Proceedings of Symposia in
PURE MATHEMATICS

Volume 98

String-Math 2016

String-Math 2016
June 27–July 2, 2016
Collège de France, Paris, France

Amir-Kian Kashani-Poor
Ruben Minasian
Nikita Nekrasov
Boris Pioline
Editors

AMERICAN MATHEMATICAL SOCIETY
Providence, Rhode Island

2010 *Mathematics Subject Classification.* Primary 14D24, 14H60, 14D21, 14J33, 58E20, 81T60, 81T30.

Library of Congress Cataloging-in-Publication Data

Names: String-Math (Conference) (2016: Paris, France) | Kashani-Poor, Amir-Kian, 1974– editor. | Minasian, Ruben, 1967– editor. | Nekrasov, Nikita, 1973– editor. | Pioline, Boris, 1972– editor.

Title: String-Math 2016: June 27–July 2, 2016, Collège de France, Paris, France / Amir-Kian Kashani-Poor, Ruben Minasian, Nikita Nekrasov, Boris Pioline, editors.

Description: Providence, Rhode Island: American Mathematical Society, [2018] | Series: Proceedings of symposia in pure mathematics; volume 98 | Includes bibliographical references.

Identifiers: LCCN 2017052181 | ISBN 9781470435158 (alk. paper)

Subjects: LCSH: Geometry, Algebraic–Congresses. | Quantum theory–Mathematics–Congresses. | AMS: Algebraic geometry – Families, fibrations – Geometric Langlands program: algebro-geometric aspects. msc | Algebraic geometry – Curves – Vector bundles on curves and their moduli. msc | Algebraic geometry – Families, fibrations – Applications of vector bundles and moduli spaces in mathematical physics (twistor theory, instantons, quantum field theory). msc | Algebraic geometry – Surfaces and higher-dimensional varieties – Mirror symmetry. msc | Global analysis, analysis on manifolds – Variational problems in infinite-dimensional spaces – Harmonic maps. msc | Quantum theory – Quantum field theory; related classical field theories – Supersymmetric field theories. msc | Quantum theory – Quantum field theory; related classical field theories – String and superstring theories; other extended objects (e.g., branes). msc

Classification: LCC QA564 .S77 2016 | DDC 516.3/5–dc23

LC record available at https://lccn.loc.gov/2017052181

DOI: http://dx.doi.org/10.1090/pspum/098

♾ The paper used in this book is acid-free and falls within the guidelines
established to ensure permanence and durability.
Visit the AMS home page at `http://www.ams.org/`

10 9 8 7 6 5 4 3 2 1 23 22 21 20 19 18

Contents

Preface

String-Math is the annual conference that was founded to reflect the most significant progress at the interface of string theory and mathematics. It is a young offspring of the annual Strings conferences, which have been gathering world experts and young researchers working in the field of string theory since 1989. The first String-Math conference was held in 2011 at the University of Pennsylvania in Philadelphia, followed by the conferences in Bonn (2012), Stony Brook (2013), Edmonton (2014), and Hainan (2015). The String-Math 2016 conference took place at Collège de France in Paris from June 27th till July 2nd, and this volume records the contributions which were presented there.

The fruitful exchange of ideas between physicists and mathematicians has a long history. The new era that began about 30 years ago with the advent of string theory and topological field theories is remarkable not only due to the increased intensity of mutually beneficial interactions between mathematicians and physicists, but also because each side is encountering the other on her own turf. Indeed, string theory has been contributing to areas deep in the heart of mathematics, far (or so we thought) from the math–physics frontier: from the classification of four manifolds using Seiberg-Witten theory, to holomorphic curve counting and other questions in enumerative geometry via topological string theory, to the study of knot invariants and their categorification via topological field theory, to progress in the geometric Langlands program inspired by the study of supersymmetric gauge theories. Conversely, mathematics developed for mathematics' sake has deeply influenced how physicists think about string theory, ranging from re-interpreting D-branes in terms of derived categories of sheaves or Lagrangian submanifolds, analyzing string compactifications using notions of generalized geometry, to the computation of elliptic genera and string amplitudes by invoking techniques from number theory. It is a safe bet that the interaction between the two fields will continue to flourish and lead to stunning new results (and this is independent of what surprises lie in store at the Large Hadron Collider of CERN in the next few years).

The String-Math 2016 conference gathered more than 200 mathematicians and physicists to discuss new, exciting developments at the interface between high energy theory and mathematics. All talks were plenary, with 17 longer talks (50 min) and 18 shorter talks (25 min) given in equal numbers by physicists and mathematicians working at the interface of high-energy theory and mathematics (with some uncertainty owing to the thin line that separates the two fields in some subdomains). The talks covered a wide range of topics, including moduli spaces of various structures such as curves, meromorphic connections and differentials, coherent sheaves, and quiver representations; BPS states and wall-crossing phenomena; topological string amplitudes in compact and non-compact geometries; exact results in supersymmetric gauge theories in various dimensions; 4D/2D correspondence; applications to the geometric Langlands program; integrability in N=4 super Yang-Mills theory; and assorted topics such as moonshine, topological quantum field theory, or analytic aspects of black hole mergers. In addition to the main scientific program, a special session comprising four lectures targeting a general scientific audience was organized in collaboration with Séminaire Poincaré on July 2nd, 2016.

All talks, including the general public session, were recorded and broadcast live. The recordings and slides can be found at the conference website:

`indico.cern.ch/e/string-math2016.`

This volume collects the written contributions from some (but unfortunately not all) of the speakers. Below we record the complete list of talks, with a one-sentence summary which hopefully will convey the scope of each talk:

Plenary talks

(1) Mina Aganagic: *Two mathematical applications of little string theory*: In this talk, new variants of the AGT correspondence and of the quantum Langlands correspondence relating q-deformed W algebra conformal blocks and K-theoretic instanton counting were presented.

(2) Benjamin Basso: *Hexagons and 3-point functions*: This talk described how the correlators of three single trace operators in $N = 4$ SYM theory can be computed at finite 't Hooft coupling, thanks to the integrability in the planar limit, by using hexagonal patches as building blocks.

(3) Mikhail Bershtein: *Plane partitions and W algebras*: This talk described a new example of W algebras labelled by three integers n, m, k, whose category of representations is equivalent to that of the product of three quantum groups $gl_{n|k}$, $gl_{k|m}$, $gl_{m|n}$.

(4) Gaetan Borot: *Chern-Simons theory on S^3/G and topological strings*: This talk discussed matrix models computing the $SU(N)$ Chern-Simons partition function on spherical Seifert manifolds.

(5) Mathew Bullimore: *Monopoles, Vortices, and Vermas*: In this talk, the notion of monopole operators creating and destroying vortices in 3d gauge theories was explored, and the implications for a finite version of the AGT correspondence discussed.

(6) Stefano Cremonesi: *3D supersymmetric gauge theories and Hilbert series*: In this talk, closely related to Nakajima's talk, the Hilbert series of 3d $N = 2$ gauge theory was obtained via the counting of dressed 't Hooft monopole operators.

(7) Thibault Damour: *Analytical Approaches to Coalescing Binary Black Holes*: This talk reviewed the analytical and numerical methods used to predict the waveform emitted by coalescing binary black holes, with particular emphasis on the Effective One-Body method.

(8) Olivia Dumitrescu: *From the Hitchin component to opers*: In this talk, a holomorphic description of the limiting oper appearing in Gaiotto's conjecture, proved by the speaker and collaborators, weas presented.

(9) Abhijit Gadde: *Conformal constraints on defects*: In this talk, the constraints imposed by conformal invariance on defects of arbitrary codimension in any conformal field theory were explored.

(10) Davide Gaiotto: *Geometric Langlands applications of boundary conditions for maximally supersymmetric Yang Mills theory*: General classes of branes in the two-dimensional sigma model on Hitchin moduli space were constructed, and their relation via the Geometric Langlands correspondence were established.

(11) Jaume Gomis: *Correlation Functions in Superconformal Field Theories*: This talk discussed new exact methods for computing correlation functions of local operators in the Coulomb branch in four-dimensional N=2 superconformal field theories.

(12) Nikolay Gromov: *Quantum Spectral Curve for AdS/CFT and its applications*: In this talk, a Riemann-Hilbert type equation known as the quantum spectral curve was introduced and applications to the computation of scaling dimensions in planar $N = 4$ SYM were discussed.

(13) Sarah Harrison: *Umbral symmetry groups and K3 CFTs*: In this talk, the extension of umbral moonshine to the case of singular CFTs and under the inclusion of worldsheet parity were discussed.

(14) André Henriques: *What Chern-Simons theory assigns to a point?*: This talk proposed that Chern-Simons theory assigns the representations of based loop groups to points and discussed the implications of this point of view.

(15) Min-xin Huang: *Derivation of modular anomaly equation in compact elliptic Calabi-Yau spaces*: In this talk, complementary to Katz' talk, the modular anomaly equation for elliptically fibered Calabi-Yau spaces was derived from the BCOV holomorphic anomaly equations.

(16) Kohei Iwaki: *Exact WKB analysis, cluster algebras and Painlevé equations*: The relation between Voros symbols in exact WKB analysis and cluster variables was described, and the notion of Voros symbols was generalized to Painlevé equations.

(17) Sheldon Katz: *Elliptically fibered Calabi-Yau threefolds: mirror symmetry and Jacobi forms*: This talk explained how to compute the topological string partition function on elliptically fibered Calabi-Yau threefolds using a combination of B-model, homological mirror symmetry, and geometric techniques.

(18) Maxim Kontsevich: *Resurgence and exact quantization via holomorphic Floer cohomology*: In this talk, the Riemann-Hilbert correspondence and wall-crossing problem was revisited by considering the Fukaya categories associated with a holomorphic symplectic manifold and a possibly singular analytic Lagrangian subvariety.

(19) Oleg Lisovyy: *Fredholm determinant and Nekrasov type representations for isomonodromic tau functions*: In this talk, Fredholm determinant representations for isomonodromic tau functions of Fuchsian systems with n regular singular points on the Riemann sphere and general monodromy in $GL(N, C)$ were derived.

(20) Marcos Mariño: *Spectral theory and topological strings*: This talk presented a conjectural correspondence between topological string theory on toric Calabi-Yau manifolds and the spectral theory of certain trace class operators, obtained by quantizing the mirror curves.

(21) Gregory Moore: *Framed BPS States In Two And Four Dimensions*: This talk gave a broad review of wall-crossing phenomena in two and four dimensions, with applications to the categorification of knot invariants.

(22) Hiraku Nakajima: *Quantized Coulomb branches of 3d N=4 gauge theories and difference operators*: This talk gave a mathematical construction of the Coulomb branch of these theories and their quantization.

(23) Rahul Pandharipande: *Moduli spaces of holomorphic and meromorphic differentials*: This talk introduced a new moduli space of holomorphic/ meromorphic differentials on Riemann surfaces and made the connection to Pixton's formulas and Witten's r-spin class.

(24) Alexander Polishchuk: *Moduli spaces of curves with non-special divisors*: This talk discussed the moduli spaces of pointed curves with possibly non-nodal singularities, in terms of a quotient of an affine scheme by a torus action.

(25) Leonardo Rastelli: *Higgs branches, vertex operator algebras and modular differential equations*: This talk discussed the relationship between the Higgs branch of a D=4, N=2 SCFT and the associated vertex operator algebra, obtaining modular equations for the Schur index.

(26) Jørgen Rennemo: *Derived equivalences from a duality of non-abelian gauge linear sigma models*: In this talk, the equivalence between the category of B-branes of GLSM duals proposed by Kentaro Hori was proved, and new examples of varieties exhibiting this duality presented.

(27) Laura Schaposnik: *Higgs bundles, branes and applications*: In this talk, Higgs bundles for complex Lie groups and the associated Hitchin fibration were introduced, a natural construction of families of subspaces corresponding to different types of branes discussed, and applications to Langlands duality and representations of 3-manifolds presented.

(28) Olivier Schiffmann: *Cohomological Hall algebra actions and Kac polynomials*: This talk related the cohomological Hall algebras associated to quivers with the Yangians constructed by Maulik and Okounkov, and showed that their Hilbert series are encoded by the Kac polynomials of the underlying quiver.

(29) Pavol Severa: *Poisson-Lie T-duality*: Poisson-Lie T-duality was reviewed and explained in terms of Chern-Simons theory and its generalizations with appropriate boundary conditions.

(30) Carlos Simpson: *Reduction for SL(3) pre-buildings*: This talk discussed the reduction of $SL(3)$ spectral curves and their relations to harmonic maps and the WKB problem.
(31) Jörg Teschner: *SUSY field theories and geometric Langlands: The other side of the coin*: In this talk, the implications of the AGT correspondence in the presence of surface operators for the geometric Langlands programme and its quantum version were discussed.
(32) Richard Thomas: *A Vafa-Witten invariant for projective surfaces*: In this talk, the Vafa-Witten invariants for algebraic surfaces were discussed and related to the DT-invariants on non-compact Calabi-Yau threefolds.
(33) Daisuke Yamakawa: *Meromorphic connections and quivers*: In this talk, the proof of Boalch's generalization of the Crawley-Boevey result relating meromorphic connections on the Riemann sphere and quivers was presented.
(34) Shing-Tung Yau: *Period integrals of algebraic manifolds and their differential equations*: This talk addressed the question of which solutions of Picard-Fuchs equations correspond to periods of algebraic manifolds.
(35) Dimitri Zvonkine: *The Chern character of the Verlinde bundle*: In this talk, the Chern character of the Verlinde bundle was computed by invoking Teleman's classification of semi-simple cohomological field theories.

General Public Session "Maths & Cordes"

(1) Hirosi Ooguri (Caltech and IPMU): *What is gravity?*: This talk introduced the general public to the challenges posed by the unification of gravity with the other fundamental forces in nature.
(2) Andrei Okounkov (Columbia U.): *Catching monodromy*: The remarkable analytic properties of a certain class of special functions which play a central role in mathematical physics were explained and generalized.
(3) Robbert Dijkgraaf (IAS Princeton): *Quantum Geometry*: The interactions between mathematics and physics over the last century were put in perspective, and ideas for unifying algebra and geometry via stringy geometry were outlined.
(4) Nima Arkani-Hamed (IAS Princeton): *Physics and Mathematics for the End of Spacetime*: The difficulty of defining observables in a theory of quantum gravity were discussed, and a new geometrical approach for computing scattering amplitudes was presented.

The 2016 conference was organized by Boris Pioline (CERN, Genève and LPTHE, Paris), Ruben Minasian (Commissariat à l'Energie Atomique, Saclay), Amir-Kian Kashani-Poor (Ecole Normale Supérieure, Paris), Nikita Nekrasov (Simons Center for Geometry and Physics, Stony Brook), Philip Boalch (Université Paris Sud-Orsay), Miranda Cheng (Amsterdam University), Alessandro Chiodo (Université Pierre et Marie Curie - Paris 6), Maxim Kontsevich (Institut des Hautes Etudes Scientifiques, Bures-sur-Yvette), and Don Zagier (Max-Planch Institut für Mathematik, Bonn).

Support from the following institutions is gratefully acknowledged: Clay Mathematics Institute, Agence Nationale de la Recherche, Centre National de la Recherche Scientifique, Fondation Mathématique Jacques Hadamard, Fédération de Recherche Interactions Fondamentales, Fondation Meyer pour le développement culturel et artistique, Institut Henri Poincaré, Institut des Hautes Etudes Scientifiques, Institut de Physique Théorique du CEA, Institut de Mathématiques de Jussieu - Paris Rive Gauche, Laboratoire de Physique Théorique et Hautes Energies, UPMC, Laboratoire de Physique Théorique de l'Ecole Normale Supérieure, National Science Foundation, Université Paris Sud - Orsay, and Université Pierre et Marie Curie - Paris 6.

After Paris, the next String-Math conferences are due to take place in Hamburg (2017), Sendai (2018), Uppsala (2019), and Stellenbosch (2020).

Amir-Kian Kashani-Poor
Ruben Minasian
Nikita Nekrasov
Boris Pioline

List of Participants

Aganagic Mina
University of California at Berkeley

Aghaei Nezhla
DESY and Hamburg University

Alexandrov Sergey
LCC, Montpellier

Anderson Louise
Imperial College London

Arkani-Hamed Nima
Institute for Advanced Study, Princeton

Artamonov Semen
Rutgers University

Ashfaque Johar
University of Liverpool

Ashmore Anthony
Imperial College London

Babalic Mirela
IBS Center for Geometry and Physics, Pohang

Bachas Constantin
Ecole Normale Supérieure, Paris

Balasubramanian Aswin
University of Hamburg and DESY

Ballard Matthew
University of South Carolina

Balzin Eduard
Laboratoire J.-A. Dieudonné, Nice

Banerjee Sibasish
IPhT, CEA, Saclay

Basso Benjamin
LPTENS, Paris

Beisert Niklas
ETH Zürich

Bena Iosif
IPhT, CEA, Saclay

Benetti Genolini Pietro
Oxford University

Benvenuti Sergio
SISSA, Trieste

Bershtein Mikhail
Landau Institute for Theoretical Physics, Moscow

Bertolini Marco
Duke University

Bettadapura Kowshik
Australian National University, Canberra

Biquard Olivier
Ecole Normale Supérieure, Paris

Boalch Philip
Université Paris Sud, Orsay

Bonechi Francesco
INFN Firenze

Borot Gaetan
MPIM, Bonn

Bossard Guillaume
CPHT, Ecole Polytechnique, Palaiseau

Bouchard Vincent
University of Alberta

Bousseau Pierrick
Imperial College London

Braghiroli Matteo
University of Rome La Sapienza

Brennan Theodore
Rutgers, the State University of New Jersey

Brini Andrea
Université de Montpellier

Bugden Mark
Australian National University, Canberra

Bulgakova Daria
IPHT, CEA, SACLAY

Bullimore Mathew
University of Oxford

Calvo Jackson Juan Carlos
University of Oxford

Caorsi Matteo
SISSA, Trieste

Cassani Davide
LPTHE, UPMC Paris 6

Cheng Miranda
University of Amsterdam

Chiodo Alessandro
IMJ-PRG, UPMC Paris 6

Choi Jinwon
Sookmyung Women's University, Seoul

Cirafici Michele
IST, Lisbon

Coimbra André
IHES, Bures-sur-Yvette

Cosnier-Horeau Charles
CPHT, Palaiseau and LPTHE, Paris 6

Cotti Giordano
SISSA, Trieste

Cremonesi Stefano
King's College London

Cromer Michael
Australian National University

Damour Thibault
IHES, Bures-sur-Yvette

Davison Ben
EPFL, Lausanne

Dijkgraaf Robbert
Institute for Advanced Study, Princeton

Donagi Ron
University of Pennsylvania

Doryn Dmitry
IBS, Center for Geometry and Physics, Pohang

Dumitrescu Olivia
MPIM, Bonn

Durand Philippe
CNAM, Paris

Elias Rebelo Jose Gustavo
SISSA, Trieste

Fang Ziqi
King's College London

FeldmanAndrey
Weizmann Institute of Science, Rehovot

Ferrari Francesca
University of Amsterdam

Fiset Marc-Antoine
University of Oxford

Franchetti Guido
Leibniz University

Franco Sebastian
City College of CUNY, New York

Frenkel Edward
University of California at Berkeley

Gaberdiel Matthias
ETH Zürich

Gadde Abhijit
Institute for Advanced Study, Princeton

Gahramanov Ilmar
Albert Einstein Institute, Potsdam

Gaiotto Davide
Perimeter Institute, Waterloo

Gautason Fridrik
IPhT, CEA, Saclay

Geipel Jakob
Leibniz Universität, Hannover

Gomis Jaume
Perimeter Institute, Waterloo

Goncharov Yegor
Lebedev Physical Institute, Moscow

Grana Mariana
IPhT, CEA Saclay

Grassi Alba
ICTP, Trieste

Gromov Nikolay
King's College London

Gu Jie
LPTENS, Paris

Guere Jeremy
Humboldt Universität, Berlin

Gukov Sergei
California Institute of Technology

Gupta Rajesh
ICTP, Trieste

Gustafsson Henrik
Chalmers University of Technology

Haehnel Philipp
Trinity College Dublin

Halmagyi Nicholas
LPTHE, UPMC Paris 6

Haouzi Nathan
University of California at Berkeley

Harrison Sarah
Harvard University

Hatsuda Yasuyuki
Université de Genève

Hemo Tamir
Technion, Haifa

Henriques André
Oxford University and Utrecht University

Huang Minxin
University of Science and Technology of China

Hulik Ondrej
Masaryk University, Brno

Hull Christopher
Imperial College London

Intriligator Kenneth
University of California at San Diego

Israel Dan
LPTHE, UPMC Paris 6

Iwaki Kohei
Nagoya University

Janda Felix
IMJ-PRG, UPMC Paris 6

Jefferson Patrick
Harvard University

Julia Bernard
LPTENS Paris

Katz Sheldon
University of Illinois at Urbana-Champaign

Keßler Enno
MPI Leipzig

Kimura Takashi
Boston University

Kiritsis Elias
APC, Université Paris 7

Kivinen Oscar
University of California at Davis

Kontsevich Maxim
IHES, Bures-sur-Yvette

Kramer Reinier
Universiteit van Amsterdam

Lasso Oscar
IFT-UAM/CSIC, Madrid

Lazaroiu Calin
IBS Center for Geometry and Physics, Pohang

Le Floch Bruno
Princeton University

Li Wei
ITP, Chinese Academy of Science, Beijing

Lisovyi Oleg
LMPT, Université de Tours

Lockhart Guglielmo
Universiteit van Amsterdam

Longhi Pietro
Uppsala University

Mariño Marcos
Université de Genève

Markou Chrysoula
LPTHE, UPMC Paris 6

Matsuo Yutaka
The University of Tokyo

Minasian Ruben
IPhT, CEA, Saclay

Miramontes J Luis
University of Santiago de Compostela

Mischi Alexander
University of Innsbruck

Mohaupt Thomas
University of Liverpool

Monnier Samuel
Université de Genève

Moore Gregory
Rutgers University

Moosavian Seyed Faroogh
Perimeter Institute, Waterloo

Morrison David
University of California at Santa Barbara

Mulase Motohico
University of California at Davis

Musaev Edvard
Albert Einstein Institute, Golm

Muteeb Muhammad Nouman
ICTP, Trieste

Nagoya Hajime
Kanazawa University

Nakajima Hiraku
RIMS, Kyoto

Nedelin Anton
University of Milano-Bicocca

Negut Andrei
MIT, Boston

Nekrasov Nikita
SCGP, Stony Brook

Nian Jun
IHES, Bures-sur-Yvette

Nieri Fabrizio
Uppsala University

Norton Thomas
California Institute of Technology

Okounkov Andrei
Columbia University

Ooguri Hirosi
California Institute of Technology

Pandharipande Rahul
ETH Zürich

Paquette Natalie
Stanford University

Pawelkiewicz Michal
IPHT, Saclay

Petrini Michela
LPTHE, UPMC, Paris 6

Pezzella Franco
INFN, Napoli

Piazzalunga Nicolo'
SISSA, Trieste

Pioline Boris
LPTHE, Paris 6 and CERN, Geneva

Plencner Daniel
LPTENS/LPTHE, Paris

Plesser M. Ronen
Duke University

Policastro Giuseppe
LPTENS, Paris

Polishchuk Alexander
University of Oregon

Pradisi Gianfranco
University of Rome Tor Vergata

Prins Daniel
IPhT, CEA, Saclay

Quigley Callum
University of Toronto

Rabinovici Eliezer
IHES, Bures-sur-Yvette

Ramirez Pedro F.
IFT Madrid

Rastelli Leonardo
YITP, Stony Brook

Reid-Edwards Ron
University of Hull

Rennemo
Jørgen All Souls College, Oxford

Ronzani
Massimiliano SISSA, Trieste

Ros Camacho Ana
IMJ-PRG, Paris 6

Rose Simon
Köbenhavns Universitet

Rossi Paolo
Université de Bourgogne

Sasmal Soumya
Université Paris Sud

Savelli Raffaele
CEA Saclay

Schaposnik Laura
University of Illinois at Chicago

Schiffmann Olivier
Université de Paris-Sud Orsay

Schmid Christian
University of California at Berkeley

Sciarappa Antonio
KIAS, Seoul

Severa Pavol
Université de Genève

Shahbazi Carlos
IPhT, CEA, Saclay

Sharpe Eric
Virginia Tech

Sheshmani Artan
Massachussetts Institute of Technology

Simpson Carlos
CNRS, Université Nice Sophia Antipolis

Solard Gautier
University Milano Bicocca

Strickland-Constable Charles
IPhT, Saclay and IHES,
Bures-sur-Yvette

Stromwall Joakim
Surrey University

Svanes Eirik
LPTHE, UPMC Paris 6

Tavakol Mehdi
IBS Center for Geometry and Physics,
Pohang

Taylor Washington
Massachusetts Institute of Technology

Teschner Joerg
DESY, Hamburg

Theisen Stefan
Albert Einstein Institute, Golm

Thomas Richard
Imperial College London

Tonita Valentin
Humboldt University, Berlin

Troost Jan
LPTENS, Paris

Turton David
IPhT, CEA, Saclay

Vaintrob Dmitry
Massachusetts Institute of Technology

van Garrel Michel
KIAS, Seoul

van Leuven Sam
University of Amsterdam

Voros André
IPhT, CEA Saclay

Wakimoto Yuki
Tokyo Metropolitan University

Weitsman Jonathan
Northeastern University, Boston

Wimmer Robert
AMS (Physical Review D), Astoria

Windey Paul
LPTHE, UPMC Paris 6

Wrase Timm
TU Vienna

Yamakawa Daisuke
Tokyo Institute of Technology

Yang Zhiwei
University of Pisa

Yang Di
SISSA, Trieste

Yau Shing Tung
Harvard University

Ye Ke
California Institute of Technology

Zabzine Maxim
Uppsala University

Zagier Don
MPIM, Bonn

Zenkevich Yegor
INR, Moscow

Zerbini Federico
MPIM, Bonn

Zhang Hong
Sogang University, Seoul

Zhang Ming
University of Michigan

Zvonkine Dimitri
IMJ-PRG, Paris 6

Proceedings of Symposia in Pure Mathematics
Volume **98**, 2018
http://dx.doi.org/10.1090/pspum/098/01729

Three-dimensional $\mathcal{N} = 4$ gauge theories in omega background

Mathew Bullimore

ABSTRACT. We review the description of three-dimensional gauge theories with $\mathcal{N} = 4$ supersymmetry in the presence of an omega background as an $\mathcal{N} = 4$ supersymmetric quantum mechanics. We will focus throughout on a simple abelian example. The Hilbert space of supersymmetric ground states is populated by generalized vortex configurations, while half-BPS monopole operators act on the Hilbert space by creating and annihilating vortices, furnishing it with the structure of a Verma module for the quantized Coulomb branch chiral ring. Furthermore, by introducing two-dimensional $\mathcal{N} = (2,2)$ boundary conditions, we find a finite version of the AGT correspondence between vortex partition functions and overlaps of Whittaker vectors for quantized Coulomb branch chiral rings.

1. Introduction

We review the description of three-dimensional gauge theories with $\mathcal{N} = 4$ supersymmetry in the presence of an omega background in the $x^{1,2}$-plane as an $\mathcal{N} = 4$ supersymmetric quantum mechanics on the x^3-axis, summarizing and illustrating results from the author's joint paper [**7**]. We focus exclusively on a simple abelian example, which is sufficient to illustrate the main points and will hopefully provide a foundation for the richer non-abelian examples treated in [**7**]. The setup is shown schematically in Figure 1.

We will provide an explicit description of the $\mathcal{N} = 4$ supersymmetric quantum mechanics on the x^3-axis, which is summarized as follows:

- The supersymmetric ground states are vortex configurations localized at the origin of the $x^{1,2}$-plane.
- Monopole operators on the x^3-axis become half-BPS operators in the supersymmetric quantum mechanics that create and destroy vortices.

The monopole operators on the x^3-axis generate a non-commutative algebra that quantizes the Coulomb branch in a given complex structure. The space of supersymmetric ground states transforms as a Verma module for this non-commutative algebra. Sending the omega background parameter $\epsilon \to 0$, we recover the exact Coulomb branch chiral ring. This provides a derivation of the proposed structure of 1-loop and non-perturbative quantum corrections to the Coulomb branch chiral

2010 *Mathematics Subject Classification.* Primary 81T60.

The author was supported by ERC Starting Grant no. 306260 'Dualities in Supersymmetric Gauge Theories, String Theory and Conformal Field Theories'.

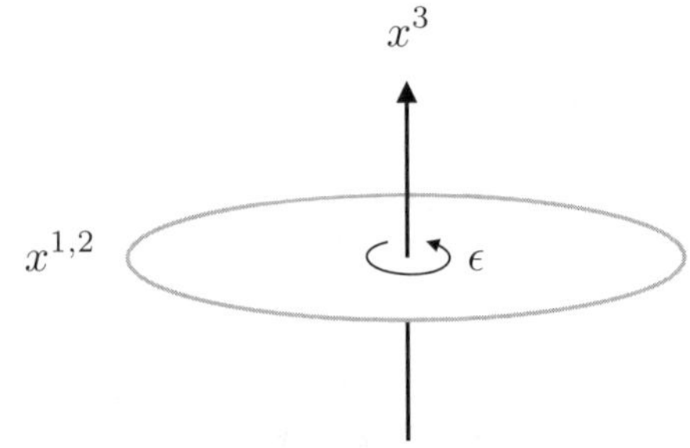

FIGURE 1. We will describe a three-dimensional gauge theory with $\mathcal{N}=4$ supersymmetry with an omega background in the $x^{1,2}$-plane as an $\mathcal{N}=4$ supersymmetric quantum mechanics on the x^3-axis.

ring developed in [**5**], and provides a complementary approach to the mathematical work of [**4, 13**].

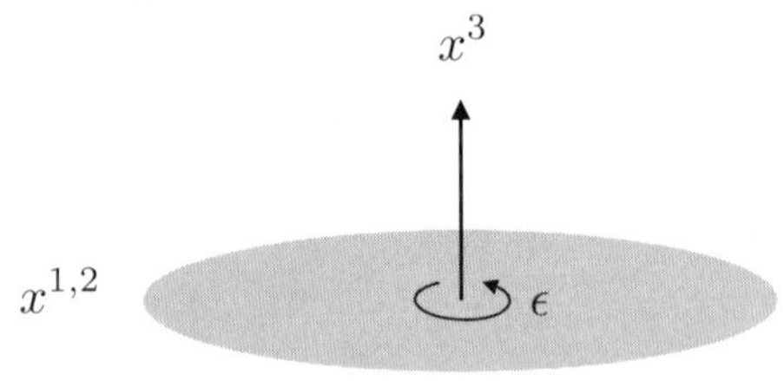

FIGURE 2. We will enrich the setup by introducing boundary conditions in the $x^{1,2}$-plane preserving $\mathcal{N}=(2,2)$ supersymmetry.

We will also enrich this setup by including boundary conditions that preserve a two-dimensional $\mathcal{N}=(2,2)$ supersymmetry in the $x^{1,2}$-plane [**6**]. Such boundary conditions define a boundary state in the Hilbert space of the $\mathcal{N}=4$ supersymmetric quantum mechanics. We show that 'Neumann' boundary conditions lead to coherent states of vortices, or generalized Whittaker vectors. Furthermore, by evaluating partition functions on an interval with $\mathcal{N}=(2,2)$ boundary conditions at each end, we provide a vast generalization and physical explanation for the 'finite' AGT correspondence introduced in [**3**].

2. Setup

2.1. 3d $\mathcal{N}=4$ Supersymmetry. We work in flat euclidean $\mathbb{R}^3$ with coordinates x^1, x^2, x^3 and spinor indices α, β for the $SU(2)_E$ isometry group. The R-symmetry is $SU(2)_H \times SU(2)_C$ and we introduce indices A, B, and $\dot{A}, \dot{B}$ for the spinor representations of $SU(2)_H$ and $SU(2)_C$ respectively. We use uniform conventions for all $SU(2)$ indices: $(\sigma_i)^\alpha{}_\beta$ are the standard Pauli matrices, while spinor indices are raised and lowered as $\psi_\alpha = \epsilon_{\alpha\beta}\psi^\beta$, $\psi^\alpha = \epsilon^{\alpha\beta}\psi_\beta$ with $\epsilon^{12} = \epsilon_{21} = 1$.

The supersymmetry generators are denoted by $Q_\alpha^{A\dot{A}}$ with

(1) $$\{Q_\alpha^{A\dot{A}}, Q_\beta^{B\dot{B}}\} = -2\epsilon^{AB}\epsilon^{\dot{A}\dot{B}}P_{\alpha\beta} + 2\epsilon_{\alpha\beta}(\epsilon^{AB}Z^{\dot{A}\dot{A}} + \epsilon^{\dot{A}\dot{B}}Z^{AB})$$

where $P_{\alpha\beta}$ is the momentum generator and Z^{AB}, $Z^{\dot{A}\dot{B}}$ are central charges in the adjoint representation of $SU(2)_H$, $SU(2)_C$.

We are primarily concerned with supersymmetric gauge theories, in which

- scalars Q^A in hypermultiplets transform in the fundamental of $SU(2)_H$,
- scalars $\varphi^{\dot{A}\dot{B}}$ in vectormultiplets transform in the adjoint of $SU(2)_C$.

In particular, $SU(2)_H$ rotates the complex structure on the Higgs branch $\mathcal{M}_H$ while $SU(2)_C$ rotates the complex structure on the Coulomb branch $\mathcal{M}_C$. The central charges have the following form:

- Z^{AB} is a linear combination conserved charges for Coulomb branch flavor symmetries, with coefficients given by FI parameters t^{AB}.
- $Z^{\dot{A}\dot{B}}$ is a linear combination of conserved charges for gauge and Higgs branch flavor symmetries, with coefficients given by vectormultiplet scalars $\varphi^{\dot{A}\dot{B}}$ and mass parameters $m^{\dot{A}\dot{B}}$ respectively.

It is often convenient to decompose fields and parameters according their charges under a fixed maximal torus $U(1)_H \times U(1)_C$ of the R-symmetry. We will make the standard choice such that the supercharges $Q^{1\dot{A}}_\alpha$, $Q^{2\dot{A}}_\alpha$ have $U(1)_H$ charge $+\frac{1}{2}$, $-\frac{1}{2}$, while the supercharges $Q^{A\dot{1}}_\alpha$, $Q^{A\dot{2}}_\alpha$ have $U(1)_C$ charge $+\frac{1}{2}$, $-\frac{1}{2}$.

2.2. Example. Throughout this note, we will focus on a simple example: $G = U(1)$ with $N \geq 1$ hypermultiplets in the fundamental representation. We decompose the bosonic fields according to their charge under the $U(1)_H \times U(1)_C$:

- Hypermultiplet scalars Q^A_j with $j = 1, \ldots, N$ decompose into complex components $(X_j, \bar{Y}_j)$ transforming with charge $+1$ under $G = U(1)$ and charge $(+\frac{1}{2}, -\frac{1}{2})$ under $U(1)_H$.
- Vectormultiplet scalars $\varphi^{\dot{A}\dot{B}}$ decompose into real and complex components $(\sigma, \varphi, \bar{\varphi})$ transforming with charge $(0, +1, -1)$ under $U(1)_C$.

There is a $G_H = PSU(N)$ flavor symmetry transforming the hypermultiplets. We can turn on mass parameters $m^{\dot{A}\dot{B}}$ corresponding to this symmetry by coupling to a background vectormultiplet and turning on a vacuum expectation value for the scalars in the Lie algebra of the maximal torus $T_H \subset G_H$. Here, we will turn on only complex masses $m^{\dot{1}\dot{1}} \sim m_{\mathbb{C}} = (m_1, \ldots, m_N)$ with $\sum_i m_i = 0$. This spontaneously breaks the $U(1)_C$ R-symmetry. The complex masses contribute to the central charges

$$Z^{\dot{1}\dot{2}} \sim \sigma \qquad Z^{\dot{1}\dot{1}} \sim \varphi + m_{\mathbb{C}} \,. \tag{2}$$

To simplify notation we omit the symmetry generators: $\varphi + m_{\mathbb{C}}$ stands for an infinitesimal complex gauge transformation with parameter φ and T_H flavor transformation with parameter $m_{\mathbb{C}}$.

In addition, there is a topological symmetry $G_C = U(1)_t$ and corresponding FI parameters t^{AB}. Here we only turn on a real FI parameter $t^{12} \sim t_{\mathbb{R}} < 0$. This preserves the $U(1)_H$ R-symmetry and contributes to the central charge

$$Z^{12} \sim t_{\mathbb{R}} \,. \tag{3}$$

This contribution vanishes on the elementary fields but acts non-trivially on monopole operators, which are charged under the $U(1)_t$ topological symmetry.

2.3. Supersymmetric Vacua. Supersymmetric vacua are determined classically by minimizing the potential and preserve all of the supercharges $Q_\alpha^{A\dot{A}}$. With generic complex masses $m_\mathbb{C} = (m_1, \ldots, m_N)$ and real FI parameter $t_\mathbb{R} < 0$, this requires that

$$\begin{aligned} \mu_\mathbb{C} &= 0 & \mu_\mathbb{R} + t_\mathbb{R} &= 0 \\ (\varphi + m_j)X_j &= 0 & \sigma X_j &= 0 \\ (-\varphi - m_j)Y_j &= 0 & -\sigma Y_j &= 0 \end{aligned} \tag{4}$$

modulo $U(1)$ gauge transformations, where

$$\mu_\mathbb{C} = \sum_{j=1}^N X_j Y_j \qquad \mu_\mathbb{R} = \sum_{j=1}^N |X_j|^2 - |Y_j|^2 \tag{5}$$

are the complex and real moment maps for the action of $G = U(1)$ on the hypermultiplets (X_j, Y_j). In the language of $\mathcal{N} = 2$ supersymmetry, they arise from F-term and D-term contributions to the lagrangian respectively.

Setting the complex masses to vanish, $(m_1, \ldots, m_N) = 0$, there is a moduli space of supersymmetric vacua known as the Higgs branch $\mathcal{M}_H$. This is protected from quantum corrections by supersymmetry and the classical description in terms of equations (4) is exact. In particular, $(\sigma, \varphi) = 0$ and $\mathcal{M}_H$ is the hyper-Kähler quotient,

$$\mathcal{M}_H = \left\{ \sum_{j=1}^N X_j Y_j = 0, \sum_{j=1}^N |X_j|^2 - |Y_j|^2 = -t_\mathbb{R} \right\} / U(1)\,. \tag{6}$$

In the complex structure where X_j, Y_J are holomorphic, this is $T^*\mathbb{CP}^{N-1}$ with compact base parameterized by X_j and Kähler parameter $-t_\mathbb{R}$. A complex algebraic description is found by replacing the real moment map equation by the stability condition $X \neq 0$ and dividing by complex gauge transformations,

$$\mathcal{M}_H = \left\{ \sum_{j=1}^N X_j Y_j = 0,\ X_j \neq 0 \right\} / \mathbb{C}^*\,. \tag{7}$$

The coordinate ring of $\mathcal{M}_H$ in this complex structure is then generated by the hypermultiplet bilinears $X_i Y_j$ subject to the complex moment map constraint $\mu_\mathbb{C} = 0$. In physical language, this is the chiral ring generated by gauge invariant local operators annihilated by half of the supercharge $Q_\alpha^{1\dot{A}}$.

On the other hand, for generic complex masses $(m_1, \ldots, m_N)$ but vanishing FI parameter $t_\mathbb{R} = 0$, there is a moduli space of supersymmetric vacua known as the Coulomb branch $\mathcal{M}_C$. Equations (4) require $X_j = Y_j = 0$ and the Coulomb branch is parametrized by the expectation values of (φ, σ) and the periodic dual photon $\gamma \sim \gamma + 2\pi$. However, the classical geometry $\mathbb{R}^3 \times S^1$ is modified by 1-loop quantum corrections to an N-centered Taub-NUT metric describing an S^1 fibration over $\mathbb{R}^3$ with singular fibers at $(\varphi, \sigma) = (-m_j, 0)$ for all $j = 1, \ldots, N$. The $U(1)_t$ topological symmetry acts by rotating the S^1 fibers.

In the complex structure where φ is a holomorphic coordinate, the coordinate ring of the Coulomb branch coincides with the chiral ring generated by operators annihilated by another half of the supercharges $Q_\alpha^{A\dot{1}}$. The chiral operators are generated by φ and monopole operators $u^\pm$ of charge ± 1 under $G_C = U(1)_t$. The monopole operators can be defined classically by $u^{\pm} \sim e^{\pm(\sigma + i\gamma)}$. However, the classical relations $u^+u^- = 1$ are modified by 1-loop quantum corrections to

$$u^+u^- = \prod_{j=1}^{N} (\varphi + m_j)\,. \tag{8}$$

This identifies the Coulomb branch in a given complex structure with $\mathbb{C}^2/\mathbb{Z}_N$ with deformation parameters $(m_1, \ldots, m_N)$. We will reproduce this quantum corrected chiral ring relation by localization to a supersymmetric quantum mechanics in the following sections.

Finally, turning on both generic complex masses $(m_1, \ldots, m_N)$ and real FI parameter $t_{\mathbb{R}} < 0$, there are N isolated massive supersymmetric vacua,

$$\nu_i : \qquad X_j = \sqrt{-t_{\mathbb{R}}}\delta_{ij} \qquad Y_j = 0 \qquad \varphi = -m_i \qquad \sigma = 0\,. \tag{9}$$

The massive vacua can be identified with fixed points of the T_H-action on $\mathcal{M}_H$ generated by $(m_1, \ldots, m_C)$: they are the coordinate hyperplanes in the $\mathbb{CP}^{N-1}$. Equivalently, since the topological symmetry rotates the S^1 fibers of $\mathcal{M}_C$, the massive vacua can be identified with fixed points of the G_C-action on $\mathcal{M}_C$. This illustrates an important theme: turning on mass parameters localizes the system to fixed points of the corresponding symmetry.

2.4. $\mathcal{N} = 4$ Quantum Mechanics. We will now identify a subalgebra of the 3d $\mathcal{N} = 4$ supersymmetry algebra (1) corresponding to an $\mathcal{N} = 4$ supersymmetric quantum mechanics on the x^3-axis. It is then convenient to introduce a complex coordinate $z = x^1 + ix^2$ in the $x^{1,2}$-plane. We will require that the system sits in a supersymmetric massive vacuum ν_i defined in (9) as $|z| \to \infty$.

First of all, let us denote by $U(1)_E \subset SU(2)_E$ the subgroup of rotations around the x^3-axis under which the supercharges $Q_1^{A\dot{A}}$, $Q_2^{A\dot{A}}$ have charge $-\frac{1}{2}$, $+\frac{1}{2}$ respectively. We now restrict attention to generators commuting with the diagonal subgroup

$$U(1)_\epsilon \subset U(1)_E \times U(1)_H\,. \tag{10}$$

The commuting supersymmetry generators are

$$Q^{\dot{A}} := Q_1^{1\dot{A}} \qquad \tilde{Q}^{\dot{A}} := Q_2^{2\dot{A}} \tag{11}$$

with

$$\begin{aligned} \{Q^{\dot{A}}, Q^{\dot{B}}\} &= 0 \\ \{Q^{\dot{A}}, \tilde{Q}^{\dot{B}}\} &= 2\epsilon^{\dot{A}\dot{B}}H + 2Z^{\dot{A}\dot{B}} \\ \{\tilde{Q}^{\dot{A}}, \tilde{Q}^{\dot{B}}\} &= 0\,. \end{aligned} \tag{12}$$

where we define $H = P_3 + Z^{12}$. This is the supersymmetry algebra of an $\mathcal{N} =$ 4 quantum mechanics on the x^3-axis with R-symmetry $U(1)_H \times SU(2)_C$ and a distinguished flavor symmetry $U(1)_\epsilon$. This type of supersymmetry algebra can also be obtained by dimensional reduction of 2d $\mathcal{N} = (2,2)$ supersymmetry.

It is straightforward to formally describe a 3d $\mathcal{N} = 4$ gauge theory as an infinite dimensional $\mathcal{N} = 4$ supersymmetric quantum mechanics on the x^3-axis with supersymmetry algebra (12). Borrowing the supermultiplet terminology from 2d $\mathcal{N}=(2,2)$ supersymmery, we have

- Chiral multiplets with complex scalar components X_j, Y_j and $D_{\bar{z}}$.
- A vectormultiplet for the gauge group $\mathcal{G}$ of maps from the $x^{1,2}$-plane into G that are constant at $|z|\to\infty$, with scalar components σ, φ and D_3.
- A superpotential

$$W \sim \sum_j \int |dz|^2 X_j D_{\bar{z}} Y_j \,. \tag{13}$$

It also also important to note that the vectormultiplet fields can be organized into a twisted chiral multiplet with bottom component give by the complex scalar φ.

The complex masses $(m_1,\ldots,m_N)$ are incorporated by coupling to a background vectormultiplet for the G_H symmetry and giving a vacuum expectation value to the bottom component of the twisted chiral multiplet in T_H. As above, this contributes a non-vanishing central charge $Z^{\dot{1}\dot{1}}$ proportional to $\varphi+m_{\mathbb{C}}$ breaking the $U(1)_C$ R-symmetry. Finally, the real FI parameter is incorporated by adding a twisted superpotential

$$\widetilde{W} \sim t_{\mathbb{R}} \int |dz|^2 \varphi \,. \tag{14}$$

which contributes a non-vanishing central charge Z^{12} proportional to $t_{\mathbb{R}}$, which preserves $U(1)_H$ R-symmetry.

2.5. $\mathcal{N}=2$ Quantum Mechanics. For many purposes, it is convenient to describe the $\mathcal{N}=4$ quantum mechanics in the language of $\mathcal{N}=2$ quantum mechanics. There are two types of $\mathcal{N}=2$ supersymmetric quantum mechanics that can be obtained from the dimensional reduction of $\mathcal{N}=(2,0)$ supersymmetry and $\mathcal{N}=(1,1)$ supersymmetry in two dimensions. Given a choice of complex structure on $\mathcal{M}_C$, we can define both an $\mathcal{N}=(2,0)$ quantum mechanics and an S^1 family of $\mathcal{N}=(1,1)$ quantum mechanics as follows.

First, we note that a complex structure on $\mathcal{M}_C$ is specified by a spinor $\xi_{\dot{A}} = (\xi_{\dot{1}},\xi_{\dot{2}})$ modulo complex rescalings, forming the homogeneous coordinates of a point on the twistor sphere $\mathbb{CP}^1$ over the Coulomb branch. It is convenient to fix the normalization $\xi_{\dot{A}}\xi^{\dagger\dot{A}} = |\xi_{\dot{1}}|^2+|\xi_{\dot{2}}|^2 = 1$ where $\xi^\dagger_{\dot{A}} = (-\bar{\xi}_{\dot{2}},\bar{\xi}_{\dot{1}})$ and so describe the twistor sphere as a quotient of S^3 along the fibers of the Hopf fibration. It is important to note that $U(1)_C$ rotates the twistor sphere with fixed points $\xi_{\dot{A}}=(1,0)$ and $(0,1)$.

Now, given a choice of complex structure $\xi_{\dot{A}}$ on $\mathcal{M}_C$, let us define

$$\begin{aligned} Q_\xi &= \xi_{\dot{A}} Q^{\dot{A}} & \widetilde{Q}_\xi &= \xi_{\dot{A}} \widetilde{Q}^{\dot{A}} \\ Q_{\xi^\dagger} &= \xi^\dagger_{\dot{A}} Q^{\dot{A}} & \widetilde{Q}_{\xi^\dagger} &= \xi^\dagger_{\dot{A}} \widetilde{Q}^{\dot{A}} \end{aligned} \tag{15}$$

Then we have

- An $\mathcal{N}=(2,0)$ quantum mechanics generated by the supercharges Q_ξ and $\widetilde{Q}_{\zeta^\dagger}$ with algebra

$$\{Q_\xi, \widetilde{Q}_{\xi^\dagger}\} = 2(H + \xi_{\dot{A}}\xi^\dagger_{\dot{B}} Z^{\dot{A}\dot{B}}) \,. \tag{16}$$

  This supersymmetric quantum mechanics has R-symmetry $U(1)_H$ and inherits the $U(1)_\epsilon$ flavor symmetry.

- An S^1 family of $\mathcal{N}=(1,1)$ supersymmetric quantum mechanics labelled by a phase ζ. This is generated by supercharges

(17) $$\mathcal{Q}_\zeta = \zeta^{-1/2} Q_\xi + \zeta^{1/2}\widetilde{Q}_\xi \qquad \widetilde{\mathcal{Q}}_\zeta = \zeta^{-1/2} Q_{\xi^\dagger} - \zeta^{1/2}\widetilde{Q}_{\xi^\dagger}$$

with

(18) $$\{\mathcal{Q}_\zeta, \widetilde{\mathcal{Q}}_\zeta\} = -2H \qquad \{\mathcal{Q}_\zeta, \mathcal{Q}_\zeta\} = 2Z_\zeta \qquad \{\widetilde{\mathcal{Q}}_\zeta, \widetilde{\mathcal{Q}}_\zeta\} = 2Z_{\zeta^\dagger}$$

where $Z_\zeta = \xi_{\dot{A}}\xi_{\dot{B}} Z^{\dot{A},\dot{B}}$. This family of supersymmetric quantum mechanics inherits the $U(1)_\epsilon$ flavor symmetry but $U(1)_H$ transformations rotate the S^1 family by $\zeta \to \zeta e^{i\theta}$.

It is straightforward to decompose our $\mathcal{N}=4$ supersymmetric quantum mechanics further in either of these two cases. For the purpose of this note, it will be convenient to phrase our computations in terms of the S^1 family of $\mathcal{N}=(1,1)$ supersymmetric quantum mechanics. Furthermore, we will fix a complex structure $\xi_{\dot{A}} = (1,0)$ on the Coulomb branch. This is left invariant by $U(1)_C$ and is therefore 'adapted' to this choice of maximal torus in $SU(2)_C$. In particular,

(19) $$\mathcal{Q}_\zeta^2 = Z^{\dot{1}\dot{1}} \sim \varphi + m_{\mathbb{C}}\,,$$

showing that Q_ζ behaves as an equivariant differential for G gauge and T_H flavor transformations, a fact that will become important later.

With this choice, the supersymmetric quantum mechanics can be conveniently described in terms of $\mathcal{N}=(1,1)$ real supermultiplets with bottom components σ, A_1, A_2 and the real and imaginary parts of X_j, Y_j together with the real superpotential

(20) $$h_\zeta = h + \mathrm{Re}(W/\zeta)$$

where

(21) $$h = \int |dz|^2\, \sigma\, (\mu_{\mathbb{R}} + t_{\mathbb{R}} + 2iF_{z\bar{z}}) \qquad W = \sum_j \int |dz|^2 X_j D_{\bar{z}} Y_j\,.$$

The supersymmetric ground states of the $\mathcal{N}=4$ quantum mechanics are configurations solving the BPS equations for all of the supercharges $Q^{\dot{A}}$, $\widetilde{Q}^{\dot{A}}$. This means they are supersymmetric ground states for every member of the S^1 family of $\mathcal{N}=(1,1)$ quantum mechanics and therefore critical points of the real superpotential h_ζ for all $|\zeta|=1$.

In Section 3, we will demonstrate that the critical points of the real superpotential h_ζ for all $|\zeta|=1$ modulo gauge transformations are generalized vortices in the $x^{1,2}$-plane. In the absence of the complex mass parameters, there is a moduli space of solutions $\mathcal{M}_n$ for each vortex number $n \in \mathbb{Z}_{\geq 0}$, which is a finite-dimensional non-compact Kähler manifold. Turning on complex mass parameters, the system is restricted to fixed points of the complex T_H transformation on $\mathcal{M}_n$ generated by $(m_1,\ldots,m_N)$. The supercharge Q_ζ descends to the T_H-equivariant differential on $\mathcal{M}_n$ and supersymmetric grounds states should be identified with the cohomology of Q_ζ.

This statement is quite subtle because the moduli spaces $\mathcal{M}_n$ and the fixed points of the complex T_H transformation generated by $m_{\mathbb{C}} = (m_1,\ldots,m_N)$ are non-compact. Standard physical considerations from supersymmetric quantum mechanics suggest one should use L^2 harmonic forms on $\mathcal{M}_n$. However, as we explain in the next section, such subtleties can be avoided by turning on a mass parameter for the $U(1)_\epsilon$ flavor symmetry.

2.6. Omega Background. Recall that the supersymmetric quantum mechanics has flavor symmetry $U(1)_\epsilon \times T_H$ but so far we have only turned on complex mass parameters $(m_1, \ldots, m_N)$ for T_H. From the point of view of supersymmetric quantum mechanics, there is no reason not to turn on a complex mass ϵ for the $U(1)_\epsilon$ flavor symmetry. From a three-dimensional perspective, this is known as an Ω-deformation in the $x^{1,2}$-plane.

The virtue of this deformation is that the combined $T_H \times U(1)_\epsilon$ action on $\mathcal{M}_n$ generated by $(m_1, \ldots, m_N)$ then has only isolated fixed points on $\mathcal{M}_n$, so that the supersymmetric quantum mechanics has only isolated massive vacua. The mass parameter ϵ makes an additional contribution to the central charge Z^{ii} so that the supersymmetry algebra is modified to

$$Q_\zeta^2 \sim \varphi + m_{\mathbb{C}} + \epsilon\,. \tag{22}$$

The Hilbert space of the supersymmetric quantum mechanics is then identified with the the standard $T_H \times U(1)_\epsilon$ equivariant cohomology of $\mathcal{M}_n$, summed over all vortex numbers $n \geq 0$, with the equivariant differential given by $\mathcal{Q}_\zeta$.

We note that the Ω-background was introduced in the context of 4d $\mathcal{N} = 2$ supersymmetry on $\mathbb{R}^4$ with coordinates x^1, x^2, x^3, x^4 and deformation parameters ϵ and ϵ' corresponding to rotations in the $x^{1,2}$-plane and $x^{3,4}$-planes respectively [**14**]. Our construction can be obtained by sending $\epsilon' \to 0$, compactifying $x^4 \sim x^4 + 2\pi R$ and sending $R \to 0$.

3. Hilbert space

3.1. Half-BPS Equations. The supersymmetric ground states of the $\mathcal{N} = 4$ supersymmetric quantum mechanics are configurations preserving all of the supercharges $Q^{\dot{A}}, \tilde{Q}^{\dot{A}}$. Such configurations are supersymmetric ground states for every member of the S^1 family of $\mathcal{N} = (1,1)$ quantum mechanics and are therefore critical points of the real superpotential h_ζ given in equation (21) for all $|\zeta| = 1$. This requires that

$$dh = 0 \qquad dW = 0\,. \tag{23}$$

Expanding these equations and grouping them into real and complex equations, we find the following half-BPS equations in the three-dimensional gauge theory for the supercharges $Q^{\dot{A}}, \tilde{Q}^{\dot{A}}$,

$$-2iF_{z\bar{z}} = \mu_{\mathbb{R}} + t_{\mathbb{R}} \qquad D_z\sigma = 0 \qquad D_{\bar{z}}\sigma = 0 \tag{24}$$

$$\mu_{\mathbb{C}} = 0 \qquad D_{\bar{z}}X_j = 0 \qquad D_{\bar{z}}Y_j = 0 \tag{25}$$

$$\sigma X_j = 0 \qquad -\sigma Y_j = 0\,. \tag{26}$$

Note that these equations are independent of the coordinate x^3. We require that solutions tend to a supersymmetric vacuum ν_i from equation (9) and divide by gauge transformations in the $x^{1,2}$-plane that tend to a constant value at $|z| \to \infty$.

In the absence of the complex masses $(m_1, \ldots, m_N, \epsilon)$, the critical point equations are further supplemented by

$$\varphi X_j = 0 \qquad -\varphi Y_j = 0\,. \tag{27}$$

As analyzed in more detail below, the solutions of these equations are labelled by a vortex number $\mathfrak{n} \in \mathbb{Z}_{\geq 0}$, which is the flux through the $x^{1,2}$-plane. For each $\mathfrak{n} \in \mathbb{Z}_{\geq 0}$ there is a corresponding non-compact Kähler moduli space $\mathcal{M}^{\mathfrak{n}}_{\nu_i}$ of solutions of complex dimension $\mathfrak{n}N$.

Turning on complex masses $(m_1, \ldots, m_N, \epsilon)$ for the $T_H \times U(1)_\epsilon$ flavor symmetry deforms the supplementary equations (27) to

$$(\varphi + m_j + \tfrac{\epsilon}{2} + \epsilon z D_z)X_j = 0 \qquad (-\varphi - m_j + \tfrac{\epsilon}{2} + \epsilon z D_z)Y_j = 0 \,. \tag{28}$$

This now requires that solutions are invariant under the combined complex gauge and flavor transformation generated by φ and $(m_1, \ldots, m_N, \epsilon)$. This corresponds to the fixed points of the corresponding $T_H \times U(1)_\epsilon$ transformation on the moduli space $\mathcal{M}_\nu$. In our example there is a single fixed point for each vortex number $\mathfrak{n} \in \mathbb{Z}_{\geq 0}$, which will contribute a single state $|\mathfrak{n}\rangle$ to the Hilbert space of supersymmetric ground states.

Mathematically, the Hilbert space of supersymmetric ground states $\mathcal{H}_{\nu_i}$ with supersymmetric vacuum ν_i at infinity is identified with the $T_H \times U(1)_\epsilon$-equivariant cohomology of the moduli space of generalized vortices,

$$\mathcal{H}_{\nu_i} = \bigoplus_{\mathfrak{n} \geq 0} H^*_{T_H \times U(1)_\epsilon}(\mathcal{M}^{\mathfrak{n}}_{\nu_i}, \mathbb{C}) \,, \tag{29}$$

with the equivariant differential realized by any of the supercharges Q_ζ with $|\zeta| = 1$. In order to compute the equivariant cohomology, we will employ a complex algebraic description of the moduli spaces $\mathcal{M}^{\mathfrak{n}}_{\nu_i}$. There is a natural basis $|\mathfrak{n}\rangle$, $n \in \mathbb{Z}_{\geq 0}$ for the equivariant cohomology in 1-1 correspondence with fixed points of $\mathcal{M}^{\mathfrak{n}}_{\nu_i}$.

3.2. General structure. We begin by studying solutions to the half-BPS equations in the absence of complex mass parameters $(m_1, \ldots, m_N, \epsilon)$. In this case, $\varphi = \sigma = 0$ everywhere. We then find a moduli space of solutions to the remaining equations,

$$\mathcal{M}_{\nu_i} = \left\{ \begin{array}{l} D_{\bar z} X_j = D_{\bar z} Y_j = 0 \\ \mu_{\mathbb{C}} = 0 \\ -2iF_{z\bar z} = \mu_{\mathbb{R}} + t_{\mathbb{R}} \end{array} \quad \Big| \; X_j, Y_j \overset{|z| \to \infty}{\longrightarrow} G \cdot \nu_i \right\} / \mathcal{G} \,, \tag{30}$$

where $\mathcal{G}$ is the infinite-dimensional group of gauge transformations in the $x^{1,2}$-plane that are constant at infinity and $G \cdot \nu_i$ denotes the $G = U(1)$ orbit of the supersymmetric vacuum ν_i on the hypermultiplet scalars.

The moduli space splits into disconnected components

$$\mathcal{M}_{\nu_i} = \bigcup_{\mathfrak{n}} \mathcal{M}^{\mathfrak{n}}_{\nu_i} \tag{31}$$

labelled by a vortex number $\mathfrak{n} \in \pi_1(G) = \mathbb{Z}$ or flux through the $x^{1,2}$-plane,

$$\mathfrak{n} = \frac{1}{2\pi} \int_{\mathbb{R}^2} F \,. \tag{32}$$

With our choice $t_{\mathbb{R}} < 0$, only the components with $\mathfrak{n} \geq 0$ are non-empty. They are Kähler manifolds of complex dimension $\mathfrak{n}N$. In order to perform explicit computations in equivariant cohomology, it is convenient to introduce a complex algebraic description of the moduli spaces $\mathcal{M}^{\mathfrak{n}}_{\nu_i}$.

3.3. Algebraic description. The complex algebraic description is obtained by dropping the real moment-map equation and instead dividing by complex gauge transformations $G_{\mathbb{C}} = \mathbb{C}^*$,

$$\mathcal{M}_{\nu_i} \simeq \left\{ \begin{array}{l} D_{\bar{z}} X_j = D_{\bar{z}} Y_j = 0 \\ \mu_{\mathbb{C}} = 0 \end{array} \mid X_j, Y_j \stackrel{|z|\to\infty}{\longrightarrow} G_{\mathbb{C}} \cdot \nu_i \right\} / \, \mathcal{G}_{\mathbb{C}} \,. \tag{33}$$

Usually, a stability condition must be imposed in the algebraic quotient. However, any solution that tends to a supersymmetric vacuum ν_i at infinity is automatically stable, so no further conditions are necessary in the algebraic quotient (33). The equivalence between the descriptions (30) and (33) is a version of the Hitchin-Kobayashi correspondence for the generalized vortex equations.

From the complex algebraic point of view, the vortex number $\mathfrak{n} \in \mathbb{Z}$ determines a complex $G_{\mathbb{C}}$-bundle $\mathcal{O}(\mathfrak{n})$ on the compactification $\mathbb{CP}^1$ of the $x^{1,2}$-plane. A point in the moduli space $\mathcal{M}^{\mathfrak{n}}_{\nu_i}$ is then specified by holomorphic sections X_j, Y_j of the associated bundle $\mathcal{O}(\mathfrak{n})^N \oplus \mathcal{O}(-\mathfrak{n})^N$ that satisfy the complex moment map constraint $\mu_{\mathbb{C}} = 0$ and lie in the complex orbit $G_{\mathbb{C}} \cdot \nu_i$ at infinity.

Using a complex gauge transformation, we can pass to a holomorphic frame in which the sections are described concretely as polynomials $X_j(z), Y_j(z)$ of degree at most $\mathfrak{n}$, $-\mathfrak{n}$ in the affine coordinate z. We then have the following description of the moduli space $\mathcal{M}^{\mathfrak{n}}_{\nu_i}$:

- If $\mathfrak{n} > 0$ then only the $X_j(z)$ are nonzero. Hitting the supersymmetric vacuum ν_i at infinity requires the leading coefficient of $X_j(z)$ with $j \neq i$ to vanish while the leading coefficient of $X_j(z)$ is nonvanishing. A constant complex gauge transformation sets the leading coefficient of $X_i(z)$ to 1, such that

$$X_j(z) = \delta_{ij} z^{\mathfrak{n}} + \sum_{l=0}^{\mathfrak{n}-1} x_{j,l} z^l \,. \tag{34}$$

  The coefficients $x_{i,l}$ are unconstrained and parameterize $\mathcal{M}^{\mathfrak{n}}_{\nu_j} \simeq \mathbb{C}^{N\mathfrak{n}}$.
- If $\mathfrak{n} = 0$, both X_j and Y_j are nonzero constants. However, the requirement that they hit the vacuum ν_i at infinity sets them equal to their vacuum values. Thus $\mathcal{M}^0_{\nu_i}$ is a point.
- If $\mathfrak{n} < 0$ then only the $Y_j(z)$ can be nonzero. This is incompatible with the vacuum ν_i, so $\mathcal{M}^{\mathfrak{n}}_{\nu_i}$ is empty.

The complex algebraic description of the moduli space is familiar in the physics literature from the work of Morrison and Plesser [**12**] and the moduli matrix construction of vortices [**9, 10**]. Mathematically, we are describing based holomorphic maps $\mathbb{CP}^1 \to [\mathcal{M}_H]$ into the Higgs branch stack $[\mathcal{M}_H] = [\mu_{\mathbb{C}}^{-1}(0)/G_{\mathbb{C}}]$, sending the point at infinity to the complex orbit $G_{\mathbb{C}} \cdot \nu_i$.

3.4. Fixed points and the Hilbert space. Turning on complex masses $(m_1, \ldots, m_N, \epsilon)$ makes the supersymmetric quantum mechanics completely massive. Equations (28) force the system to the $T_H \times U(1)_\epsilon$ fixed points on $\mathcal{M}_{\nu_i}$. The Hilbert space of supersymmetric ground states is identified with the $T_H \times U(1)_\epsilon$ equivariant cohomology of $\mathcal{M}_{\nu_i}$ with a natural basis labelled by the equivariant fixed points of $\mathcal{M}_{\nu_i}$.

The equivariant fixed points are straightforward to identify using the algebraic description of the moduli spaces $\mathcal{M}^{\mathfrak{n}}_{\nu_i}$. Let us consider an infinitesimal combined gauge and $T_H \times U(1)_\epsilon$ flavor transformation generated by $(\varphi, m_1, \ldots, m_N, \epsilon)$. This sends

(35) $$X_j(z) \mapsto (\varphi + m_j + \frac{\epsilon}{2} + \epsilon z\partial_z)X_j(z)\,.$$

For $\mathfrak{n} \geq 0$, there is a unique fixed point $X_j(z) = \delta_{ij} z^{\mathfrak{n}}$ with

(36) $$\varphi = -m_i - (\mathfrak{n} + \tfrac{1}{2})\epsilon\,.$$

which is simply origin of $\mathcal{M}^{\mathfrak{n}}_{\nu_j} = \mathbb{C}^{N\mathfrak{n}}$. Denoting the corresponding state in the quantum mechanics as $|\mathfrak{n}\rangle$, we find that

(37) $$\mathcal{H} = \bigoplus_{\mathfrak{n}\geq 0} \mathbb{C}\,|\mathfrak{n}\rangle\,.$$

The Hilbert space has a natural inner product from the supersymmetric quantum mechanics: $\langle\mathfrak{n}'|\mathfrak{n}\rangle$ is given by computing the path integral of the supersymmetric quantum mechanics with $|\mathfrak{n}\rangle$ at $x^3 \to -\infty$ and $\langle\mathfrak{n}'|$ at $x^3 \to \infty$. The path integral is zero unless $\mathfrak{n} = \mathfrak{n}'$, in which case it is given by the equivariant integral $\int_{\mathcal{M}^{\mathfrak{n}}_{\nu}}$ of the product of equivarint cohomology classes representing $\langle\mathfrak{n}'|$ and $|\mathfrak{n}\rangle$.

At this stage, there is a slight ambiguity in the normalization of the states $|\mathfrak{n}\rangle$. A natural choice is that $|\mathfrak{n}\rangle$ is the Poincaré dual of the fundamental class of the fixed point of $\mathcal{M}^{\mathfrak{n}}_{\nu_i} = \mathbb{C}^{\mathfrak{n}N}$, in other words an equivariant δ-function supported at the origin. In this case, we would find

(38) $$\langle\mathfrak{n}'|\mathfrak{n}\rangle = \delta_{\mathfrak{n}',\mathfrak{n}}\,\omega_{\mathfrak{n}}$$

where $\omega_{\mathfrak{n}}$ is the equivariant weight of the tangent space to $\mathcal{M}^{\mathfrak{n}}_{\nu_i} = \mathbb{C}^{\mathfrak{n}N}$ at the origin. Alternatively, we could normalize $|\mathfrak{n}\rangle$ by the equivariant weight of the tangent space at the fixed point, so that

(39) $$\langle\mathfrak{n}'|\mathfrak{n}\rangle = \frac{\delta_{\mathfrak{n}',\mathfrak{n}}}{\omega_{\mathfrak{n}}}$$

From a physical perspective, neither normalization is especially preferred. Here we choose the latter normalization (39).

The only remaining task is to compute the equivariant weight of the tangent space to the origin in $\mathcal{M}^{\mathfrak{n}}_{\nu_i} = \mathbb{C}^{\mathfrak{n}N}$. This is parameterized by the subleading coefficients $x_{j,l}$ in the expansion (34), which transform as

(40) $$\begin{aligned} x_{j,l} &\mapsto (\varphi + m_j + (l + \tfrac{1}{2})\epsilon)x_{j,l} \\ &= (m_j - m_i + (l - \mathfrak{n})\epsilon)x_{j,l} \end{aligned}$$

where we evaluate $\varphi = -m_i - (\mathfrak{n} + \frac{1}{2})\epsilon$ at the origin in $\mathcal{M}^{\mathfrak{n}}_{\nu_i} = \mathbb{C}^{\mathfrak{n}N}$. Therefore, the inner product on the Hilbert space is given by

(41) $$\langle\mathfrak{n}'|\mathfrak{n}\rangle = \delta_{\mathfrak{n}',\mathfrak{n}} \prod_{j=1}^{N} \prod_{l=0}^{\mathfrak{n}-1} \frac{1}{m_j - m_i + (l - \mathfrak{n})\epsilon}\,.$$

It is often convenient to introduce a characteristic polynomial for the Higgs branch flavor symmetry, $P(x) = \prod_{j=1}^{N}(x + m_j)$, and write the inner product as

(42) $$\langle\mathfrak{n}'|\mathfrak{n}\rangle = \delta_{\mathfrak{n}',\mathfrak{n}} \prod_{l=0}^{\mathfrak{n}-1} \frac{1}{P(-m_i + (l - \mathfrak{n})\epsilon)}\,.$$

4. Monopole operators

4.1. Monopole Operators. We now consider half-BPS operators in the $\mathcal{N}=4$ supersymmetric quantum mechanics preserving the supercharges $Q^{\mathfrak{i}}$, $\tilde{Q}^{\mathfrak{i}}$. Such operators arise from Coulomb branch chiral ring operators in the original three-dimensional theory annihilated by $Q^{1\dot{A}}_{\alpha}$.

In an abelian theory, one such operator is the complex scalar φ which acts on the vortex state $|\mathfrak{n}\rangle$ by evaluation at the corresponding fixed point

$$\varphi|\mathfrak{n}\rangle = (-m_i - (\mathfrak{n}+\tfrac{1}{2})\epsilon)|\mathfrak{n}\rangle\,. \tag{43}$$

However, there are also monopole operators v_A labelled by an integer magnetic charge A, which are defined by removing a small S^2_p around a point p and imposing singular boundary conditions in the path integral,

$$F = A\sin\theta\, d\theta\wedge d\phi+\cdots \qquad \sigma = -\frac{A}{2r}+\cdots \tag{44}$$

where (r,θ,ϕ) are spherical coordinates around the point p [**2**]. In this section, we explain how these monopole operators can be understood from our $\mathcal{N}=4$ supersymmetric quantum mechanics and how they act on the Hilbert space of supersymmetric ground states of the quantum mechanics.

As a preliminary observation, we note that the monopole operator v_A creates A units of flux on a small sphere S^2_p surrounding the point p where it is inserted,

$$\frac{1}{2\pi}\int_{S^2} F = A\in\mathbb{Z}\,. \tag{45}$$

Therefore, by topological considerations alone, we must have

$$v_A|\,\mathfrak{n}\,\rangle = \begin{cases} c_{A,\mathfrak{n}}|\,\mathfrak{n}+A\,\rangle & \text{if } \mathfrak{n}+A\geq 0\\ 0 & \text{if } \mathfrak{n}+A<0\end{cases}. \tag{46}$$

In other words, monopole operators create and annihilate vortices. Our task is therefore reduced to computing explicitly the coefficients $c_{A,\mathfrak{n}}$. Since the vortex states $|\mathfrak{n}\rangle$ are orthogonal, this is equivalent to computing the non-zero correlation functions $\langle\mathfrak{n}+A|v_A|\mathfrak{n}\rangle$.

4.2. Quarter BPS Equations. As noted above, the monopole operators preserve the supercharges $Q^{\mathfrak{i}}$, $\tilde{Q}^{\mathfrak{i}}$ of the $\mathcal{N}=4$ supersymmetric quantum mechanics. This is equivalent to preserving the supercharge Q_ζ for all phases ζ. They should therefore correspond to singular solutions of the instanton equations for every member of the S^1 family of $\mathcal{N}=(1,1)$ quantum mechanics.

The instanton equations for the supercharge Q_ζ are gradient flow equations for the real superpotential h_ζ,

$$D_3\Phi = -\frac{\delta h_\zeta}{\delta\Phi} \tag{47}$$

Imposing the instanton equations for all $|\zeta|=1$ we find

$$\frac{\delta W}{\delta\Phi}=0 \qquad D_3\sigma = -\frac{\delta h}{\delta\Phi} \tag{48}$$

where Φ stands for the scalar components of real $\mathcal{N}=(1,1)$ supermultiplets, namely σ, A_1, A_2 and the real and imaginary parts of X_j, Y_j. Expanding and grouping into real and complex equations, we find the following quarter-BPS equations in the three-dimensional gauge theory for the supercharges Q^{i}, $\tilde{Q}^{\mathrm{i}}$,

$$-2iF_{z\bar{z}} - D_3\sigma = \mu_{\mathbb{R}} + t_{\mathbb{R}} \qquad F_{3z} - iD_z\sigma = 0 \qquad F_{3\bar{z}} + iD_{\bar{z}}\sigma = 0 \tag{49}$$

$$\mu_{\mathbb{C}} = 0 \qquad D_{\bar{z}}X_j = 0 \qquad D_{\bar{z}}Y_j = 0 \tag{50}$$

$$(D_3 + \sigma)X = 0 \qquad (D_3 - \sigma)Y = 0\,. \tag{51}$$

These equations are again supplemented by

$$(\varphi + m_j + \tfrac{\epsilon}{2} + \epsilon z D_z)X_j = 0 \qquad (-\varphi - m_j + \tfrac{\epsilon}{2} + \epsilon z D_z)Y_j = 0\,. \tag{52}$$

Note that x^3-independent solutions of these equations reduce in axial gauge $A_3 = 0$ to supersymmetric ground states preserving all of the supercharges of the $\mathcal{N} = 4$ quantum mechanics.

In principle, a correlation function $\langle \mathfrak{n}+A|v_A|\mathfrak{n}\rangle$ in the supersymmetric quantum mechanics localizes to an equivariant integral over the moduli space of solutions to (49)-(51) that tend to vortex solutions as $x^3 \to \pm\infty$ with $\mathfrak{n}$ and $\mathfrak{n}+A$ units of flux in the $x^{1,2}$-plane, with a monopole singularity generating A units of flux at some point on the x^3-axis. Instead we will compute the action of monopole operators $v_A|\mathfrak{n}\rangle$ directly using a complex algebraic description of the solutions to equations (49)-(51).

4.3. Algebraic Approach. Let us recall from section the complex algebraic description of the moduli space $\mathcal{M}^{\mathfrak{n}}_{\nu_i}$ of solutions to the x^3-independent half-BPS equations for $Q^{\dot{A}}$, $\tilde{Q}^{\dot{A}}$ with vacuum ν_i at $|z| \to \infty$. A point in $\mathcal{M}_{\nu_i}$ is specified by the following 'holomorphic data':

- A complex line bundle $E \cong \mathcal{O}(\mathfrak{n})$.
- Holomorphic sections (X_j, Y_j) of the associated bundle $\mathcal{O}(\mathfrak{n})^N \oplus \mathcal{O}(-\mathfrak{n})^N$ obeying $\sum_j X_jY_j = 0$ and lying in the complex orbit $G_{\mathbb{C}} \cdot \nu_i$ at $|z| \to \infty$.

This description was sufficient to build an explicit description of the Hilbert space of supersymmetric ground states $\mathcal{H}_{\nu_i}$ as the $U(1)_\epsilon \times T_H$-equivariant cohomology of $\mathcal{M}_{\nu_i}$ with equivariant parameters $(m_1, \ldots, m_N, \epsilon)$.

Let us now assume that at some point $x^3 = s_0$ we have solution specified by a point in $\mathcal{M}^{\mathfrak{n}}_{\nu_i}$. We will now ask how the holomorphic data evolves as a function of $x^3 \geq s_0$ by solving the quarter-BPS equations for Q^{i}, $\tilde{Q}^{\mathrm{i}}$. Choosing axial gauge $A_3 = 0$, it follows from the quarter-BPS equations (49)-(51) that

$$\partial_3 A_{\bar{z}} = -iD_{\bar{z}}\sigma \qquad \partial_3 X = -\sigma X \qquad \partial_3 Y = -\sigma Y\,. \tag{53}$$

This shows that evolution in the x^3-direction is a complex gauge transformation with parameter $i\sigma$. Therefore provided σ is smooth, the holomorphic type of the bundle $E \simeq \mathcal{O}(\mathfrak{n})$ cannot change. Together with $D_{\bar{z}}X = 0$, $D_{\bar{z}}Y = 0$ and $\mu_{\mathbb{C}} = 0$, this ensures that the holomorphic data are constant in the x^3-direction. More precisely, the holomorphic data at nearby s and s' are related by a globally invertible, holomorphic gauge transformation $g(z; s, s')$.

However, at a collection of points $\{s_i\}$ the holomorphic data can jump due to the presence of a monopole operator on the x^3-axis with a singularity for σ. The holomorphic data at $x^3 < s_i$ and $x^3 > s_i$ are then related by a 'singular' complex gauge transformation $g(z)$ that is only invertible in the complement of the origin $z = 0$. In our example, we consider the singular gauge transformations,

$$g(z) \sim z^A \qquad A \in \mathbb{Z}\,, \tag{54}$$

corresponding to the insertion of a monopole operator v_A at $x^3 = s_i$ and $z = 0$. This is known as a 'Hecke modification' of the holomorphic data. Such modifications were analyzed by Kapustin and Witten [**11**] in a four-dimensional lift of our current setup.

4.4. Action on Hilbert Space. The action of the singular gauge transformation $g(z) = z^A$ on the holomorphic data is summarized as follows:

- If $A \geq 0$, the gauge transformation sends $X_i(z) \mapsto z^A X_i(z)$. This creates A vortices at the origin of the z-plane.
- If $A < 0$, the transformation sends $X_i(z) \mapsto z^{-|A|} X_i(z)$. Regularity of this modification requires that $X_i(z)$ have a zero of order A at $z = 0$. In other words, there must exist A vortices at the origin of the z-plane to be destroyed by the monopole operator.

To determine the coefficients $c_{A,\mathfrak{n}}$ in equation (46), we examine the action of the singular gauge transformation in a neighborhood of the fixed points of $\mathcal{M}^{\mathfrak{n}}_{\nu_i}$ and $\mathcal{M}^{\mathfrak{n}+A}_{\nu_i}$. Note that if $A > 0$, the singular gauge transformation z^A is a composition of A singular gauge transformations z. In terms of monopole operators, we therefore write $v_A = (v_+)^A$. Similarly, if $A < 0$ we write $v_A = (v_-)^{|A|}$. Thus it suffices to determine the action of v_+ and v_-.

Let us therefore consider the action of the monopole operator v_+ on the state $|\mathfrak{n}-1\rangle$. A vortex configuration in a neighborhood of the origin of $\mathcal{M}^{\mathfrak{n}-1}_{\nu_i}$ has the general form

$$X_j(z) = z^{\mathfrak{n}-1}\delta_{ij} + \sum_{l=0}^{\mathfrak{n}-2} x_{j,l+1} z^l\,. \tag{55}$$

This is mapped by the singular gauge transformation $g(z) = z$ to

$$g(z)X_j(z) = z^{\mathfrak{n}}\delta_{ij} + \sum_{l=1}^{\mathfrak{n}-1} x_{j,l} z^l\,. \tag{56}$$

Thus the image of $g(z)$ is the subspace of $\mathcal{M}^{\mathfrak{n}}_{\nu_i} \cong \mathbb{C}^{\mathfrak{n}N}$ with $x_{j,0} = 0$ for all $j = 1, \ldots, N$. This means that $|\mathfrak{n}-1\rangle$ is mapped to $|\mathfrak{n}\rangle$, times an equivariant δ-function imposing the constraints $x_{j,0} = 0$. We therefore multiply by the equivariant weights of the coordinates $x_{j,0}$ for $j = 1, \ldots, N$. The result is

$$v_+|\mathfrak{n}-1\rangle = P(-m_i - \mathfrak{n}\epsilon)|\mathfrak{n}\rangle\,, \tag{57}$$

where $P(u) = \prod_{j=1}^N (u + m_j)$.

On the other hand, to compute the action of the monopole operator v_- on the vortex state $|\mathfrak{n}\rangle$ we consider the singular gauge transformation $g(z) = z^{-1}$. This time the subspace of $\mathcal{M}^{\mathfrak{n}}_{\nu_i}$ defined by $x_{j,0} = 0$ maps isomorphically onto $\mathcal{M}^{\mathfrak{n}-1}_{\nu_i}$. We therefore have $v_-|\mathfrak{n}\rangle = |\mathfrak{n}-1\rangle$ for $\mathfrak{n} > 0$, and $v_-|0\rangle = 0$. An alternative perspective on this computation in terms of correspondences can be found in [**7**].

We can therefore summarize the action of Coulomb branch chiral ring operators on the Hilbert space of supersymmetric ground states $\mathcal{H}_{\nu_i}$ by

$$\begin{aligned} \varphi|\mathfrak{n}\rangle &= (-m_i - (\mathfrak{n}+\tfrac{1}{2})\epsilon)|\mathfrak{n}\rangle \\ v_+|\mathfrak{n}\rangle &= P\big(\varphi+\tfrac{1}{2}\epsilon\big)\big|\mathfrak{n}+1\big\rangle \\ v_-|\mathfrak{n}\rangle &= |\mathfrak{n}-1\rangle\,. \end{aligned} \tag{58}$$

A short computation shows that the monopole operators obey the algebra

$$\begin{gathered} v_+v_- = P(\varphi+\tfrac{1}{2}\epsilon) \qquad v_-v_+ = P(\varphi-\tfrac{1}{2}\epsilon) \\ [\varphi, v_\pm] = \mp\epsilon v_\pm\,. \end{gathered} \tag{59}$$

This is a non-commutative deformation of the Coulomb branch chiral ring (8). It is a deformation quantization of the Coulomb branch with holomorphic symplectic form $d\varphi\wedge d\log u^+$. The complex masses $(m_1,\dots,m_N)$ are the period of the quantization.

4.5. Some Representation Theory. The deformation quantization (59) is a spherical rational Cherednik algebra in the mathematical literature. It is graded by the topological symmetry $G_C \simeq U(1)_t$ under which φ, u^+, u^- have charge $0, -1, +1$. For generic complex masses $(m_1,\dots,m_N)$, every vortex state $|\mathfrak{n}\rangle \in \mathcal{H}_{\nu_i}$ can be obtained by acting on $|0\rangle$ with the monopole operators v_+ of negative grading. The Hilbert spaces of supersymmetric ground states $\mathcal{H}_{\nu_i}$ transform as in equation (58) as highest-weight Verma modules of the spherical rational Cherednik algebra with respect to this grading.

In the special case $N = 2$, the deformation quantization is isomorphic to a central quotient of the universal enveloping algebra $U(\mathfrak{sl}_2)$, with the quadratic Casimir element fixed by the complex masses $(m_1,\dots,m_N)$. In particular, defining the generators

$$h = 2\varphi \qquad e = -v_- \qquad f = v_+ \tag{60}$$

we find

$$[h,e] = 2\epsilon e\,, \qquad [h,f] = -2\epsilon f\,, \qquad [e,f] = \epsilon h\,, \tag{61}$$

and

$$C_2 = \frac{1}{2}h^2 + ef + fe = \frac{1}{2}((m_1-m_2)^2 - \epsilon^2)\,. \tag{62}$$

The enveloping algebra $U(\mathfrak{sl}_2)$ at a generic value of the central charge admits two irreducible Verma modules, which can be identified with the Hilbert spaces of supersymmetric ground states $\mathcal{H}_{\nu_1}$, $\mathcal{H}_{\nu_2}$ associated to the two isolated massive vacua.

5. Boundary conditions and overlaps

In this section, we enrich the setup considered previously by adding boundary conditions $\mathcal{B}$ that preserve a 2d $\mathcal{N} = (2,2)$ supersymmetry algebra in the $x^{1,2}$-plane with vector R-symmetry $U(1)_H$. Large families of boundary conditions of this type were introduced in [**6**]. Such boundary conditions preserve the supercharges Q^{i}, $\tilde{Q}^{\mathrm{i}}$ of the $\mathcal{N} = 4$ supersymmetric quantum mechanics. Correlation functions involving such boundary conditions can then be performed by localization to the appropriate solutions of the quarter-BPS equations (49)-(51).

Boundary condition of this type that are compatible with a real FI parameter $t_{\mathbb{R}} < 0$ and generic complex masses $(m_1,\dots,m_N)$ will define a state in the Hilbert space $\mathcal{H}_{\nu_i}$ of supersymmetric ground states. In section 5.1, we will construct this boundary state for a class of Neumann boundary conditions in our abelian example. In this case, the boundary state is a coherent state of vortices, or equivalently a generalized eigenvector of the monopole operators $u^{\pm}$. Mathematically, it defines a generalized Whittaker vector in $\mathcal{H}_\nu$.

Compactifying the three-dimensional theory on an interval with such boundary conditions at either end leads to a 2d $\mathcal{N}=(2,2)$ gauge theory. In section 5.2, we show that the vortex partition function of this 2d $\mathcal{N}=(2,2)$ theory in Ω-background is an inner product of the corresponding boundary states in $\mathcal{H}_{\nu_i}$. This can be viewed as a finite version of the AGT correspondence, vastly extending and providing the correct physical setup for the beautiful mathematical work [**3**].

5.1. Neumann Boundary Conditions. We will focus here on boundary conditions that involve Neumann boundary conditions for the gauge field and therefore preserve the gauge symmetry at the boundary. The boundary conditions for a $G=U(1)$ vectormultiplet are [**6**]

$$F_{3j}|=0 \qquad \partial_3\varphi|=0 \qquad \sigma+i\gamma|=\tau_{2d} \tag{63}$$

where γ is the dual photon and $\tau_{2d}=t_{2d}+i\theta_{2d}$ is a combination of a boundary FI parameter and theta angle. The exponential $\xi=e^{\tau_{2d}}$ transforms as the bottom component of a 2d $\mathcal{N}=(2,2)$ twisted chiral multiplet.

The remaining boundary conditions for the N hypermultiplets are labelled by a sign vector $\varepsilon=(\varepsilon_1,\ldots,\varepsilon_N)$ with

$$\begin{aligned} \varepsilon_j=+ \quad &: \quad D_sX_j|=0 \qquad Y_j|=0 \\ \varepsilon_j=- \quad &: \quad D_sY_j|=0 \qquad X_j|=0\,. \end{aligned} \tag{64}$$

We therefore label Neumann boundary conditions by $\mathcal{N}_{\varepsilon,\xi}$.

Recall that monopole operators are given semi-classically by

$$v_\pm \sim e^{\pm(\sigma+i\gamma)} \tag{65}$$

and one might therefore expcect that $v_\pm| \sim \xi^{\pm}$ for a monopole operator brought to the Neumann boundary condition. However, 1-loop quantum corrections modify this relation so that the boundary Ward identity for the action of bulk monopole operators $v_\pm$ on the boundary state is given by [**6**]

$$\begin{aligned} v_+|\mathcal{N}_\varepsilon\rangle &= \xi \prod_{i \text{ s.t. } \varepsilon_i=+} (\varphi+m_i+\tfrac{\epsilon}{2})\,|\mathcal{N}_\varepsilon\rangle\,, \\ v_-|\mathcal{N}_\varepsilon\rangle &= \xi^{-1} \prod_{i \text{ s.t. } \varepsilon_i=-} (-\varphi-m_i+\tfrac{\epsilon}{2})\,|\mathcal{N}_\varepsilon\rangle\,. \end{aligned} \tag{66}$$

Note that the factors appearing on the right are the equivariant weights of the chiral fields with Neumann boundary conditions: X_j if $\varepsilon_j=+$ and Y_j if $\varepsilon_j=-$. It is easy to check that this is compatible with the algebra (58). The states $|\mathcal{N}_{\varepsilon,\zeta}\rangle$ are known as generalized Whittaker vectors.

One way to derive equation (66) directly would be to compute the overlaps $\langle \mathfrak{n}|\mathcal{N}_{\varepsilon,\xi}\rangle$ from the path integral with Neumann boundary condition at $x^3\to-\infty$ and the vortex configuration corresponding to the fixed point $X_j(z)=\delta_{ij}z^{\mathfrak{n}}$ at $x^3\to+\infty$. This would reduce to an equivariant integral over solutions to the quarter BPS equations for $\widetilde{Q}^{\mathrm{i}}$, Q^{i} with these boundary conditions.

Let us examine in more detail the Neumann boundary condition with $\varepsilon = (+,\ldots,+)$ where X_j all have Neumann boundary conditions and Y_j all have Dirichlet boundary conditions. In this case, the boundary state obeys

$$\begin{aligned} v_+|\mathcal{N}_{\varepsilon,\xi}\rangle &= \xi\, P(\varphi+\tfrac{\epsilon}{2})|\mathcal{N}_{\varepsilon,\xi}\rangle \\ v_-|\mathcal{N}_{\varepsilon,\xi}\rangle &= \xi^{-1}|\mathcal{N}_{\varepsilon,\xi}\rangle\,. \end{aligned} \tag{67}$$

with solution

$$|\mathcal{N}_{(+,\ldots,+),\xi}\rangle = \xi^{\varphi/\epsilon}\sum_{\mathfrak{n}\geq 0}|\mathfrak{n}\rangle\,. \tag{68}$$

This state can be characterized as a coherent state of vortices: it is an eigenvector of the annihilation operator v_- and has non-vanishing overlap with all vortex states $|\mathfrak{n}\rangle$, $\mathfrak{n}\geq 0$. This to be expected as this Neumann boundary condition is compatible with all vortex configurations for $t_{\mathbb{R}}<0$.

In the opposite case, $\varepsilon = (-,\ldots,-)$, there is no non-trivial solution of the boundary Ward identities (66) in $\mathcal{H}_{\nu_j}$ and therefore $|\mathcal{N}_{\varepsilon,\xi}\rangle = 0$. This is compatible with the observation that this Neumann boundary condition is incompatible with solutions of the vortex equations for $t_{\mathbb{R}}<0$ so that $\langle\mathfrak{n}|\mathcal{N}_{\varepsilon,\xi}\rangle = 0$ for all $\mathfrak{n}\geq 0$. The intermediate cases are discussed in [**6**].

5.2. Overlaps. With the above results, we can now compute the partition function of our theory on an interval with Neumann boundary condition $\mathcal{N}_{\varepsilon,\xi}$ and $\mathcal{N}_{\varepsilon',\xi'}$ at either end - see figure. Let us denote the partition function of this system by $Z_{\nu_i}(q)$ where we define $q=\xi/\xi'$. This partition function can be computed in two ways:

1) In the $\mathcal{N}=4$ supersymmetric quantum mechanics, the Neumann boundary conditions defines states $|\mathcal{N}_{\varepsilon,\xi}\rangle$ and $|\mathcal{N}_{\varepsilon',\xi'}\rangle$ in $\mathcal{H}_{\nu_i}$ and the partition function Z_{ν_i} is the overlap $\langle\mathcal{N}_{\varepsilon',\xi'}|\mathcal{N}_{\varepsilon,\xi}\rangle$.
2) Since the partition function is independent of the length of the interval, we can send this length to zero to obtain a 2d $\mathcal{N}=(2,2)$ theory $\mathcal{T}_{2d}$. The partition function $Z_{\nu_i}(q)$ is then identified with the vortex partition function of $\mathcal{T}_{2d}$.

The equivalence of these computations can be viewed as a finite analogue of the AGT correspondence, providing a vast generalization and the correct physical setup of the beautiful mathematical work [**3**].

We consider the case of Neumann boundary conditions with $\varepsilon=\varepsilon'=(+,...,+)$. The overlap of boundary states (68) is

$$Z_{\nu_i}(q) = \langle\mathcal{N}_{\varepsilon,\xi}|\mathcal{N}_{\varepsilon,\xi'}\rangle = \sum_{\mathfrak{n}\geq 0}\frac{q^{\frac{m_j}{\epsilon}-\frac{1}{2}-\mathfrak{n}}}{\prod_{l=0}^{\mathfrak{n}-1}P(-m_i+(l-\mathfrak{n})\epsilon)} \tag{69}$$

This is exactly vortex partition function of the 2d $\mathcal{N}=(2,2)$ theory $\mathcal{T}_{2d}$ with gauge group $U(1)$ and N chiral multiplets X_j of charge $+1$ and an exponentiated complexified FI parameter $q=\xi/\xi'$. [**1**, **8**]. Mathematically, it is the equivariant J-function of $\mathbb{CP}^{N-1}$.

5.3. Differential equations. The vortex partition functions (69) are generalized hypergeometric functions, which satisfy an N-th order differential equation in the parameter q. This differential equation can be explicitly derived from the relation $\mathcal{Z}_{\nu_i}(q) = \langle \mathcal{N}_{\varepsilon',\zeta'} | \mathcal{N}_{\varepsilon,\zeta} \rangle$ and the defining properties of the boundary states. As above, we focus in the case $\varepsilon = \varepsilon' = (+, \dots, +)$.

Our starting point is the differential equation

$$\epsilon\, \xi \frac{d}{d\xi} |\mathcal{N}_{\varepsilon,\xi}\rangle = \varphi\, |\mathcal{N}_{\varepsilon,\xi}\rangle\,, \tag{70}$$

which follows immediately from equation (68). Recalling that $v_+ v_- = P(\varphi + \frac{\epsilon}{2})$ we now have

$$\begin{aligned} P\Big(\epsilon\, q \frac{\partial}{\partial q} + \frac{\epsilon}{2}\Big) \mathcal{Z}_{\nu_i}(q) &= \langle \mathcal{N}_{\varepsilon,\xi'} | P(\varphi + \tfrac{\epsilon}{2}) | \mathcal{N}_{\varepsilon,\xi} \rangle \\ &= \langle \mathcal{N}_{\varepsilon,\xi'} | v_+ v_- | \mathcal{N}_{\varepsilon,\xi} \rangle \\ &= q^{-1}\, \mathcal{Z}_{\nu_i}(q)\,, \end{aligned} \tag{71}$$

which is the N-th order generalized hypergeometric equation satisfied by (69). Note that the derivation did not depend on the choice of vacuum ν_i: the N different choices of vacuum produce a basis linearly independent solutions.

6. Vortex quantum mechanics

In sections 2 and 3, we argued that a 3d $\mathcal{N} = 4$ gauge theory in an Ω-background in the $x^{1,2}$-plane localizes to an $\mathcal{N} = 4$ supersymmetric quantum mechanics on the x^3-axis. The space of supersymmetric vacua decomposed as a direct sum

$$\mathcal{H}_\nu = \bigoplus_{\mathfrak{n} \geq 0} \mathcal{H}_\nu^{\mathfrak{n}}\,, \tag{72}$$

where each summand $\mathcal{H}_{\nu_i}^{\mathfrak{n}}$ is given by the equivariant cohomology of a moduli space of vortices $\mathcal{M}_\nu^{\mathfrak{n}}$ with vortex number $\mathfrak{n}$.

An alternative approach is to describe each summand in isolation as a massive gauged supersymmetric quantum mechanics $Q(\nu, \mathfrak{n})$, whose Higgs branch is the moduli space of vortices $\mathcal{M}_{\nu_i}^{\mathfrak{n}}$ and whose space of supersymmetric vacua is $\mathcal{H}_{\nu_i}$. The matter content of each quantum mechanics is known from the brane construction of $\mathcal{M}_\nu^{\mathfrak{n}}$. The monopole operators v_A are realized as a family of half-BPS interfaces between quantum mechanics $Q(\nu, \mathfrak{n})$ and $Q(\nu, \mathfrak{n} + A)$. This approach is explored in section 6 of reference [**7**] and shown to reproduce the results that we have presented here.

Acknowledgements

I would like to thank Tudor Dimofte, Davide Gaiotto, Justin Hilburn and Hee-Cheol Kim for the fruitful collaboration on reference [**7**].

References

[1] Luis F. Alday, Davide Gaiotto, Sergei Gukov, Yuji Tachikawa, and Herman Verlinde, *Loop and surface operators in $\mathcal{N} = 2$ gauge theory and Liouville modular geometry*, J. High Energy Phys. **1** (2010), 113, 50. MR2660780

[2] Vadim Borokhov, Anton Kapustin, and Xinkai Wu, *Monopole operators and mirror symmetry in three dimensions*, J. High Energy Phys. **12** (2002), 044, 23. MR1955002

[3] Alexander Braverman, Boris Feigin, Michael Finkelberg, and Leonid Rybnikov, *A finite analog of the AGT relation I: Finite W-algebras and quasimaps' spaces*, Comm. Math. Phys. **308** (2011), no. 2, 457–478. MR2851149

[4] Hiraku Nakajima, *Towards a mathematical definition of Coulomb branches of 3-dimensional $\mathcal{N} = 4$ gauge theories, I*, Adv. Theor. Math. Phys. **20** (2016), no. 3, 595–669. MR3565863

[5] Mathew Bullimore, Tudor Dimofte, and Davide Gaiotto, *The Coulomb Branch of 3d $\mathcal{N} = 4$ Theories*, (2015).

[6] Mathew Bullimore, Tudor Dimofte, Davide Gaiotto, and Justin Hilburn, *Boundaries, mirror symmetry, and symplectic duality in 3d $\mathcal{N} = 4$ gauge theory*, J. High Energy Phys. **10** (2016), 108, front matter+191. MR3578533

[7] Mathew Bullimore, Tudor Dimofte, Davide Gaiotto, Justin Hilburn, and Hee-Cheol Kim, *Vortices and Vermas*, (2016).

[8] Tudor Dimofte, Sergei Gukov, and Lotte Hollands, *Vortex counting and Lagrangian 3-manifolds*, Lett. Math. Phys. **98** (2011), no. 3, 225–287. MR2852983

[9] Minoru Eto, Youichi Isozumi, Muneto Nitta, Keisuke Ohashi, and Norisuke Sakai, *Moduli space of non-abelian vortices*, Phys. Rev. Lett. **96** (2006), no. 16, 161601, 4. MR2221048

[10] Minoru Eto, Youichi Isozumi, Muneto Nitta, Keisuke Ohashi, and Norisuke Sakai, *Solitons in the Higgs phase: the moduli matrix approach*, J. Phys. A **39** (2006), no. 26, R315–R392. MR2238500

[11] Anton Kapustin and Edward Witten, *Electric-magnetic duality and the geometric Langlands program*, Commun. Number Theory Phys. **1** (2007), no. 1, 1–236. MR2306566

[12] David R. Morrison and M. Ronen Plesser, *Summing the instantons: quantum cohomology and mirror symmetry in toric varieties*, Nuclear Phys. B **440** (1995), no. 1-2, 279–354. MR1336089

[13] Hiraku Nakajima, *Towards a mathematical definition of Coulomb branches of 3-dimensional $\mathcal{N} = 4$ gauge theories, I*, Adv. Theor. Math. Phys. **20** (2016), no. 3, 595–669. MR3565863

[14] Nikita A. Nekrasov, *Seiberg-Witten prepotential from instanton counting*, Adv. Theor. Math. Phys. **7** (2003), no. 5, 831–864. MR2045303

Mathematical Institute, University of Oxford, Woodstock Road, Oxford, OX2 6GG, United Kingdom

Email address: mathew.bullimore@maths.ox.ac.uk

Proceedings of Symposia in Pure Mathematics
Volume **98**, 2018
http://dx.doi.org/10.1090/pspum/098/01728

3d supersymmetric gauge theories and Hilbert series

Stefano Cremonesi

ABSTRACT. The Hilbert series is a generating function that enumerates gauge invariant chiral operators of a supersymmetric field theory with four supercharges and an R-symmetry. In this article I review how counting dressed 't Hooft monopole operators leads to a formula for the Hilbert series of a 3d $\mathcal{N} \geq 2$ gauge theory, which captures precious information about the chiral ring and the moduli space of supersymmetric vacua of the theory. (Conference paper based on a talk given at String-Math 2016, Collège de France, Paris.)

1. Introduction

There is a long and illustrious tradition of fruitful interplay between supersymmetric quantum field theory and geometry [**51, 55, 55, 63**]. The main bridge between the two topics is the concept of the moduli space of supersymmetric vacua, the set of zero energy configurations of the field theory, which in the context of supersymmetric field theories with at least four supercharges is a complex algebraic variety equipped with a Kähler metric.

Moduli spaces of vacua of quantum field theories with four supercharges in four spacetime dimensions have been studied in great detail since the 1990's, and their algebro-geometric structure is well understood. Less understood are their counterparts in three dimensions, due to new scalar fields which are obtained by dualizing vectors in three dimensions. Interesting results on the moduli spaces of vacua of three-dimensional theories with four supercharges were obtained by means of semiclassical analysis [**1, 29**], but a precise understanding of the underlying algebraic geometry was lacking, except for a few simple theories [**11**].

In this article I will review recent developments that allow one to make exact statements on the algebraic geometry of the moduli spaces of supersymmetric vacua of three-dimensional gauge theories with four or more supercharges [**22–24, 26, 27**]. The key idea is to count the gauge invariant chiral operators that parametrize the moduli space of supersymmetric vacua, using a generating function called the Hilbert series. In the context of three-dimensional supersymmetric field theories, the gauge invariant chiral operators are dressed 't Hooft monopole operators: I will describe their properties and how to count them, leading to a formula for the Hilbert series. A peculiarity of 't Hooft monopole operators, that hindered the understanding of the algebraic geometry of moduli spaces of vacua of three-dimensional supersymmetric gauge theories, is that they obey relations that arise

2010 *Mathematics Subject Classification.* Primary 81T60, 14D21.

in the quantum theory. However, by applying plethystic techniques to the Hilbert series that counts dressed monopole operators, one can deduce information on the charges of generators and relations of the chiral ring of the field theory, which is the coordinate ring of its moduli space of vacua. The formalism is therefore capable of producing predictions on the quantum relations between monopole operators, without making any assumptions on them.

The article is organized as follows. In section 2 I recall the concepts of moduli space of vacua and of chiral ring in the familiar ground of four-dimensional theories with four supercharges ($4d$ $\mathcal{N}=1$ theories). In section 3 I introduce the Hilbert series in that context and give a few useful examples for what follows. In section 4 I introduce three-dimensional theories with four supercharges ($3d$ $\mathcal{N}=2$ theories) and contrast them with $4d$ $\mathcal{N}=1$ theories. In section 5 I discuss supersymmetric 't Hooft monopole operators and some of their properties. In section 6 I present the main formula (6.3) for the Hilbert series, that counts dressed monopole operators. In section 7 I apply these ideas to Coulomb branches of $3d$ $\mathcal{N}=4$ theories, and in section 8 to moduli spaces of vacua of $3d$ $\mathcal{N}=2$ Yang-Mills and Chern-Simons theories. I conclude with a few remarks and open questions in section 9.

2. Moduli space of supersymmetric vacua and chiral ring

Let us first recall some well-known facts about four-dimensional gauge theories with four Poincaré supercharges ($4d$ $\mathcal{N}=1$ theories). Most of the structure of $4d$ $\mathcal{N}=1$ supersymmetric theories carries over to $3d$ $\mathcal{N}=2$ supersymmetric theories, which will be our main focus in the following.

The $4d$ $\mathcal{N}=1$ supersymmetry algebra consists of the following generators: the Lorentz generators $M_{\mu\nu}=-M_{\nu\mu}$ ($\mu,\nu=0,\dots,3$) for rotations and boosts, which generate the Lorentz group $SO(1,3)$; the momentum P_μ, an $SO(1,3)$ vector which generates translations in the Minkowski spacetime $\mathbb{R}^{1,3}$; the complex supercharges $\mathcal{Q}_\alpha$ and $\overline{\mathcal{Q}}^{\dot\alpha}=(\mathcal{Q}_\alpha)^\dagger$ ($\alpha=1,2$ and $\dot\alpha=1,2$), left-handed and right-handed Weyl spinors transforming in the two-dimensional representations $[1;0]$ and $[0;1]$ of the double cover of the Lorentz group $\mathrm{Spin}(1,3)=SL(2,\mathbb{C})\cong SL(2,\mathbb{R})\times SL(2,\mathbb{R})$, which generate translations along the Grassmann odd directions of superspace; and possibly R, a Lorentz scalar which generates a $U(1)_R$ symmetry that acts non-trivially on the supercharges. The commutation relations are those required by the Lorentz properties of the generators recalled above, together with

$$\begin{gathered}\{\mathcal{Q}_\alpha,\overline{\mathcal{Q}}_{\dot\alpha}\}=2(\sigma^\mu)_{\alpha\dot\alpha}P_\mu\\ [R,\mathcal{Q}_\alpha]=-\mathcal{Q}_\alpha\ ,\qquad [R,\overline{\mathcal{Q}}_{\dot\alpha}]=\overline{\mathcal{Q}}_{\dot\alpha}\ ,\end{gathered}\tag{2.1}$$

where $(\sigma^\mu)_{\alpha\dot\alpha}=(-\mathbf{1},\sigma_i)_{\alpha\dot\alpha}$ and σ_i are the Pauli matrices that satisfy $\sigma_h\sigma_j=\delta_{hj}+i\epsilon_{hjk}\sigma_k$. In addition there might be a global non-R symmetry algebra, often called flavour symmetry, generated by scalar charges which commute with the generators of the supersymmetry algebra.

Fields in a $4d$ $\mathcal{N}=1$ supersymmetric field theory fit in irreducible representations of the $4d$ $\mathcal{N}=1$ superalgebra, which in turn constrains the form of interactions. Altogether, a $4d$ $\mathcal{N}=1$ supersymmetric Lagrangian gauge theory is specified by the following data:

(1) **Gauge group**: a compact semisimple Lie group G, to which one associates real vector multiplets V^a, with $a=1,\dots,\mathrm{rk}(G)$;

(2) **Matter content**: a representation $\mathcal{R}$ of G, to which one associates complex chiral multiplets X^i, with $i = 1, \dots, \dim(\mathcal{R})$;
(3) **Superpotential**: a G-invariant holomorphic polynomial $W(X)$ in the matter fields.

The scalar fields in the chiral multiplets, which we also denote as X^i with a common abuse of notation, interact through a potential

$$V = \sum_i |F_i|^2 + \frac{g^2}{2} \sum_a (D^a)^2 \ . \tag{2.2}$$

Here $F_i(X) = \frac{\partial W}{\partial X^i}$ are the F-terms, which equal the derivatives of the superpotential, $D^a(X, X^\dagger) = \sum_i X_i^\dagger (T^a)^i{}_j X^j$ are the D-terms, which equal the moment maps of the action of the gauge group G on the matter representation $\mathcal{R}$, and g is the Yang-Mills coupling constant.[1] For $U(1)$ gauge factors, the moment map can be shifted by a constant, the Fayet-Iliopoulos (FI) parameter ξ, so that $D^{U(1)} = \sum_i q_i |X^i|^2 - \xi$, with q_i the $U(1)$ charge of X^i.

One can associate to a supersymmetric gauge theory specified by these data an object of great physical and mathematical interest: its moduli space of supersymmetric vacua $\mathcal{M}$ [**63**]. Physically, $\mathcal{M}$ controls the low energy behaviour of the quantum field theory. Many of the impressive results on the dynamics of supersymmetric field theories obtained in the 1990's were indeed rooted in the analysis of their moduli spaces of vacua [**55**]. Mathematically, $\mathcal{M}$ provides a natural bridge between supersymmetric field theories and (differential and algebraic) geometry. Physically based results on moduli spaces of vacua of supersymmetric field theories can thus lead to interesting mathematical predictions.

Concretely, the **moduli space of supersymmetric vacua** $\mathcal{M}$ is defined as the set of constant field configurations that minimize the potential (2.2), modulo gauge equivalence:

$$\begin{aligned} \mathcal{M} &= \{(X, X^\dagger)|\ F_i(X) = 0\ \forall i,\ D^a(X, X^\dagger) = 0\ \forall a\}/G = \mathcal{F}//G \\ &\cong \{(X)|\ \partial W(X) = 0\}/G^{\mathbb{C}} = \mathcal{F}/G^{\mathbb{C}} \ . \end{aligned} \tag{2.3}$$

$4d$ $\mathcal{N} = 1$ supersymmetry implies that the moduli space of vacua $\mathcal{M}$ is a (possibly singular) Kähler manifold. The first line of (2.3) expresses $\mathcal{M}$ as a symplectic (in fact Kähler) quotient of $\mathcal{F}$ by the gauge group G, whereas the second line expresses $\mathcal{M}$ equivalently as a holomorphic (GIT) quotient by the complexified gauge group $G^{\mathbb{C}}$. Here

$$\mathcal{F} = \{(X)|\ \partial W(X) = 0\} \tag{2.4}$$

is the space of solutions of the F-term equations, often called F-flat moduli space or master space [**35**]. Algebraically, it is a complex affine variety defined by the vanishing of the F-term relations $\partial W(X) = 0$.

In the following I will adopt the holomorphic viewpoint in the second line of (2.3) and view the moduli space of vacua $\mathcal{M}$ as a complex algebraic variety. $\mathcal{M}$ is typically an affine variety. Of particular interest are superconformal field theories, whose moduli spaces of vacua are cones. The $\mathbb{C}^*$ action whose radial part dilates the cone is the complexification of the $U(1)_R$ symmetry of the field theory.

[1] In the D^2 term I have used a basis of the Lie algebra that diagonalizes the Killing form. If the gauge group G is semisimple there is one Yang-Mills coupling constant per simple factor.

Closely related to the moduli space of supersymmetric vacua are the concepts of chiral operators and chiral ring [**62**]. Local gauge invariant **chiral operators** $\mathcal{O}_i(x)$ form a subset of observables in a 4d $\mathcal{N}=1$ field theory which are protected from quantum corrections. These are $\frac{1}{2}$-BPS operators that are annihilated by all the supercharges of positive R-charge:

$$\overline{\mathcal{Q}}_{\dot{\alpha}}\mathcal{O}_i(x) = 0 \qquad \forall\, \dot{\alpha} = 1,2\ . \tag{2.5}$$

A crucial property of chiral operators is that their spacetime derivatives are $\overline{\mathcal{Q}}$-exact and therefore vanish in expectation values, provided supersymmetry is unbroken. It follows that a product of chiral operators is free of short distance divergences and that its expectation value factorizes into the product of spacetime constant one-point functions: $\langle \mathcal{O}_{i_1}(x_1)\ldots\mathcal{O}_{i_n}(x_n)\rangle = \langle \mathcal{O}_{i_1}\rangle\ldots\langle\mathcal{O}_{i_n}\rangle$.

Chiral operators form a commutative ring, the **chiral ring** $\mathcal{R}$, with product

$$\mathcal{O}_i\mathcal{O}_j = c_{ij}{}^k\mathcal{O}_k + \overline{\mathcal{Q}}_{\dot{\alpha}}(\ldots)^{\dot{\alpha}}\ , \tag{2.6}$$

where the only spacetime dependence is in the $\overline{\mathcal{Q}}$-exact term, and repeated indices are summed over. Since we are physically interested in taking expectation values, we will work at the level of $\overline{\mathcal{Q}}$-cohomology and omit $\overline{\mathcal{Q}}$-exact terms in (2.6) from now on. The chiral ring is then specified once a basis of chiral operators $\{\mathcal{O}_i\}$ and the structure constants $c_{ij}{}^k$ are provided.

The expectation values $\langle\mathcal{O}_i\rangle$ of gauge invariant chiral operators, or equivalently $\overline{\mathcal{Q}}$-cohomology classes, are holomorphic functions on the moduli space of vacua $\mathcal{M}$. It is generally expected, though not proven to the best of my knowledge, that the correspondence between expectation values of chiral operators and holomorphic functions on $\mathcal{M}$ is one-to-one, once relations are taken into account. With this physically motivated assumption, the chiral ring $\mathcal{R}$ of the supersymmetric field theory is identified with the coordinate ring of its moduli space of vacua $\mathcal{M}$. We would then like to characterize the chiral ring as a quotient ring

$$\mathcal{R} = \mathbb{C}[\mathcal{O}_1,\ldots,\mathcal{O}_n]/I\ , \tag{2.7}$$

determine the generators $\mathcal{O}_1,\ldots,\mathcal{O}_n$ of the polynomial ring and the defining relations of the ideal I.

In a 4d $\mathcal{N}=1$ gauge theory, the chiral operators are G-invariant polynomials in the matter fields X.[2] If there is no gauge symmetry, the chiral ring is just the Jacobian ring of the superpotential W. In a gauge theory, however, the quotient by the gauge group in (2.3) makes it often hard to explicitly determine generators and relations of the chiral ring, and therefore the defining equations of the moduli space $\mathcal{M}$ as an algebraic variety.

3. The Hilbert series

Since determining generators and relations of the chiral ring of a supersymmetric gauge theory is in general a difficult task, it helps to exploit as much as possible the symmetries of the theory. A very useful tool in this respect is the **Hilbert series** [**8**],[3] a generating function that counts scalar gauge invariant chiral operators,

[2]We neglect here glueball operators [**17**], since they do not play a role in three dimensions.
[3]See also [**70**] for an early incarnation of this concept.

graded by their charges under a Cartan subalgebra of the global symmetry:

$$H(t,\widehat{x}) = \mathrm{Tr}_{\mathcal{H}}\left(t^R \prod_{\widehat{i}} \widehat{x}_{\widehat{i}}^{\widehat{Q}_{\widehat{i}}}\right). \tag{3.1}$$

Here $\mathcal{H} = \{\mathcal{O}_i | \overline{\mathcal{Q}}_{\dot{\alpha}}\mathcal{O}_i = 0,\ M_{\mu\nu}\mathcal{O}_i = 0\}$ denotes the vector space of gauge invariant scalar chiral operators, that parametrize the moduli space of supersymmetric vacua.[4] It can be decomposed into common eigenspaces of the $U(1)_R$ generator R and the generators $\widehat{Q}_{\widehat{i}}$ of the Cartan subalgebra of the flavour symmetry. For a superconformal field theory the R-charges can be taken positive, thus (3.1) is a Taylor series in t (and a Laurent series in $\widehat{x}_{\widehat{i}}$), and the eigenspaces are finite dimensional. The Hilbert series is a character on the vector space of scalar chiral operators: the coefficients of the series are the dimensions of the common eigenspaces of the global symmetry.

In light of the correspondence between the space $\mathcal{H}$ of scalar chiral operators of the supersymmetric field theory in (3.1) and the space $\mathcal{H}^0(\mathcal{M})$ of holomorphic functions on its moduli space of vacua $\mathcal{M}$, the Hilbert series (3.1) can be interpreted geometrically as a character of the action of the global symmetry group on $\mathcal{H}^0(\mathcal{M})$. For a superconformal field theory, whose moduli space of vacua is a cone, the Hilbert series equals the equivariant index of the Dolbeault operator on $\mathcal{M}$

$$H(t,\widehat{x}) = \mathrm{Tr}\left(t^R \prod_{\widehat{i}} \widehat{x}_{\widehat{i}}^{\widehat{Q}_{\widehat{i}}} \,\Big|\, \mathcal{H}^0(\mathcal{M})\right) = \sum_p (-1)^p\, \mathrm{Tr}\left(t^R \prod_{\widehat{i}} \widehat{x}_{\widehat{i}}^{\widehat{Q}_{\widehat{i}}} \,\Big|\, \mathcal{H}^p(\mathcal{M})\right), \tag{3.2}$$

dubbed index-character in [**65**], because higher Dolbeault cohomology groups $\mathcal{H}^p(\mathcal{M})$ vanish for $p > 1$. In (3.1)-(3.2) we have distinguished the R-charge, which generates a $\mathbb{C}^*$ action that rescales the holomorphic top form of $\mathcal{M}$, from the flavour charges $\widehat{Q}_{\widehat{i}}$, which generate a torus action that leave the holomorphic top form invariant.

Useful information on the moduli space of vacua $\mathcal{M}$ can be extracted from the Hilbert series [**8**]. For instance, the complex dimension d of $\mathcal{M}$ is the order of the pole at $t = 1$ of the unrefined Hilbert series $H(t,1)$, and the coefficient of $(1-t)^{-d}$ is proportional to the volume of the $(d-1)$-dimensional base of $\mathcal{M}$. Most importantly, the charges of the generators and relations can be extracted using plethystic techniques, if higher syzygies can be disentangled.[5] Once this is achieved, the problem of presenting the moduli space $\mathcal{M}$ as an algebraic variety (or equivalently the chiral ring as a quotient ring (2.7)) is reduced to determining a finite number of coefficients, that specify which linear combinations of the chiral operators having the appropriate charges appear as generators or relations.

After this general discussion, let us see how the Hilbert series is computed in practice for a sample of $4d$ $\mathcal{N} = 1$ supersymmetric quantum field theories. We start by considering theories with no gauge group, so that the moduli space (2.3) coincides with the F-flat space (2.4). For the theory of a free chiral multiplet X of R-charge $R[X] = r$, the moduli space is the complex plane, the chiral ring is the

[4] An alternative count of protected operators is provided by the "superconformal" index [**72**], which also counts fermionic and short non-chiral operators, and depends on the superpotential only through the R-charges of matter fields. Since our interest is in the moduli space of vacua and the chiral ring, we focus on the Hilbert series rather than the superconformal index.

[5] This is often possible with some physical input, such as an independent determination of the dimension of the moduli space, and the help of computer algebra such as `Macaulay2` [**42**].

polynomial ring in one complex variable $\mathbb{C}[X]$ and the Hilbert series is simply the geometric series counting powers of X:[6]

$$H = 1 + \tau + \tau^2 + \cdots = \frac{1}{1-\tau} = \mathrm{PE}[\tau] , \qquad \tau = t^r . \tag{3.3}$$

If X is subject to a superpotential $W(X) = X^{N+1}$, the Hilbert series becomes

$$H = 1 + \tau + \tau^2 + \cdots + \tau^{N-1} = \frac{1-\tau^N}{1-\tau} = \mathrm{PE}[\tau - \tau^N] , \qquad \tau = t^{\frac{2}{N+1}} . \tag{3.4}$$

The chiral ring is $\mathbb{C}[X]/\langle X^N \rangle$ and the moduli space $\mathcal{M}$ consists of a point of multiplicity N. The generator X and the relation $X^N = 0$ are respectively associated to the positive term $+\tau$ and the negative term $-\tau^N$ in the argument of the plethystic exponential. In general, the plethystic logarithm of the Hilbert series terminates for theories whose moduli spaces are complete intersections, the dimension of which is the number of generators minus the number of relations.

The case of a generic polynomial superpotential $W(X) = X^{N+1} + \sum_{i=1}^N c_i X^{N-i}$ of degree $N+1$ can be treated similarly: even though the $U(1)_R$ symmetry is explicitly broken by the subleading terms in the superpotential, it can be restored by assigning R-charges to the parameters c_i, as is common practice in the analysis of supersymmetric field theories [**73**]. The parameters c_i are not dynamical and are not counted by the Hilbert series, which is insensitive to them and remains (3.4), but they may (and do) appear in the relations. The Hilbert series only constrains the charges of the relations, which in this case must be of the form $X^N + \sum_i \alpha_i c_i X^{N-i} = 0$, but does not fix the coefficients, which in this case we know to be $\alpha_i = (N-i)/(N+1)$. Of course there is no need to invoke the Hilbert series to study the chiral ring of such a simple theory, but this example makes it clear which information can be extracted from the Hilbert series (*i.e.* charges of operators, generators and relations) and which cannot (*i.e.* the precise form of the relations, unless they are entirely fixed by symmetry). Even when there are coefficients in the relations that cannot be determined by symmetry alone, the Hilbert series is a very useful tool for deducing the most general form of the chiral ring relations that is consistent with symmetry.

Another simple but more interesting example of moduli space is provided by the XYZ model, a theory of three chiral multiplets X, Y and Z with the trilinear superpotential $W = XYZ$. From the F-term relations $\partial W = 0$ we deduce that the chiral ring is $\mathbb{C}[X,Y,Z]/\langle YZ, ZX, XY \rangle$. The moduli space $\mathcal{M}$ consists of three 1-dimensional components (in physical jargon "branches") parametrized by X, Y

---

[6] PE is the plethystic exponential, the generating function of symmetric powers. For a multivariate function $f(x_1, \ldots, x_n)$ that vanishes at the origin,

$$\mathrm{PE}[f(x_1,\ldots,x_n)] = \exp\left(\sum_{p=1}^{\infty} \frac{1}{p} f(x_1^p, \ldots, x_n^p) \right) .$$

This implies that $\mathrm{PE}[\sum_i a_i \prod_j x_j^{b_{ij}}] = \prod_i (1 - \prod_j x_j^{b_{ij}})^{-a_i}$. The inverse of the plethystic exponential is the plethystic logarithm PL. For a multivariate function g that equals 1 at the origin,

$$\mathrm{PL}[g(x_1,\ldots,x_n)] = \sum_{k=1}^{\infty} \frac{\mu(k)}{k} \log g(x_1^k, \ldots, x_n^k) ,$$

where $\mu(k)$ is the Möbius function.

and Z respectively, meeting at a point. The Hilbert series reads

$$\begin{aligned} H &= \frac{1}{1-\tau x} + \frac{1}{1-\tau y} + \frac{1}{1-\tau z} - 2 = \qquad (\tau = t^{2/3}\ , \quad xyz = 1) \\ &= \frac{1-\tau^2(yz+xz+xy)+2\tau^3 xyz}{(1-\tau x)(1-\tau y)(1-\tau z)} \ . \\ \mathrm{PL}[H] &= (x+y+z)\tau - (yz+xz+xy)\tau^2 + 2\tau^3 - (x+y+z)\tau^4 + \dots \ . \end{aligned} \tag{3.5}$$

The first line shows that the moduli space $\mathcal{M}$ consists of three copies of $\mathbb{C}$ meeting at a point; the second line shows the three generators in the denominator, and that the moduli space is not a complete intersection because the numerator does not factorize. The plethystic logarithm in the last line allows us to extract information about the ring: the generators X, Y, Z of the polynomial ring appear at order τ, the generators of the ideal of relations by which we quotient $\partial_X W, \partial_Y W, \partial_Z W$ at order τ^2, and then we see higher order syzygies: $X\partial_X W - Y\partial_Y W$ and $X\partial_X W - Z\partial_Z W$ at order τ^3, $\partial_X W \partial_Y W - Z^2\partial_Z W$ and cyclic permutations at order τ^4, and so on. The plethystic logarithm is a series which does not terminate: this is the general structure for theories with non-complete intersection moduli spaces.

Next we consider gauge theories. **Gauging** a subgroup G of the flavour symmetry leads in the holomorphic description of the moduli space in (2.3) to the quotient $\mathcal{M} = \mathcal{F}/G^{\mathbb{C}}$ of the F-flat moduli space by the complexified gauge group. At the level of the Hilbert series, the projection to gauge singlets is achieved by averaging the Hilbert series $H_{\mathcal{F}}$ of the ungauged theory (whose moduli space is the F-flat moduli space $\mathcal{F}$) over the gauge group

$$H(t,\widehat{x}) = \oint d\mu_G(x) H_{\mathcal{F}}(t,x,\widehat{x}) \tag{3.6}$$

using the Haar measure

$$\oint d\mu_G(x) = \left(\prod_{j=1}^{r} \oint \frac{dx_j}{2\pi i x_j}\right) \prod_{\alpha\in\Delta_+} (1-x^\alpha) \ . \tag{3.7}$$

The integral is over the maximal torus of G, $r = \mathrm{rk}(G)$ is the rank of the gauge group, Δ_+ is the set of positive roots of its Lie algebra, and I have used the shorthand $x^\alpha = \prod_{i=1}^r x_i^{\alpha_i}$. Gauge fugacities are denoted by x and ungauged flavour fugacities by $\widehat{x}$.

A simple class of examples, that will be useful in the following, is provided by theories with gauge group G and a chiral multiplet Φ in the adjoint representation. (This is also the vector multiplet sector of 4d $\mathcal{N} = 2$ theories with gauge group G. The branch of the moduli space of vacua where the vector multiplet scalar Φ takes expectation values is called the Coulomb branch.) Setting $\tau = t^{R[\Phi]}$, the Hilbert series reads

$$H(\tau) = \oint d\mu_G(x)\, \mathrm{PE}[\tau\chi^G_{ad}(x)] = \prod_{i=1}^{r} \frac{1}{1-\tau^{d_i(G)}} \ , \tag{3.8}$$

where $\chi^G_{ad}(x)$ is the character of the adjoint representation of the gauge group. The result expresses the well-known fact that the ring of invariants of the adjoint representation is freely generated by Casimir invariants u_i of degrees $d_i(G)$. Hence $\mathcal{R} = \mathbb{C}[g]^G = \mathbb{C}[\phi_1,\dots,\phi_r]/\mathcal{W}_G = \mathbb{C}[u_1,\dots,u_r]$. *E.g.* for $G = SU(N)$ the Casimir invariants are $d_i(SU(N)) = 2, 3, \dots, N$.

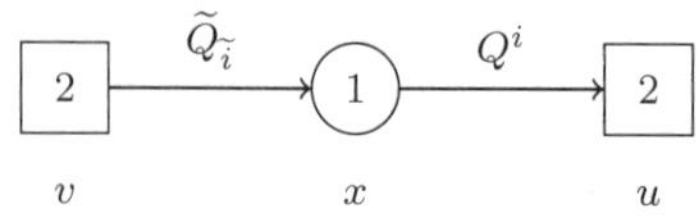

FIGURE 1. Quiver diagram for SQED with two flavours.

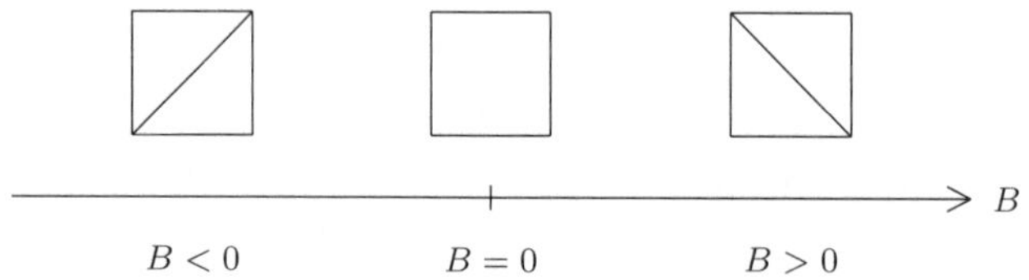

FIGURE 2. Line bundles on the conifold and resolutions.

Another example is a $U(1)$ gauge theory with two matter fields Q^i of charge 1 and two matter fields $\widetilde{Q}_{\widetilde{i}}$ of charge -1, also known as SQED with two flavours. See figure 1 for the quiver diagram of this theory. It turns out to be interesting to compute the Hilbert series in the presence of a background electric charge $-B$ for the $U(1)$ gauge symmetry, even though this may seem artificial from the perspective of four-dimensional gauge theory. (We will see that B has a more natural interpretation in three dimensions.) This modified Hilbert series, often called baryonic Hilbert series [**36**], counts polynomials in the matter fields Q and $\widetilde{Q}$ of total electric charge B to compensate the background electric charge $-B$, and is computed by the formula [**36**]

$$\begin{aligned} H_{-B}(\tau,u,v) \equiv g_1(\tau,u,v;B) &= \oint \frac{dx}{2\pi i x} x^{-B}\,\mathrm{PE}\left[\tau x\left(u+\frac{1}{u}\right)+\tau\frac{1}{x}\left(v+\frac{1}{v}\right)\right] \\ &= \begin{cases} \sum_{n=0}^{\infty}[n+B;n]_{u,v}\tau^{2n+B} & B\geq 0 \\ \sum_{n=0}^{\infty}[n;n-B]_{u,v}\tau^{2n-B} & B\leq 0 \end{cases} . \end{aligned} \tag{3.9}$$

Here u and v are fugacities for the $SU(2)_u \times SU(2)_v$ flavour symmetries that rotate Q^i and $\widetilde{Q}_{\widetilde{i}}$ respectively, and $[n;m]_{u,v}$ denotes the character of the representation $[n;m]$ of $SU(2)_u \times SU(2)_v$.

The background electric charge or "baryonic charge" B in (3.9) is a discrete analogue of the Fayet-Iliopoulos parameter ξ introduced after (2.2), which leads to a resolution of the conical moduli space of vacua in the symplectic reduction in the first line of (2.3). The theory that we are discussing is nothing but the gauged linear sigma model for the conifold, albeit viewed as a four-dimensional theory: its moduli space of vacua $\mathcal{M}$ is the conifold if $\xi=0$, and the resolved conifold if $\xi \neq 0$, with the resolutions at $\xi>0$ and $\xi<0$ being related by a flop transition. In the holomorphic language, the Hilbert series (3.9) with insertion of the baryonic charge B counts holomorphic sections of the line bundle $\mathcal{O}(BD)$, where D is the toric divisor associated to Q fields, and $-D$ the toric divisor associated to $\widetilde{Q}$ fields. See figure 2 for a summary. The baryonic Hilbert series counts operators of the schematic form $Q^B(Q\widetilde{Q})^n$ for $B\geq 0$ and $\widetilde{Q}^{-B}(Q\widetilde{Q})^n$ for $B\leq 0$. For $B=0$ we obtain the Hilbert series of the conifold

$$H_0(\tau,u,v) = \mathrm{PE}[\tau[1;1]_{u,v} - \tau^2] \,, \tag{3.10}$$

corresponding to the ring $\mathbb{C}[M_1^1, M_2^1, M_1^2, M_2^2]/\langle M_1^1 M_2^2 - M_2^1 M_1^2\rangle$ generated by the four mesons $M_{\tilde{i}}^{i} = Q^i \widetilde{Q}_{\tilde{i}}$ subject to a singlet relation $\det(M) = 0$. We have thus recovered the algebraic description of the conifold [**18**].

4. $3d$ $\mathcal{N} = 2$ gauge theories vs $4d$ $\mathcal{N} = 1$ gauge theories

Moduli spaces of supersymmetric vacua of four-dimensional supersymmetric gauge theories, which have been described so far and analysed with the help of the Hilbert series, are of mathematical and physical interest and many nontrivial results have been obtained [**8, 34, 41, 45, 47**]. However, the construction is limited on both fronts. Mathematically, even though supersymmetry only requires the moduli space of vacua $\mathcal{M}$ of $4d$ $\mathcal{N} = 1$ theories to be a Kähler manifold (or hyperkähler for the Higgs branch of 4d $\mathcal{N} = 2$ theories), the latter is actually a Kähler quotient (or hyperkähler quotient for the Higgs branch of 4d $\mathcal{N} = 2$ theories) for a Lagrangian supersymmetric gauge theory. On the physical side, the computation of the Hilbert series is essentially classical and reduces to counting gauge invariant polynomials in the matter fields appearing in the UV Lagrangian.[7]

Moduli space of vacua which are not (hyper)Kähler quotients occur for non-Lagrangian supersymmetric theories in four dimensions, and string theory or M-theory constructions have been used to study some of their properties, along the lines of [**37**]. We will pursue here an alternative way to overcome the limitations explained above, by considering three-dimensional supersymmetric gauge theories with at least four supercharges ($3d$ $\mathcal{N} \geq 2$) instead of four-dimensional theories.

The $3d$ $\mathcal{N} = 2$ supersymmetry algebra can be obtained by dimensional reduction of the $4d$ $\mathcal{N} = 1$ supersymmetry algebra and is therefore very similar to (2.1). The main modifications are that the supercharges are complex conjugate $3d$ Dirac spinors $\mathcal{Q}_\alpha$ and $\overline{\mathcal{Q}}_\beta$ transforming as doublets of $SL(2, \mathbb{R})$, and that the anticommutator of two supercharges is

$$\{\mathcal{Q}_\alpha, \overline{\mathcal{Q}}_\beta\} = 2(\gamma^\mu)_{\alpha\beta} P_\mu + 2i\epsilon_{\alpha\beta} Z \ . \tag{4.1}$$

Here $\gamma^\mu = (-\mathbf{1}, \sigma_1, \sigma_3)$, and Z is a real generator of the centre of the supersymmetry algebra. The central charge Z originates from the momentum along the reduced dimension. In the same vein, the vector and chiral multiplets of $3d$ $\mathcal{N} = 2$ supersymmetry can be obtained by dimensional reduction of the vector and chiral multiplets of $4d$ $\mathcal{N} = 1$ supersymmetry. Supersymmetric actions in three dimensions take the same form as in four dimensions, except for the possibility, peculiar to odd spacetime dimensions, to add supersymmetric Chern-Simons (CS) terms

$$S_{CS} = \frac{k_{ab}}{4\pi} \int d^3x \, d^4\theta \, \Sigma_a V_b = \frac{k_{ab}}{4\pi} \int (A_a dA_b + \sigma_a D_b + \text{superpartners}) \ , \tag{4.2}$$

where $\Sigma_a = \epsilon^{\alpha\beta} \overline{D}_\alpha D_\beta V_a$ is the field strength multiplet and A_a is the gauge connection for a $U(1)$ gauge group. The real scalar σ_a is the lowest component of the field strength multiplet Σ_a, and originates from the component A_4 of the gauge connection in the dimensional reduction from four dimensions. (4.2) is a Chern-Simons interaction between abelian gauge groups, but can be generalized to nonabelian

[7]Quantum corrected moduli spaces of vacua of several four-dimensional supersymmetric gauge theories were intensively studied in the 1990's [**55**], but the Hilbert series is to a large extent insensitive to the complex structure deformations induced by quantum effects.

gauge groups G in a straightforward way using the Killing form as the symmetric pairing.

Despite the similarities between $4d\,\mathcal{N}=1$ and $3d\,\mathcal{N}=2$ supersymmetric gauge theories at the classical level, they behave very differently quantum-mechanically. In what follows I will describe how the quantum physics of three-dimensional $\mathcal{N}\geq 2$ supersymmetric gauge theories leads to moduli spaces of supersymmetric vacua $\mathcal{M}$ which are (hyper)Kähler, as required by supersymmetry, though not (hyper)Kähler quotients. Unlike most of the previous results on moduli spaces of vacua of $3d$ supersymmetric gauge theories which were based on semiclassical analysis (see [**1**, **28**, **29**, **32**, **54**, **56**, **75**] for a partial sample), the aim here will be to understand these moduli spaces as algebraic varieties and to develop general methods to characterize the chiral rings of such theories. In particular, I will review explicit formulae for the Hilbert series of the Coulomb branch of the moduli space of vacua of $3d\,\mathcal{N}=4$ gauge theories and for the Hilbert series of the moduli space of vacua of $3d\,\mathcal{N}=2$ gauge theories, which have been obtained in a series of recent works [**22**, **26**, **27**] (see also [**43**]).

The count of holomorphic functions in the Hilbert series encodes information about the moduli space of vacua and hints at a new construction of (hyper)Kähler moduli spaces which is alternative to the standard (hyper)Kähler quotient. In the context of $3d\,\mathcal{N}=4$ gauge theories, the Hilbert series formula of [**26**] has spurred recent activity both on the physical [**14**–**16**, **19**, **23**–**25**, **30**, **31**, **44**, **49**, **50**, **66**] and on the mathematical front [**12**, **13**, **60**, **67**–**69**], leading eventually to a mathematical definition of the Coulomb branch and to several other interesting developments (see also Nakajima's and Bullimore's talk at this conference).

The novelty compared to four-dimensional theories is that three-dimensional supersymmetric gauge theories contain chiral 't Hooft monopole operators, a new class of gauge invariant chiral operators which are not polynomials in the matter fields. Monopole operators are subject to relations that arise quantum-mechanically and cannot be obtained by differentiating a superpotential. It is difficult to directly determine these relations for general theories, although impressive results have been obtained by direct computation for simple theories using conformal field theory techniques [**11**]. Nevertheless, we will see that physical arguments lead to a general group-theoretic formula for the Hilbert series that completely bypasses this problem. Once the Hilbert series is computed, exactly or in a Taylor expansion, we can learn about the quantum relations among monopole operators by means of plethystic methods, as was sketched in the four-dimensional examples described above.

5. 't Hooft monopole operators

Before we can explain how to compute the Hilbert series of $3d\,\mathcal{N}\geq 2$ supersymmetric gauge theories, we need to understand some of the properties of supersymmetric 't Hooft monopole operators.

In recent years it has been increasingly appreciated that local operators in a quantum field theory need not be expressible as polynomials in the microscopic fields that are used to write down the Lagrangian [**58**]. One also needs to include "disorder" (or defect) operators, which may be defined by prescribing appropriate singular boundary conditions in the path integral. While the definition of local disorder operators appears to put them on a different footing from standard local "order" operators, this difference is an artefact of our choice of field variables in the description of the quantum field theory. All local operators are on the same footing in the quantum theory. There are by now numerous examples of dualities (quantum equivalences between classically different field theories) that map standard order operators to disorder operators and vice versa, going from sine-Gordon – massive Thirring duality [**21**, **64**], T-duality and mirror symmetry [**39**, **52**] in two spacetime dimensions to Intriligator-Seiberg mirror symmetry (supersymmetric particle-vortex duality) [**56**] and Aharony duality [**2**] in three dimensions.

In the context of three-dimensional gauge theories, the relevant local disorder operators are 't Hooft monopole operators (monopole operators henceforth) [**74**], which are introduced in the Euclidean formulation of the theory and may be obtained by dimensionally reducing 't Hooft loop operators in four dimensions. To insert a monopole operator $V_m(x)$ in a correlation function, one path integrates over gauge field configurations with a Dirac monopole singularity at the insertion point x. In a $3d$ $\mathcal{N}=2$ gauge theory, the monopole operator can be supersymmetrized by imposing singular boundary conditions for all the bosonic fields in the $3d$ $\mathcal{N}=2$ vector multiplet. Using spherical coordinates (r,θ,φ) centred at x, we define a $\frac{1}{2}$-BPS **bare monopole operator** by imposing the following singular boundary conditions as $r\to 0$ [**10**, **11**]:

$$
\begin{aligned}
A_\pm &\sim \frac{m}{2}(\pm 1-\cos\theta)d\varphi \\
\sigma &\sim \frac{m}{2r}\,.
\end{aligned}
\tag{5.1}
$$

In the first line of (5.1), $A_\pm$ is the gauge connection in the north/south patch of a trivialization of the G-bundle over the S^2 surrounding the insertion point x. The Dirac monopole singularity is given by an embedding $U(1)\hookrightarrow G$, specified by the **magnetic charge** $m\in\mathfrak{h}/\mathcal{W}$, a constant element of the Cartan subalgebra $\mathfrak{h}$ of the gauge Lie algebra $\mathfrak{g}$, defined modulo Weyl reflections. Well-definedness of the gauge bundle requires the Dirac quantization condition [**33**, **40**]

$$
e^{2\pi i m}=\mathbf{1}_G \qquad \Longrightarrow \qquad m\in\Gamma_{G^\vee}/\mathcal{W}\,,
\tag{5.2}
$$

hence the magnetic charge m is an element of the magnetic weight lattice, the weight lattice of the Langlands [**61**] or GNO [**40**] dual group $G^\vee$ of the gauge group G, modulo Weyl reflections. m can be viewed as the highest weight of an irreducible representation of the dual group $G^\vee$, or equivalently as specifying a cocharacter in $\mathrm{Hom}(U(1),G)$.

The boundary condition for the gauge connection in the first line of (5.1) defines a monopole operator which does not preserve any supersymmetry. By further imposing the boundary condition in the second line of (5.1) for the real scalar σ in the $3d$ $\mathcal{N}=2$ vector multiplet (coming from the component of the $4d$ gauge field in the reduced dimension), we define a $\frac{1}{2}$-BPS monopole operator that sits in a chiral multiplet, like all the matter fields in the theory. It is a crucial fact [**11**] that there exists a single half-BPS bare monopole operator for each choice of magnetic charge $m\in\Gamma_{G^\vee}/\mathcal{W}$.

The bare monopole operator V_m defined by the boundary conditions (5.1) is made gauge invariant by averaging over the Weyl group, if there are no gauge Chern-Simons terms. It is called bare because, as will be explained below, monopole operators can also be dressed by matter fields before they are made gauge invariant.

Note that in this construction a vector multiplet, containing the bosonic fields A, σ appearing in (5.1) and their supersymmetric partners, is traded for a tower of chiral multiplets V_m, the monopole operators, labelled by their magnetic charges m. As gauge invariant chiral operators, the monopole operators V_m can take expectation value on the moduli space of vacua $\mathcal{M}$ of the $3d$ $\mathcal{N}=2$ gauge theory. In a semiclassical description, this corresponds to the fact that the real scalar σ in the vector multiplet can take expectation value and contribute to the moduli space. At a generic point of the Coulomb branch, where the adjoint σ takes expectation value breaking G to its maximal torus, the Cartan components σ_i of σ are complexified by dual photons τ_i defined by $d\tau_i = *F_i$. For large expectation values monopole operators can be expressed as $V_m \simeq \exp\left(m \cdot (\frac{\sigma}{g^2} + i\tau)\right)$ up to quantum corrections, but this semiclassical expression for monopole operators breaks down at loci of enhanced gauge symmetry [**1**, **29**]. We will use instead the path integral definition of monopole operators (5.1), which is more implicit but is valid even in strongly coupled regions of the moduli space of vacua with unbroken nonabelian gauge symmetry.

For later purposes, it is useful to repeat the previous construction in the presence of **background magnetic charges** for background vector multiplets associated to the global non-R symmetries of the $3d$ $\mathcal{N}=2$ gauge theory. These include flavour symmetries acting on the matter fields, as in $4d$ $\mathcal{N}=1$ field theories, but also topological symmetries special to three dimensions that only act on monopole operators. Vector multiplets associated to gauge symmetries are fluctuating dynamical fields, whereas vector multiplets associated to global symmetries are frozen external parameters. We label these generalized monopole operators as $V_{m;\widehat{m},B}$, where m is the dynamical magnetic charge for the gauge symmetry G, and $\widehat{m}$ and B are fixed background magnetic charges for the global flavour and topological symmetries. Supersymmetry relates magnetic charges and real scalars in (5.1), leading to the following correspondence between quantized charges that characterize monopole operators and continuous real scalar fields or masses:

$$\begin{array}{lcccl} \text{Gauge:} & m & \longleftrightarrow & \sigma & \text{real scalar} \\ \text{Flavour:} & \widehat{m} & \longleftrightarrow & \widehat{\sigma} & \text{real mass} \\ \text{Topological:} & B & \longleftrightarrow & \xi & \text{FI parameter} \end{array} \tag{5.3}$$

We have encountered the "baryonic charge" B earlier in the context of $4d$ $\mathcal{N}=1$ gauge theories in (3.9). There B was introduced as a background electric charge for the gauge symmetry, the counterpart in the holomorphic quotient of the Fayet-Iliopoulos parameter ξ, whose effect was to resolve the conical moduli space of vacua in the symplectic quotient description. In the context of $3d$ $\mathcal{N}=2$ gauge theories, the FI parameter ξ can be interpreted as a background real scalar in the vector multiplet for the topological symmetry, and B is the associated background magnetic charge.

5.1. Charges of monopole operators. We have defined a set of new chiral operators, the monopole operators V_m, which together with the standard gauge invariant polynomials in the matter fields parametrize the moduli space of supersymmetric vacua of a $3d$ $\mathcal{N} \geq 2$ gauge theory. Since we are going to count monopole operators in the Hilbert series, we need to know how they are (electrically) charged under the global symmetries of the theory.

The first symmetry under which monopole operators are charged is the topological symmetry that was mentioned above. For a theory with gauge group G, the topological symmetry group is $G_J = Z(G)$, the centre of the gauge group.

The magnetic charge m of a monopole operator is an element of the magnetic weight lattice of the gauge group G, $\Gamma_{G^\vee}$, up to Weyl symmetry. The topological charge of the monopole operator $J(m)$ is the magnetic charge modulo elements of the coroot lattice of G. For example, if $G = U(N)$, the magnetic charge is $m = \text{diag}(m_1, \dots, m_N) \in \mathbb{Z}^N/S_N$, the topological symmetry group is $G_J = U(1)$ and the topological charge is $J(m) = \text{Tr}(m) = \sum_i m_i \in \mathbb{Z}$.

Monopole operators also carry electric charges under other global (as well as gauge) symmetries at the classical level, if the theory has Chern-Simons couplings (4.2). In general, the classical charges of monopole operators under the Cartan generators are given by

$$Q_A^{cl}(M) = -\sum_B k_{AB} M_B \ , \tag{5.4}$$

where I labelled by $\{M_A\} = \{m_i, \widehat{m}_{\widehat{i}}, B_i, 0\}$ all the magnetic charges allowed by supersymmetry: m_i for the Cartan subalgebra of the gauge symmetry, $\widehat{m}_{\widehat{i}}$ for the flavour symmetry, B_i for the topological symmetry, and 0 for the R-symmetry. Formula (5.4) is just the statement that magnetic charges lead to electric charges in the presence of Chern-Simons interactions (4.2). The charges of monopole operators under continuous abelian subgroups of the topological symmetry can be included in (5.4) by means of appropriate mixed gauge-topological Chern-Simons couplings.

The classical charges (5.4) receive quantum corrections of the form [**7**, **11**, **53**]

$$Q_A^q(M) = -\frac{1}{2}\sum_{\psi_a} Q_A[\psi_a] |m_a^{\text{eff}}(M)| \ , \tag{5.5}$$

where the sum runs over all fermionic fields ψ_a in matter chiral multiplets and vector multiplets, of charge $Q_A[\psi_a]$ and "effective mass"[8]

$$m_a^{\text{eff}}(M) = \sum_A Q_A[\psi_a] M_A \ . \tag{5.6}$$

Adding up the classical contribution (5.4) and the quantum contribution (5.5), the total electric charges of monopole operators are

$$Q_A^{\text{eff}}(M) = -\sum_B k_{AB}^{\text{eff}}(M) M_B \ , \tag{5.7}$$

where the **effective Chern-Simons levels**

$$k_{AB}^{\text{eff}}(M) = k_{AB} + \frac{1}{2}\sum_{\psi_a} Q_A[\psi_a] Q_B[\psi_a] \ \text{sign}(m_a^{\text{eff}}(M)) \tag{5.8}$$

must be integer for gauge invariance. This in turn constrains the values of the bare CS levels k_{AB}.

[8]We are slightly abusing terminology here. The effective real mass of a chiral multiplet, which equal its central charge Z appearing in the supersymmetry algebra (4.1), is actually (5.6) with the magnetic charges M_A replaced by the real scalars or masses Σ_A, according to the correspondence (5.3). In the background of a half-BPS monopole operator, magnetic charges and real scalars are related as in the second line of (5.1), therefore the effective real mass is proportional to (5.6) and inversely proportional to the distance r from the point where the monopole operator is inserted.

5.2. Dressed monopole operators. So far we have discussed bare monopole operators, chiral operators which are defined in terms of vector multiplets. The boundary condition (5.1) breaks the gauge group G to a subgroup G_m, the stabilizer of the magnetic charge: $G_m \cdot m = 0$. We call G_m the **residual gauge group**. (A similar discussion applies to the flavour symmetry group $\widehat{G}$ and the topological symmetry group G_J.) In the background of the monopole operator V_m, the vector multiplets for G_m, associated to roots α such that $\alpha(m) = 0$, are massless. The vector multiplets for G/G_m, associated to roots α such that $\alpha(m) \neq 0$, are massive by the Higgs mechanism. When integrated out, they correct the R-charge of V_m according to the formula (5.5), where all fermions in the vector multiplet have R-charge 1.

Translating the previous discussion in a mathematical formula, the contribution of a dynamical vector multiplet to the Hilbert series of a $3d$ $\mathcal{N} = 2$ gauge theory is

$$\prod_{\alpha \in \Delta_+} t^{-|\alpha(m)|} (1 - x^\alpha)^{\delta_{\alpha(m),0}} \ . \tag{5.9}$$

The first factor accounts for the correction to the R-charge of the monopole operator V_m due to the massive vector multiplets; the second factor is the contribution to the Hilbert series of the residual gauge group G_m, whose fugacities x are eventually integrated over.

A similar discussion applies to matter fields, which transform in the representation $(\mathcal{R}, \widehat{\mathcal{R}}, 1)$ of $G \times \widehat{G} \times G_J$ with weights $(\rho, \widehat{\rho}, 0)$. In the background of the monopole operator, the matter fields neutral under the $U(1)$ subgroup of $G \times \widehat{G}$ singled out by m have vanishing "effective mass"

$$m^{\text{eff}}_{\rho,\widehat{\rho}}(m, \widehat{m}) = \rho(m) + \widehat{\rho}(m) \ . \tag{5.10}$$

They can take expectation value and dress the bare monopole operator without spoiling its supersymmetry [**22**, **26**, **27**]. We call these massless matter fields **residual matter fields**. On the other hand, the matter fields with nonvanishing (5.10) are massive, cannot take expectation value and are integrated out. Their only effect is to correct the charges of the bare monopole operator quantum-mechanically according to formula (5.5).

In formulae, the contribution to the Hilbert series of a $3d$ $\mathcal{N} = 2$ gauge theory of a matter chiral multiplet X of R-charge r transforming in the representation $(\mathcal{R}, \widehat{\mathcal{R}}, 1)$ of $G \times \widehat{G} \times G_J$ is

$$\prod_{\rho, \widehat{\rho}} (t^{r-1} x^\rho \widehat{x}^{\widehat{\rho}})^{-\frac{1}{2}|\rho(m) + \widehat{\rho}(\widehat{m})|} \, \mathrm{PE}[\delta_{\rho(m)+\widehat{\rho}(\widehat{m}),0} \ t^r x^\rho \widehat{x}^{\widehat{\rho}}] \ , \tag{5.11}$$

where we have assumed that the matter field is not subject to F-term relations descending from a superpotential in order to simplify the formula (see [**22**, **27**] for the generalization). The first factor accounts for the quantum correction to the charges of the monopole operator V_m due to the massive matter fields; the second factor is the contribution of the massless residual matter fields.

In conclusion, we can dress a bare gauge-variant monopole operator by a polynomial in the residual matter fields to construct an operator that is invariant under the residual gauge group G_m. The resulting dressed monopole operator is then made G-invariant by averaging over the action of the Weyl group $\mathcal{W}_G / \mathcal{W}_{G_m}$.

6. Monopole formula for the Hilbert series of $3d$ $\mathcal{N} \geq 2$ gauge theories

We have collected all the necessary ingredients to write down a formula for the Hilbert series of a $3d$ $\mathcal{N} = 2$ supersymmetric gauge theory. The Hilbert series counts the gauge invariant chiral operators of the theory, which are dressed monopole operators, possibly in the presence of background magnetic charges $(\widehat{m}, B)$ for the flavour and topological symmetries:

$$H(t, \widehat{x}, z; \widehat{m}, B) = \mathrm{Tr}_{\mathcal{H}_{\widehat{m},B}} \left(t^R z^J \prod_{\widehat{i}} \widehat{x}_{\widehat{i}}^{\widehat{Q}_{\widehat{i}}} \right) . \tag{6.1}$$

$\mathcal{H}_{\widehat{m},B}$ denotes the vector space of scalar chiral monopole operators of fixed background magnetic charges $(\widehat{m}, B)$ for the flavour and topological symmetries $\widehat{G} \times G_J$. When $\widehat{m} = B = 0$, (6.1) is the standard Hilbert series that counts gauge invariant chiral operators that parametrize the moduli space $\mathcal{M}_0$ of the superconformal field theory to which the gauge theory flows at low energy. The background magnetic charges $(\widehat{m}, B)$ correspond to turning on real masses for the flavour symmetry and Fayet-Iliopoulos parameters, that lead to a resolution of the moduli space $\mathcal{M}_0$ of the SCFT. The generalized Hilbert series with background magnetic charges then counts holomorphic sections of line bundles rather than holomorphic functions on $\mathcal{M}_0$, analogously to the baryonic Hilbert series (3.9).

The dressing of monopole operators is summarized in the data of a **residual gauge theory** $T_{m;\widehat{m},B}$ of massless fields in the background of a monopole operator of magnetic charges $(m; \widehat{m}, B)$:

(1) A residual gauge group G_m (and flavour group $\widehat{G}_{\widehat{m}}$);
(2) Residual matter fields transforming in representations of $G_m \times \widehat{G}_{\widehat{m}}$;
(3) A residual superpotential W_m, which is obtained by setting to zero all the massive matter fields in the original superpotential W;
(4) Background electric charges $Q_i(m, \widehat{m}, B)$ under the Cartan generators of G_m.

Equipped with these data, we can write down the Hilbert series

$$H^{T_{m;\widehat{m},B}}_{Q(m,\widehat{m},B)}(t, \widehat{x}) \tag{6.2}$$

that counts chiral operators of electric charges $-Q(m, \widehat{m}, B)$ in the residual gauge theory, according to the rules of section 3.

The Hilbert series (6.1) counts dressed monopole operators labelled by their magnetic charges m for the gauge group G, which are summed over, and $(\widehat{m}; B, 0)$ for the global symmetry group $\widehat{G} \times G_J \times U(1)_R$, which are held fixed. Putting together all the ingredients discussed so far leads to the **monopole formula** for the Hilbert series [**22, 27**]

$$H(t, \widehat{x}, z; \widehat{m}; B) = \sum_{m \in \Gamma_q} t^{R(m,\widehat{m},B)} z^{J(m,\widehat{m},B)} \prod_{\widehat{i}} \widehat{x}_{\widehat{i}}^{\widehat{Q}_{\widehat{i}}(m,\widehat{m},B)} \cdot H^{T_{m;\widehat{m},B}}_{Q(m,\widehat{m},B)}(t, \widehat{x}) . \tag{6.3}$$

The sum is over the quantum lattice of magnetic charges Γ_q, which is $\Gamma_{G^\vee}/\mathcal{W}$ or a sublattice thereof if there are nonperturbative effects that lift part of the semiclassical Coulomb branch (see [**22**] and section 8 for details). The powers of the fugacities t, z and $\widehat{x}$ in the prefactor keep track of the charges (5.7) of a bare monopole operator under the global symmetry, and the Hilbert series $H^{T_{m;\widehat{m},B}}_{Q(m,\widehat{m},B)}$

of the residual gauge theory is the **dressing factor** that keeps track of the charges of the residual matter fields that dress the bare monopole operator.

We will see some applications of the monopole formula (6.3) for the Hilbert series of the moduli space of vacua of three-dimensional $\mathcal{N} \geq 2$ gauge theories in the following sections.

7. Coulomb branch of $3d$ $\mathcal{N} = 4$ gauge theories

The monopole formula (6.3) becomes particularly simple when it is applied to the vector multiplet sector of three-dimensional theories with eight supercharges, *i.e.* $3d$ $\mathcal{N} = 4$ supersymmetry. We will view $3d$ $\mathcal{N} = 4$ gauge theories as special cases of $3d$ $\mathcal{N} = 2$ gauge theories, fixing once and for all an $\mathcal{N} = 2$ subalgebra of the $\mathcal{N} = 4$ supersymmetry algebra. This is equivalent to choosing a particular complex structure out of a $\mathbb{P}^1$ worth of them.

Like $4d$ $\mathcal{N} = 2$ theories, $3d$ $\mathcal{N} = 4$ supersymmetric gauge theories are completely specified by the following data:

(1) **Gauge group**: a compact semisimple Lie group G, to which one associates $\mathcal{N} = 4$ vector multiplets $\mathcal{V}^a$, with $a = 1, \ldots, \mathrm{rk}(G)$, which contains a gauge connection, three real scalars and fermionic partners. The $\mathcal{N} = 4$ vector multiplet $\mathcal{V}$ decomposes into an $\mathcal{N} = 2$ vector multiplet V (containing a gauge connection A with curvature F, a real scalar σ and fermions) and an $\mathcal{N} = 2$ chiral multiplet Φ in the adjoint representation of G (containing a complex scalar that we also call Φ and fermions).

(2) **Matter content**: a (quaternionic) representation $\mathcal{R}$ of G, to which one associates hypermultiplets Y^i, with $i = 1, \ldots, \dim(\mathcal{R})$. The hypermultiplets Y^i decompose into a pair of $\mathcal{N} = 2$ chiral multiplets X^i and $\widetilde{X}_i$, each containing a complex scalar plus fermions, and transforming in complex conjugate representations $\mathcal{R}$ and $\overline{\mathcal{R}}$.[9]

The interactions are completely determined by $\mathcal{N} = 4$ supersymmetry. In particular, the superpotential takes the form $W = \widetilde{X}\Phi X$, where Φ acts on X in the representation $\mathcal{R}$ and the projection to the gauge singlet in $\overline{\mathcal{R}} \otimes \mathcal{R}$ is implied.

The $\mathcal{N} = 4$ supersymmetry algebra admits an R-symmetry automorphism $SU(2)_C \times SU(2)_H$. $SU(2)_C$ acts on vector multiplets, rotating the three real scalars as a triplet. $SU(2)_H$ acts on hypermultiplets, rotating the two complex scalars X^i and $(\widetilde{X}_i)^\dagger$ as a doublet. The moduli space of supersymmetric vacua of a $3d$ $\mathcal{N} = 4$ theory is locally of the form $\mathcal{M}_C \times \mathcal{M}_H$: on the Coulomb branch $\mathcal{M}_C$ the scalars in the vector multiplet take expectation value, whereas on the Higgs branch $\mathcal{M}_H$ the scalars in the hypermultiplets take expectation value. Although Hilbert series methods can be easily applied to the total moduli space of vacua, which includes mixed branches (see for instance [**19**]), we restrict here to the maximal-dimensional components of the Coulomb and Higgs branches.

Higgs branches of 3d $\mathcal{N} = 4$ supersymmetric gauge theories in three dimensions are identical to the Higgs branches of 4d $\mathcal{N} = 2$ supersymmetric gauge theories with the same gauge group and matter content, being protected against quantum corrections by a non-renormalization theorem [**4**]. In particular they are hyperkähler

[9] If the representation $\mathcal{R}$ is pseudoreal, one can introduce a half-hypermultiplet, which contains half as many degrees of freedom as a standard hypermultiplet.

quotients. Hilbert series of Higgs branches of gauge theories with eight supercharges have been computed in [**9**, **46**, **48**].

More interesting results are obtained by applying the logic of this section to Coulomb branches of $3d$ $\mathcal{N}=4$ gauge theories, which are hyperkähler manifolds of quaternionic dimension $r = \mathrm{rk}(G)$. Hypermultiplet scalars vanish on the Coulomb branch, which is parametrized by monopole operators dressed by scalars Φ in the vector multiplet of the residual gauge group. Specializing formula (5.5) to the Cartan generator of the $SU(2)_C$ R-symmetry, which assigns charge 2 to Φ and 0 to hypermultiplet scalars, one obtains the following R-charge for bare monopole operators [**11**, **38**]:

$$R(m,\widehat{m}) = -\sum_{\alpha\in\Delta_+} |\alpha(m)| + \frac{1}{2}\sum_{\rho,\widehat{\rho}} |\rho(m)+\widehat{\rho}(\widehat{m})| \; . \tag{7.1}$$

Taking into account the dressing of monopole operators by the adjoint scalar Φ in the vector multiplet and the absence of nonperturbative corrections to the superpotential, which is a consequence of $\mathcal{N}=4$ supersymmetry, the monopole formula (6.3) takes the simple form [**24**, **26**]

$$H(\tau, z; \widehat{m}) = \sum_{m\in\Gamma_G/\mathcal{W}_G} z^{J(m)} \tau^{2R(m,\widehat{m})} P_G(\tau^2; m) \; , \tag{7.2}$$

where the fugacity $\tau = t^{1/2}$ is introduced to have integer powers, and the dressing factor

$$P_G(t;m) = \prod_{i=1}^{r} \frac{1}{1-t^{d_i(G_m)}} \tag{7.3}$$

counts polynomials in the Casimir invariants of the residual gauge group G_m, as in formula (3.8) with G replaced by G_m. Note that the background magnetic charge B for the topological symmetry was set to zero in (7.2), otherwise there are no gauge invariant monopole operators dressed by Φ only; this agrees with the well known fact that a Fayet-Iliopoulos parameter lifts the Coulomb branch.

It should be noted that the monopole formula (7.2) yields a well defined Taylor series in τ provided the theory is *good* or *ugly* in the terminology of [**38**].[10] Then the conformal dimension of chiral gauge invariant operators in the infrared superconformal field theory is equal to their R-charge.

Recently, a mathematical interpretation of the monopole formula (7.2) for the Hilbert series of the Coulomb branch of 3d $\mathcal{N}=4$ gauge theories has been provided by Nakajima [**68**], leading to a number of interesting mathematical developments, for which I refer to Nakajima's talk at this conference.

A few additional remarks on (7.2) are in order. First of all, $H(t,z;0)$ is the Hilbert series that counts holomorphic function on the Coulomb branch of the low energy superconformal field theory, which is a cone. Conversely, $\widehat{m}\neq 0$ corresponds to turning on real masses that resolve the singularity.

[10] For *bad* theories (7.2) is not a convergent Taylor series, because there are infinitely many monopole operators with the same charges. This problem can be bypassed by adding extra hypermultiplets to make the theory good or ugly, with a large background magnetic charge for the flavour symmetry that acts on them only. The background charge ensures that the common eigenspaces of the Cartan subalgebra of the global symmetry have finite dimension and serves at the same time as a cut-off.

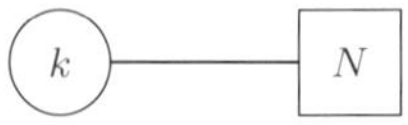

FIGURE 3. 8-supercharge quiver diagram for $U(k)$ SQCD with N flavours.

Secondly, it has been shown in [**71**] that the Hilbert series of the Coulomb branch given by (7.2) can also be obtained as a particular (and more easily computable) limit of the superconformal index. This means that in this limit the index only receives contribution from the scalar chiral operators which are counted by the Hilbert series, and not by other protected operators.

Thirdly, as we will see in some of the examples below, one can often deduce the charges of generators and relations of the Coulomb branch chiral ring by resumming (7.2) and applying plethystic techniques. In this respect the Hilbert series provides complementary information to the more recent physical description of the Coulomb branch in [**14**]: the latter allows to construct the relations explicitly, but in practice determining the generators of the chiral ring and of the ideal of relations can be difficult if the Coulomb branch is not a complete intersection. Instead the Hilbert series can be computed as easily for complete as for non-complete intersections.

Finally, the Hilbert series (7.2) is sensitive to resolutions of the singularity (Kähler deformations) through the dependence on $\widehat{m}$ and B, but is insensitive to complex structure deformations. On the other hand the formalism of [**14**] is sensitive to complex structure deformations but insensitive to resolutions.

7.1. Examples. We conclude this section with a few examples of Hilbert series of $3d$ $\mathcal{N}=4$ theories with interesting Coulomb branches, restricting for simplicity to zero background magnetic charges. Examples with $\widehat{m}\neq 0$ can be found in [**24, 25**].

It is well known that the Coulomb branch of $3d$ $\mathcal{N}=4$ SQED, with $G=U(1)$ and N flavours of hypermultiplets of charge 1, is the A_{N-1} singularity $\mathbb{C}^2/\mathbb{Z}_N$ [**56**]. This result is easily recovered by computing the Hilbert series [**26**]

$$H(\tau,z)=\frac{1}{1-\tau^2}\sum_{m\in\mathbb{Z}} z^m\tau^{N|m|}=\mathrm{PE}[\tau^2+(z+z^{-1})\tau^N-\tau^{2N}]\,, \tag{7.4}$$

which shows that the Coulomb branch chiral ring is generated by three operators Φ, $V_+\equiv V_1$ and $V_-\equiv V_{-1}$, subject to a single relation $V_+V_-=\Phi^N$. This is the well-known algebraic description of the A_{N-1} singularity.

The computation is easily generalized to SQCD theories with $G=U(k)$ and $N\geq 2k-1$ flavours of fundamental hypermultiplets, which are summarized by "quiver diagram" in figure 3, which uses the the eight-supercharge notation where a circular node denotes a unitary gauge group, a square node denotes a unitary flavour group, and edges denote hypermultiplets in the bifundamental representation. The Hilbert series of the Coulomb branch is [**26**]

$$\begin{aligned} H(\tau,z)&=\sum_{m_1\geq\cdots\geq m_k} P_{U(k)}(\tau^2;m)z^{\sum_i m_i}\,\tau^{-2\sum_{i<j}(m_i-m_j)+N\sum_i|m_i|}\\ &=\mathrm{PE}\left[\sum_{j=1}^{k}\left(\tau^{2j}+(z+z^{-1})\tau^{N+2(j-k)}-\tau^{2(N+j-k)}\right)\right]. \end{aligned} \tag{7.5}$$

The dressing factor $P_{U(k)}$ can be written in terms of a partition of k that encodes how many magnetic charges m_i are equal, see appendix A of [**26**] for the explicit

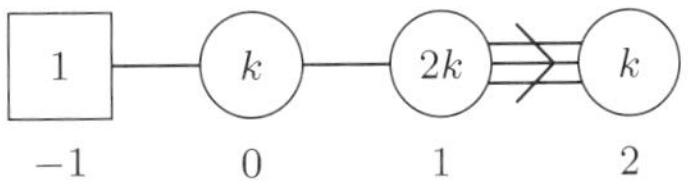

FIGURE 4. Generalized affine Dynkin quiver for k G_2 instantons.

form. The final expression in (7.5) shows that the Coulomb branch is a complete intersection: there are $3k$ generators (k Casimirs and $k+k$ dressed monopole operators of topological charges ± 1) subject to k relations, whose explicit form was later obtained in [**14**].

A very interesting application is to theories whose Coulomb branches are moduli spaces $\mathcal{M}_{\mathcal{G},k}$ of k $\mathcal{G}$-instantons on $\mathbb{C}^2$ [**23**]. These theories are described by (generalized) quiver diagrams which are affine Dynkin diagrams of $\mathcal{G}$, with a $U(1)$ flavour node attached to the zeroth root. See figure 4 for the example of $\mathcal{G} = G_2$. If $\mathcal{G}$ has a non-simply-laced Lie algebra, the field theory has no known Lagrangian description, but a monopole formula for the Hilbert series of their Coulomb branches was conjectured in [**23**] based on brane constructions and the action of outer automorphisms of simply laced Lie algebras. The proposal involves a minimal modification of (7.2) that deals with multiple bonds in the Dynkin diagrams. For instance, the triple bond in the affine Dynkin diagram of G_2 leads to the contribution

$$\Delta R(m) = -\frac{1}{2}\sum_{i=1}^{2k}\sum_{j=1}^{k}|3m_i^{(1)} - m_j^{(2)}| \tag{7.6}$$

to the R-charge of monopole operators, where the factor of 3 accounts for the triple bond directed from node 1 to node 2.

This conjectured modification ensures that the affine Dynkin quiver without the flavour node is balanced, which is crucial for the enhancement of the topological symmetry to a nonabelian gauge group $\mathcal{G} \times SU(2)$, and passes several other nontrivial consistency checks. Most importantly, it provides a uniform description of instanton moduli spaces which is purely based on group theoretic data and is alternative to the ADHM construction [**5**] that applies to classical gauge groups $\mathcal{G}$. See [**12**] for a further mathematical exploration of this modified monopole formula.

Using this general description, it was shown in [**23**] that the Hilbert series of the moduli space $\mathcal{M}_{\mathcal{G},k}$ of k $\mathcal{G}$-instantons has the following perturbative expansion, that holds for any $\mathcal{G}$:

$$H_{\mathcal{M}_{\mathcal{G},k}}(t,x,y) = \mathrm{PE}\left[\sum_{p=0}^{k-1}\left([p+1;0]_{x,y}t^{p+1} + [p;\mathrm{Ad}]_{x,y}t^{p+2}\right) - \dots t^{k+2} + \dots\right]. \tag{7.7}$$

Here x is an $SU(2)$ fugacity, y collectively denotes $\mathcal{G}$ fugacities, and Ad denotes the adjoint representation of $\mathcal{G}$. The first $2k$ positive terms in the argument of the plethystic exponential are generators. In particular $[1;0]_{x,y}t$ corresponds to the centre of mass of the k instantons, and $([2;0]_{x,y} + [0;\mathrm{Ad}]_{x,y})t^2$ correspond to the moment maps of the $SU(2) \times \mathcal{G}$ symmetry. The first relation appears at order t^{k+2}. It would be interesting to see whether this uniform analysis of instanton moduli spaces can be pushed further to fully determine generators and relations.

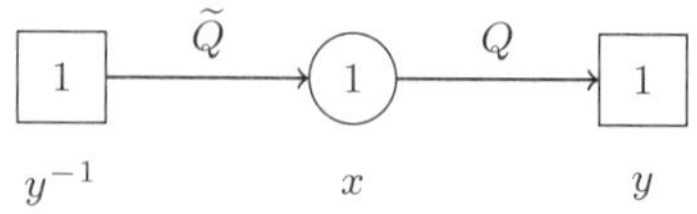

FIGURE 5. Quiver diagram and fugacities for SQED with one flavour.

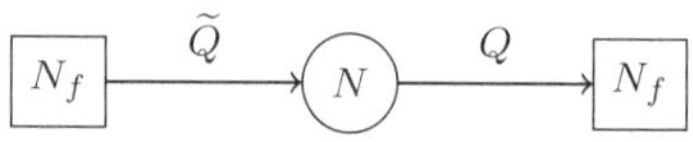

FIGURE 6. Quiver diagram for $U(N)$ SQCD with N_f flavours.

8. Moduli spaces of $3d$ $\mathcal{N}=2$ theories and Hilbert series

The rationale for counting dressed monopole operators presented in section 6 was initially developed to study Coulomb branches of $3d$ $\mathcal{N}=4$ theories, but it was soon realized that it applies just as well to the moduli space of vacua of theories with $\mathcal{N} \geq 2$ supersymmetry. In this section we present a few examples involving Maxwell, Yang-Mills [**22**] and Chern-Simons theories [**27**].

8.1. Maxwell and Yang-Mills theories. We begin with CP-invariant theories, with no Chern-Simons interactions. For simplicity we set all background magnetic charges $\widehat{m}$ and B to zero and take the superpotential to vanish (see [**27**] for the generalization). Then the Hilbert series that counts gauge invariant chiral operators, or equivalently holomorphic functions on the moduli space of supersymmetric vacua $\mathcal{M}$, takes the general form [**22, 43**]

$$
\begin{aligned}
H(t,z,\widehat{x}) = &\sum_{m\in\Gamma_q} z^{J(m)} \left(\prod_{i=1}^{r}\oint \frac{dx_i}{2\pi i x_i}\right) \prod_{\alpha\in\Delta_+} (1-x^\alpha)^{\delta_{\alpha(m),0}} t^{-|\alpha(m)|} \cdot \\
&\cdot \prod_{\rho,\widehat{\rho}} \left(t^{r_{\rho,\widehat{\rho}}-1} x^\rho \widehat{x}^{\widehat{\rho}}\right)^{-\frac{1}{2}|\rho(m)|} \mathrm{PE}\left[\delta_{\rho(m),0} t^{r_{\rho,\widehat{\rho}}} x^\rho \widehat{x}^{\widehat{\rho}}\right] ,
\end{aligned}
\tag{8.1}
$$

where the product over ρ and $\widehat{\rho}$ runs over all weights of the representation $\mathcal{R}\times\widehat{\mathcal{R}}$ of the matter fields under the gauge and flavour symmetry, and $r_{\rho,\widehat{\rho}}$ are the R-charges of the matter fields, which are constant in each irreducible representation. As anticipated below (6.3), due to nonperturbative effects that lift part of the classical Coulomb branch [**29**], the sum over magnetic charges is restricted to a sublattice of $\Gamma_{G^\vee}/\mathcal{W}$, the quantum lattice of magnetic charges Γ_q, which in this case is

$$
\Gamma_q = \{m\in\Gamma_{G^\vee}/\mathcal{W} \mid \sum_\rho \rho(\alpha_i^\vee)\mathrm{sign}\rho(m) \neq 0 \ \vee \ \alpha_i(m)=0 \ \forall i=1,\dots,r\} ,
\tag{8.2}
$$

where α_i and $\alpha_i^\vee$ are simple roots and coroots of the gauge group G.

For instance, applying formula (8.1) to $3d$ $\mathcal{N}=2$ SQED with one flavour, whose quiver diagram is in figure 5, one obtains the Hilbert series

$$
\begin{aligned}
H(t,z,y) &= \sum_{m\in\mathbb{Z}} z^m t^{(1-r)|m|} y^{-|m|} \oint \frac{dx}{2\pi i x} \mathrm{PE}\left[\delta_{m,0} t^r y (x+\frac{1}{x})\right] \\
&= \frac{1}{1-y^2t^{2r}} + \frac{1}{1-zy^{-1}t^{1-r}} + \frac{1}{1-z^{-1}y^{-1}t^{1-r}} - 2 ,
\end{aligned}
\tag{8.3}
$$

where y is a fugacity for the axial $U(1)$ symmetry under which the matter fields Q and $\widetilde{Q}$ have charge 1, and equal R-charge r has been assigned to Q and $\widetilde{Q}$. The final result shows that the moduli space consists of three-one dimensional components meeting at a point, generated by the meson $M = \widetilde{Q}Q$ and the monopole operators $V_\pm \equiv V_{\pm 1}$, subject to the relations $V_+V_- = V_+M = V_-M = 0$. This reproduces the moduli space and chiral ring of the XYZ model (3.5), of which SQED with one flavour is known to be dual [**1**, **29**].

Even more instructive is the case of $U(N)$ SQED with N_f flavours of fundamental and antifundamental matter, see figure 6. The nonabelian gauge dynamics induces nonperturbative corrections to the superpotential, which make the quantum lattice (8.2) two-dimensional:

$$\Gamma_q = \{m = (m_1, 0, \ldots, 0, m_N) \in \mathbb{Z}^N \mid m_1 \geq 0 \geq m_N\} \ . \tag{8.4}$$

The Hilbert series can be computed straightforwardly [**22**], but we do not report the result here. Suffices to say that the chiral ring of the theory follows immediately from the structure of the residual gauge theories associated to lattice points in (8.4) and the knowledge of their moduli spaces. Over the point $m_1 = m_N = 0$, the residual theory is nothing but $U(N)$ SQCD with N_f flavours; over the 1-dimensional boundary the residual theory is $U(N-1)$ SQCD with N_f flavours (plus a free $U(1)$ theory); and over the two-dimensional interior the residual theory is $U(N-2)$ SQCD with N_f flavours (plus two free $U(1)$ theories). It follows almost immediately that the chiral ring of the theory is generated by the $N_f \times N_f$ meson matrix $M = \widetilde{Q}Q$ and the two monopole operators $V_+ \equiv V_{(1,0,\ldots,0)}$ and $V_- \equiv V_{(0,\ldots,0,-1)}$, subject to the chiral ring relations

$$\mathrm{minor}_{N+1}(M) = 0 \ , \quad V_\pm \mathrm{minor}_N(M) = 0 \ , \quad V_+V_-\mathrm{minor}_{N-1}(M) = 0 \tag{8.5}$$

which constrain the rank of the meson matrix and monopole operators. This result is confirmed by explicit computation of the Hilbert series.

This structure of the moduli space of vacua was argued earlier in [**2**] by invoking an extra nonperturbative superpotential involving mesons and monopole operators, which however is singular at certain subloci of the moduli space. We learn from this example that the moduli spaces and the chiral rings of the nonabelian $\mathcal{N} = 2$ Yang-Mills theories can be successfully analysed by looking at dressed monopole operators and using well-defined nonperturbative superpotentials that partially lift the Coulomb branch, without resorting to singular nonperturbative superpotentials.

8.2. Chern-Simons theories. The formalism presented in section 6 can also be applied to CP-violating gauge theories with Chern-Simons interactions (4.2), which were studied intensively following [**3**] since they often appear on the worldvolume of M2-branes in M-theory.

As an example, we consider here the $3d\,\mathcal{N} = 2$ gauge theory on the worldvolume of M2-branes probing the Calabi-Yau cone over the toric Sasaki-Einstein 7-fold $Q^{1,1,1}$, which is a circle fibration over $\mathbb{P}^1 \times \mathbb{P}^1 \times \mathbb{P}^1$. The toric diagram of the Calabi-Yau Gorenstein singularity $C(Q^{1,1,1})$ is a convex polytope in three dimensions, which is shown in figure 7.

The gauge theory on the worldvolume of M2-branes is specified by the quiver diagram in figure 8 and by the superpotential [**6**, **57**]

$$W = \mathrm{Tr}(A_1B_1A_2B_2 - A_1B_2A_2B_1) + p_1B_1q_1 + p_2B_2q_2 \ . \tag{8.6}$$

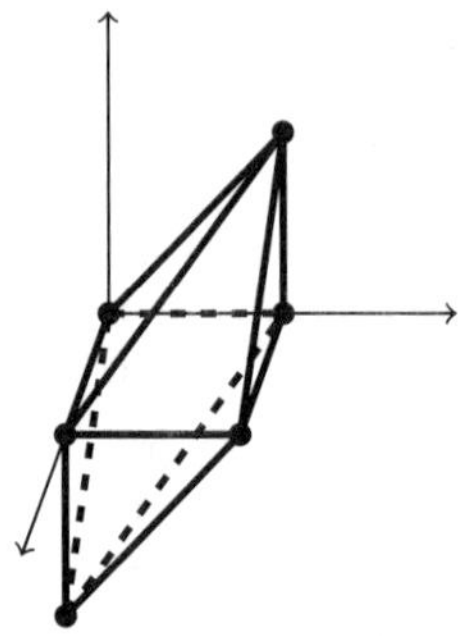

FIGURE 7. Toric diagram of $C(Q^{1,1,1})$.

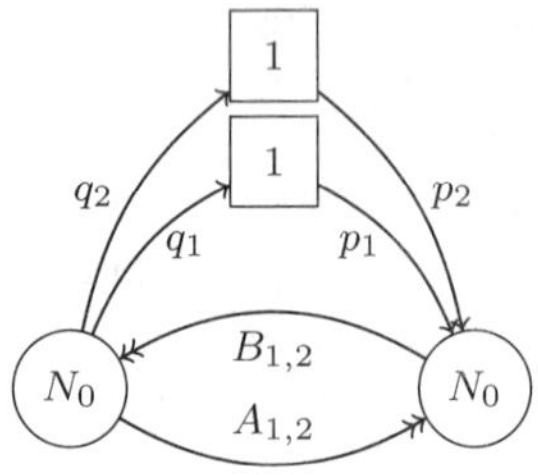

FIGURE 8. Quiver diagram of the worldvolume theory on N M2-branes probing $C(Q^{1,1,1})$. The subscripts denote bare CS levels.

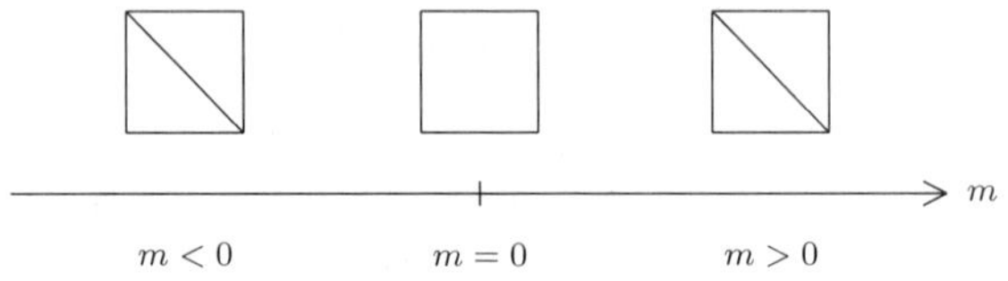

FIGURE 9. Line bundles and resolutions of the conifold from holomorphic functions on $C(Q^{1,1,1})$. The toric diagrams of the resolved conifolds can be obtained by projecting the toric diagram of $C(Q^{1,1,1})$ of figure 7 in the vertical direction.

The bare Chern-Simons levels for the two gauge groups in the Lagrangian vanish, $k_1 = k_2 = 0$, but effective Chern-Simons levels (5.8) are radiatively generated once the fundamental flavours p_i and q_i gain a real mass and are integrated out.

The Hilbert series of the moduli space of vacua is easily computed for the abelian theory on $N = 1$ M2-brane. Setting the background magnetic charges to zero and the dynamical magnetic charges equal $m_1 = m_2 = m$ (there are no gauge invariant dressed monopole operators if $m_1 \neq m_2$), the Hilbert series of the gauge theory (6.3) becomes [**27**]

(8.7)

$$\begin{aligned} H(t,u,v,z) &= \sum_{m\in\mathbb{Z}} z^m t^{\frac{1}{2}|m|} \oint \frac{dx}{2\pi i x} x^{-|m|}\, \mathrm{PE}\left[t^{\frac{1}{2}} x \left(u+\frac{1}{u}\right) + t^{\frac{1}{2}} \frac{1}{x}\left(v+\frac{1}{v}\right)\right] \\ &= \sum_{m\in\mathbb{Z}} z^m t^{\frac{1}{2}|m|} g_1(t^{1/2},u,v;|m|) = \sum_{n=0}^{\infty} [n;n;n]_{\alpha,\beta,\gamma} t^n \end{aligned}$$

where we have used (3.9) in the second line. The final result is nothing but the Hilbert series of the cone over $Q^{1,1,1}$, which can be computed alternatively using the toric description of the geometry: the moduli space of the gauge theory on the M2-branes is the transverse space that they probe. The integral in the first line shows that the residual theory when $m_1 = m_2 = m$ is the worldvolume theory for a D-brane on the conifold [**59**], which up to a decoupled $U(1)$ is the SQED theory with 2 flavours discussed at the end of section 3. The conifold arises here as the Kähler quotient of the cone over $Q^{1,1,1}$ by the $U(1)$ action that corresponds to the vertical direction in the toric diagram 7. Going from the first to the second line we have used the baryonic Hilbert series (3.9) that counts holomorphic sections of line bundles on the conifold, with baryonic charge $B = |m|$, due to the effective Chern-Simons levels $k_1^{\text{eff}} = -k_2^{\text{eff}} = \text{sign}(m)$. The last equality relates the sum over holomorphic sections of line bundles $\oplus_m \mathcal{O}(|m|D)$ on the conifold to holomorphic functions on the cone $C(Q^{1,1,1})$ (see figure 9 and compare with figure 2). The final expression in (8.7), involving $SU(2)^3$ characters in terms of the fugacities $\alpha = u$, $\beta = (v/z)^{1/2}$ and $\beta = (vz)^{1/2}$, shows a global symmetry enhancement to $SU(2)^3 \times U(1)_R$ and reproduces the Hilbert series of the cone over $Q^{1,1,1}$. Generators and relations of the ring can be easily extracted from the final formula [**27**].

The computation (8.7) can be easily generalized to include non-zero background magnetic charges: the Hilbert series then counts holomorphic section of toric line bundles on $C(Q^{1,1,1})$ [**27**], which are in correspondence with resolutions of the cone [**7**]. It can also be extended to the case of $N > 1$ M2-branes, in which case one can show by means of the Hilbert series that the moduli space is $\mathcal{M} = \text{Sym}^N C(Q^{1,1,1})$, as expected from string theory considerations [**27**].

Many more examples of M2-brane theories with $\mathcal{N} = 2$ and $\mathcal{N} = 3$ supersymmetries have been studied using the Hilbert series formalism in [**27**]. Crucially, the Hilbert series can be computed by counting gauge invariant dressed monopole operators, and information on the chiral ring can be extracted from it, with no need to assume the form of the quantum relation between monopole operators as was originally done in [**6, 57**].

9. Conclusion

I have presented here a general formalism to count gauge invariant chiral operators that parametrize moduli spaces of supersymmetric vacua of three-dimensional $\mathcal{N} \geq 2$ gauge theories. The formalism enumerates gauge invariant dressed monopole operators and packages the information in a generating function called the Hilbert series. By applying plethystic techniques to the Hilbert series, one can extract the charges of the generators of the chiral ring and of the ideal of relations. The formalism simplifies considerably the study of the chiral ring of three-dimensional supersymmetric gauge theories: in favourable cases it determines it completely, and more generally it reduces the problem to determining a finite number of coefficients.

A number of interesting open questions are raised by these results:

(1) For the Coulomb branch of $3d$ $\mathcal{N} = 4$ gauge theories, a procedure to determine the chiral ring relations, including all coefficients, has been put forward by Bullimore, Dimofte and Gaiotto [**14**]. Can that construction be generalized to theories with lower supersymmetry, for which I have reviewed how to compute the Hilbert series here?

(2) In a parallel development, a mathematical definition of the Coulomb branch of $3d$ $\mathcal{N} = 4$ gauge theories has been introduced by Nakajima and collaborators [**12, 13, 68**], building on the monopole formula (7.2) for the Hilbert series. Can one similarly define moduli spaces of vacua of $3d$ $\mathcal{N} \geq 2$ gauge theories mathematically, building on formula (6.3)?

(3) A path integral definition of the Hilbert series of the Coulomb branch as an index has been provided for $3d$ $\mathcal{N} = 4$ gauge theories [**20, 71**]. Can the Hilbert series of the moduli space of vacua of $3d$ $\mathcal{N} \geq 2$ theory be defined as an index, at least when the F-flat moduli space $\mathcal{F}$ is a complete intersection?

I hope that some of these questions can be answered in the near future.

Acknowledgements

It is a pleasure to thank Giulia Ferlito, Amihay Hanany, Noppadol Mekareeya and Alberto Zaffaroni for a fruitful and enjoyable collaboration on most of the work reported here, and the organizers of String–Math 2016 for giving me the opportunity to present these results in a highly stimulating environment.

References

[1] O. Aharony, A. Hanany, K. Intriligator, N. Seiberg, and M. J. Strassler, *Aspects of $N = 2$ supersymmetric gauge theories in three dimensions*, Nuclear Phys. B **499** (1997), no. 1-2, 67–99. MR1468698

[2] Ofer Aharony, *IR duality in $d = 3$ $N = 2$ supersymmetric* $\mathrm{USp}(2N_c)$ *and* $\mathrm{U}(N_c)$ *gauge theories*, Phys. Lett. B **404** (1997), no. 1-2, 71–76. MR1456971

[3] Ofer Aharony, Oren Bergman, Daniel Louis Jafferis, and Juan Maldacena, *$\mathcal{N} = 6$ superconformal Chern-Simons-matter theories, M2-branes and their gravity duals*, J. High Energy Phys. **10** (2008), 091, 38. MR2452954

[4] Philip C. Argyres, M. Ronen Plesser, and Nathan Seiberg, *The moduli space of vacua of $N = 2$ SUSY QCD and duality in $N = 1$ SUSY QCD*, Nuclear Phys. B **471** (1996), no. 1-2, 159–194. MR1401295

[5] M. F. Atiyah, N. J. Hitchin, V. G. Drinfel′d, and Yu. I. Manin, *Construction of instantons*, Phys. Lett. A **65** (1978), no. 3, 185–187. MR598562

[6] Francesco Benini, Cyril Closset, and Stefano Cremonesi, *Chiral flavors and M2-branes at toric CY4 singularities*, JHEP **02** (2010), 036.

[7] ———, *Quantum moduli space of Chern-Simons quivers, wrapped D6-branes and AdS4/CFT3*, JHEP **09** (2011), 005.

[8] Sergio Benvenuti, Bo Feng, Amihay Hanany, and Yang-Hui He, *Counting BPS operators in gauge theories: quivers, syzygies and plethystics*, J. High Energy Phys. **11** (2007), 050, 48. MR2362099

[9] Sergio Benvenuti, Amihay Hanany, and Noppadol Mekareeya, *The Hilbert series of the one instanton moduli space*, J. High Energy Phys. **6** (2010), 100, 40. MR2680314

[10] Vadim Borokhov, *Monopole operators in three-dimensional $\mathcal{N} = 4$ SYM and mirror symmetry*, J. High Energy Phys. **3** (2004), 008, 21. MR2062369

[11] Vadim Borokhov, Anton Kapustin, and Xinkai Wu, *Monopole operators and mirror symmetry in three dimensions*, J. High Energy Phys. **12** (2002), 044, 23. MR1955002

[12] Alexander Braverman, Michael Finkelberg, and Hiraku Nakajima, *Coulomb branches of 3d $\mathcal{N} = 4$ quiver gauge theories and slices in the affine Grassmannian (with appendices by Alexander Braverman, Michael Finkelberg, Joel Kamnitzer, Ryosuke Kodera, Hiraku Nakajima, Ben Webster, and Alex Weekes)*, (2016).

[13] ———, *Towards a mathematical definition of Coulomb branches of 3-dimensional $\mathcal{N} = 4$ gauge theories, II*, (2016).

[14] Mathew Bullimore, Tudor Dimofte, and Davide Gaiotto, *The Coulomb Branch of 3d $\mathcal{N} = 4$ Theories*, (2015).

[15] Mathew Bullimore, Tudor Dimofte, Davide Gaiotto, and Justin Hilburn, *Boundaries, mirror symmetry, and symplectic duality in 3d $\mathcal{N} = 4$ gauge theory*, J. High Energy Phys. **10** (2016), 108, front matter+191. MR3578533

[16] Mathew Bullimore, Tudor Dimofte, Davide Gaiotto, Justin Hilburn, and Hee-Cheol Kim, *Vortices and Vermas*, (2016).

[17] Nathan Seiberg, *Adding fundamental matter to: "Chiral rings and anomalies in supersymmetric gauge theory" [J. High Energy Phys.* ***2002****, no. 12, 071, 56 pp; MR1960462] by F. Cachazo, M. R. Douglas, Seiberg and E. Witten*, J. High Energy Phys. **1** (2003), 061, 12. MR1969863

[18] Philip Candelas and Xenia C. de la Ossa, *Comments on conifolds*, Nuclear Phys. B **342** (1990), no. 1, 246–268. MR1068113

[19] Federico Carta and Hirotaka Hayashi, *Hilbert series and mixed branches of $T[SU(N)]$ theory*, (2016).

[20] Cyril Closset and Heeyeon Kim, *Comments on twisted indices in 3d supersymmetric gauge theories*, J. High Energy Phys. **8** (2016), 059, front matter+84. MR3555623

[21] Sidney R. Coleman, *The Quantum Sine-Gordon Equation as the Massive Thirring Model*, Phys. Rev. **D11** (1975), 2088.

[22] Stefano Cremonesi, *The Hilbert series of 3d $\mathcal{N} = 2$ Yang-Mills theories with vectorlike matter*, J. Phys. A **48** (2015), no. 45, 455401, 18. MR3418004

[23] Stefano Cremonesi, Giulia Ferlito, Amihay Hanany, and Noppadol Mekareeya, *Coulomb Branch and The Moduli Space of Instantons*, JHEP **1412** (2014), 103.

[24] Stefano Cremonesi, Amihay Hanany, Noppadol Mekareeya, and Alberto Zaffaroni, *Coulomb branch Hilbert series and Hall-Littlewood polynomials*, J. High Energy Phys. **9** (2014), 178, front matter+57. MR3267902

[25] ———, *$T^{\sigma}_{\rho}(G)$ Theories and Their Hilbert Series*, (2014).

[26] Stefano Cremonesi, Amihay Hanany, and Alberto Zaffaroni, *Monopole operators and Hilbert series of Coulomb branches of 3d $\mathcal{N} = 4$ gauge theories*, JHEP **1401** (2014), 005.

[27] Stefano Cremonesi, Noppadol Mekareeya, and Alberto Zaffaroni, *The moduli spaces of 3d $\mathcal{N} \geq 2$ Chern-Simons gauge theories and their Hilbert series*, J. High Energy Phys. **10** (2016), 046, front matter+91. MR3577515

[28] Jan de Boer, Kentaro Hori, Hirosi Ooguri, and Yaron Oz, *Mirror symmetry in three-dimensional gauge theories, quivers and D-branes*, Nuclear Phys. B **493** (1997), no. 1-2, 101–147. MR1454291

[29] Jan de Boer, Kentaro Hori, and Yaron Oz, *Dynamics of $N = 2$ supersymmetric gauge theories in three dimensions*, Nuclear Phys. B **500** (1997), no. 1-3, 163–191. MR1471654

[30] Michele Del Zotto and Amihay Hanany, *Complete Graphs, Hilbert Series, and the Higgs branch of the 4d $\mathcal{N} = 2$ (A_n, A_m) SCFTs*, Nucl. Phys. **B894** (2015), 439–455.

[31] Anindya Dey, Amihay Hanany, Peter Koroteev, and Noppadol Mekareeya, *Mirror Symmetry in Three Dimensions via Gauged Linear Quivers*, JHEP **1406** (2014), 059.

[32] Nick Dorey and David Tong, *Mirror symmetry and toric geometry in three dimensional gauge theories*, J. High Energy Phys. **5** (2000), Paper 18, 16. MR1768734

[33] F. Englert and P. Windey, *Quantization condition for 't Hooft monopoles in compact simple Lie groups*, Phys. Rev D (3) **14** (1976), no. 10, 2728–2731. MR0495850

[34] Bo Feng, Amihay Hanany, and Yang-Hui He, *Counting gauge invariants: the plethystic program*, J. High Energy Phys. **3** (2007), 090, 42. MR2313959

[35] Davide Forcella, Amihay Hanany, Yang-Hui He, and Alberto Zaffaroni, *The Master Space of N=1 Gauge Theories*, JHEP **0808** (2008), 012.

[36] Davide Forcella, Amihay Hanany, and Alberto Zaffaroni, *Baryonic generating functions*, J. High Energy Phys. **12** (2007), 022, 43. MR2366534
[37] Davide Gaiotto, *$N = 2$ dualities*, J. High Energy Phys. **8** (2012), 034, front matter + 57. MR3006961
[38] Davide Gaiotto and Edward Witten, *S-duality of boundary conditions in $\mathcal{N} = 4$ super Yang-Mills theory*, Adv. Theor. Math. Phys. **13** (2009), no. 3, 721–896. MR2610576
[39] A. Giveon, M. Porrati, and E. Rabinovici, *Target space duality in string theory*, Phys. Rep. **244** (1994), no. 2-3, 77–202. MR1288936
[40] P. Goddard, J. Nuyts, and D. Olive, *Gauge theories and magnetic charge*, Nuclear Phys. B **125** (1977), no. 1, 1–28. MR0443696
[41] James Gray, Amihay Hanany, Yang-Hui He, Vishnu Jejjala, and Noppadol Mekareeya, *SQCD: A Geometric Apercu*, JHEP **0805** (2008), 099.
[42] Daniel R. Grayson and Michael E. Stillman, *Macaulay2, a software system for research in algebraic geometry*, Available at http://www.math.uiuc.edu/Macaulay2/.
[43] Amihay Hanany, Chiung Hwang, Hyungchul Kim, Jaemo Park, and Rak-Kyeong Seong, *Hilbert series for theories with Aharony duals*, J. High Energy Phys. **11** (2015), 132, front matter+29. MR3454940
[44] Amihay Hanany and Rudolph Kalveks, *Quiver theories for moduli spaces of classical group nilpotent orbits*, J. High Energy Phys. **6** (2016), 130, front matter+60. MR3538768
[45] Amihay Hanany and Noppadol Mekareeya, *Counting gauge invariant operators in SQCD with classical gauge groups*, J. High Energy Phys. **10** (2008), 012, 27. MR2453033
[46] ———, *Complete Intersection Moduli Spaces in N=4 Gauge Theories in Three Dimensions*, JHEP **1201** (2012), 079.
[47] Amihay Hanany, Noppadol Mekareeya, and Giuseppe Torri, *The Hilbert series of adjoint SQCD*, Nuclear Phys. B **825** (2010), no. 1-2, 52–97. MR2552386
[48] Amihay Hanany and Christian Römelsberger, *Counting BPS operators in the chiral ring of $\mathcal{N} = 2$ supersymmetric gauge theories or $\mathcal{N} = 2$ brane surgery*, Adv. Theor. Math. Phys. **11** (2007), no. 6, 1091–1112. MR2368942
[49] Amihay Hanany and Marcus Sperling, *Algebraic properties of the monopole formula*, (2016).
[50] ———, *Coulomb branches for rank 2 gauge groups in 3d $\mathcal{N} = 4$ gauge theories*, JHEP **08** (2016), 016.
[51] Nigel J. Hitchin, A. Karlhede, U. Lindström, and M. Roček, *HyperKahler Metrics and Supersymmetry*, Commun.Math.Phys. **108** (1987), 535.
[52] Kentaro Hori and Cumrun Vafa, *Mirror symmetry*, (2000).
[53] Yosuke Imamura and Shuichi Yokoyama, *Index for three dimensional superconformal field theories with general R-charge assignments*, J. High Energy Phys. **4** (2011), 007, 21. MR2833291
[54] Kenneth Intriligator and Nathan Seiberg, *Aspects of 3d $\mathcal{N} = 2$ Chern-Simons-matter theories*, J. High Energy Phys. **7** (2013), 079, front matter+62. MR3106220
[55] K. Intriligator and N. Seiberg, *Lectures on supersymmetric gauge theories and electric-magnetic duality*, Nuclear Phys. B Proc. Suppl. **45BC** (1996), 1–28. String theory, gauge theory and quantum gravity (Trieste, 1995). MR1410545
[56] K. Intriligator and N. Seiberg, *Mirror symmetry in three-dimensional gauge theories*, Phys. Lett. B **387** (1996), no. 3, 513–519. MR1413696
[57] Daniel Louis Jafferis, *Quantum corrections to $\mathcal{N} = 2$ Chern-Simons theories with flavor and their* AdS_4 *duals*, J. High Energy Phys. **8** (2013), 046, front matter+22. MR3106392
[58] Anton Kapustin, *Wilson-'t Hooft operators in four-dimensional gauge theories and S-duality*, Phys. Rev. D (3) **74** (2006), no. 2, 025005, 14. MR2249977
[59] Igor R. Klebanov and Edward Witten, *Superconformal field theory on threebranes at a Calabi-Yau singularity*, Nuclear Phys. B **536** (1999), no. 1-2, 199–218. MR1666725
[60] Ryosuke Kodera and Hiraku Nakajima, *Quantized Coulomb branches of Jordan quiver gauge theories and cyclotomic rational Cherednik algebras*, (2016).
[61] Robert Langlands, *Letter to Andre Weil*, unpublished, January 1967.
[62] Wolfgang Lerche, Cumrun Vafa, and Nicholas P. Warner, *Chiral rings in $N = 2$ superconformal theories*, Nuclear Phys. B **324** (1989), no. 2, 427–474. MR1025424
[63] Markus A. Luty and Washington Taylor IV, *Varieties of vacua in classical supersymmetric gauge theories*, Phys. Rev. D (3) **53** (1996), no. 6, 3399–3405. MR1380943

[64] S. Mandelstam, *Soliton operators for the quantized sine-Gordon equation*, Phys. Rev. D (3) **11** (1975), 3026–3030. MR0456086

[65] Dario Martelli, James Sparks, and Shing-Tung Yau, *Sasaki-Einstein manifolds and volume minimisation*, Comm. Math. Phys. **280** (2008), no. 3, 611–673. MR2399609

[66] Noppadol Mekareeya, *The moduli space of instantons on an ALE space from 3d $\mathcal{N} = 4$ field theories*, J. High Energy Phys. **12** (2015), 174, front matter+29pp. MR3464628

[67] Hiraku Nakajima, *Questions on provisional Coulomb branches of 3-dimensional $\mathcal{N} = 4$ gauge theories*, (2015).

[68] ———, *Towards a mathematical definition of Coulomb branches of 3-dimensional $\mathcal{N} = 4$ gauge theories, I*, (2015).

[69] Hiraku Nakajima and Yuuya Takayama, *Cherkis bow varieties and Coulomb branches of quiver gauge theories of affine type A*, (2016).

[70] Philippe Pouliot, *Molien function for duality*, J. High Energy Phys. **1** (1999), Paper 21, 6. MR1684510

[71] Shlomo S. Razamat and Brian Willett, *Down the rabbit hole with theories of class $\mathcal{S}$*, J. High Energy Phys. **10** (2014), 099, front matter+35. MR3324825

[72] Christian Römelsberger, *Counting chiral primaries in $\mathcal{N} = 1$, $d = 4$ superconformal field theories*, Nuclear Phys. B **747** (2006), no. 3, 329–353. MR2241553

[73] Nathan Seiberg, *Naturalness versus supersymmetric nonrenormalization theorems*, Phys. Lett. B **318** (1993), no. 3, 469–475. MR1248357

[74] G. 't Hooft, *On the phase transition towards permanent quark confinement*, Nuclear Phys. B **138** (1978), no. 1, 1–25. MR0489444

[75] David Tong, *Dynamics of $N = 2$ supersymmetric Chern-Simons theories*, J. High Energy Phys. **7** (2000), Paper 19, 26. MR1780358

Department of Mathematics, King's College London, The Strand, London WC2R 2LS, United Kingdom

Current address: Department of Mathematical Sciences, Durham University, Durham DH1 3LE, United Kingdom

Email address: `stefano.cremonesi@durham.ac.uk`

Proceedings of Symposia in Pure Mathematics
Volume **98**, 2018
http://dx.doi.org/10.1090/pspum/098/01720

Quantized Coulomb branches of Jordan quiver gauge theories and cyclotomic rational Cherednik algebras

Ryosuke Kodera and Hiraku Nakajima

Abstract. We study quantized Coulomb branches of quiver gauge theories of Jordan type. We prove that the quantized Coulomb branch is isomorphic to the spherical graded Cherednik algebra in the unframed case, and is isomorphic to the spherical cyclotomic rational Cherednik algebra in the framed case. We also prove that the quantized Coulomb branch is a deformation of a subquotient of the Yangian of the affine $\mathfrak{gl}(1)$.

1. Introduction

1(i). Let G be a complex reductive group and $\mathbf{M}$ be its finite dimensional symplectic representation. To such a pair, physicists consider a $3d$ $\mathcal{N} = 4$ supersymmetric gauge theory, and associate a hyper-Kähler manifold possibly with a singularity with an $\mathrm{SU}(2)$-action rotating complex structures, called the *Coulomb branch* of the gauge theory. Its physical definition involves quantum corrections, hence is difficult to be justified in a mathematically rigorous way. The second named author proposed an approach towards a mathematically rigorous definition of the Coulomb branch [**Nak16**]. When the symplectic representation $\mathbf{M}$ is of a form $\mathbf{N} \oplus \mathbf{N}^*$, its definition as an affine variety with a symplectic form on the regular locus, together with a $\mathbb{C}^\times$-action was firmly established in [**BFN16a**] written by the second named author with Braverman and Finkelberg.

As a byproduct of the definition in [**Nak16, BFN16a**], the Coulomb branch $\mathcal{M}$ has a natural quantization, i.e., we have a noncommutative algebra $\mathcal{A}_\hbar$ over $\mathbb{C}[\hbar]$ such that its specialization $\mathcal{A}_\hbar \otimes_{\mathbb{C}[\hbar]} (\mathbb{C}[\hbar]/\hbar\mathbb{C}[\hbar])$ at $\hbar = 0$ is isomorphic to the coordinate ring $\mathbb{C}[\mathcal{M}]$, and the Poisson bracket associated with the symplectic form is given by the formula $\{\ ,\ \} = \frac{[\ ,\ ]}{\hbar} \bmod \hbar$. We call $\mathcal{A}_\hbar$ the *quantized Coulomb branch*.

By a general result proved in [**BFN16a**], the quantized Coulomb branch $\mathcal{A}_\hbar$ is embedded into a localization of the quantized Coulomb branch associated with the maximal torus T of G and the zero representation. The latter quantized Coulomb branch is the ring of difference operators on the Lie algebra of T, and the localization is the complement of a union of hyperplanes in $\operatorname{Lie} T$.

2010 *Mathematics Subject Classification.* Primary 20C08; Secondary 16G20, 17B67.

The first author's research was supported by JSPS Kakenhi Grant Number 25220701.
The second author's research was supported by National Science Foundation grant PHY-1066293 and by JSPS Kakenhi Grant Numbers 24224001, 25220701, and 16H06335.

This embedding was further studied for a framed quiver gauge theory of type ADE in two appendices of [**BFN16b**], written by the present authors with Braverman, Finkelberg, Kamnitzer, Webster and Weekes. In particular, generators of the quantized Coulomb branch are given by explicit difference operators, and it was shown that the quantized Coulomb branch is a quotient of shifted Yangian introduced in [**KWWY14**] (when the weight μ determined by dimension vectors of a framed quiver representation is dominant).

In this paper, we study the quantized Coulomb branch of a framed quiver gauge theory of Jordan type, i.e., the gauge group is $\mathrm{GL}(N)$, and its representation $\mathbf{N}$ is the direct sum of the adjoint representation $\mathfrak{gl}(N)$ and l copies of the vector representation $\mathbb{C}^N$, where l is a nonnegative integer. The Coulomb branch of the gauge theory was known in [**dBHOO97**]: it is the Nth symmetric power of the surface $\mathcal{S}_l$ in $\mathbb{C}^3$ given by the equation $xy = z^l$. (For example, $\mathcal{S}_0$ is $\mathbb{C} \times \mathbb{C}^\times$.) The definition in [**Nak16, BFN16a**] reproduces this answer [**BFN16b**, §3(vii)]. We have its quantization, namely the spherical part $\mathbf{SH}^{\mathrm{cyc}}_{N,l}$ of the cyclotomic rational Cherednik algebra $\mathbf{H}^{\mathrm{cyc}}_{N,l}$ associated with the wreath product $\mathfrak{S}_N \ltimes (\mathbb{Z}/l\mathbb{Z})^N$ if $l > 0$, and the spherical part $\mathbf{SH}^{\mathrm{gr}}_N$ of the graded Cherednik algebra $\mathbf{H}^{\mathrm{gr}}_N$ associated with $\mathfrak{S}_N$ if $l = 0$. Here the cyclotomic rational Cherednik algebra $\mathbf{H}^{\mathrm{cyc}}_{N,l}$ with $l = 1$ is understood as the rational Cherednik algebra associated with $\mathfrak{S}_N$, which is the rational degeneration of $\mathbf{H}^{\mathrm{gr}}_N$.

Our first main result is to show that the quantized Coulomb branch is isomorphic to this quantization:

THEOREM 1.1. *The quantized Coulomb branch $\mathcal{A}_\hbar$ of the gauge theory $(G, \mathbf{N}) = (\mathrm{GL}(N), \mathfrak{gl}(N) \oplus (\mathbb{C}^N)^{\oplus l})$ is isomorphic to $\mathbf{SH}^{\mathrm{cyc}}_{N,l}$ if $l > 0$, and to $\mathbf{SH}^{\mathrm{gr}}_N$ if $l = 0$. The parameters $\mathbf{t}$, $\hbar$ of the quantized Coulomb branch and the Cherednik algebra are the same, and others are matched by*

$$z_k = -\frac{1}{l}\left((l-k)\hbar + \sum_{m=1}^{l-1} \frac{c_m(1-\varepsilon^{mk})}{1-\varepsilon^m}\right)$$

when $l > 0$.

REMARK 1.2. The quantized Coulomb branch has the parameter $z_1, \ldots, z_l$ corresponding to equivariant variables for the additional $(\mathbb{C}^\times)^l$-action. However the overall shift $z_1 \to z_1 + c, \ldots, z_l \to z_l + c$ is irrelevant. Therefore our convention $z_l = 0$ does not loose generality.

While we were preparing the paper, we noticed that Losev shows that any filtered quantization of $\mathrm{Sym}^N \mathcal{S}_l$ is $\mathbf{SH}^{\mathrm{cyc}}_{N,l}\Big|_{\hbar=1}$ for some choice of parameters [**Los16**]. In order to calculate parameters, we can use localization of the quantized Coulomb branch as in [**BFN16b**, §3(iii)] to reduce the cases $l = 0$ or $N = 1$. (We learn the argument through discussion with Losev.) The case $l = 0$ is easy to handle, as we will do in §2. The proof for the case $N = 1$ requires only Proposition 5.3, thus our calculation is reduced about to the half. We think that our proof is elementary and interesting its own way.

The second main result is the relation of $\mathcal{A}_\hbar$ to the Yangian $Y(\widehat{\mathfrak{gl}}(1))$ of the affine $\mathfrak{gl}(1)$, which appeared in cohomology of moduli spaces of framed torsion free sheaves on $\mathbb{C}^2$ ([**MO12, SV13**]). We will use its presentation in [**AS13**] (see also [**Tsy17**]) by generators and relations.

THEOREM 1.3 (Theorem 6.14 for detail). *The quantized Coulomb branch $\mathcal{A}_\hbar$ is a deformation of a subquotient of $Y(\widehat{\mathfrak{gl}}(1))$. A little more precisely, we consider a subalgebra of $Y(\widehat{\mathfrak{gl}}(1))$ generated by elements $D_{0,m}$ $(m \geq 1)$, e_n, f_{n+l} $(n \geq 0)$ and deform it by replacing the relation* (6.1h) *by* (6.13)*. Let $Y_l(\vec{z})$ denote the resulting algebra. We have a surjective homomorphism $Y_l(\vec{z}) \to \mathcal{A}_\hbar$.*

This result for $l = 0$ is a consequence of Theorem 1.1 and [**SV13**], where $Y(\widehat{\mathfrak{gl}}(1))$ is introduced as a limit of $\mathbf{SH}_N^{\mathrm{gr}}$ as $N \to \infty$. But we will give a self-contained proof starting from the presentation in [**AS13**] so that it also gives the result with $l > 0$. We call $Y_l(\vec{z})$ the *shifted Yangian* of $\widehat{\mathfrak{gl}}(1)$, as it is an analog of the shifted Yangian of $\mathfrak{gl}_n$ [**BK06, BK08**] and a finite dimensional simple Lie algebra [**KWWY14**].

1(ii). All necessary computation for the quantized Coulomb branch in the proofs of two main theorems is already given in appendices of [**BFN16b**]. The remaining steps are to relate the spherical cyclotomic rational Cherednik algebra and the affine Yangian of $\mathfrak{gl}(1)$ with the ring of difference operators on $T = (\mathbb{C}^\times)^N$. These steps are completely independent of the quantized Coulomb branch. Let us formulate them as results on those algebras.

Let $\mathcal{A}_\hbar(T, 0)$ be the ring of $\hbar$-difference operators on the Lie algebra of the torus T. Taking coordinates, we represent it as the $\mathbb{C}[\hbar]$-algebra with generators w_i, $\mathsf{u}_i^{\pm 1}$ $(1 \leq i \leq \dim T)$ with relations

$$[w_i, w_j] = 0 = [\mathsf{u}_i, \mathsf{u}_j], \qquad \mathsf{u}_i^{-1}\mathsf{u}_i = 1 = \mathsf{u}_i\mathsf{u}_i^{-1}, \qquad [\mathsf{u}_i^{\pm 1}, w_j] = \pm\delta_{i,j}\hbar\mathsf{u}_i^{\pm 1}.$$

Let $\mathbf{t}$, $z_1, \dots, z_l$ be other variables. We introduce difference operators

(1.4)

$$E_n[f] \stackrel{\text{def.}}{=} \sum_{\substack{I \subset \{1,\dots,N\} \\ \#I = n}} f(w_I) \prod_{i \in I, j \notin I} \frac{w_i - w_j - \mathbf{t}}{w_i - w_j} \prod_{i \in I} \mathsf{u}_i,$$

$$F_n[f] \stackrel{\text{def.}}{=} \sum_{\substack{I \subset \{1,\dots,N\} \\ \#I = n}} f(w_I - \hbar) \prod_{i \in I, j \notin I} \frac{w_i - w_j + \mathbf{t}}{w_i - w_j} \prod_{i \in I} \left(\prod_{k=1}^{l} (w_i - \hbar - z_k) \cdot \mathsf{u}_i^{-1} \right),$$

where $1 \leq n \leq N$ and f is a symmetric polynomial in n variables, and $f(w_I)$, $f(w_I - \hbar)$ mean substitution of $(w_i)_{i \in I}$, $(w_i - \hbar)_{i \in I}$ to f respectively. These are elements in the localized ring

$$\tilde{\mathcal{A}}_\hbar = \mathbb{C}[\hbar, \mathbf{t}, z_1, \dots, z_l]\langle w_i, \mathsf{u}_i^{\pm 1}, (w_i - w_j + m\hbar)^{-1} (m \in \mathbb{Z})\rangle/\text{relations}.$$

Then [**BFN16b**, Th. A.8] says the quantized Coulomb branch $\mathcal{A}_\hbar$ is the subalgebra of $\tilde{\mathcal{A}}_\hbar$ generated by operators $E_n[f]$, $F_n[f]$, and symmetric polynomials in $(w_1, \dots, w_N)$. This result gives us an algebraic characterization of $\mathcal{A}_\hbar$. We will use it as a starting point, and will not use the original definition of $\mathcal{A}_\hbar$ in [**Nak16, BFN16a**].

If $f \equiv 1$, $E_n[1]$ is a rational version of the nth Macdonald operator, once we understand u_i as the $\hbar$-difference operator $f(w_1, \dots, w_N) \mapsto f(w_1, \dots, w_i + \hbar, \dots, w_N)$. The same is true for $F_n[1]$ with $l = 0$. These observation will give us a link between $\mathcal{A}_\hbar$ and the graded Cherednik algebra with $l = 0$, and lead to a proof of Theorem 1.1 with $l = 0$. (See §2.) The proof of Theorem 1.1 with $l > 0$ is given in the same way, by relating $E_n[f]$, $F_n[f]$ with the cyclotomic rational Cherednik algebra.

THEOREM 1.5. *There is a faithful embedding of* $\mathbf{SH}^{\mathrm{cyc}}_{N,l}$ *to the ring* $\tilde{\mathcal{A}}_\hbar$ *of localized difference operators such that*

$$e_{\Gamma_N}\left(\sum_{i=1}^N (l^{-1}\xi_i\eta_i + \mathbf{t}\sum_{j<i} s_{ji})^n\right)e_{\Gamma_N} = \sum_{i=1}^N w_i^n \quad (n>0),$$

$$e_{\Gamma_N}\left(\sum_{i=1}^N \xi_i^l\right)e_{\Gamma_N} = E_1[1], \qquad e_{\Gamma_N}\left(\sum_{i=1}^N (l^{-1}\eta_i)^l\right)e_{\Gamma_N} = F_1[1],$$

where e_{Γ_N} *is the idempotent for the group* $\Gamma_N = \mathfrak{S}_N \ltimes (\mathbb{Z}/l\mathbb{Z})^N$, *and* ξ_i, η_i *are generators of* $\mathbf{H}^{\mathrm{cyc}}_{N,l}$. *(See §5 for the presentation of* $\mathbf{H}^{\mathrm{cyc}}_{N,l}$*).*

This embedding is given by the composition of an embedding $e_\Gamma\mathbf{H}^{\mathrm{cyc}}_{N,l}e_\Gamma \to \mathbf{H}^{\mathrm{gr}}_N[c_1,\dots,c_{l-1}]$ in [**Obl07**] (also [**Suz05**] for $l=1$) and the faithful embedding of $\mathbf{H}^{\mathrm{gr}}_N$ by rational Demazure-Lusztig operators. Here e_Γ is the idempotent for the subgroup $(\mathbb{Z}/l\mathbb{Z})^N$.

For Yangian $Y(\hat{\mathfrak{gl}}(1))$, we define a homomorphism to $\tilde{\mathcal{A}}_\hbar$ by setting images of generators $D_{0,m}$, e_n, f_n as explicit operators. This is an analog of representations of Yangian of finite type by difference operators [**GKLO05, KWWY14**].

The paper is organized as follows. We give a proof of Theorem 1.1 with $l=0$ in §2, assuming some results on the graded Cherednik algebra. In §3 we compute some Poisson brackets in order to reduce generators of the quantized Coulomb branch. In §4 we recall the definitions and some properties of the graded and rational Cherednik algebras. It includes two embeddings of the graded Cherednik algebra, one is into the ring of differential operators via Dunkl operators and the other is into the ring of difference operators via rational Demazure-Lusztig operators. In §5 we recall the definition of the cyclotomic rational Cherednik algebra and an embedding of its partially symmetrized subalgebra into the graded Cherednik algebra due to Oblomkov. Then some calculations yield Theorem 1.5, and hence Theorem 1.1 is proved for general l. We relate the quantized Coulomb branch to the affine Yangian in §6, §A, §B.

Acknowledgments. This research was originally started together with A. Braverman and M. Finkelberg as continuation of the appendix in [**BFN16b**]. P. Etingof continuously helped us from the beginning. It became soon clear that we only need to calculate the image of certain operators in the partially symmetrized $\mathbf{H}^{\mathrm{cyc}}_{N,l}$ under the embedding to $\mathbf{H}^{\mathrm{gr}}_N$ given in [**Obl07**]. The authors of this paper took this computational approach, while the other three consider another route through non-spherical Cherednik algebras. Hence the papers are written separately as this paper and [**BEF16**]. The authors thank them for sharing many insights. We also thank I. Losev for discussing another derivation of Theorem 1.1 from [**Los16**].

H.N. thanks A. Kirillov for teaching him basics on Cherednik algebras, T. Suzuki for telling him the reference [**Obl07**], and A. Tsymbaliuk for discussion on affine Yangian.

This work was started when H.N. was staying at the Simons Center for Geometry and Physics, and was completed when he was visiting at the Aspen Center for Physics. H.N. thanks both institutes for hospitality and nice research atmosphere.

2. The case with no framing

Let us start with the case $l = 0$ to illustrate our strategy. We assume some results on the graded Cherednik algebra. The proof of the case $l > 0$ will be given after preparing necessary results for cyclotomic rational Cherednik algebras in §4.

Consider the commutator of the class $E_n[f]$ and the kth power sum $p_k(\vec{w}) = \sum_{i=1}^N w_i^k$. It is

$$[p_k(\vec{w}), E_n[f]] \\ = - \sum_{I\subset\{1,\dots,N\},\#I=n} (p_k((w_i+\hbar)_{i\in I}) - p_k(w_I))\, f(w_I) \prod_{i\in I, j\notin I} \frac{w_i - w_j - \mathsf{t}}{w_i - w_j} \prod_{i\in I} \mathsf{u}_i,$$

thanks to the commutation relation $[\prod_{i\in I} \mathsf{u}_i, w_j] = \hbar \prod_{i\in I}\mathsf{u}_i$ if $j \in I$ and 0 otherwise. Therefore we can get elements $E_n[f]$ for general f inductively from $E_n[1]$ and symmetric functions in w_i by taking commutators divided by $\hbar$. The same is true for $F_n[f]$. (In particular, $\mathcal{A}$ is generated by $E_n[1]$, $F_n[1]$ and symmetric functions in $\vec{w}$ *as a Poisson algebra.*)

Proof of Theorem 1.1 with $l = 0$. Consider the embedding of $\mathbf{SH}_N^{\mathrm{gr}}$ to the ring of rational difference operators on $\mathsf{t} = \mathbb{C}^N$, obtained as the trigonometric degeneration of the usual embedding of the spherical part of the double affine Hecke algebra $\mathbf{H}_N$. Symmetric functions in w_i are considered as functions on t. (The detail will be reviewed in §4(iii) for a reader who is unfamiliar with Cherednik algebras.)

Since $E_n[1]$, $F_n[1]$ are trigonometric degeneration of Macdonald operators, they are contained in $\mathbf{SH}_N^{\mathrm{gr}}$. (In the notation in §4, $E_n[1]$ and $F_n[1]$ correspond to $S(e_n(X))S$ and $S(e_n(X^{-1}))S$ respectively, where S is the symmetrizer and e_n is the nth elementary symmetric polynomial. cf. Example 4.5.)

By the explanation preceding the proof, all operators $E_n[f]$, $F_n[f]$ in (1.4) are also contained in $\mathbf{SH}_N^{\mathrm{gr}}$. Therefore, by [**BFN16b**, Th. A.8], the image of $\mathcal{A}_\hbar$ in $\tilde{\mathcal{A}}_\hbar$ is contained in $\mathbf{SH}_N^{\mathrm{gr}}$. Thus we have an injective homomorphism $\mathcal{A}_\hbar \to \mathbf{SH}_N^{\mathrm{gr}}$. We know that both $\mathcal{A}_\hbar$, $\mathbf{SH}_N^{\mathrm{gr}}$ degenerate to $\mathrm{Sym}^N(\mathbb{C}\times\mathbb{C}^\times)$ at $\hbar = 0 = \mathsf{t}$, we are done. □

3. Poisson brackets

Let us continue computation of Poisson brackets in order to reduce the number of Poisson generators further.

Recall the Coulomb branch is the Nth symmetric power $\mathrm{Sym}^N \mathcal{S}_l$ of the surface $\mathcal{S}_l = \{xy = w^l\}$ in $\mathbb{C}^3$. Let us introduce x_i, y_i, w_i $(1 \le i \le N)$ for functions on $(\mathcal{S}_l)^N$. The Poisson brackets are given by

$$\{x_i, y_j\} = \delta_{ij} w_i^{l-1}, \qquad \{w_i, x_j\} = -\delta_{ij} x_i, \qquad \{w_i, y_j\} = \delta_{ij} y_j.$$

Hence

$$\left\{\sum_{i=1}^N w_i^2, \sum_{j=1}^N y_j\right\} = 2\sum_{i=1}^N w_i y_i, \qquad \left\{\sum_{i=1}^N w_i y_i, \sum_{j=1}^N y_j^n\right\} = n\sum_{i=1}^N y_i^{n+1}$$

for $n \in \mathbb{Z}_{>0}$. Note that $F_n[1]$ is specialized to the nth elementary symmetric polynomial in y_i at $\mathsf{t} = \hbar = z_k = 0$. Therefore the above implies that we can obtain elements $F_n[1]$ inductively from $F_1[1]$ and symmetric polynomials in w_i by taking commutators divided by $\hbar$. The same is true for $E_n[1]$. Therefore to obtain an isomorphism $\mathcal{A}_\hbar \cong \mathbf{SH}^{\mathrm{cyc}}_{N,l}$, it is enough to construct an algebra embedding of $\mathbf{SH}^{\mathrm{cyc}}_{N,l}$ to the ring $\tilde{\mathcal{A}}_\hbar$ of rational difference operators on $\operatorname{Lie} T$ such that $E_1[1]$, $F_1[1]$ are contained in $\mathbf{SH}^{\mathrm{cyc}}_{N,l}$.

4. Cherednik algebras

We recall definitions of various versions of Cherednik algebras and their faithful representations by differential (Dunkl) difference operators. Our basic reference is [**Kir97**]. All the results are due to Cherednik.

4(i). Definitions. The *graded Cherednik algebra* (alias the *trigonometric double affine Hecke algebra*) $\mathbf{H}^{\mathrm{gr}}_N$ for $\mathfrak{gl}_N$ is the $\mathbb{C}[\hbar,\mathbf{t}]$-algebra generated by $\pi^{\pm 1}$, s_0, …, s_{N-1}, w_1, …, w_N with the relations

(4.1a) $$[w_i, w_j] = 0 \quad (i,j = 1,\dots,N),$$

(4.1b) π, s_i $(i = 0,\dots,N-1)$ satisfy the relation of the extended affine Weyl group of $\mathfrak{gl}_N$, e.g., $\pi s_i = s_{i+1}\pi$ $(i = 0,\dots,N-1 \bmod N)$, etc,

(4.1c) $$\pi w_i = w_{i+1}\pi \quad (i = 1,\dots,N-1), \qquad \pi w_N = (w_1+\hbar)\pi,$$

(4.1d) $$s_i w_i = w_{i+1}s_i - \mathbf{t}, \quad s_i w_{i+1} = w_i s_i + \mathbf{t},$$
$$s_i w_j = w_j s_i \quad (i = 1,\dots,N-1, j \neq i, i+1),$$

(4.1e) $$s_0 w_1 = (-\hbar + w_N)s_0 + \mathbf{t}, \quad s_0 w_N = (\hbar + w_1)s_0 - \mathbf{t}, \quad s_0 w_i = w_i s_0 \quad (i \neq 1, N).$$

This is the presentation obtained from one for the original Cherednik algebra (see e.g., [**Kir97**, Def. 4.1]) by setting $X_i = \exp(\boldsymbol{\beta} w_i)$, $q^2 = \exp(\boldsymbol{\beta}\hbar)$, $t^2 = \exp(-\boldsymbol{\beta}\mathbf{t})$ and taking the limit $\boldsymbol{\beta} \to 0$. We will not use the original Cherednik algebra considered in [**Kir97**], in particular we will use the notation X_i for a different object below.

The graded Cherednik algebra $\mathbf{H}^{\mathrm{gr}}_N$ has another presentation with generators s_1, …, s_{N-1}, $X_i^{\pm 1}$, w_i $(i = 1,\dots,N)$ and relations

(4.2a) $$[w_i, w_j] = 0 = [X_i, X_j] \quad (i,j = 1,\dots,N),$$

(4.2b) s_i $(i = 1,\dots,N-1)$ are the standard generators of the symmetric group $\mathfrak{S}_N$,

(4.2c) the same as (4.1d),

(4.2d) $$sX_i^{\pm 1} = X_{s(i)}^{\pm 1} s \quad (s \in \mathfrak{S}_N),$$

(4.2e) $$[w_i, X_j] = \begin{cases} -\mathbf{t}X_j s_{ji} & \text{if } i > j, \\ -\mathbf{t}X_i s_{ij} & \text{if } i < j, \\ -\hbar X_i + \mathbf{t}\sum_{k<i} X_k s_{ki} + \mathbf{t}\sum_{k>i} X_i s_{ik} & \text{if } i = j, \end{cases}$$

where s_{ij} is the transposition (ij). The isomorphism of two presentations is given by setting $X_1 = \pi s_{N-1}\dots s_2 s_1$, $X_2 = s_1 X_1 s_1 = s_1 \pi s_{N-1}\dots s_3 s_2$, etc. (See [**AST98**, Section 1] for detail.) The inverse is given by $\pi = X_1 s_1 s_2 \dots s_{N-1}$, $s_0 = \pi^{-1}s_1\pi$. This presentation matches with one in [**SV13**, §1.1] by setting $\mathbf{t} = \kappa$, $\hbar = -1$.

The *rational Cherednik algebra* $\mathbf{H}_N^{\mathrm{rat}}$ for $\mathfrak{gl}_N$ is the quotient of the algebra $\mathbb{C}[\hbar, \mathbf{t}]\langle x_1, \dots, x_N, y_1, \dots, y_N\rangle \rtimes \mathfrak{S}_N$ by the relations

$$[x_i, x_j] = 0 = [y_i, y_j] \quad (i, j = 1, \dots N), \tag{4.3a}$$

$$[y_i, x_j] = \begin{cases} -\hbar + \mathbf{t}\sum_{k\neq i} s_{ik} & \text{if } i = j, \\ -\mathbf{t}s_{ij} & \text{if } i \neq j. \end{cases} \tag{4.3b}$$

Suzuki [**Suz05**] introduced an embedding $\iota\colon \mathbf{H}_N^{\mathrm{rat}} \to \mathbf{H}_N^{\mathrm{gr}}$ given by

$$\begin{aligned} \iota(w) &= w \quad (w \in \mathfrak{S}_N), \\ \iota(x_i) &= X_i \quad (i = 1, \dots, N), \\ \iota(y_i) &= X_i^{-1}\left(w_i - \mathbf{t}\sum_{j<i} s_{ji}\right) \quad (i = 1, \dots, N). \end{aligned} \tag{4.4}$$

It will be clear that this is an algebra embedding thanks to trigonometric Dunkl operators recalled in the next subsection.

4(ii). Dunkl operators. Consider the (Laurent) polynomial ring

$$\mathbb{C}[\hbar, \mathbf{t}, X_1^{\pm 1}, \dots, X_N^{\pm 1}].$$

We define a representation of $\mathbf{H}_N^{\mathrm{gr}}$ as follows: $X_i^{\pm 1}$, s_i act in the standard way and w_i acts by the trigonometric Dunkl operator

$$w_i \mapsto -\hbar X_i \frac{\partial}{\partial X_i} + \mathbf{t}\sum_{k\neq i} \frac{X_i}{X_i - X_k}(1 - s_{ik}) + \mathbf{t}\sum_{k<i} s_{ik}.$$

See [**Opd00**, Th. 3.7] for the proof. It is not difficult to check the defining relations directly.

The corresponding representation of the rational Cherednik algebra $\mathbf{H}_N^{\mathrm{rat}}$ is given by the rational Dunkl operator

$$y_i \mapsto -\hbar\frac{\partial}{\partial x_i} + \mathbf{t}\sum_{k\neq i} \frac{1}{x_i - x_k}(1 - s_{ik}).$$

See [**EG02**, Prop. 4.5] for the proof. It is even simpler to check the defining relations than the trigonometric case. It is known that this is a faithful representation. (See [**EG02**, Prop. 4.5].) Now trigonometric and rational Dunkl operators are compatible, hence (4.4) indeed gives an embedding of algebras.

4(iii). Rational Demazure-Lusztig operators. Let us consider the polynomial ring $\mathbb{C}[\hbar, \mathbf{t}, w_1, \dots, w_N]$. Let s_i^w denote the ordinary simple reflection on $\mathbb{C}[w_1, \dots, w_N]$ for $i \neq 0$, and $s_0^w(w_1) = w_N - \hbar$, $s_0^w(w_N) = w_1 + \hbar$, $s_0^w w_i = w_i$ $(i \neq 1, N)$. We extend them linearly in $\hbar$, $\mathbf{t}$. We define a representation of $\mathbf{H}_N^{\mathrm{gr}}$ as follows: w_i acts by multiplication, π is as above, and

$$s_i \mapsto s_i^w + \frac{\mathbf{t}}{w_i - w_{i+1}}(s_i^w - 1) \quad (i \neq 0), \qquad s_0 \mapsto s_0^w - \frac{\mathbf{t}}{\hbar + w_1 - w_N}(s_0^w - 1).$$

Note that $\pi s_{N-1}^w \cdots s_2^w s_1^w$ is the difference operator $\mathsf{u}_1(w_i) = w_i + \hbar\delta_{i1}$.

It is known that this is a faithful representation [**Che05**, Prop. 1.6.3 (a)]. We can prove it by an argument using the notion of leading term, which will be introduced in §5. The argument is similar to one for the case of double affine Hecke algebra (See [**Kir97**, Th. 5.7 and Cor. 5.8]).

Let us also consider the restriction Res to the space of symmetric polynomials in $\vec{w}$. (See [**Kir97**, (4.5)].) The spherical subalgebra $\mathbf{SH}_N^{\mathrm{gr}}$ preserves the space of symmetric polynomials, and Res gives a faithful representation of $\mathbf{SH}_N^{\mathrm{gr}}$.

EXAMPLE 4.5. Consider the $N = 2$ case. We have

$$X_1 = \pi s_1 = \left(1 + \frac{\mathsf{t}}{w_2 - w_1 - \hbar}\right) \mathsf{u}_1 - \frac{\mathsf{t}}{w_2 - w_1 - \hbar}\pi,$$
$$X_2 = s_1 X_1 s_1 = \left(1 + \frac{\mathsf{t}}{w_1 - w_2}\right) \mathsf{u}_2 - \frac{\mathsf{t}}{w_1 - w_2}\pi.$$

Note $\operatorname{Res}\pi = \mathsf{u}_1$ as $\pi = \mathsf{u}_1 s_1^w$. Therefore

$$\operatorname{Res} X_1 + \operatorname{Res} X_2 = \left(1 - \frac{\mathsf{t}}{w_1 - w_2}\right) \mathsf{u}_1 + \left(1 + \frac{\mathsf{t}}{w_1 - w_2}\right) \mathsf{u}_2.$$

This is nothing but $E_1[1]$ in (1.4).

We also have

$$X_1^{-1} = s_1 \pi^{-1} = \left(1 + \frac{\mathsf{t}}{w_1 - w_2}\right) \mathsf{u}_1^{-1} - \frac{\mathsf{t}}{w_1 - w_2}\pi^{-1},$$
$$X_2^{-1} = \pi^{-1} s_1 = \left(1 + \frac{\mathsf{t}}{w_2 - \hbar - w_1}\right) \mathsf{u}_2^{-1} - \frac{\mathsf{t}}{w_2 - \hbar - w_1}\pi^{-1}.$$

Since $\operatorname{Res}\pi^{-1} = \mathsf{u}_2^{-1}$, we get

$$\operatorname{Res} X_1^{-1} + \operatorname{Res} X_2^{-1} = \left(1 + \frac{\mathsf{t}}{w_1 - w_2}\right) \mathsf{u}_1^{-1} + \left(1 - \frac{\mathsf{t}}{w_1 - w_2}\right) \mathsf{u}_2^{-1}.$$

This is nothing but $F_1[1]$ in (1.4) with $l = 0$. By the discussion in §3, we thus proved that $\mathcal{A}_\hbar \cong \mathbf{SH}_N^{\mathrm{gr}}$ for $N = 2$, $l = 0$. This proof is nothing but the detail of one in §2, as we have just computed Macdonald operators explicitly.

We also have

$$X_1^{-1} w_1 = \left(1 + \frac{\mathsf{t}}{w_1 - w_2}\right) (w_1 - \hbar)\mathsf{u}_1^{-1} - \frac{\mathsf{t}}{w_1 - w_2}(w_2 - \hbar)\pi^{-1},$$
$$X_2^{-1} w_2 = \left(1 + \frac{\mathsf{t}}{w_2 - w_1 - \hbar}\right) (w_2 - \hbar)\mathsf{u}_2^{-1} - \frac{\mathsf{t}}{w_2 - w_1 - \hbar} w_1 \pi^{-1}.$$

Hence

$$\begin{aligned}&\operatorname{Res} X_1^{-1} w_1 + \operatorname{Res} X_2^{-1} w_2 \\ &= \left(1 + \frac{\mathsf{t}}{w_1 - w_2}\right) (w_1 - \hbar)\mathsf{u}_1^{-1} + \left(1 - \frac{\mathsf{t}}{w_1 - w_2}\right) (w_2 - \hbar)\mathsf{u}_2^{-1} + \mathsf{t}\mathsf{u}_2^{-1}.\end{aligned}$$

Then $X_2^{-1} s_1 = \pi^{-1}$, hence

$$\begin{aligned}&\operatorname{Res} X_1^{-1} w_1 + \operatorname{Res} X_2^{-1} (w_2 - \mathsf{t} s_1) \\ &= \left(1 + \frac{\mathsf{t}}{w_1 - w_2}\right) (w_1 - \hbar)\mathsf{u}_1^{-1} + \left(1 - \frac{\mathsf{t}}{w_1 - w_2}\right) (w_2 - \hbar)\mathsf{u}_2^{-1}.\end{aligned}$$

This is nothing but $F_1[1]$ in (1.4) with $l = 1$, $z_k = 0$. Under Suzuki's embedding (4.4), we have $y_1 = X_1^{-1} w_1$, $y_2 = X_2^{-1}(w_2 - \mathsf{t} s_1)$. Therefore $\operatorname{Res} y_1 + \operatorname{Res} y_2 = F_1[1]$.

By the discussion in §3, we thus proved that $\mathcal{A}_\hbar \cong \mathbf{SH}_N^{\mathrm{rat}}$ for $N = 2$, $l = 1$.

5. Cyclotomic rational Cherednik algebras

Let $\Gamma_N = \mathfrak{S}_N \ltimes (\mathbb{Z}/l\mathbb{Z})^N$ be the wreath product of the symmetric group and the cyclic group of order l. Denote a fixed generator of the ith factor of $(\mathbb{Z}/l\mathbb{Z})^N$ by α_i. The group Γ_N acts on $\mathbb{C}\langle \xi_1, \dots, \xi_N, \eta_1, \dots, \eta_N\rangle$ by

$$\alpha_i(\xi_i) = \varepsilon\xi_i,\ \alpha_i(\xi_j) = \xi_j,\ \alpha_i(\eta_i) = \varepsilon^{-1}\eta_i,\ \alpha_i(\eta_j) = \eta_j \quad (i \neq j),$$

with the obvious $\mathfrak{S}_N$-action. Here ε denotes a primitive lth root of unity.

The *cyclotomic rational Cherednik algebra* $\mathbf{H}^{\mathrm{cyc}}_{N,l}$ for $\mathfrak{gl}_N$ is the quotient of the algebra $\mathbb{C}[\hbar, \mathbf{t}, c_1, \dots, c_{l-1}]\langle \xi_1, \dots, \xi_N, \eta_1, \dots, \eta_N\rangle \rtimes \Gamma_N$ by the relations

$$[\xi_i, \xi_j] = 0 = [\eta_i, \eta_j] \quad (i, j = 1, \dots, N), \tag{5.1a}$$

$$[\eta_i, \xi_j] = \begin{cases} -\hbar + \mathbf{t}\sum_{k\neq i}\sum_{m=0}^{l-1} s_{ik}\alpha_i^m\alpha_k^{-m} + \sum_{m=1}^{l-1} c_m\alpha_i^m & \text{if } i = j, \\ -\mathbf{t}\sum_{m=0}^{l-1} s_{ij}\varepsilon^m\alpha_i^m\alpha_j^{-m} & \text{if } i \neq j. \end{cases} \tag{5.1b}$$

Let e_{Γ_N} be the idempotent for the group $\Gamma_N = \mathfrak{S}_N \ltimes (\mathbb{Z}/l\mathbb{Z})^N$. The spherical part of the cyclotomic rational Cherednik algebra $\mathbf{H}^{\mathrm{cyc}}_{N,l}$ is defined as

$$\mathbf{SH}^{\mathrm{cyc}}_{N,l} = e_{\Gamma_N}\mathbf{H}^{\mathrm{cyc}}_{N,l}e_{\Gamma_N}.$$

Let e_Γ be the idempotent for the group $(\mathbb{Z}/l\mathbb{Z})^N$, that is,

$$e_\Gamma = \frac{1}{l^N}\sum_{g\in(\mathbb{Z}/l\mathbb{Z})^N} g = \frac{1}{l^N}\prod_{i=1}^{N}\left(\sum_{m_i=0}^{l-1}\alpha_i^{m_i}\right).$$

Oblomkov [**Obl07**] introduced an embedding $\mathbf{H}^{\mathrm{rat}}_N \to \left(e_\Gamma\mathbf{H}^{\mathrm{cyc}}_{N,l}e_\Gamma\right)_{(\xi_i^l)}$ given by

$$x_i \mapsto e_\Gamma\xi_i^l e_\Gamma \quad (i = 1, \dots, N),$$
$$y_i \mapsto l^{-1}e_\Gamma\xi_i^{1-l}\eta_i e_\Gamma \quad (i = 1, \dots, N).$$

This induces an embedding $\iota\colon e_\Gamma\mathbf{H}^{\mathrm{cyc}}_{N,l}e_\Gamma \to \mathbf{H}^{\mathrm{gr}}_N[c_1, \dots, c_{l-1}]$ such that

$$\iota(e_\Gamma\xi_i^l e_\Gamma) = X_i \quad (i = 1, \dots, N),$$

$$\iota(e_\Gamma\xi_i\eta_i e_\Gamma) = l\left(w_i - \mathbf{t}\sum_{j<i} s_{ji}\right) \quad (i = 1, \dots, N).$$

When $l = 1$, this coincides with Suzuki's embedding (4.4). We restrict ι to the spherical part and obtain the embedding $\iota\colon \mathbf{SH}^{\mathrm{cyc}}_{N,l} \to \mathbf{SH}^{\mathrm{gr}}_N[c_1, \dots, c_{l-1}]$.

Fix $i = 1, \dots, N$ and define for each $k = 0, 1, \dots, l-1$

$$e_\Gamma[k] = \frac{1}{l^N}\left(\sum_{m_i=0}^{l-1}(\varepsilon^{-k}\alpha_i)^{m_i}\right)\prod_{j\neq i}\left(\sum_{m_j=0}^{l-1}\alpha_j^{m_j}\right).$$

We have

$$\alpha_i^m e_\Gamma[k] = \varepsilon^{km}e_\Gamma[k], \quad \alpha_j^m e_\Gamma[k] = e_\Gamma[k] \quad (j \neq i),$$

and hence

$$e_\Gamma[k]e_\Gamma = e_\Gamma e_\Gamma[k] = 0 \quad (k \neq 0), \quad e_\Gamma[k]^2 = e_\Gamma[k].$$

We also have

$$\xi_i e_\Gamma[k] = e_\Gamma[k+1]\xi_i, \quad e_\Gamma[k]\eta_i = \eta_i e_\Gamma[k+1].$$

These imply

$$\iota(e_\Gamma \xi_i^k e_\Gamma) = \iota(e_\Gamma \eta_i^k e_\Gamma) = 0$$

for $k = 1, \dots, l-1$.

LEMMA 5.2. *For $k = 1, \dots, l$, we have*

$$\iota(e_\Gamma \eta_i^k \xi_i^k e_\Gamma) = (lw_i - \hbar + \sum_{m=1}^{l-1} c_m + l\mathbf{t}\sum_{i<j} s_{ij})(lw_i - 2\hbar + \sum_{m=1}^{l-1}(1+\varepsilon^m)c_m + l\mathbf{t}\sum_{i<j} s_{ij}) \cdots (lw_i - k\hbar + \sum_{m=1}^{l-1}(1 + \varepsilon^m + \cdots + \varepsilon^{(k-1)m})c_m + l\mathbf{t}\sum_{i<j} s_{ij}).$$

PROOF. For $k = 1$, we have

$$\begin{aligned}
\iota(e_\Gamma \eta_i \xi_i e_\Gamma) &= \iota(e_\Gamma(\xi_i\eta_i - [\xi_i, \eta_i])e_\Gamma) \\
&= l\left(w_i - \mathbf{t}\sum_{j<i} s_{ji}\right) + \left(-\hbar + l\mathbf{t}\sum_{j\neq i} s_{ij} + \sum_{m=1}^{l-1} c_m\right) \\
&= lw_i - \hbar + \sum_{m=1}^{l-1} c_m + l\mathbf{t}\sum_{i<j} s_{ij},
\end{aligned}$$

hence the assertion holds. For $k \geq 1$, we have

$$\begin{aligned}
(e_\Gamma \eta_i^k \xi_i^k e_\Gamma)(e_\Gamma \eta_i \xi_i e_\Gamma) &= e_\Gamma(\eta_i^k \xi_i^k)(\eta_i \xi_i) e_\Gamma \\
&= e_\Gamma \eta_i^{k+1} \xi_i^{k+1} e_\Gamma + e_\Gamma \eta_i^k [\xi_i^k, \eta_i] \xi_i e_\Gamma.
\end{aligned}$$

We calculate the second term as

$$\begin{aligned}
e_\Gamma \eta_i^k [\xi_i^k, \eta_i] \xi_i e_\Gamma &= e_\Gamma \eta_i^k (\sum_{p=1}^{k} \xi_i^{k-p}[\xi_i, \eta_i]\xi_i^{p-1}) \xi_i e_\Gamma \\
&= \sum_{p=1}^{k} e_\Gamma \eta_i^k \xi_i^{k-p}[\xi_i, \eta_i] e_\Gamma[p] \xi_i^p e_\Gamma \\
&= \sum_{p=1}^{k} (\hbar - \sum_{m=1}^{l-1} \varepsilon^{pm} c_m) e_\Gamma \eta_i^k \xi_i^k e_\Gamma.
\end{aligned}$$

Therefore

$$\begin{aligned}
&\iota(e_\Gamma \eta_i^{k+1} \xi_i^{k+1} e_\Gamma) = \iota(e_\Gamma \eta_i^k \xi_i^k e_\Gamma)\iota(e_\Gamma \eta_i \xi_i e_\Gamma) - \iota(e_\Gamma \eta_i^k [\xi_i^k, \eta_i] \xi_i e_\Gamma) \\
&= \iota(e_\Gamma \eta_i^k \xi_i^k e_\Gamma)\left((lw_i - \hbar + \sum_{m=1}^{l-1} c_m + l\mathbf{t}\sum_{i<j} s_{ij}) - (k\hbar - \sum_{m=1}^{l-1}(\varepsilon^m + \cdots + \varepsilon^{km})c_m)\right) \\
&= \iota(e_\Gamma \eta_i^k \xi_i^k e_\Gamma)\left(lw_i - (k+1)\hbar + \sum_{m=1}^{l-1}(1 + \varepsilon^m + \cdots + \varepsilon^{km})c_m + l\mathbf{t}\sum_{i<j} s_{ij}\right)
\end{aligned}$$

and the assertion follows by induction on k. □

By Lemma 5.2 and $\iota(e_\Gamma \xi_i^l e_\Gamma) = X_i$, we obtain the following.

PROPOSITION 5.3. *We have*

$$\iota(e_\Gamma(l^{-1}\eta_i)^l e_\Gamma) = (w_i - \hbar - z_1 + \mathbf{t}\sum_{i<j} s_{ij}) \cdots (w_i - \hbar - z_l + \mathbf{t}\sum_{i<j} s_{ij}) X_i^{-1}$$

where

$$z_k = -l^{-1}\Big((l-k)\hbar + \sum_{m=1}^{l-1}(1 + \varepsilon^m + \cdots + \varepsilon^{(k-1)m})c_m\Big).$$

Remark that $z_l = 0$.

LEMMA 5.4. *If $i < j$ then the following identities hold in $\mathbf{H}_N^{\mathrm{gr}}$:*

$$s_{ij} w_i = w_j s_{ij} - \mathbf{t}(1 + \sum_{k=i+1}^{j-1} s_{ik} s_{kj}), \tag{5.5}$$

$$s_{ij}(w_i + \mathbf{t}\sum_{i<k} s_{ik}) = (w_j + \mathbf{t}\sum_{j<k} s_{jk}) s_{ij}. \tag{5.6}$$

In particular we have

$$\begin{aligned} & s_{i_{p-1},i_p} s_{i_{p-2},i_{p-1}} \cdots s_{i_1,i_2} s_{i_0,i_1} (w_{i_0} + \mathbf{t}\sum_{i_0<k} s_{i_0 k}) \\ &= (w_{i_p} + \mathbf{t}\sum_{i_p<k} s_{i_p k}) s_{i_{p-1},i_p} s_{i_{p-2},i_{p-1}} \cdots s_{i_1,i_2} s_{i_0,i_1} \end{aligned} \tag{5.7}$$

for $i_0 < i_1 < \cdots < i_p$.

PROOF. We prove (5.5) by induction on j. The case $j = i+1$ is nothing but the relation (4.1d) in $\mathbf{H}_N^{\mathrm{gr}}$. Assume it holds for j. Then

$$\begin{aligned} s_{i,j+1} w_i &= s_j s_{ij} s_j w_i = s_j (s_{ij} w_i) s_j \\ &= s_j (w_j s_{ij} - \mathbf{t}(1 + \sum_{k=i+1}^{j-1} s_{ik} s_{kj})) s_j \\ &= (w_{j+1} s_j - \mathbf{t}) s_{ij} s_j - \mathbf{t}(1 + \sum_{k=i+1}^{j-1} s_j s_{ik} s_{kj} s_j) \\ &= w_{j+1} s_{i,j+1} - \mathbf{t} s_{ij} s_j - \mathbf{t}(1 + \sum_{k=i+1}^{j-1} s_{ik} s_{k,j+1}) \\ &= w_{j+1} s_{i,j+1} - \mathbf{t}(1 + \sum_{k=i+1}^{j} s_{ik} s_{k,j+1}). \end{aligned}$$

Then we use (5.5) to prove (5.6):

$$\begin{aligned}
s_{ij}(w_i + \mathbf{t}\sum_{i<k} s_{ik}) &= w_j s_{ij} - \mathbf{t}(1 + \sum_{k=i+1}^{j-1} s_{ik}s_{kj}) \\
&\quad + \mathbf{t}\sum_{i<k} s_{ij}s_{ik} \\
&= w_j s_{ij} - \mathbf{t}(1 + \sum_{k=i+1}^{j-1} s_{ik}s_{kj}) \\
&\quad + \mathbf{t}(\sum_{k=i+1}^{j-1} s_{ij}s_{ik} + 1 + \sum_{j<k} s_{ij}s_{ik}) \\
&= (w_j + \mathbf{t}\sum_{j<k} s_{jk})s_{ij}.
\end{aligned}$$

□

LEMMA 5.8. *We have*

$$\begin{aligned}
&(w_i - \hbar - z_1 + \mathbf{t}\sum_{i<j} s_{ij}) \cdots (w_i - \hbar - z_l + \mathbf{t}\sum_{i<j} s_{ij}) \\
&= \sum_{p=0}^{l} \mathbf{t}^p \sum_{0=k_0<k_1<\cdots<k_p\le l} \; \sum_{i=i_0<i_1<\cdots<i_p} \prod_{k=k_0+1}^{k_1-1} (w_{i_0} - \hbar - z_k) \prod_{k=k_1+1}^{k_2-1} (w_{i_1} - \hbar - z_k) \\
&\qquad \cdots \prod_{k=k_p+1}^{l} (w_{i_p} - \hbar - z_k) s_{i_{p-1},i_p} s_{i_{p-2},i_{p-1}} \cdots s_{i_1,i_2} s_{i_0,i_1}.
\end{aligned}$$

PROOF. We set $A_k = w_i - \hbar - z_k + \mathbf{t}\sum_{i<j} s_{ij}$ and

$$\begin{aligned}
B_s = \sum_{p=0}^{s} \mathbf{t}^p \sum_{0=k_0<k_1<\cdots<k_p\le s} \; \sum_{i=i_0<i_1<\cdots<i_p} \prod_{k=k_0+1}^{k_1-1} (w_{i_0} - \hbar - z_k) \prod_{k=k_1+1}^{k_2-1} (w_{i_1} - \hbar - z_k) \\
\cdots \prod_{k=k_p+1}^{s} (w_{i_p} - \hbar - z_k) s_{i_{p-1},i_p} s_{i_{p-2},i_{p-1}} \cdots s_{i_1,i_2} s_{i_0,i_1}.
\end{aligned}$$

We prove $A_1 \cdots A_s = B_s$ by induction on s. For the case $s = 1$, $A_1 = B_1$ follows by definition. By the induction hypothesis and (5.7) in Lemma 5.4, we have

$$\begin{aligned}
&A_1 \cdots A_{s+1} \\
&= \Big\{ \sum_{p=0}^{s} \mathbf{t}^p \sum_{0=k_0<k_1<\cdots<k_p\le s} \sum_{i=i_0<i_1<\cdots<i_p} \prod_{k=k_0+1}^{k_1-1} (w_{i_0} - \hbar - z_k) \prod_{k=k_1+1}^{k_2-1} (w_{i_1} - \hbar - z_k) \\
&\qquad \cdots \prod_{k=k_p+1}^{s} (w_{i_p} - \hbar - z_k) s_{i_{p-1},i_p} s_{i_{p-2},i_{p-1}} \cdots s_{i_1,i_2} s_{i_0,i_1} \Big\} \\
&\qquad \times (w_i - \hbar - z_{s+1} + \mathbf{t} \sum_{i<j} s_{ij}) \\
&= \sum_{p=0}^{s} \mathbf{t}^p \sum_{0=k_0<k_1<\cdots<k_p\le s} \sum_{i=i_0<i_1<\cdots<i_p} \prod_{k=k_0+1}^{k_1-1} (w_{i_0} - \hbar - z_k) \prod_{k=k_1+1}^{k_2-1} (w_{i_1} - \hbar - z_k) \\
&\qquad \cdots \prod_{k=k_p+1}^{s} (w_{i_p} - \hbar - z_k)(w_{i_p} - \hbar - z_{s+1} + \mathbf{t} \sum_{i_p<j} s_{i_p j}) \\
&\qquad \times s_{i_{p-1},i_p} s_{i_{p-2},i_{p-1}} \cdots s_{i_1,i_2} s_{i_0,i_1} \\
&= B'_{s+1} + B''_{s+1},
\end{aligned}$$

where

$$\begin{aligned}
B'_{s+1} &= \sum_{p=0}^{s} \mathbf{t}^p \sum_{0=k_0<k_1<\cdots<k_p\le s} \sum_{i=i_0<i_1<\cdots<i_p} \prod_{k=k_0+1}^{k_1-1} (w_{i_0} - \hbar - z_k) \\
&\times \prod_{k=k_1+1}^{k_2-1} (w_{i_1} - \hbar - z_k) \cdots \prod_{k=k_p+1}^{s+1} (w_{i_p} - \hbar - z_k) s_{i_{p-1},i_p} s_{i_{p-2},i_{p-1}} \cdots s_{i_1,i_2} s_{i_0,i_1}
\end{aligned}$$

and

$$\begin{aligned}
B''_{s+1} &= \sum_{p=0}^{s} \mathbf{t}^{p+1} \sum_{0=k_0<k_1<\cdots<k_p\le s} \sum_{i=i_0<i_1<\cdots<i_{p+1}} \prod_{k=k_0+1}^{k_1-1} (w_{i_0} - \hbar - z_k) \\
&\times \prod_{k=k_1+1}^{k_2-1} (w_{i_1} - \hbar - z_k) \cdots \prod_{k=k_p+1}^{s} (w_{i_p} - \hbar - z_k) s_{i_p,i_{p+1}} s_{i_{p-1},i_p} \cdots s_{i_1,i_2} s_{i_0,i_1}.
\end{aligned}$$

The term B'_{s+1} is equal to the contribution of the terms satisfying $k_p < s+1$ to B_{s+1}. The term B''_{s+1} after replacing $p+1$ by p is equal to the contribution of the terms satisfying $k_p = s+1$ to B_{s+1}. Hence the assertion follows. □

By Lemma 5.8, we obtain

(5.9) $\iota(e_\Gamma (l^{-1}\eta_i)^l e_\Gamma)$

$$\begin{aligned}
&= \sum_{p=0}^{l} \mathbf{t}^p \sum_{0=k_0<k_1<\cdots<k_p\le l} \sum_{i=i_0<i_1<\cdots<i_p} \prod_{k=k_0+1}^{k_1-1} (w_{i_0} - \hbar - z_k) \prod_{k=k_1+1}^{k_2-1} (w_{i_1} - \hbar - z_k) \\
&\qquad\qquad \cdots \prod_{k=k_p+1}^{l} (w_{i_p} - \hbar - z_k) X_{i_p}^{-1} s_{i_{p-1},i_p} s_{i_{p-2},i_{p-1}} \cdots s_{i_1,i_2} s_{i_0,i_1}.
\end{aligned}$$

Put $Y_i = \iota(e_\Gamma (l^{-1}\eta_i)^l e_\Gamma)$.

Recall the representation of the graded Cherednik algebra $\mathbf{H}_N^{\mathrm{gr}}$ on the polynomial ring $\mathbb{C}[\hbar, \mathbf{t}, w_1, \dots, w_N]$ given in §4(iii). We extend the scalar to $\mathbb{C}[\hbar, \mathbf{t}, c_1, \dots, c_{l-1}]$ and consider the action of $e_\Gamma \mathbf{H}_{N,l}^{\mathrm{cyc}} e_\Gamma$ on $\mathbb{C}[\hbar, \mathbf{t}, c_1, \dots, c_{l-1}][w_1, \dots, w_N]$ via the embedding $\iota\colon e_\Gamma \mathbf{H}_{N,l}^{\mathrm{cyc}} e_\Gamma \to \mathbf{H}_N^{\mathrm{gr}}[c_1, \dots, c_{l-1}]$.

We introduce the notion of leading term of operators acting on $\mathbb{C}[\hbar, \mathbf{t}, c_1, \dots, c_{l-1}][w_1, \dots, w_N]$ as in [**Kir97**, Def. 5.1]. Let $P = \bigoplus_{i=1}^N \mathbb{Z}\varepsilon_i$ be the weight lattice of $\mathfrak{gl}_N$. Fix positive roots $R^+ = \{\varepsilon_i - \varepsilon_j \mid i < j\}$ and denote by P^+ the set of dominant integral weights. The Weyl group is denoted by $W = \mathfrak{S}_N$. For $\lambda, \mu \in P$, we define $\lambda \le \mu$ if $\mu - \lambda$ is a sum of positive roots with coefficients in $\mathbb{Z}_{\ge 0}$. Let us define another partial order $\lessdot$ on P. Given $\lambda \in P$, we denote by λ^+ the unique element in $P^+ \cap W\lambda$. For $\lambda, \mu \in P$, we define $\lambda \lessdot \mu$ if

$$\lambda^+ < \mu^+ \text{ or}$$
$$\lambda^+ = \mu^+ \text{ and } \lambda > \mu.$$

EXAMPLE 5.10. We have $\varepsilon_1 > \varepsilon_2 > \cdots > \varepsilon_{N-1} > \varepsilon_N$ and they are all in the same W-orbit. Hence we have $\varepsilon_1 \lessdot \varepsilon_2 \lessdot \cdots \lessdot \varepsilon_{N-1} \lessdot \varepsilon_N$. Moreover we see that $\{\lambda \in P \mid \lambda \lessdot \varepsilon_i\} = \{\varepsilon_j \mid j < i\}$ since ε_1 is a minuscule dominant weight and $\varepsilon_1, \dots, \varepsilon_N$ exhaust its W-orbit. Similarly we see that $\{\lambda \in P \mid \lambda \lessdot -\varepsilon_i\} = \{-\varepsilon_j \mid j > i\}$.

We denote by $\mathsf{u}^\lambda = \mathsf{u}_1^{\lambda_1} \cdots \mathsf{u}_N^{\lambda_N}$ for $\lambda = \sum_{i=1}^N \lambda_i \varepsilon_i \in P$.

DEFINITION 5.11. Let T be an operator on $\mathbb{C}[\hbar, \mathbf{t}, c_1, \dots, c_{l-1}][w_1, \dots, w_N]$ of the form

$$T = \sum_{\lambda \in P, w \in W} g_{\lambda,w} \mathsf{u}^\lambda w$$

for some $g_{\lambda,w} = g_{\lambda,w}(w_1, \dots, w_N) \in \mathbb{C}[\hbar, \mathbf{t}, c_1, \dots, c_{l-1}][w_1, \dots, w_N][(w_i - w_j)^{-1}]$. If it can be written as

$$T = \sum_{w \in W} g_{\lambda_0,w} \mathsf{u}^{\lambda_0} w + \sum_{\lambda \lessdot \lambda_0, w' \in W} g_{\lambda,w'} \mathsf{u}^\lambda w'$$

for some λ_0 and at least one of $g_{\lambda_0,w} \neq 0$, we say that $\sum_{w \in W} g_{\lambda_0,w} \mathsf{u}^{\lambda_0} w$ is the leading term of T.

EXAMPLE 5.12. In Example 4.5, we have

$$\operatorname{Res} X_1 = \mathsf{u}_1, \quad \operatorname{Res} X_2 = \left(1 + \frac{\mathbf{t}}{w_1 - w_2}\right) \mathsf{u}_2 - \frac{\mathbf{t}}{w_1 - w_2} \mathsf{u}_1,$$
$$\operatorname{Res} X_1^{-1} = \left(1 + \frac{\mathbf{t}}{w_1 - w_2}\right) \mathsf{u}_1^{-1} - \frac{\mathbf{t}}{w_1 - w_2} \mathsf{u}_2^{-1}, \quad \operatorname{Res} X_2^{-1} = \mathsf{u}_2^{-1}.$$

Therefore the leading terms are

$$\mathsf{u}_1, \quad \left(1 + \frac{\mathbf{t}}{w_1 - w_2}\right) \mathsf{u}_2, \quad \left(1 + \frac{\mathbf{t}}{w_1 - w_2}\right) \mathsf{u}_1^{-1}, \quad \mathsf{u}_2^{-1},$$

respectively. We also have

$$\operatorname{Res} y_1 = \left(1 + \frac{\mathbf{t}}{w_1 - w_2}\right) (w_1 - \hbar) \mathsf{u}_1^{-1} - \frac{\mathbf{t}}{w_1 - w_2} (w_2 - \hbar) \mathsf{u}_2^{-1}, \quad \operatorname{Res} y_2 = (w_2 - \hbar) \mathsf{u}_2^{-1}.$$

Therefore the leading terms are

$$\left(1 + \frac{\mathbf{t}}{w_1 - w_2}\right) (w_1 - \hbar) \mathsf{u}_1^{-1}, \quad (w_2 - \hbar) \mathsf{u}_2^{-1}$$

respectively.

PROPOSITION 5.13. (1) *We have*

$$\operatorname{Res} X_i = \sum_{j \le i} g_j \mathsf{u}_j$$

for some rational functions $g_j = g_j(w_1, \ldots, w_N)$*, hence the leading term of* $\operatorname{Res} X_i$ *is* $g_i \mathsf{u}_i$*. Moreover the leading term of* $\operatorname{Res} X_N$ *is*

$$\prod_{j \neq N} \frac{w_N - w_j - \mathbf{t}}{w_N - w_j} \mathsf{u}_N.$$

(2) *We have*

$$\operatorname{Res} X_i^{-1} = \sum_{j \ge i} g'_j \mathsf{u}_j^{-1}$$

for some rational functions $g'_j = g'_j(w_1, \ldots, w_N)$*, hence the leading term of* $\operatorname{Res} X_i^{-1}$ *is* $g'_i \mathsf{u}_i^{-1}$*. Moreover the leading term of* $\operatorname{Res} X_1^{-1}$ *is*

$$\prod_{j \neq 1} \frac{w_1 - w_j + \mathbf{t}}{w_1 - w_j} \mathsf{u}_1^{-1}.$$

(3) *We have*

$$\operatorname{Res} Y_i = \sum_{j \ge i} g''_j \mathsf{u}_j^{-1}$$

for some rational functions $g''_j = g''_j(w_1, \ldots, w_N)$*, hence the leading term of* $\operatorname{Res} Y_i$ *is* $g''_i \mathsf{u}_i^{-1}$*. Moreover the leading term of* $\operatorname{Res} Y_1$ *is*

$$\prod_{k=1}^{l} (w_1 - \hbar - z_k) \prod_{j \neq 1} \frac{w_1 - w_j + \mathbf{t}}{w_1 - w_j} \mathsf{u}_1^{-1}.$$

PROOF. We follow [**Kir97**, Lecture 5] and modify arguments for our degenerate setting. For an affine root $\alpha = \varepsilon_i - \varepsilon_j + k\delta$, we set

$$G(\alpha) = 1 + \frac{\mathbf{t}}{w_i - w_j - k\hbar}(s_\alpha^w - 1).$$

Take $\lambda = \sum_{i=1}^N \lambda_i \varepsilon_i \in P$ and define $X^\lambda = X_1^{\lambda_1} \cdots X_N^{\lambda_N}$. Let t_λ be the corresponding translation in the extended affine Weyl group. Let $t_\lambda = \pi^m s_{i_r} \cdots s_{i_1}$ be a reduced expression and put $\beta_1 = \alpha_{i_1}, \beta_2 = s_{i_1}\alpha_{i_2}, \ldots, \beta_r = s_{i_1} \cdots s_{i_{r-1}} \alpha_{i_r}$. Then X^λ acts on $\mathbb{C}[\hbar, \mathbf{t}, c_1, \ldots, c_{l-1}][w_1, \ldots, w_N]$ as an operator $\mathsf{u}^\lambda G(\beta_r) \cdots G(\beta_1)$. The assertions (1) and (2) follow from a similar argument as in [**Kir97**, Th. 5.6 and Ex. 5.4]. Note that X_N and X_1^{-1} correspond to anti-dominant weights ε_N and $-\varepsilon_1$ respectively. Hence their leading terms can be calculated explicitly as in [**Kir97**, Ex. 5.4].

We can slightly generalize the assertion (2) (and (1), clearly): given $w \in W$ we have

$$\operatorname{Res}(X_i^{-1} w) = \sum_{j \ge i} g'_j \mathsf{u}_j^{-1}$$

for some rational functions $g'_j = g'_j(w_1, \ldots, w_N)$. Then by (5.9), we see that the term $p = 0$ only contributes to the leading term of $\operatorname{Res} Y_i$. This implies the assertion (3). □

By Proposition 5.13 (1), we see that the term containing u_N in $\sum_{i=1}^N \operatorname{Res} X_i$ only comes from $\operatorname{Res} X_N$. Hence the term is

$$\prod_{j\neq N} \frac{w_N - w_j - \mathbf{t}}{w_N - w_j} \mathsf{u}_N.$$

Since $\sum_{i=1}^N \operatorname{Res} X_i$ is W-invariant, we conclude that

$$\sum_{i=1}^N \operatorname{Res} X_i = \sum_{i=1}^N \prod_{j\neq i} \frac{w_i - w_j - \mathbf{t}}{w_i - w_j} \mathsf{u}_i.$$

This coincides with $E_1[1]$ in (1.4). By the same argument we conclude that

$$\sum_{i=1}^N \operatorname{Res} X_i^{-1} = \sum_{i=1}^N \prod_{j\neq i} \frac{w_i - w_j + \mathbf{t}}{w_i - w_j} \mathsf{u}_i^{-1}$$

and

$$\sum_{i=1}^N \operatorname{Res} Y_i = \sum_{i=1}^N \prod_{j\neq i} \frac{w_i - w_j + \mathbf{t}}{w_i - w_j} \prod_{k=1}^{l} (w_i - \hbar - z_k)\mathsf{u}_i^{-1}.$$

These coincide with $F_1[1]$ for $l = 0$ and $l \geq 1$ in (1.4) respectively. Thus we obtain a complete proof of Theorem 1.1 for general l.

6. Affine Yangian of $\mathfrak{gl}(1)$

6(i). Presentation. We use the presentation of the affine Yangian $Y(\widehat{\mathfrak{gl}}(1))$ in [**AS13**], given based on [**SV13**]. See also [**Tsy17**].

Let us first prepare some functions. Let $\hbar$, $\mathbf{t}$ be indeterminate as before. We set

$$G_0(x) = -\log x, \quad G_n(x) = \frac{x^{-n} - 1}{n} \quad (n \geq 1),$$
$$\varphi_n(x) = \sum_{q=-\hbar,\hbar+\mathbf{t},-\mathbf{t}} x^n (G_n(1 - qx) - G_n(1 + qx)),$$
$$\phi_n(x) = x^n G_n(1 - (\hbar + \mathbf{t})x).$$

The affine Yangian $Y(\widehat{\mathfrak{gl}}(1))$ of $\mathfrak{gl}(1)$ is a $\mathbb{C}[\omega, \hbar, \mathbf{t}]$-algebra generated by $D_{0,m}$ $(m \geq 1)$, e_n, f_n $(n \geq 0)$ with relations

(6.1a) $$[D_{0,m}, D_{0,n}] = 0 \quad (m, n \geq 1),$$

(6.1b) $$[D_{0,m}, e_n] = -\hbar e_{m+n-1} \quad (m \geq 1, n \geq 0),$$

(6.1c) $$[D_{0,m}, f_n] = \hbar f_{m+n-1} \quad (m \geq 1, n \geq 0),$$

(6.1d) $$3[e_2, e_1] - [e_3, e_0] + (\hbar^2 + \mathbf{t}(\hbar + \mathbf{t}))[e_1, e_0] + \hbar\mathbf{t}(\hbar + \mathbf{t})e_0^2 = 0,$$

(6.1e) $$3[f_2, f_1] - [f_3, f_0] + (\hbar^2 + \mathbf{t}(\hbar + \mathbf{t}))[f_1, f_0] - \hbar\mathbf{t}(\hbar + \mathbf{t})f_0^2 = 0,$$

(6.1f) $$[e_0, [e_0, e_1]] = 0 = [f_0, [f_0, f_1]],$$

(6.1g) $$[e_m, f_n] = \hbar h_{m+n} \quad (m, n \geq 0),$$

where elements h_{m+n} are determined through the formula

(6.1h)

$$1 - \mathbf{t}(\hbar + \mathbf{t}) \sum_{n\geq 0} h_n x^{n+1} = \frac{(1 - (\hbar + \mathbf{t})x)(1 + \omega\mathbf{t}x)}{1 - (\hbar + (1 - \omega)\mathbf{t})x} \exp\left(-\sum_{n\geq 0} \frac{D_{0,n+1}}{\hbar} \varphi_n(x) \right).$$

Note that the right hand side is 1 at $\mathbf{t} = 0$ or $\hbar + \mathbf{t} = 0$. (φ_n vanishes if $\hbar = 0$, $\mathbf{t} = 0$ or $\hbar + \mathbf{t} = 0$.) Therefore h_n is a polynomial in $D_{0,m}$ ($m \geq 0$) with coefficients in $\mathbb{C}[\omega, \hbar, \mathbf{t}]$. For example,

$$\begin{gathered} h_0 = 0, \quad h_1 = \omega, \quad h_2 = 2D_{0,1} + \omega(\hbar + (1-\omega)\mathbf{t}), \\ h_{l+1} = l(l+1)D_{0,l} + (\text{lower order term}), \end{gathered} \tag{6.2}$$

where the lower order term means a polynomial in $\hbar$, $\mathbf{t}$, ω and $D_{0,m}$ with $m < l$.

If we set $\hbar = -1$, $\mathbf{t} = \kappa$, $\xi = 1 - \kappa = 1 - \mathbf{t}$, $D_{1,n} = e_n$, $D_{-1,n} = f_n$, these are the defining relations in [**AS13**] with $c_0 = 0$, $c_n = -\mathbf{t}^n\omega^n$ ($n > 0$) as $\exp(\sum_{n\geq 0}(-1)^{n+1}\mathbf{c}_n\phi_n(x)) = (1-(\hbar+\mathbf{t})x)(1+\omega\mathbf{t}x)/(1-(\hbar+(1-\omega)\mathbf{t})x)$.[1]

REMARK 6.3. Applying $[D_{0,n+1}, [D_{0,m+1}, \bullet]] + \hbar[D_{0,m+n+1}, \bullet]$ to (6.1e), we get

$$\begin{aligned} \text{(6.1e')}\quad & 3[f_{m+2}, f_{n+1}] - 3[f_{m+1}, f_{n+2}] - [f_{m+3}, f_n] + [f_m, f_{n+3}] \\ & + (\hbar^2 + \mathbf{t}(\hbar+\mathbf{t}))([f_{m+1}, f_n] - [f_m, f_{n+1}]) - \hbar\mathbf{t}(\hbar+\mathbf{t})(f_m f_n + f_n f_m) = 0. \end{aligned}$$

(6.1e) is the special case $m = n = 0$. Similarly (6.1f) implies

$$\begin{aligned} \text{(6.1f')}\quad & [f_i, [f_j, f_{k+1}]] + [f_i, [f_k, f_{j+1}]] + [f_j, [f_i, f_{k+1}]] \\ & + [f_j, [f_k, f_{i+1}]] + [f_k, [f_i, f_{j+1}]] + [f_k, [f_j, f_{i+1}]] = 0. \end{aligned}$$

We have the corresponding formula for e_n. See [**Tsy17**].

REMARK 6.4. The relation (6.1b) is replaced by

(6.1b')

$$\begin{gathered} [h_{m+3}, e_n] - 3[h_{m+2}, e_{n+1}] + 3[h_{m+1}, e_{n+2}] - [h_m, e_{n+3}] \\ - (\hbar^2 + \mathbf{t}(\hbar+\mathbf{t}))([h_{m+1}, e_n] - [h_m, e_{n+1}]) - \hbar\mathbf{t}(\hbar+\mathbf{t})(h_m e_n + e_n h_m) = 0, \\ [h_0, e_n] = 0 = [h_1, e_n], \quad [h_2, e_n] = -2\hbar e_n \end{gathered}$$

for m, $n \geq 0$ in [**Tsy17**]. One can check two relations are equivalent as follows. First observe that (6.1b') is equivalent to

$$h(x)e(y) = \left(e(y)h(x)\frac{(1-(y^{-1}+\hbar)x)(1-(y^{-1}+\mathbf{t})x)(1-(y^{-1}-\hbar-\mathbf{t})x)}{(1-(y^{-1}-\hbar)x)(1-(y^{-1}-\mathbf{t})x)(1-(y^{-1}+\hbar+\mathbf{t})x)}\right)_+,$$

where

$$h(x) = 1 - \mathbf{t}(\hbar+\mathbf{t})\sum_{n=0}^{\infty} h_n x^{n+1}, \quad e(y) = \sum_{n=0}^{\infty} e_n y^{n+1}$$

and $(\ )_+$ denotes the part with positive powers in y. Here

$$\frac{(1-(y^{-1}+\hbar)x)(1-(y^{-1}+\mathbf{t})x)(1-(y^{-1}-\hbar-\mathbf{t})x)}{(1-(y^{-1}-\hbar)x)(1-(y^{-1}-\mathbf{t})x)(1-(y^{-1}+\hbar+\mathbf{t})x)}$$

is regarded as an element in $\mathbb{C}[\hbar, \mathbf{t}, y^{-1}][[x]]$. Then this is equivalent to

$$[\log h(x), e(y)] = \left(e(y)\log\frac{(1-(y^{-1}+\hbar)x)(1-(y^{-1}+\mathbf{t})x)(1-(y^{-1}-\hbar-\mathbf{t})x)}{(1-(y^{-1}-\hbar)x)(1-(y^{-1}-\mathbf{t})x)(1-(y^{-1}+\hbar+\mathbf{t})x)}\right)_+.$$

Now we observe that

$$\log\frac{(1-(y^{-1}+\hbar)x)(1-(y^{-1}+\mathbf{t})x)(1-(y^{-1}-\hbar-\mathbf{t})x)}{(1-(y^{-1}-\hbar)x)(1-(y^{-1}-\mathbf{t})x)(1-(y^{-1}+\hbar+\mathbf{t})x)} = \sum_{n=0}^{\infty} y^{-n}\varphi_n(x).$$

[1]The sign of $\hbar\mathbf{t}(\hbar+\mathbf{t})e_0^2$, $\hbar\mathbf{t}(\hbar+\mathbf{t})f_0^2$ are opposite, and $\mathbf{t}$ is missing in the definition of h_n. We believe that they are typo.

Similarly we have the corresponding relation (6.1c') on h_m and f_n equivalent to (6.1c). We also have the obvious relation

$$(6.1a') \qquad [h_m, h_n] = 0 \quad (m, n \geq 0)$$

instead of (6.1a). Thus $Y(\widehat{\mathfrak{gl}}(1))$ is generated by h_{n+2}, e_n, f_n ($n \geq 0$) with relations (6.1a', 6.1b', 6.1c', 6.1d, 6.1e, 6.1f, 6.1g) and $h_0 = 0$, $h_1 = \omega$. This is the presentation in [**Tsy17**].

6(ii). From Yangian to difference operators. Let $(B_n(x))_{n\geq 1}$ be the Bernoulli polynomials:

(1) $B_n(x+1) - B_n(x) = nx^{n-1}$,
(2) $\int_0^1 B_n(x)dx = 0$.

We also set $B_0(x) \equiv 1$. We then set $\bar{B}_n(w) = (-\hbar)^n B_n(-w/\hbar)/n$ so that $\bar{B}_n(w-\hbar) - \bar{B}_n(w) = -\hbar w^{n-1}$.

THEOREM 6.5. *Let $\mathcal{A}_\hbar$ be the quantized Coulomb branch for the quiver gauge theory for the Jordan quiver with* $\dim V = N$, $\dim W = 0$. *Then we have a surjective homomorphism of algebras* $\Phi\colon Y(\widehat{\mathfrak{gl}}(1)) \to \mathcal{A}_\hbar$ *given by*

$$D_{0,m} \mapsto \sum_{i=1}^{N} \bar{B}_m(w_i - (N-1)\mathbf{t}) - \bar{B}_m(-(i-1)\mathbf{t}) \quad (m \geq 1),$$

$$e_n \mapsto E_1[(w + \hbar - (N-1)\mathbf{t})^n], \qquad f_n \mapsto F_1[(w + \hbar - (N-1)\mathbf{t})^n] \quad (n \geq 0),$$

where $\omega = N$.

This follows from Theorem 1.1 and [**SV13**], where $Y(\widehat{\mathfrak{gl}}(1))$ is defined as the limit of $\mathbf{SH}_N^{\mathrm{gr}}$ as $N \to \infty$. The equivalence of the definition in [**SV13**] and the presentation (6.1) is given afterwards in [**AS13**]. We present a self-contained proof using the presentation (6.1).

6(iii). Proof. We check (6.1a, 6.1b, 6.1c) and (6.1g, 6.1h) in this subsection. A proof of (6.1d, 6.1e, 6.1f) will be given in §A.

A proof of (6.1b', 6.1c') will be given in §B. Hence a reader who prefers the presentation in Remark 6.4 should read this subsection until Lemma 6.8 for (6.1a, 6.1g), and then §A, §B.

It is obvious that (6.1a) is satisfied.

Let us set $\bar{w}_i = w_i - (N-1)\mathbf{t}$, $\bar{w} = w - (N-1)\mathbf{t}$.

It is clear that we have

LEMMA 6.6. *Let $f = f(w)$ be a polynomial in one variable w. Then we have*

$$\left[\sum_{i=1}^{N} \bar{B}_n(\bar{w}_i), E_1[f]\right] = -\hbar E_1[f(\bar{w}+\hbar)^{n-1}], \quad \left[\sum_{i=1}^{N} \bar{B}_n(\bar{w}_i), F_1[f]\right] = \hbar F_1[f(\bar{w}+\hbar)^{n-1}].$$

This checks (6.1b, 6.1c).

REMARK 6.7. In [**SV13**] a polynomial denoted by $B_m^{(N)}$ is introduced, and $\Phi(D_{0,m})$ is introduced as $SB_m^{(n)}(w_1 - N\mathbf{t}, \dots, w_N - N\mathbf{t})S$, where S is the complete idempotent. See [**SV13**, Lem. 1.9 and (1.29)]. One can directly check that $B_m^{(N)}$ is given by the Bernoulli polynomial, and our definition of $\Phi(D_{0,m})$ coincides with [**SV13**]. Also $\Phi(D_{\pm1,0})$ is defined as $\sum_i \operatorname{Res} X_i^{\pm1}$. See [**SV13**, (1.32)]. Therefore it coincides with our $E_1[1] = \Phi(e_0)$, $F_1[1] = \Phi(f_0)$. Together with the relations (6.1b, 6.1c), we see that our homomorphism is exactly the same as one in [**SV13**].

In order to check (6.1g) we start with the following:

LEMMA 6.8. *For $m, n \geq 0$*

$$\begin{aligned}&[E_1[(\bar{w}+\hbar)^m], F_1[(\bar{w}+\hbar)^n]]\\ &= -\frac{\hbar}{\mathbf{t}(\hbar+\mathbf{t})}[x^{m+n+1}]\left[\prod_{i=1}^N \frac{(1-(\bar{w}_i-\mathbf{t})x)(1-(\bar{w}_i+\hbar+\mathbf{t})x)}{(1-\bar{w}_i x)(1-(\bar{w}_i+\hbar)x)}\right],\end{aligned}$$

where $[x^{m+n+1}]$ denotes the coefficient of x^{m+n+1}.

PROOF. The left hand side is

$$\sum_{i,j=1}^N \left[(\bar{w}_i+\hbar)^m \prod_{s\neq i} \frac{\bar{w}_i-\bar{w}_s-\mathbf{t}}{\bar{w}_i-\bar{w}_s}\mathsf{u}_i, \bar{w}_j^n \prod_{u\neq j}\frac{\bar{w}_j-\bar{w}_u+\mathbf{t}}{\bar{w}_j-\bar{w}_u}\mathsf{u}_j^{-1}\right].$$

(Note that $w_i - w_s = \bar{w}_i - \bar{w}_s$.) It is easy to check that terms with $i \neq j$ vanish. Next consider the sum over $i = j$. We have

$$\begin{aligned}\sum_{i=1}^N \Bigg((\bar{w}_i+\hbar)^{m+n} \prod_{s\neq i} \frac{\bar{w}_i-\bar{w}_s-\mathbf{t}}{\bar{w}_i-\bar{w}_s}\frac{\bar{w}_i+\hbar-\bar{w}_s+\mathbf{t}}{\bar{w}_i+\hbar-\bar{w}_s}\\ - \bar{w}_i^{m+n}\prod_{s\neq i}\frac{\bar{w}_i-\bar{w}_s+\mathbf{t}}{\bar{w}_i-\bar{w}_s}\frac{\bar{w}_i-\hbar-\bar{w}_s-\mathbf{t}}{\bar{w}_i-\hbar-\bar{w}_s}\Bigg).\end{aligned}$$

This expression was appeared in [**SV13**, Cor. B.6]. Let us use the same technique to compute this: Consider

$$\begin{aligned}\sum_{i=1}^N \Bigg(\frac{x\mathbf{t}}{1-(\bar{w}_i+\hbar)x} \prod_{s\neq i} \frac{\bar{w}_i-\bar{w}_s-\mathbf{t}}{\bar{w}_i-\bar{w}_s}\frac{\bar{w}_i+\hbar-\bar{w}_s+\mathbf{t}}{\bar{w}_i+\hbar-\bar{w}_s}\\ - \frac{x\mathbf{t}}{1-\bar{w}_i x}\prod_{s\neq i}\frac{\bar{w}_i-\bar{w}_s+\mathbf{t}}{\bar{w}_i-\bar{w}_s}\frac{\bar{w}_i-\hbar-\bar{w}_s-\mathbf{t}}{\bar{w}_i-\hbar-\bar{w}_s}\Bigg).\end{aligned}$$

This is a rational function in x, vanishes at $x = 0$, regular at $x = \infty$, and with at most simple poles. Compare it with

$$-\frac{\hbar}{\hbar+\mathbf{t}}\left(\prod_{i=1}^N \frac{(1-(\bar{w}_i-\mathbf{t})x)(1-(\bar{w}_i+\hbar+\mathbf{t})x)}{(1-\bar{w}_i x)(1-(\bar{w}_i+\hbar)x)} - 1\right),$$

which has the same properties and the equal residues. Therefore two are equal thanks to the maximal principle. Since we are taking coefficients of x^{m+n+1}, we can ignore the constant term -1. □

Finally let us rewrite

$$\begin{aligned}&\prod_{i=1}^N \frac{(1-(\bar{w}_i-\mathbf{t})x)(1-(\bar{w}_i+\hbar+\mathbf{t})x)}{(1-\bar{w}_i x)(1-(\bar{w}_i+\hbar)x)}\\ &= \exp\left[\sum_{n=1}^\infty\sum_{i=1}^N \left(\bar{w}_i^n - (\bar{w}_i-\mathbf{t})^n + (\bar{w}_i+\hbar)^n - (\bar{w}_i+\hbar+\mathbf{t})^n\right)\frac{x^n}{n}\right]\end{aligned} \tag{6.9}$$

in terms of the normalized Bernoulli polynomials $\bar{B}_n(\bar{w}_i)$.

We use the following formula for Bernoulli polynomials:

$$w^n = \frac{1}{n+1}\sum_{k=0}^{n}\binom{n+1}{k}B_k(w), \qquad B_k(w+v) = \sum_{i=0}^{k}\binom{k}{i}B_i(w)v^{k-i}.$$

A direct calculation shows

$$\begin{aligned}\sum_{n=1}^{\infty}(\bar{w}_i - \mathbf{t})^n\frac{x^n}{n} &= 1 + \frac{1+\mathbf{t}x}{\hbar x}\log(1+\mathbf{t}x) - \frac{1+(\hbar+\mathbf{t})x}{\hbar x}\log(1+(\mathbf{t}+\hbar)x)\\ &\quad + \frac{\bar{B}_1(\bar{w}_i)}{\hbar}\log\left(\frac{1+x(\hbar+\mathbf{t})}{1+x\mathbf{t}}\right)\\ &\quad - \sum_{k=2}^{\infty}\frac{\bar{B}_k(\bar{w}_i)}{(k-1)\hbar}\left(\left(\frac{x}{1+x(\hbar+\mathbf{t})}\right)^{k-1} - \left(\frac{x}{1+x\mathbf{t}}\right)^{k-1}\right).\end{aligned}$$

Taking difference with the same expression with $\bar{w}_i = -(i-1)\mathbf{t}$, we get

(6.10)

$$\begin{aligned}\sum_{n=1}^{\infty}(\bar{w}_i - \mathbf{t})^n\frac{x^n}{n} &= -\log(1+ix\mathbf{t}) + \frac{\bar{B}_1(\bar{w}_i) - \bar{B}_1(-(i-1)\mathbf{t})}{\hbar}\log\left(\frac{1+x(\hbar+\mathbf{t})}{1+x\mathbf{t}}\right)\\ &\quad - \sum_{k=2}^{\infty}\frac{\bar{B}_k(\bar{w}_i) - \bar{B}_k(-(i-1)\mathbf{t})}{(k-1)\hbar}\left(\left(\frac{x}{1+x(\hbar+\mathbf{t})}\right)^{k-1} - \left(\frac{x}{1+x\mathbf{t}}\right)^{k-1}\right).\end{aligned}$$

Similarly we have

$$\begin{aligned}\sum_{n=1}^{\infty}\bar{w}_i^n\frac{x^n}{n} &= -\log(1+x(i-1)\mathbf{t}) + \frac{\bar{B}_1(\bar{w}_i) - \bar{B}_1(-(i-1)\mathbf{t})}{\hbar}\log(1+x\hbar)\\ &\quad - \sum_{k=2}^{\infty}\frac{\bar{B}_k(\bar{w}_i) - \bar{B}_k(-(i-1)\mathbf{t})}{(k-1)\hbar}\left(\left(\frac{x}{1+x\hbar}\right)^{k-1} - x^{k-1}\right),\\ \sum_{n=1}^{\infty}(\bar{w}_i + \hbar)^n\frac{x^n}{n} &= -\log(1+x((i-1)\mathbf{t}-\hbar))\\ &\quad + \frac{\bar{B}_1(\bar{w}_i) - \bar{B}_1(-(i-1)\mathbf{t})}{\hbar}\log\left(\frac{1}{1-x\hbar}\right)\\ &\quad - \sum_{k=2}^{\infty}\frac{\bar{B}_k(\bar{w}_i) - \bar{B}_k(-(i-1)\mathbf{t})}{(k-1)\hbar}\left(x^{k-1} - \left(\frac{x}{1-x\hbar}\right)^{k-1}\right),\\ \sum_{n=1}^{\infty}(\bar{w}_i + \hbar + \mathbf{t})^n\frac{x^n}{n} &= -\log(1+x((i-2)\mathbf{t}-\hbar))\\ &\quad + \frac{\bar{B}_1(\bar{w}_i) - \bar{B}_1(-(i-1)\mathbf{t})}{\hbar}\log\left(\frac{1-x\mathbf{t}}{1-x(\hbar+\mathbf{t})}\right)\\ &\quad - \sum_{k=2}^{\infty}\frac{\bar{B}_k(\bar{w}_i) - \bar{B}_k(-(i-1)\mathbf{t})}{(k-1)\hbar}\left(\left(\frac{x}{1-x\mathbf{t}}\right)^{k-1} - \left(\frac{x}{1-x(\hbar+\mathbf{t})}\right)^{k-1}\right).\end{aligned}$$

The log terms give us

$$\sum_{i=1}^{N}(\log(1+xi\mathbf{t})-\log(1+x(i-1)\mathbf{t})-\log(1+x((i-1)\mathbf{t}-\hbar))+\log(1+x((i-2)\mathbf{t}-\hbar)))$$
$$=\log(1+Nx\mathbf{t})-\log(1+x((N-1)\mathbf{t}-\hbar))+\log(1-x(\mathbf{t}+\hbar)).$$

The alternating sum for $\bar{B}_1(\bar{w}_i)-\bar{B}_1(-(i-1)\mathbf{t})/\hbar$ is

$$\log(1-x\hbar)-\log(1-x(\hbar+\mathbf{t}))+\log(1-x\mathbf{t})$$
$$-\log(1+x\hbar)-\log(1+x\mathbf{t})+\log(1+x(\hbar+\mathbf{t})).$$

This is nothing but $\varphi_0(x)$.

Finally the alternating sum for $\bar{B}_k(\bar{w}_i)-\bar{B}_k(-(i-1)\mathbf{t})/\hbar$ is

$$\frac{1}{k-1}\left[-\left(\frac{x}{1-\hbar x}\right)^{k-1}+\left(\frac{x}{1-(\hbar+\mathbf{t})x}\right)^{k-1}-\left(\frac{x}{1-\mathbf{t}x}\right)^{k-1}\right.$$
$$\left.+\left(\frac{x}{1+\hbar x}\right)^{k-1}+\left(\frac{x}{1+\mathbf{t}x}\right)^{k-1}-\left(\frac{x}{1+(\hbar+\mathbf{t})x}\right)^{k-1}\right].$$

This is $\varphi_{k-1}(x)$. Therefore (6.9) is equal to

$$\frac{(1-(\hbar+\mathbf{t})x)(1+N\mathbf{t}x)}{1-(\hbar+(1-N)\mathbf{t})x}\exp\left(-\sum_{k=1}^{\infty}\sum_{i=1}^{N}\frac{\bar{B}_k(\bar{w}_i)-\bar{B}_k(-(i-1)\mathbf{t})}{\hbar}\varphi_{k-1}(x)\right).$$

This shows (6.1g) with $\omega=N$.

The proof of the remaining relations is given in §A.

6(iv). Automorphism. Let a be a complex number. We define an automorphism τ_a of $Y(\widehat{\mathfrak{gl}}(1))$ by

$$\tau_a(e_n)=\sum_{k=0}^{n}\binom{n}{k}a^{n-k}e_k,\qquad \tau_a(f_n)=\sum_{k=0}^{n}\binom{n}{k}a^{n-k}f_k,$$
$$\tau_a(h_n)=\sum_{k=0}^{n}\binom{n}{k}a^{n-k}h_k.$$

A direct computation shows that relations (6.1d, 6.1e, 6.1f, 6.1g) are preserved.

It is also easy to check (6.1a', 6.1b', 6.1c') in Remark 6.4. Since those are equivalent to (6.1a, 6.1b, 6.1c), the automorphism τ_a is well-defined.

Let us give another proof of (6.1a, 6.1b, 6.1c).

LEMMA 6.11. *We have*

$$\tau_a(D_{0,m})\equiv\sum_{k=1}^{m}\binom{m-1}{k-1}a^{m-k}D_{0,k}$$

modulo a central element.

PROOF. By the binomial theorem, we have

$$\tau_a\left(1-\mathbf{t}(\hbar+\mathbf{t})\sum_{n=0}^{\infty}h_nx^{n+1}\right)=1-\mathbf{t}(\hbar+\mathbf{t})\sum_{n=0}^{\infty}h_n\left(\frac{x}{1-ax}\right)^{n+1}.$$

On the other hand

$$\exp\left(-\sum_{n=0}^{\infty}\sum_{k=1}^{n+1}\binom{n}{k-1}a^{n+1-k}\frac{D_{0,k}}{\hbar}\varphi_n(x)\right) = \exp\left(-\sum_{n=0}^{\infty}\frac{D_{0,n+1}}{\hbar}\varphi_n\left(\frac{x}{1-ax}\right)\right)$$

follows from the identity

$$\sum_{n=k}^{\infty}\binom{n}{k}a^{n-k}\varphi_n(x) = \varphi_k\left(\frac{x}{1-ax}\right) \qquad (n \geq 0).$$

Therefore

$$C_m \overset{\text{def.}}{=} \tau_a(D_{0,m}) - \sum_{k=1}^{m}\binom{m-1}{k-1}a^{m-k}D_{0,k}$$

is given by

$$\begin{aligned}\frac{(1-(\hbar+\mathbf{t})x)(1+\omega\mathbf{t}x)}{1-(\hbar+(1-\omega)\mathbf{t})x}\exp\left(-\sum_{n=0}^{\infty}\frac{C_{n+1}}{\hbar}\varphi_n(x)\right)\\ = \frac{(1-(a+\hbar+\mathbf{t})x)(1-(a-\omega\mathbf{t})x)}{(1-ax)(1-(a+\hbar+(1-\omega)\mathbf{t})x)}.\end{aligned}$$

Assuming $\omega = N$ is a positive integer, we substitute $\bar{w}_i = a - (i-1)\mathbf{t}$ to (6.10) and the subsequent three equations. We have

$$\begin{aligned}\exp\left(-\sum_{k=1}^{\infty}\sum_{i=1}^{N}\frac{\bar{B}_k(a-(i-1)\mathbf{t}) - \bar{B}_k(-(i-1)\mathbf{t})}{\hbar}\varphi_{k-1}(x)\right)\\ = \frac{(1-(a+\hbar+\mathbf{t})x)(1-(a-N\mathbf{t})x)(1-(\hbar+(1-N)\mathbf{t})x)}{(1-ax)(1-(a+\hbar+(1-N)\mathbf{t})x)(1-(\hbar+\mathbf{t})x)(1+N\mathbf{t}x)}.\end{aligned}$$

Note that $\sum_{i=1}^{N}\bar{B}_k(a-(i-1)\mathbf{t}) - \bar{B}_k(-(i-1)\mathbf{t})$ is a polynomial in a, $\hbar$, $\mathbf{t}$ and N. Therefore C_k is the central element obtained by replacing N by ω. □

From the proof, we can remove the shift $-(N-1)\mathbf{t}$ in Theorem 6.5.

PROPOSITION 6.12. *Let $\mathcal{A}_\hbar$ be the quantized Coulomb branch for* $\dim V = N$, $\dim W = 0$. *Then we have a surjective homomorphism of algebras* $\Psi\colon Y(\widehat{\mathfrak{gl}}(1)) \to \mathcal{A}_\hbar$ *given by*

$$\begin{aligned}D_{0,m} &\mapsto \sum_{i=1}^{N}\bar{B}_m(w_i) - \bar{B}_m(-(i-1)\mathbf{t}) \quad (m \geq 1),\\ e_n &\mapsto E_1[(w+\hbar)^n], \qquad f_n \mapsto F_1[(w+\hbar)^n] \quad (n \geq 0),\end{aligned}$$

where $\omega = N$.

In fact, looking at the proof in §6(iii), §A, we find that the argument go through when we use w_i, w instead of $\bar{w}_i$, $\bar{w}$. It gives a direct proof without using the automorphism τ_a.

6(v). Shifted Yangian. Now we consider the case $\dim W = l > 0$. Let us compare operators F_1 in (1.4) for $l > 0$ and $l = 0$. In order to distinguish them, let us denote them by $F_1^{(l)}$, $F_1^{(0)}$ respectively. They are related by

$$F_1^{(l)}[(w+\hbar)^n] = F_1^{(0)}[(w+\hbar)^n \prod_{k=1}^{l}(w-z_k)] = \sum_{i=0}^{l}(-1)^{l-i}e_{l-i}(\vec{z}+\hbar)F_1^{(0)}[(w+\hbar)^{i+n}],$$

where $e_{l-i}(\vec{z}+\hbar)$ is the $(l-i)$th elementary symmetric function in variables $z_1+\hbar$, …, $z_l+\hbar$. Thus the commutation relations on $F_1^{(l)}[(w+\hbar)^n]$ are deduced from those on $F_1^{(0)}[(w+\hbar)^n]$.

Let $Y_l(\vec{z})$ be a $\mathbb{C}[\omega, \hbar, \mathbf{t}]$-algebra generated by $D_{0,m}$ $(m \geq 1)$, e_n, f_{n+l} $(n \geq 0)$ with relations (6.1) where (6.1h) is replaced by

$$\text{(6.13)}\quad 1 - \mathbf{t}(\hbar+\mathbf{t})\sum_{n\geq 0} h_n x^{n+1}$$
$$= \prod_{k=1}^{l}(1-(z_k+\hbar)x) \times \frac{(1-(\hbar+\mathbf{t})x)(1+\omega\mathbf{t}x)}{1-(\hbar+(1-\omega)\mathbf{t})x}\exp\left(-\sum_{n\geq 0}\frac{D_{0,n+1}}{\hbar}\varphi_n(x)\right).$$

The right hand side is *not* 1 at $\mathbf{t}=0$ nor $\hbar+\mathbf{t}=0$. Nevertheless we only need h_n with $n \geq l$ in (6.1g), and they are well-defined as $\prod_{k=1}^{l}(1-(z_k+\hbar)x)$ is of degree l.

THEOREM 6.14. *Let $\mathcal{A}_\hbar$ be the quantized Coulomb branch for $\dim V = N$, $\dim W = l$. Then we have a surjective homomorphism of algebras $\Psi\colon Y_l(\vec{z}) \to \mathcal{A}_\hbar$ given by*

$$D_{0,m} \mapsto \sum_{i=1}^{N}\bar{B}_m(w_i) - \bar{B}_m(-(i-1)\mathbf{t}) \quad (m \geq 1),$$
$$e_n \mapsto E_1[(w+\hbar)^n], \qquad f_{n+l} \mapsto F_1^{(l)}[(w+\hbar)^n] \quad (n \geq 0),$$

where $\omega = N$.

REMARK 6.15. Let us switch to the presentation in Remark 6.4. Let Y_l be the subalgebra of $Y(\hat{\mathfrak{gl}}(1))$ generated by h_n, e_n, f_{n+l}. Then we can 'formally' define a homomorphism $Y_l \to \mathcal{A}_\hbar$ by

$$h(x) \mapsto \prod_{k=1}^{l}(1-(z_k+\hbar)x)\prod_{i=1}^{N}\frac{(1-(w_i-\mathbf{t})x)(1-(w_i+\hbar+\mathbf{t})x)}{(1-w_ix)(1-(w_i+\hbar)x)},$$
$$e_n \mapsto E_1[(w+\hbar)^n], \qquad f_{n+l} \mapsto F_1^{(l)}[(w+\hbar)^n] \quad (n \geq 0).$$

(A proof is given in §B.) However the target of $h(x)$ is not 1 at $\mathbf{t}=0$ nor $\hbar+\mathbf{t}=0$. Therefore the image of h_n $(n < l)$ is contained in $\frac{1}{\mathbf{t}(\hbar+\mathbf{t})}\mathcal{A}_\hbar$, but not in $\mathcal{A}_\hbar$. Similarly $Y_l(\vec{z})$ is *almost* isomorphic to Y_l, which is independent of parameters z_k, but not quite yet.

This problem does not arise for finite type shifted Yangian: Y_l and $Y_l(\vec{z})$ are isomorphic in this case. See [**KWWY14**, 3G].

Appendix A

Let us prove (6.1e) for $f_n = F_1[(\bar{w}+\hbar)^n]$. Put

$$C_i = \prod_{j\neq i} \frac{\bar{w}_i - \bar{w}_j + \mathbf{t}}{\bar{w}_i - \bar{w}_j}$$

so that $F_1[(\bar{w}+\hbar)^n] = \sum_{i=1}^N \bar{w}_i^n C_i \mathsf{u}_i^{-1}$. We define C_i' and $C_j^{(i)}$ $(i\neq j)$ by

$$C_i' = \prod_{j\neq i} \frac{\bar{w}_i - \bar{w}_j - \hbar + \mathbf{t}}{\bar{w}_i - \bar{w}_j - \hbar},$$

$$C_j^{(i)} = C_j \frac{\bar{w}_j - \bar{w}_i}{\bar{w}_j - \bar{w}_i + \mathbf{t}} = \prod_{k\neq i,j} \frac{\bar{w}_j - \bar{w}_k + \mathbf{t}}{\bar{w}_j - \bar{w}_k}$$

so that

$$\mathsf{u}_i^{-1} C_i = C_i' \mathsf{u}_i^{-1}, \quad \mathsf{u}_i^{-1} C_j = C_j^{(i)} \frac{\bar{w}_j - \bar{w}_i + \hbar + \mathbf{t}}{\bar{w}_j - \bar{w}_i + \hbar} \mathsf{u}_i^{-1} \ (i\neq j).$$

We have

$$\begin{aligned}
[f_m, f_n] &= \sum_{i,j} [\bar{w}_i^m C_i \mathsf{u}_i^{-1}, \bar{w}_j^n C_j \mathsf{u}_j^{-1}] \\
&= \sum_i \left(\bar{w}_i^m (\bar{w}_i - \hbar)^n - \bar{w}_i^n (\bar{w}_i - \hbar)^m\right) C_i C_i' \mathsf{u}_i^{-2} \\
&\quad + \sum_{i\neq j} \bar{w}_i^m \bar{w}_j^n \left(C_i C_j^{(i)} \frac{\bar{w}_j - \bar{w}_i + \hbar + \mathbf{t}}{\bar{w}_j - \bar{w}_i + \hbar} - C_j C_i^{(j)} \frac{\bar{w}_i - \bar{w}_j + \hbar + \mathbf{t}}{\bar{w}_i - \bar{w}_j + \hbar} \right) \mathsf{u}_i^{-1} \mathsf{u}_j^{-1}.
\end{aligned}$$

The second term is calculated as

$$\begin{aligned}
&C_i C_j^{(i)} \frac{\bar{w}_j - \bar{w}_i + \hbar + \mathbf{t}}{\bar{w}_j - \bar{w}_i + \hbar} - C_j C_i^{(j)} \frac{\bar{w}_i - \bar{w}_j + \hbar + \mathbf{t}}{\bar{w}_i - \bar{w}_j + \hbar} \\
&= -C_i^{(j)} C_j^{(i)} \frac{2\hbar \mathbf{t}(\hbar + \mathbf{t})}{(\bar{w}_i - \bar{w}_j)(\bar{w}_i - \bar{w}_j + \hbar)(\bar{w}_i - \bar{w}_j - \hbar)}.
\end{aligned}$$

Therefore

$$\begin{aligned}
[f_m, f_n] &= \sum_i \left(\bar{w}_i^m (\bar{w}_i - \hbar)^n - \bar{w}_i^n (\bar{w}_i - \hbar)^m\right) C_i C_i' \mathsf{u}_i^{-2} \\
-2\hbar\mathbf{t}(\hbar{+}\mathbf{t}) &\sum_{i<j} (\bar{w}_i^m \bar{w}_j^n {-} \bar{w}_i^n \bar{w}_j^m) C_i^{(j)} C_j^{(i)} \frac{1}{(\bar{w}_i{-}\bar{w}_j)(\bar{w}_i{-}\bar{w}_j{+}\hbar)(\bar{w}_i{-}\bar{w}_j{-}\hbar)} \mathsf{u}_i^{-1} \mathsf{u}_j^{-1}.
\end{aligned}$$

We also have

$$\begin{aligned}
f_0^2 &= \sum_i C_i C_i' \mathsf{u}_i^{-2} \\
&\quad + \sum_{i<j} C_i^{(j)} C_j^{(i)} \left(\frac{\bar{w}_i{-}\bar{w}_j{+}\mathbf{t}}{\bar{w}_i{-}\bar{w}_j} \frac{\bar{w}_j{-}\bar{w}_i{+}\hbar{+}\mathbf{t}}{\bar{w}_j{-}\bar{w}_i{+}\hbar} + \frac{\bar{w}_j{-}\bar{w}_i{+}\mathbf{t}}{\bar{w}_j{-}\bar{w}_i} \frac{\bar{w}_i{-}\bar{w}_j{+}\hbar{+}\mathbf{t}}{\bar{w}_i{-}\bar{w}_j{+}\hbar} \right) \mathsf{u}_i^{-1} \mathsf{u}_j^{-1} \\
&= \sum_i C_i C_i' \mathsf{u}_i^{-2} + 2 \sum_{i<j} C_i^{(j)} C_j^{(i)} \frac{(\bar{w}_i - \bar{w}_j)^3 - (\hbar^2 + \mathbf{t}(\hbar + \mathbf{t}))(\bar{w}_i - \bar{w}_j)}{(\bar{w}_i - \bar{w}_j)(\bar{w}_i - \bar{w}_j + \hbar)(\bar{w}_i - \bar{w}_j - \hbar)} \mathsf{u}_i^{-1} \mathsf{u}_j^{-1}.
\end{aligned}$$

The coefficient of u_i^{-2} in the left hand side of (6.1e) is

$$C_iC_i' \\ \times \left(3\left(\bar{w}_i^2(\bar{w}_i-\hbar)-\bar{w}_i(\bar{w}_i-\hbar)^2\right)-\left(\bar{w}_i^3-(\bar{w}_i-\hbar)^3\right)+(\hbar^2+\mathbf{t}(\hbar+\mathbf{t}))\left(\bar{w}_i-(\bar{w}_i-\hbar)\right) - \hbar\mathbf{t}(\hbar+\mathbf{t})\right)=0.$$

The coefficient of $\mathsf{u}_i^{-1}\mathsf{u}_j^{-1}$ $(i<j)$ in the left hand side of (6.1e) is

$$C_i^{(j)}C_j^{(i)}\frac{2\hbar\mathbf{t}(\hbar+\mathbf{t})}{(\bar{w}_i-\bar{w}_j)(\bar{w}_i-\bar{w}_j+\hbar)(\bar{w}_i-\bar{w}_j-\hbar)} \\ \times\left(-\left(3\left(\bar{w}_i^2\bar{w}_j-\bar{w}_i\bar{w}_j^2\right)-\left(\bar{w}_i^3-\bar{w}_j^3\right)+(\hbar^2+\mathbf{t}(\hbar+\mathbf{t}))\left(\bar{w}_i-\bar{w}_j\right)\right) - \left((\bar{w}_i-\bar{w}_j)^3-(\hbar^2+\mathbf{t}(\hbar+\mathbf{t}))(\bar{w}_i-\bar{w}_j)\right)\right)=0.$$

The proof of (6.1d) is the same, hence is omitted.

Let us prove (6.1f) for f_n. The proof for e_n is similar. We define C_i'' and $C_j'^{(i)}$ $(i\neq j)$ by

$$C_i''=\prod_{j\neq i}\frac{\bar{w}_i-\bar{w}_j+\mathbf{t}-2\hbar}{\bar{w}_i-\bar{w}_j-2\hbar},$$
$$C_j'^{(i)}=C_j'\frac{\bar{w}_j-\bar{w}_i-\hbar}{\bar{w}_j-\bar{w}_i-\hbar+\mathbf{t}}=\prod_{k\neq i,j}\frac{\bar{w}_j-\bar{w}_k-\hbar+\mathbf{t}}{\bar{w}_j-\bar{w}_k-\hbar}$$

so that

$$\mathsf{u}_i^{-1}C_i'=C_i''\mathsf{u}_i^{-1},\quad \mathsf{u}_i^{-1}C_j'=C_j'^{(i)}\frac{\bar{w}_j-\bar{w}_i+\mathbf{t}}{\bar{w}_j-\bar{w}_i}\mathsf{u}_i^{-1}\ (i\neq j).$$

We have

$$\text{(A.1)}\quad [f_0,[f_0,f_1]]=-\hbar\sum_{i,j}[C_i\mathsf{u}_i^{-1},C_jC_j'\mathsf{u}_j^{-2}] \\ +2\hbar\mathbf{t}(\hbar+\mathbf{t})\sum_i\sum_{j<k}[C_i\mathsf{u}_i^{-1},C_j^{(k)}C_k^{(j)}\frac{1}{(\bar{w}_j-\bar{w}_k+\hbar)(\bar{w}_j-\bar{w}_k-\hbar)}\mathsf{u}_j^{-1}\mathsf{u}_k^{-1}].$$

Consider the case $i=j$ in the first sum of (A.1). The summand is

$$(C_iC_i'C_i''-C_iC_i'C_i'')\mathsf{u}_i^{-3}=0.$$

For the case $i\neq j$ in the first sum of (A.1), we have

$$C_i\mathsf{u}_i^{-1}C_jC_j'\mathsf{u}_j^{-2}=C_iC_j^{(i)}\frac{\bar{w}_j-\bar{w}_i+\hbar+\mathbf{t}}{\bar{w}_j-\bar{w}_i+\hbar}C_j'^{(i)}\frac{\bar{w}_j-\bar{w}_i+\mathbf{t}}{\bar{w}_j-\bar{w}_i}\mathsf{u}_i^{-1}\mathsf{u}_j^{-2} \\ =C_i^{(j)}C_jC_j'^{(i)}\frac{\bar{w}_i-\bar{w}_j+\mathbf{t}}{\bar{w}_i-\bar{w}_j}\frac{\bar{w}_i-\bar{w}_j-\hbar-\mathbf{t}}{\bar{w}_i-\bar{w}_j-\hbar}\mathsf{u}_i^{-1}\mathsf{u}_j^{-2}.$$

Furthermore,

$$C_jC_j'\mathsf{u}_j^{-2}C_i\mathsf{u}_i^{-1}=C_jC_j'C_i^{(j)}\frac{\bar{w}_i-\bar{w}_j+2\hbar+\mathbf{t}}{\bar{w}_i-\bar{w}_j+2\hbar}\mathsf{u}_i^{-1}\mathsf{u}_j^{-2} \\ =C_jC_j'^{(i)}C_i^{(j)}\frac{\bar{w}_j-\bar{w}_i-\hbar+\mathbf{t}}{\bar{w}_j-\bar{w}_i-\hbar}\frac{\bar{w}_i-\bar{w}_j+2\hbar+\mathbf{t}}{\bar{w}_i-\bar{w}_j+2\hbar}\mathsf{u}_i^{-1}\mathsf{u}_j^{-2}.$$

Thus we have

$$\begin{aligned}&[C_i\mathsf{u}_i^{-1}, C_jC'_j\mathsf{u}_j^{-2}]\\ &= C_i^{(j)}C_jC_j'^{(i)}\left(\frac{\bar{w}_i-\bar{w}_j+\mathbf{t}}{\bar{w}_i-\bar{w}_j}\frac{\bar{w}_i-\bar{w}_j-\hbar-\mathbf{t}}{\bar{w}_i-\bar{w}_j-\hbar}-\frac{\bar{w}_j-\bar{w}_i-\hbar+\mathbf{t}}{\bar{w}_j-\bar{w}_i-\hbar}\frac{\bar{w}_i-\bar{w}_j+2\hbar+\mathbf{t}}{\bar{w}_i-\bar{w}_j+2\hbar}\right)\mathsf{u}_i^{-1}\mathsf{u}_j^{-2}\end{aligned}$$

and denote this by A_{ij}. Consider the case $i = j < k$ in the second sum of (A.1). We have

$$\begin{aligned}&C_j\mathsf{u}_j^{-1}C_j^{(k)}C_k^{(j)}\frac{1}{(\bar{w}_j-\bar{w}_k+\hbar)(\bar{w}_j-\bar{w}_k-\hbar)}\mathsf{u}_j^{-1}\mathsf{u}_k^{-1}\\ &= C_jC_j'^{(k)}C_k^{(j)}\frac{1}{(\bar{w}_j-\bar{w}_k)(\bar{w}_j-\bar{w}_k-2\hbar)}\mathsf{u}_j^{-2}\mathsf{u}_k^{-1},\end{aligned}$$

and

$$\begin{aligned}&C_j^{(k)}C_k^{(j)}\frac{1}{(\bar{w}_j-\bar{w}_k+\hbar)(\bar{w}_j-\bar{w}_k-\hbar)}\mathsf{u}_j^{-1}\mathsf{u}_k^{-1}C_j\mathsf{u}_j^{-1}\\ &= C_j^{(k)}C_k^{(j)}\frac{1}{(\bar{w}_j-\bar{w}_k+\hbar)(\bar{w}_j-\bar{w}_k-\hbar)}C_j'^{(k)}\frac{\bar{w}_j-\bar{w}_k+\mathbf{t}}{\bar{w}_j-\bar{w}_k}\mathsf{u}_j^{-2}\mathsf{u}_k^{-1}\\ &= C_jC_k^{(j)}C_j'^{(k)}\frac{1}{(\bar{w}_j-\bar{w}_k+\hbar)(\bar{w}_j-\bar{w}_k-\hbar)}\mathsf{u}_j^{-2}\mathsf{u}_k^{-1}.\end{aligned}$$

Thus we have

$$\begin{aligned}&[C_j\mathsf{u}_j^{-1}, C_j^{(k)}C_k^{(j)}\frac{1}{(\bar{w}_j-\bar{w}_k+\hbar)(\bar{w}_j-\bar{w}_k-\hbar)}\mathsf{u}_j^{-1}\mathsf{u}_k^{-1}]\\ &= C_jC_j'^{(k)}C_k^{(j)}\left(\frac{1}{(\bar{w}_j-\bar{w}_k)(\bar{w}_j-\bar{w}_k-2\hbar)}-\frac{1}{(\bar{w}_j-\bar{w}_k+\hbar)(\bar{w}_j-\bar{w}_k-\hbar)}\right)\mathsf{u}_j^{-2}\mathsf{u}_k^{-1}\end{aligned}$$

and denote this by B_{jk}. Similarly the case $j < k = i$ in the second sum of (A.1) is given by

$$\begin{aligned}&[C_k\mathsf{u}_k^{-1}, C_j^{(k)}C_k^{(j)}\frac{1}{(\bar{w}_j-\bar{w}_k+\hbar)(\bar{w}_j-\bar{w}_k-\hbar)}\mathsf{u}_j^{-1}\mathsf{u}_k^{-1}]\\ &= C_kC_j^{(k)}C_k'^{(j)}\left(\frac{1}{(\bar{w}_j-\bar{w}_k+2\hbar)(\bar{w}_j-\bar{w}_k)}-\frac{1}{(\bar{w}_j-\bar{w}_k+\hbar)(\bar{w}_j-\bar{w}_k-\hbar)}\right)\mathsf{u}_j^{-1}\mathsf{u}_k^{-2}\\ &= B_{kj}.\end{aligned}$$

Hence the coefficient of $\mathsf{u}_j^{-2}\mathsf{u}_k^{-1}$ $(j < k)$ in (A.1) is given by

$$\begin{aligned}&-\hbar A_{kj} + 2\hbar\mathbf{t}(\hbar+\mathbf{t})B_{jk}\\ &= -\hbar C_k^{(j)}C_jC_j'^{(k)}\left(\frac{\bar{w}_k-\bar{w}_j+\mathbf{t}}{\bar{w}_k-\bar{w}_j}\frac{\bar{w}_k-\bar{w}_j-\hbar-\mathbf{t}}{\bar{w}_k-\bar{w}_j-\hbar}-\frac{\bar{w}_j-\bar{w}_k-\hbar+\mathbf{t}}{\bar{w}_j-\bar{w}_k-\hbar}\frac{\bar{w}_k-\bar{w}_j+2\hbar+\mathbf{t}}{\bar{w}_k-\bar{w}_j+2\hbar}\right)\\ &\quad + 2\hbar\mathbf{t}(\hbar+\mathbf{t})C_jC_j'^{(k)}C_k^{(j)}\left(\frac{1}{(\bar{w}_j-\bar{w}_k)(\bar{w}_j-\bar{w}_k-2\hbar)}-\frac{1}{(\bar{w}_j-\bar{w}_k+\hbar)(\bar{w}_j-\bar{w}_k-\hbar)}\right).\end{aligned}$$

Set $x = \bar{w}_j - \bar{w}_k$. A direct calculation shows the identity

$$\begin{aligned}&-\hbar\left(\frac{(x-\mathbf{t})(x+\hbar+\mathbf{t})}{x(x+\hbar)}-\frac{(x-\hbar+\mathbf{t})(x-2\hbar-\mathbf{t})}{(x-\hbar)(x-2\hbar)}\right)\\ &\qquad + 2\hbar\mathbf{t}(\hbar+\mathbf{t})\left(\frac{1}{x(x-2\hbar)}-\frac{1}{(x+\hbar)(x-\hbar)}\right) = 0.\end{aligned}$$

Hence the above vanishes. The same argument shows that the coefficient of $\mathsf{u}_j^{-1}\mathsf{u}_k^{-2}$ $(j<k)$ in (A.1) is

$$-\hbar A_{jk} + 2\hbar\mathbf{t}(\hbar+\mathbf{t})B_{kj} = 0.$$

We consider the coefficient of $\mathsf{u}_i^{-1}\mathsf{u}_j^{-1}\mathsf{u}_k^{-1}$ $(i<j<k)$ in (A.1). Define $C_j^{(k,i)}$ by

$$C_j^{(k,i)} = C_j^{(k)}\frac{\bar{w}_j-\bar{w}_i}{\bar{w}_j-\bar{w}_i+\mathbf{t}} = \prod_{l\neq i,j,k}\frac{\bar{w}_j-\bar{w}_l+\mathbf{t}}{\bar{w}_j-\bar{w}_l}$$

so that

$$\mathsf{u}_i^{-1}C_j^{(k)} = C_j^{(k,i)}\frac{\bar{w}_j-\bar{w}_i+\hbar+\mathbf{t}}{\bar{w}_j-\bar{w}_i+\hbar}\mathsf{u}_i^{-1}.$$

Define $C_k^{(j,i)}$ and $C_i^{(j,k)}$ similarly. Then we have

$$\begin{aligned}
&C_i\mathsf{u}_i^{-1}C_j^{(k)}C_k^{(j)}\frac{1}{(\bar{w}_j-\bar{w}_k+\hbar)(\bar{w}_j-\bar{w}_k-\hbar)}\mathsf{u}_j^{-1}\mathsf{u}_k^{-1}\\
&= C_iC_j^{(k,i)}\frac{\bar{w}_j-\bar{w}_i+\hbar+\mathbf{t}}{\bar{w}_j-\bar{w}_i+\hbar}\\
&\qquad\times C_k^{(j,i)}\frac{\bar{w}_k-\bar{w}_i+\hbar+\mathbf{t}}{\bar{w}_k-\bar{w}_i+\hbar}\frac{1}{(\bar{w}_j-\bar{w}_k+\hbar)(\bar{w}_j-\bar{w}_k-\hbar)}\mathsf{u}_i^{-1}\mathsf{u}_j^{-1}\mathsf{u}_k^{-1}\\
&= C_i^{(k,j)}C_j^{(k,i)}C_k^{(j,i)}\frac{\bar{w}_i-\bar{w}_j+\mathbf{t}}{\bar{w}_i-\bar{w}_j}\frac{\bar{w}_i-\bar{w}_k+\mathbf{t}}{\bar{w}_i-\bar{w}_k}\frac{\bar{w}_j-\bar{w}_i+\hbar+\mathbf{t}}{\bar{w}_j-\bar{w}_i+\hbar}\frac{\bar{w}_k-\bar{w}_i+\hbar+\mathbf{t}}{\bar{w}_k-\bar{w}_i+\hbar}\\
&\qquad\times\frac{1}{(\bar{w}_j-\bar{w}_k+\hbar)(\bar{w}_j-\bar{w}_k-\hbar)}\mathsf{u}_i^{-1}\mathsf{u}_j^{-1}\mathsf{u}_k^{-1}.
\end{aligned}$$

Also

$$\begin{aligned}
&C_j^{(k)}C_k^{(j)}\frac{1}{(\bar{w}_j-\bar{w}_k+\hbar)(\bar{w}_j-\bar{w}_k-\hbar)}\mathsf{u}_j^{-1}\mathsf{u}_k^{-1}C_i\mathsf{u}_i^{-1}\\
&= C_j^{(k)}C_k^{(j)}\frac{1}{(\bar{w}_j-\bar{w}_k+\hbar)(\bar{w}_j-\bar{w}_k-\hbar)}\\
&\qquad\times C_i^{(j,k)}\frac{\bar{w}_i-\bar{w}_j+\hbar+\mathbf{t}}{\bar{w}_i-\bar{w}_j+\hbar}\frac{\bar{w}_i-\bar{w}_k+\hbar+\mathbf{t}}{\bar{w}_i-\bar{w}_k+\hbar}\mathsf{u}_i^{-1}\mathsf{u}_j^{-1}\mathsf{u}_k^{-1}\\
&= C_j^{(k,i)}C_k^{(j,i)}C_i^{(j,k)}\frac{1}{(\bar{w}_j-\bar{w}_k+\hbar)(\bar{w}_j-\bar{w}_k-\hbar)}\\
&\qquad\times\frac{\bar{w}_j-\bar{w}_i+\mathbf{t}}{\bar{w}_j-\bar{w}_i}\frac{\bar{w}_k-\bar{w}_i+\mathbf{t}}{\bar{w}_k-\bar{w}_i}\frac{\bar{w}_i-\bar{w}_j+\hbar+\mathbf{t}}{\bar{w}_i-\bar{w}_j+\hbar}\frac{\bar{w}_i-\bar{w}_k+\hbar+\mathbf{t}}{\bar{w}_i-\bar{w}_k+\hbar}\mathsf{u}_i^{-1}\mathsf{u}_j^{-1}\mathsf{u}_k^{-1}.
\end{aligned}$$

Thus we have

$$
\begin{aligned}
&[C_i \mathsf{u}_i^{-1}, C_j^{(k)} C_k^{(j)} \frac{1}{(\bar{w}_j - \bar{w}_k + \hbar)(\bar{w}_j - \bar{w}_k - \hbar)} \mathsf{u}_j^{-1}\mathsf{u}_k^{-1}] \\
&= C_i^{(k,j)} C_j^{(k,i)} C_k^{(j,i)} \frac{1}{(\bar{w}_i - \bar{w}_j)(\bar{w}_i - \bar{w}_k)(\bar{w}_j - \bar{w}_k + \hbar)(\bar{w}_j - \bar{w}_k - \hbar)} \\
&\quad \left(\frac{(\bar{w}_i - \bar{w}_j + \mathbf{t})(\bar{w}_i - \bar{w}_j - \hbar - \mathbf{t})(\bar{w}_i - \bar{w}_k + \mathbf{t})(\bar{w}_i - \bar{w}_k - \hbar - \mathbf{t})}{(\bar{w}_i - \bar{w}_j - \hbar)(\bar{w}_i - \bar{w}_k - \hbar)} \right. \\
&\qquad \left. - \frac{(\bar{w}_i - \bar{w}_j - \mathbf{t})(\bar{w}_i - \bar{w}_j + \hbar + \mathbf{t})(\bar{w}_i - \bar{w}_k - \mathbf{t})(\bar{w}_i - \bar{w}_k + \hbar + \mathbf{t})}{(\bar{w}_i - \bar{w}_j + \hbar)(\bar{w}_i - \bar{w}_k + \hbar)} \right) \mathsf{u}_i^{-1}\mathsf{u}_j^{-1}\mathsf{u}_k^{-1} \\
&= C_i^{(k,j)} C_j^{(k,i)} C_k^{(j,i)} \frac{1}{x_{ij}x_{jk}x_{ki}(x_{ij}+\hbar)(x_{ij}-\hbar)(x_{jk}+\hbar)(x_{jk}-\hbar)(x_{ki}+\hbar)(x_{ki}-\hbar)} \\
&\quad \times x_{jk}\Big((x_{ij}+\hbar)(x_{ij}+\mathbf{t})(x_{ij}-\hbar-\mathbf{t})(x_{ki}-\hbar)(x_{ki}-\mathbf{t})(x_{ki}+\hbar+\mathbf{t}) \\
&\qquad - (x_{ij}-\hbar)(x_{ij}-\mathbf{t})(x_{ij}+\hbar+\mathbf{t})(x_{ki}+\hbar)(x_{ki}+\mathbf{t})(x_{ki}-\hbar-\mathbf{t}) \Big) \mathsf{u}_i^{-1}\mathsf{u}_j^{-1}\mathsf{u}_k^{-1},
\end{aligned}
$$

where we set $x_{ij} = \bar{w}_i - \bar{w}_j$ and so on. Put

$$
\begin{aligned}
A_{ijk} = x_{jk}\Big(&(x_{ij}+\hbar)(x_{ij}+\mathbf{t})(x_{ij}-\hbar-\mathbf{t})(x_{ki}-\hbar)(x_{ki}-\mathbf{t})(x_{ki}+\hbar+\mathbf{t}) \\
&- (x_{ij}-\hbar)(x_{ij}-\mathbf{t})(x_{ij}+\hbar+\mathbf{t})(x_{ki}+\hbar)(x_{ki}+\mathbf{t})(x_{ki}-\hbar-\mathbf{t}) \Big).
\end{aligned}
$$

It is enough to show $A_{ijk} + A_{jki} + A_{kij} = 0$. We have

$$
\begin{aligned}
A_{ijk} &= 2\hbar\mathbf{t}(\hbar+\mathbf{t})x_{jk}(x_{ij}^3 - x_{ki}^3 - (\hbar^2 + \mathbf{t}(\hbar+\mathbf{t}))(x_{ij} - x_{ki})), \\
A_{jki} &= 2\hbar\mathbf{t}(\hbar+\mathbf{t})x_{ki}(x_{jk}^3 - x_{ij}^3 - (\hbar^2 + \mathbf{t}(\hbar+\mathbf{t}))(x_{jk} - x_{ij})), \\
A_{kij} &= 2\hbar\mathbf{t}(\hbar+\mathbf{t})x_{ij}(x_{ki}^3 - x_{jk}^3 - (\hbar^2 + \mathbf{t}(\hbar+\mathbf{t}))(x_{ki} - x_{jk})).
\end{aligned}
$$

A straightforward calculation shows

$$
\begin{aligned}
x_{jk}(x_{ij}^3 - x_{ki}^3) + x_{ki}(x_{jk}^3 - x_{ij}^3) + x_{ij}(x_{ki}^3 - x_{jk}^3) = 0, \\
x_{jk}(x_{ij} - x_{ki}) + x_{ki}(x_{jk} - x_{ij}) + x_{ij}(x_{ki} - x_{jk}) = 0.
\end{aligned}
$$

This completes the proof of (6.1f).

Appendix B

We check that the relation (6.1b') holds for

$$
\begin{gathered}
h(x) = \prod_{k=1}^{l} (1 - (z_k + \hbar)x) \prod_{i=1}^{N} \frac{(1 - (w_i - \mathbf{t})x)(1 - (w_i + \hbar + \mathbf{t})x)}{(1 - w_i x)(1 - (w_i + \hbar)x)}, \\
e_n = E_1[(w+\hbar)^n].
\end{gathered}
$$

We have

$$
\begin{aligned}
&[h(x), e_n] \\
&= \sum_{i=1}^{N} (w_i+\hbar)^n \prod_{j\neq i} \frac{w_i-w_j-\mathbf{t}}{w_i-w_j} \prod_{k=1}^{l}(1-(z_k+\hbar)x) \prod_{j\neq i} \frac{(1-(w_j-\mathbf{t})x)(1-(w_j+\hbar+\mathbf{t})x)}{(1-w_jx)(1-(w_j+\hbar)x)} \\
&\quad \times \left(\frac{(1-(w_i-\mathbf{t})x)(1-(w_i+\hbar+\mathbf{t})x)}{(1-w_ix)(1-(w_i+\hbar)x)} - \frac{(1-(w_i+\hbar-\mathbf{t})x)(1-(w_i+2\hbar+\mathbf{t})x)}{(1-(w_i+\hbar)x)(1-(w_i+2\hbar)x)} \right) \mathsf{u}_i \\
&= \sum_{i=1}^{N} (w_i+\hbar)^n \prod_{j\neq i} \frac{w_i-w_j-\mathbf{t}}{w_i-w_j} \prod_{k=1}^{l}(1-(z_k+\hbar)x) \prod_{j\neq i} \frac{(1-(w_j-\mathbf{t})x)(1-(w_j+\hbar+\mathbf{t})x)}{(1-w_jx)(1-(w_j+\hbar)x)} \\
&\quad \times \frac{2\hbar\mathbf{t}(\hbar+\mathbf{t})x^3}{(1-w_ix)(1-(w_i+\hbar)x)(1-(w_i+2\hbar)x)} \mathsf{u}_i.
\end{aligned}
$$

This shows $[h_0, e_n] = [h_1, e_n] = 0$ and $[h_2, e_n] = -2\hbar e_n$. Then it is enough to prove that the term with positive powers in y of

$$
\begin{aligned}
\text{(B.1)} \quad \left((x^{-1} - y^{-1})^3 - (\hbar^2 + \mathbf{t}(\hbar + \mathbf{t}))(x^{-1} - y^{-1})\right) [h(x), e(y)] \\
- \hbar\mathbf{t}(\hbar + \mathbf{t})(h(x)e(y) + e(y)h(x))
\end{aligned}
$$

vanishes. We have

$$
\begin{aligned}
h(x)e(y) + e(y)h(x) &= \sum_{n\geq 0} y^{n+1} \sum_{i=1}^{N} (w_i + \hbar)^n \\
&\times \prod_{j\neq i} \frac{w_i - w_j - \mathbf{t}}{w_i - w_j} \prod_{k=1}^{l}(1 - (z_k + \hbar)x) \prod_{j\neq i} \frac{(1 - (w_j - \mathbf{t})x)(1 - (w_j + \hbar + \mathbf{t})x)}{(1 - w_jx)(1 - (w_j + \hbar)x)} \\
&\times \frac{2P_i(x)}{(1 - w_ix)(1 - (w_i + \hbar)x)(1 - (w_i + 2\hbar)x)} \mathsf{u}_i,
\end{aligned}
$$

where

$$
\begin{aligned}
P_i(x) = 1 - 3(w_i + \hbar)x + \left(3(w_i + \hbar)^2 - (\hbar^2 + \mathbf{t}(\hbar + \mathbf{t}))\right) x^2 \\
- \left((w_i + \hbar)^3 - (\hbar^2 + \mathbf{t}(\hbar + \mathbf{t}))(w_i + \hbar)\right) x^3.
\end{aligned}
$$

Consider the coefficient of y^{n+1} $(n \geq 0)$ in (B.1). Since we have

$$
\begin{aligned}
(w_i + \hbar)^n - 3(w_i + \hbar)^{n+1}x + 3(w_i + \hbar)^{n+2}x^2 - (w_i + \hbar)^{n+3}x^3 \\
- (\hbar^2 + \mathbf{t}(\hbar + \mathbf{t}))((w_i + \hbar)^n x^2 - (w_i + \hbar)^{n+1}x^3) - (w_i + \hbar)^n P_i(x) = 0,
\end{aligned}
$$

the assertion is proved. The relation (6.1c') can be checked similarly.

References

[AS13] N. Arbesfeld and O. Schiffmann, *A presentation of the deformed $W_{1+\infty}$ algebra*, Symmetries, integrable systems and representations, Springer Proc. Math. Stat., vol. 40, Springer, Heidelberg, 2013, pp. 1–13, DOI 10.1007/978-1-4471-4863-0_1. MR3077678

[AST98] T. Arakawa, T. Suzuki, and A. Tsuchiya, *Degenerate double affine Hecke algebra and conformal field theory*, Topological field theory, primitive forms and related topics (Kyoto, 1996), Progr. Math., vol. 160, Birkhäuser Boston, Boston, MA, 1998, pp. 1–34, DOI 10.1007/978-1-4612-0705-4_1. MR1653020

[BEF16] A. Braverman, P. Etingof, and M. Finkelberg, *Cyclotomic double affine Hecke algebras*, ArXiv e-prints (2016), `arXiv:1611.10216 [math.RT]`.

[BFN16a] A. Braverman, M. Finkelberg, and H. Nakajima, *Towards a mathematical definition of Coulomb branches of 3-dimensional $\mathcal{N} = 4$ gauge theories, II*, ArXiv e-prints (2016), arXiv:1601.03586 [math.RT].

[BFN16b] ———, *Coulomb branches of 3d $\mathcal{N} = 4$ quiver gauge theories and slices in the affine Grassmannian (with appendices by Alexander Braverman, Michael Finkelberg, Joel Kamnitzer, Ryosuke Kodera, Hiraku Nakajima, Ben Webster, and Alex Weekes)*, ArXiv e-prints (2016), arXiv:1604.03625 [math.RT].

[BK06] J. Brundan and A. Kleshchev, *Shifted Yangians and finite W-algebras*, Adv. Math. **200** (2006), no. 1, 136–195, DOI 10.1016/j.aim.2004.11.004. MR2199632

[BK08] ———, *Representations of shifted Yangians and finite W-algebras*, Mem. Amer. Math. Soc. **196** (2008), no. 918, viii+107. MR2456464 (2009i:17020)

[Che05] I. Cherednik, *Double affine Hecke algebras*, London Mathematical Society Lecture Note Series, vol. 319, Cambridge University Press, Cambridge, 2005. MR2133033

[dBHOO97] J. de Boer, K. Hori, H. Ooguri, and Y. Oz, *Mirror symmetry in three-dimensional gauge theories, quivers and D-branes*, Nuclear Phys. B **493** (1997), no. 1-2, 101–147, DOI 10.1016/S0550-3213(97)00125-9. MR1454291

[EG02] P. Etingof and V. Ginzburg, *Symplectic reflection algebras, Calogero-Moser space, and deformed Harish-Chandra homomorphism*, Invent. Math. **147** (2002), no. 2, 243–348, DOI 10.1007/s002220100171. MR1881922

[GKLO05] A. Gerasimov, S. Kharchev, D. Lebedev, and S. Oblezin, *On a class of representations of the Yangian and moduli space of monopoles*, Comm. Math. Phys. **260** (2005), no. 3, 511–525, DOI 10.1007/s00220-005-1417-3. MR2182434

[Kir97] A. A. Kirillov Jr., *Lectures on affine Hecke algebras and Macdonald's conjectures*, Bull. Amer. Math. Soc. (N.S.) **34** (1997), no. 3, 251–292, DOI 10.1090/S0273-0979-97-00727-1. MR1441642

[KWWY14] J. Kamnitzer, B. Webster, A. Weekes, and O. Yacobi, *Yangians and quantizations of slices in the affine Grassmannian*, Algebra Number Theory **8** (2014), no. 4, 857–893, DOI 10.2140/ant.2014.8.857. MR3248988

[Los16] I. Losev, *Deformations of symplectic singularities and Orbit method for semisimple Lie algebras*, ArXiv e-prints (2016), arXiv:1605.00592 [math.RT].

[MO12] D. Maulik and A. Okounkov, *Quantum Groups and Quantum Cohomology*, ArXiv e-prints (2012), arXiv:1211.1287 [math.AG].

[Nak16] H. Nakajima, *Towards a mathematical definition of Coulomb branches of 3-dimensional $\mathcal{N} = 4$ gauge theories, I*, Adv. Theor. Math. Phys. **20** (2016), no. 3, 595–669, DOI 10.4310/ATMP.2016.v20.n3.a4. MR3565863

[Obl07] A. Oblomkov, *Deformed Harish-Chandra homomorphism for the cyclic quiver*, Math. Res. Lett. **14** (2007), no. 3, 359–372, DOI 10.4310/MRL.2007.v14.n3.a1. MR2318640

[Opd00] E. M. Opdam, *Lecture notes on Dunkl operators for real and complex reflection groups*, MSJ Memoirs, vol. 8, Mathematical Society of Japan, Tokyo, 2000. With a preface by Toshio Oshima. MR1805058

[Suz05] T. Suzuki, *Rational and trigonometric degeneration of the double affine Hecke algebra of type A*, Int. Math. Res. Not. **37** (2005), 2249–2262, DOI 10.1155/IMRN.2005.2249. MR2181264

[SV13] O. Schiffmann and E. Vasserot, *Cherednik algebras, W-algebras and the equivariant cohomology of the moduli space of instantons on $\mathbf{A}^2$*, Publ. Math. Inst. Hautes Études Sci. **118** (2013), 213–342, DOI 10.1007/s10240-013-0052-3. MR3150250

[Tsy17] A. Tsymbaliuk, *The affine Yangian of $\mathfrak{gl}_1$ revisited*, Adv. Math. **304** (2017), 583–645, DOI 10.1016/j.aim.2016.08.041. MR3558218

Department of Mathematics, Kyoto University, Kyoto 606-8502, Japan
Email address: rkodera@math.kyoto-u.ac.jp

Research Institute for Mathematical Sciences, Kyoto University, Kyoto 606-8502, Japan
Email address: nakajima@kurims.kyoto-u.ac.jp

Proceedings of Symposia in Pure Mathematics
Volume **98**, 2018
http://dx.doi.org/10.1090/pspum/098/01723

Supersymmetric field theories and geometric Langlands: The other side of the coin

Aswin Balasubramanian and Jörg Teschner

ABSTRACT. This note announces results on the relations between the approach of Beilinson and Drinfeld to the geometric Langlands correspondence based on conformal field theory, the approach of Kapustin and Witten based on $N = 4$ SYM, and the AGT-correspondence. The geometric Langlands correspondence is described as the Nekrasov-Shatashvili limit of a generalisation of the AGT-correspondence in the presence of surface operators. Following the approaches of Kapustin - Witten and Nekrasov - Witten we interpret some aspects of the resulting picture using an effective description in terms of two-dimensional sigma models having Hitchin's moduli spaces as target-manifold.

1. Introduction

Some remarkable connections between supersymmetric gauge theories and conformal field theory (CFT) have been discovered in the last few years. One of the most explicit connections was discovered by Alday, Gaiotto and Tachikawa [**3**], nowadays often referred to as AGT-correspondence. It overlaps with another family of results for which Nekrasov and collaborators have introduced the name BPS/CFT-correspondence, see [**40**] for the first in a series of papers on this subject and references to previous work in this direction. An older development, the relations between the $N = 4$ super-Yang-Mills theory (SYM) and the geometric Langlands correspondence exhibited by Kapustin and Witten [**34**], naturally fits into the emerging picture [**39, 41**]. One may note, however, that the approach of Beilinson and Drinfeld to the geometric Langlands correspondence has close connections to CFT, see [**15**] for a review and further references, which are not obvious in the approach of Kapustin and Witten. Some connections between these subjects have been discussed in [**26, 39, 41, 45**], and the recent work [**1, 35**] indicates that at least a part of these relations admit a further deformation, motivated by supersymmetric gauge and string theory. However, the picture still seems to be incomplete in many respects.

This note, prepared for the proceedings of String-Math 2016, announces results shedding some light on the relations between the approaches to the geometric Langlands correspondence of Beilinson and Drinfeld, Kapustin and Witten, and the AGT-correspondence. We will mostly restrict attention to the cases where the underlying Lie-algebra is $\mathfrak{sl}_2$ in order to keep the length of this note within reasonable

2010 *Mathematics Subject Classification.* Primary 81T60, 22E57, 14D21.

bounds. The results described here are part of a larger project being pursued by the authors together with Ioana Coman-Lohi. A series of publications containing more details, an extended discussion of mathematical aspects, and a discussion of the higher rank cases is in preparation.

In the main text we will freely use several standard definitions and results concerning Hitchin's moduli spaces. A very brief summary is collected in Appendix A in the form of a glossary. If a glossary entry exists for a term, its first occurrence will appear with a superscript as in term$^{g)}$.

2. Review

2.1. What is the geometric Langlands correspondence? The geometric Langlands correspondence is often loosely formulated as a correspondence which assigns $\mathcal{D}$-modules on Bun_G to ${}^LG$-local systems$^{g)}$on a Riemann surface C. ${}^LG$ is the Langlands dual group of a simple complex Lie group G. The ${}^LG$-local systems appearing in this correspondence can be represented by pairs $(\mathcal{E}, \nabla')$ composed of a holomorphic ${}^LG$-bundle $\mathcal{E}$ with a holomorphic connection ∇', or equivalently by the representations ρ of the fundamental group $\pi_1(C)$ defined from the holonomies of $(\mathcal{E}, \nabla')$. We will mostly be interested in the case of *irreducible* ${}^LG$-local systems. The corresponding $\mathcal{D}$-modules on Bun_G can be described more concretely as systems of partial differential equations taking the form of eigenvalue equations $D_i f = E_i f$ for a family of differential operators D_i on Bun_G obtained by quantising the Hamiltonians of Hitchin's integrable system. A more ambitious version of the geometric Langslands correspondence has been formulated in [**6**] in which it becomes necessary to extend it to certain classes of reducible local systems.

Some of the original approaches to the geometric Langlands correspondence start from the cases where the ${}^LG$-local systems are *opers*$^{g)}$, pairs $(\mathcal{E}, \nabla')$ in which ∇' is gauge-equivalent to a certain standard form. The space of opers forms a Lagrangian subspace in the moduli space of all local systems. The CFT-based approach of Beilinson and Drinfeld constructs for each oper an object in the category of $\mathcal{D}$-modules on Bun_G as conformal blocks of the affine Lie algebra $\hat{\mathfrak{g}}_k$ at the critical level $k = -h^\vee$. The Ward-identities characterising the conformal blocks equip the sheaves of conformal blocks with a $\mathcal{D}$-module structure. The universal enveloping algebra $\mathcal{U}(\hat{\mathfrak{g}}_k)$ has a large center at $k = -h^\vee$, isomorphic to the space of ${}^L\mathfrak{g}$-opers on the formal disc [**13**]. This can be used to show that the $\mathcal{D}$-module structure coming from the Ward identities can be described by the system of eigenvalue equations $D_i f = E_i f$ for the quantised Hitchin Hamiltonians, with eigenvalues E_i parameterising the choice of opers [**15**].

There exists an extension of the Beilinson-Drinfeld construction of the geometric Langlands correspondence described in [**15**, Section 9.6] from the case of opers to general irreducible local systems. It is based on the fact that such local systems are always gauge-equivalent to opers with certain extra singularities [**5**]. The construction of Beilinson and Drinfeld associates to such opers conformal blocks of $\hat{\mathfrak{g}}_{-h^\vee}$ with certain degenerate representations induced from the finite-dimensional representations of $\mathfrak{g}$ inserted at the extra singularities. We may in this sense regard the geometric Langlands correspondence for general irreducible local systems as an extension of the correspondence that exists for ordinary, non-singular opers. This point of view will turn out to be natural from the perspective we will propose in this paper. Let us also remark that the construction of Beilinson-Drinfeld plays an

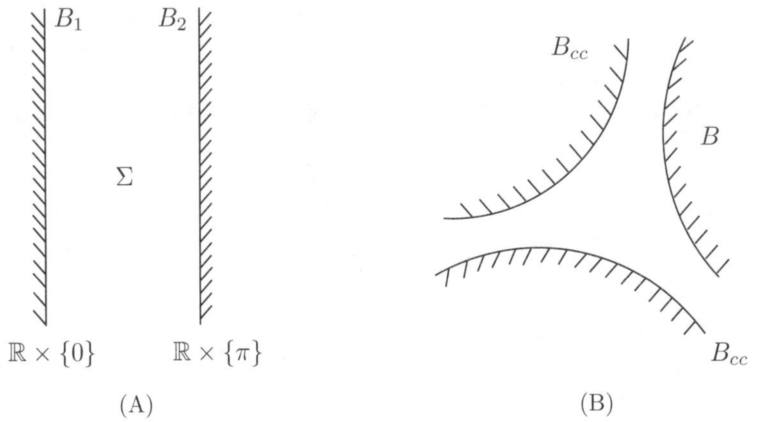

FIGURE 1. The 2d TQFT setup to compute (A) $\mathrm{Hom}_{\mathcal{M}_H}(B_1, B_2)$ and (B) the left action of the algebra $\mathcal{A}$ on $\mathrm{Hom}_{\mathcal{M}_H}(B_{cc}, B)$.

important role in the outline given in [**27**] for a proof of the strengthened geometric Langlands conjecture formulated in [**6**].

For future reference let us note that the Beilinson-Drinfeld construction describes the $\mathcal{D}$-modules appearing in the geometric Langlands correspondence as spaces of conformal blocks which, naturally being fibered over Bun_G on the one hand, are furthermore getting fibered over the spaces of irreducible ${}^LG$-local systems on the other hand.

2.2. Geometric Langlands correspondence - approach of Kapustin-Witten. The variant of the geometric Langlands correspondence proposed by Kapustin and Witten [**34**] is based on the consideration of $N = 4$ SYM theory with gauge group G_c, a compact real form of G, on four-manifolds of the form $\Sigma \times C$, where C is a Riemann surface. Compactification on C allows one to represent the topologically twisted version of $N = 4$ SYM effectively by a topologically twisted $2d$ sigma-model with target being the Hitchin moduli space$^{(g)}$ $M_H(G)$ on Σ. The complete integrability of the Hitchin moduli space, as is manifest in the description of $M_H(G)$ as a torus fibration, allows one to describe the consequences of the S-duality of $N = 4$ SYM theory as the SYZ mirror symmetry relating the $2d$ sigma-models with target $M_H(G)$ and $M_H({}^LG)$, respectively.

In order to relate this to the geometric Langlands correspondence, Kapustin and Witten consider the cases when $\Sigma = R \times I$, $I = [0, \pi]$. Supersymmetric boundary condition of $N = 4$ SYM theory will upon compactification on C define boundary conditions of the $\mathcal{N} = (4, 4)$ sigma model with target $\mathcal{M}_H(G)$ on Σ. Let $\mathfrak{B}$ be the category having as objects boundary conditions B called branes preserving the maximal number of supersymmetries, with morphisms being the spaces of "open string" states $\mathrm{Hom}_{\mathcal{M}_H(G)}(\mathsf{B}_1, \mathsf{B}_2)$ of the sigma model on the strip $\mathbb{R} \times I$, having associated boundary conditions B_1 and B_2 to the boundaries $\mathbb{R} \times \{0\}$ and $\mathbb{R} \times \{\pi\}$, respectively. The set-up is illustrated in Figure 1(A).

A distinguished role is played by the so-called canonical coisotropic brane $\mathsf{B}_{\rm cc}$ [**34, 41**]. The vector space $\mathcal{A}_{\rm cc} = \mathrm{Hom}_{\mathcal{M}_H(G)}(\mathsf{B}_{\rm cc}, \mathsf{B}_{\rm cc})$ has a natural algebra structure with product defined by "joining open strings". The spaces $\mathcal{H}(\mathsf{B}) =$

$\mathrm{Hom}_{\mathcal{M}_H(G)}(\mathsf{B}_{\mathrm{cc}}, \mathsf{B})$ are left modules over the algebra $\mathcal{A}_{\mathrm{cc}}$ with action defined by "joining open strings" from $\mathcal{A}_{\mathrm{cc}}$ on the left boundary of the strip I, as depicted in Figure 1(B). Kapustin and Witten argue that the algebra $\mathcal{A}_{\mathrm{cc}}$ contains the algebra of global differential operators on Bun_G. It follows that the spaces $\mathcal{H}(\mathsf{B})$ represent $\mathcal{D}$-modules on Bun_G.

The reduction of Wilson- and 't Hooft line operators with support on $\mathbb{R} \times \{x\} \times P$, $x \in I$, $P \in C$, to the two dimensional TQFT defines natural functors on the category of branes, inducing modifications of the spaces $\mathcal{H}(\mathsf{B})$. The functors defined in this way are identified in [**34**] with the Hecke functors in the geometric Langlands correspondence. For some branes B one may represent for each fixed $P \in C$ the resulting modification as the tensor product of $\mathcal{H}(\mathsf{B})$ with a finite-dimensional representation V of ${}^LG$. One says that the brane B satisfies the Hecke eigenvalue property if the family of modifications obtained by varying the point $P \in C$ glues into a local system.

A family of branes F_μ is identified in [**34**] having this property. The branes F_μ are supported on fibers of the Hitchin's torus fibration. S-duality of $N = 4$ SYM gets represented within the sigma model with target $\mathcal{M}_H(G)$ as a variant of SYZ mirror symmetry, relating the branes F_μ to branes in the dual sigma model with target $\mathcal{M}_H({}^LG)$ represented by skyscraper sheaves $\check{\mathsf{F}}_\mu$ having pointlike support at $\mu \in \mathcal{M}_H({}^LG)$.

2.3. AGT-correspondence - approach of Nekrasov-Witten. Alday, Gaiotto and Tachikawa discovered a relation between the instanton partition functions of certain $N = 2$ supersymmetric gauge theories and conformal blocks of the Virasoro algebra [**3**]. This discovery has stimulated a lot of work leading in particular to various generalisations of such relations. In an attempt to explain the relations discovered in [**3**], Nekrasov and Witten considered four-dimensional $N = 2$ supersymmetric gauge theories of class $\mathcal{S}$ obtained from the maximally supersymmetric six-dimensional QFT on manifolds of the form $\mathcal{M}^4 \times C$ by compactification on the Riemann surface C. For the case associated to the Lie algebra $\mathfrak{g} = \mathfrak{sl}_2$ one has weakly coupled Lagrangian descriptions of the resulting theory associated to the choice of a pants decomposition σ of C. For four-manifolds $\mathcal{M}^4$ which can be represented as a fibered product locally of the form $\mathbb{R} \times I \times S^1 \times S^1$ it is argued in [**41**] that (i) an Ω-deformation with parameters ϵ_1, ϵ_2 can be defined, and (ii) an effective representation is obtained by compactification on $S^1 \times S^1$ in terms of a sigma-model with target $\mathcal{M}_H(G)$ on $R \times I$. The coupling parameter of this sigma model is ϵ_1/ϵ_2.

The end points of the interval I in the representation $\mathcal{M}^4 \simeq \mathbb{R} \times I \times S^1 \times S^1$ correspond to points where $\mathcal{M}^4$ is perfectly regular. One must therefore have distinguished boundary condition in the sigma-model with target $\mathcal{M}_H(G)$ on $R \times I$ describing the compactification of a class $\mathcal{S}$ theory on $\mathcal{M}^4$. When the compactification yields a sigma model with target $\mathcal{M}_H(G)$, it is argued in [**41**] that the corresponding boundary conditions are described by a variant B_{cc} of the canonical coisotropic brane at $\mathbb{R} \times \{0\}$, and a new type of brane called the "brane of opers", here denoted by B_{op}, respectively.[1] The brane B_{op} is the mirror dual of B_{cc}, and it is proposed in [**41**] that the brane B_{op} is a Lagrangian brane supported on the variety of opers within $\mathcal{M}_H(G)$.

[1]The branes denoted B_{cc} in this context are similar but not identical with the brane considered in [**34**]. The paper [**41**] used the notation $B_{N'}$ for the brane denoted B_{op} here

In [**41**] it is furthermore proposed that the space $\mathcal{H} = \mathrm{Hom}_{\mathcal{M}_H(G)}(\mathsf{B}_{\mathrm{cc}}, \mathsf{B}_{\mathrm{op}})$ can be identified with the space of Virasoro conformal blocks. In order to motivate this identification, Nekrasov and Witten note that the algebra $\mathcal{A}^{\hbar}_{\mathrm{cc}}(G) = \mathrm{Hom}_{\mathcal{M}_H(G)}(\mathsf{B}_{\mathrm{cc}}, \mathsf{B}_{\mathrm{cc}})$ with $\hbar = \epsilon_1/\epsilon_2$ is isomorphic to the algebra of Verlinde line operators acting on the space of Virasoro conformal blocks. Mirror symmetry produces a dual description of $\mathcal{H}(G) \simeq \mathcal{H}({}^{\mathrm{L}}G)$ as the space $\mathrm{Hom}_{\mathcal{M}_H({}^LG)}(\mathsf{B}'_{\mathrm{op}}, \mathsf{B}'_{\mathrm{cc}})$, with $\mathsf{B}'_{\mathrm{op}}$ and $\mathsf{B}'_{\mathrm{cc}}$ being close relatives of B_{op} and B_{cc}, respectively, with modified SUSY invariance properties. In the dual representation one has an obvious right action of the algebra $\check{\mathcal{A}}^{1/\hbar}_{\mathrm{cc}}({}^LG) = \mathrm{Hom}_{\mathcal{M}_H({}^LG)}(\mathsf{B}'_{\mathrm{cc}}, \mathsf{B}'_{\mathrm{cc}})$ with action defined by "joining open strings" on the right boundary of the strip I. The existence of (almost) commuting actions of $\mathcal{A}^{\hbar}_{\mathrm{cc}}(G)$ and $\check{\mathcal{A}}^{1/\hbar}_{\mathrm{cc}}({}^LG)$ is a characteristic feature of the space of Virasoro conformal blocks.

2.4. The other way around. It is no accident that the work of Nekrasov and Witten [**41**] has many elements in common with the approach Kapustin and Witten [**34**]. A common root can be found in the fact that both the class $\mathcal{S}$-theories and $N = 4$ SYM [**47**] can be obtained as compactifications of the six-dimensional $(2,0)$-theory on six-manifolds $\mathcal{M}^6 = \mathcal{M}^4 \times C$, where C is a Riemann surface, and $\mathcal{M}^4$ is a four-manifold locally represented as a circle fibration locally of the form $\mathbb{R} \times I \times S^1 \times S^1$. Compactification on C yields class $\mathcal{S}$-theories [**23**], while compactification on $S^1 \times S^1$ yields $N = 4$ SYM on $\mathbb{R} \times I \times C$, the set-up considered in [**34**] as was further discussed in [**47**].

One should note, however, that different topological twists are used in the two compactifications, making the comparison of the results somewhat subtle. This fact can nevertheless be used to relate supersymmetric boundary conditions in the 2d sigma model with target $\mathcal{M}_H$ arising from compactification of class $\mathcal{S}$-theories to boundary conditions in $N = 4$ SYM on C. These boundary conditions have been classified in the work of Gaiotto and Witten [**25**]. In this way, one may establish a relationship between the canonical coisotropic brane and pure Neumann boundary conditions in $N = 4$ SYM [**26, 34**]. Exchanging the two circles in $S^1 \times S^1$ gets related to the S-duality of $N = 4$ SYM which implies relations between its boundary conditions studied in [**24**]. This led [**26**] to relate the brane B_{op}, the mirror dual of the canonical coisotropic brane in [**41**], to the boundary condition descending from the so-called Nahm pole boundary conditions in $N = 4$ SYM, as will briefly be discussed in Subsection 3.2 below.

2.5. Other approaches. As noted above, one needs to use different twists in the two reductions from six to four dimensions considered above. In order to get the set-up studied in [**41**], for example, one needs to twist the $(2,0)$ theory on $\mathcal{M}^4 \times C$ in such a way that it becomes topological on $\mathcal{M}^4$. A different twist is obviously needed to get the topologically twisted $N = 4$ SYM on $\mathbb{R} \times I \times C$ studied in [**34**] from the six-dimensional $(2,0)$ theory.

In order to describe the dimensional reduction of topologically twisted $N = 4$ SYM on $\mathbb{R} \times I \times C$ one may find it natural to consider boundary conditions that are purely topological, not depending on the complex structure on C. This point of view motivated Ben-Zvi and Nadler [**7**] to propose the Betti geometric Langlands conjecture as a purely topological variant of the geometric Langlands correspondence formulated in [**6**] that is naturally adapted to the four-dimensional TQFT's studied in [**34**].

Yet another approach towards understanding the AGT-correspondence was proposed by Cordova and Jafferis in [**10**]. It starts from six-manifolds of the form $\mathcal{M}^4 \times C$, with $\mathcal{M}^4$ being a squashed four-sphere, represented as three-sphere fibration over an interval I_0, locally $\mathcal{M}^4 \sim I_0 \times S^3_{\epsilon_1\epsilon_2}$ with ϵ_1, ϵ_2 now being the squashing parameters. Considering partial topological twisting on C, it is argued that the consequence of Weyl invariance of the $(2,0)$ theory with the twist on C is the fact that partition functions depend neither on the overall radius of $\mathcal{M}^4$ nor on the volume of C. The effective description at small volume of C will be given by the partition functions of class S theories on (squashed) four-spheres studied in the work of Pestun [**43**] and many subsequent publications. It is argued that an equivalent description in terms of a CFT on C is obtained by first compactifying on the three spheres $S^3_{\epsilon_1\epsilon_2}$ appearing in the representation $\mathcal{M}^4 \sim I_0 \times S^3_{\epsilon_1\epsilon_2}$. Using five-dimensional SYM theory as an intermediate step, Cordova and Jafferis found an effective description in terms of complex Chern-Simons theory on $I_0 \times C$. The boundary conditions at the ends of C are determined by the Nahm pole boundary conditions studied in [**24–26**]. A generalised version of the usual correspondence between Chern-Simons theory and two-dimensional conformal field theory can then be invoked, with Nahm pole boundary conditions translating into the constraints reducing the WZNW models to the conformal Toda theories.

3. Defects of co-dimension two and surface operators

M theory suggests a natural extension of the set-up considered above. Viewing the six-dimensional $(2,0)$ theory as the effective theory on a stack of M5-branes in M-theory leads to a natural extension of the set-up by an additional stack of M5'-branes sharing part of the support with the original M5-branes, as indicated in the following table:

Brane	0	1	2	3	4	5	6	7	8	9	10
M5	x	x	x	x			x				x
M5'	x	x					x		x	x	x

Depending on the type of compactification considered one gets surface operators in class $\mathcal{S}$ theories, or line operators in $N = 4$ SYM theory. This will be the first type of modification turning out to be crucial for our story.

3.1. Surface operators in class $\mathcal{S}$ theories. By now there exists a fair amount of evidence that an effective description of a system of M5-M5'-branes obtained by compactification on C is provided by theories of class $\mathcal{S}$ modified by the presence of a certain type of surface operator. The relevant surface operators can be described by prescribing a certain type of singular behaviour of the gauge fields along a two-dimensional submanifold $\mathcal{M}^2 \subset \mathcal{M}^4$ characterised by a set of parameters $x = (x_1, \dots, x_{3g-3+n})$ if C has genus g and n punctures. A generalisation $\mathcal{Z}^{\rm inst}_\sigma(a; x, \tau; \epsilon_1, \epsilon_2)$ of the instanton partition functions can be defined in the presence of such surface operators, carrying an additional dependence on the parameters x on top of the dependencies on the scalar zero mode values a, the UV gauge couplings associated to complex structure moduli τ of C, and the parameters ϵ_1, ϵ_2 associated to the Omega-deformation. The subscript σ refers to the pants decomposition determining the Lagrangian representation being used. Explicit calculations in [**4, 8, 36–38**] gave evidence for a generalisation of the AGT-correspondence, identifying $\mathcal{Z}^{\rm inst}_\sigma(a; x, \tau; \epsilon_1, \epsilon_2)$ with the chiral partition functions

$\mathcal{Z}_\sigma^{\rm WZW}(a;x,\tau;k)$ associated to conformal blocks of the affine Lie algebras $\hat{\mathfrak{g}}_k$ at level $k=-h^\vee-\epsilon_2/\epsilon_1$ defined by the gluing construction. The parameters x get identified with coordinates on the moduli space Bun_G of holomorphic G-bundles on C under this correspondence. A proof of this generalisation of the AGT-correspondence will follow from the result announced in [**40**] that $\mathcal{Z}_\sigma^{\rm inst}$ satisfies the KZB equations.[2]

Our goal in this section will be to generalise the approach of [**26, 41**] to theories of class $\mathcal{S}$ in the presence of surface operators of co-dimension two. We will propose an effective description in terms of the two-dimensional sigma model with target $\mathcal{M}_H(G)$ in which the brane of opers $\mathsf{B}_{\rm op}$ gets replaced by a family of Lagrangian branes $\mathsf{L}_x^{(2)}$ supported on the fibers of Hitchin's second fibration over bundles $\mathcal{E}_x$ labelled by coordinates x for Bun_G. The relevance of the branes $\mathsf{L}_x^{(2)}$ in this context and the relation to conformal blocks has first been proposed by E. Frenkel in [**16**], as has been pointed out to one of us (J.T.) in 2012. Our goal in the following will be to offer additional support for this proposal.

3.2. Nahm pole boundary conditions. Gaiotto and Witten have classified 1/2 BPS boundary conditions of $N=4$ SYM in [**25**] using the data (ρ,H,T), where $\rho:\mathfrak{sl}_2\to\mathfrak{g}$ is an embedding of $\mathfrak{sl}_2$ into the Lie algebra $\mathfrak{g}$ of the gauge group G_c, H is a subgroup of the commutant in G_c of the image of ρ, and T is a three-dimensional SCFT with $N=4$ supersymmetry and at least H global symmetry. We will only need two of the simplest of these boundary conditions. In the following we will first briefly review the so-called Nahm pole boundary condition studied in [**26**] which is associated to a triple $(\rho,\mathrm{Id},T_\emptyset)$, where ρ is a *principal* $\mathfrak{sl}_2$-embedding, and $T_\emptyset$ stands for the trivial three-dimensional SCFT. We will then discuss the even simpler case where ρ is replaced by the trivial embedding mapping $\mathfrak{sl}_2$ to $0\in\mathfrak{g}$, which will be of particular interest for us.

It is for our purposes sufficient to describe the Nahm pole boundary conditions for the solutions of the BPS-equations [**34**] characterising field configuration in $N=4$ SYM preserving certain supersymmetries. Restricting attention to solutions to the BPS-equations on $\mathbb{R}\times\mathbb{R}_+\times C$ which are invariant under translations along $\mathbb{R}$, one gets a system of differential equations of the form

$$[\mathcal{D}_z\,,\mathcal{D}_{\bar z}]=0\,,\quad[\mathcal{D}_y\,,\mathcal{D}_z]=0\,,\quad[\mathcal{D}_y\,,\mathcal{D}_{\bar z}]=0\,,\tag{3.1a}$$

$$\sum_{i=1}^3[\mathcal{D}_i\,,\mathcal{D}_i^\dagger]=0\,.,\tag{3.1b}$$

where the notations $z=x_2+ix_3$ and $y=x_1$ have been used, and the differential operators $\mathcal{D}_i$ are of the form[3]

$$\begin{aligned}\mathcal{D}_z&=\zeta\partial_z+\mathcal{A}_z\,, & \mathcal{A}_z&=\zeta A_z+\phi_z\,,\\ \mathcal{D}_{\bar z}&=\partial_{\bar z}+\mathcal{A}_{\bar z}\,, & \mathcal{A}_{\bar z}&=A_{\bar z}+\zeta\phi_{\bar z}\,,\end{aligned}\qquad \mathcal{D}_y=\partial_y+A_y-i\phi_y\,.\tag{3.2}$$

The parameter ζ determines the supersymmetries that are preserved. It is proposed in [**26**] that the space of solutions to (3.1) modulo compact gauge transformations is isomorphic to the moduli space of the solutions to the "F-term" equations (3.1a) modulo complex gauge transformations. Equations $[\mathcal{D}_z,\mathcal{D}_{\bar z}]=0$ determine a flat

[2] see also [**38**] for an earlier result in this direction.

[3] Our conventions differ slightly from [**26**].

complex connection on C at each fixed y. The remaining equations in (3.1a) imply that the y-dependence of this flat connection is represented by complex gauge transformations.

Boundary conditions of Nahm pole type are defined in [**26**] by demanding that the solutions to (3.1) have a singular behaviour of the form

$$(3.3)\quad \mathcal{A}_z \underset{y\to 0}{\sim} \mathfrak{t}_- y^{-1} + \mathcal{O}(y^0), \qquad \mathcal{A}_{\bar{z}} \underset{y\to 0}{\sim} \mathcal{O}(y^0), \qquad \mathcal{A}_1 \underset{y\to 0}{\sim} \mathfrak{t}_3\, y^{-1} + \mathcal{O}(y^0),$$

with $\mathfrak{t}_+ = \mathfrak{t}_1 + i\mathfrak{t}_2$, and $\mathfrak{t}_i$, $i = 1, 2, 3$, being the generators of a principal $\mathfrak{sl}_2$ subalgebra of $\mathfrak{g}$. By a gauge transformation we may always set $\mathcal{A}_{\bar{z}}$ to zero, allowing us to represent the flat connection on C we get at each y as a local system $(\mathcal{E}_y, \nabla'_y)$ consisting of a holomorphic bundle and a holomorphic connection $\nabla'_y = dz(\partial_z + \mathcal{A}_z(z; y))$. In the case $\mathfrak{g} = \mathfrak{sl}_2$, we may reformulate the first condition in (3.3) as the condition that there exists a basis of sections $s = \{s_1, s_2\}$ with respect to which $\mathcal{A}$ has the form $\mathcal{A} = g\tilde{\mathcal{A}}g^{-1} + gdg^{-1}$, with

$$(3.4)\qquad \tilde{\mathcal{A}}_z \underset{y\to 0}{\sim} \begin{pmatrix} 0 & t \\ 1 & 0 \end{pmatrix} + \mathcal{O}(y^1), \qquad g \underset{y\to 0}{\sim} \begin{pmatrix} y^{1/2} & 0 \\ 0 & y^{-1/2} \end{pmatrix} + \mathcal{O}(y^0).$$

Horizontal sections $(d + \mathcal{A})s = 0$ will then have a first component s_1 vanishing as $y^{1/2}$. As explained in [**26**], this implies that the local system $\lim_{y\to 0}(\tilde{\mathcal{E}}_y, \tilde{\nabla}'_y)$ on C is an oper.

The Nahm pole boundary condition has the feature that it breaks G_c maximally since the commutant of the principal $\mathfrak{sl}_2$-embedding is trivial. At the opposite extreme, associated to the trivial $\mathfrak{sl}_2$-embedding, one gets a similar boundary condition associated to a triple $(0, \mathrm{Id}, T_\emptyset)$ by fixing the boundary value of the gauge field $\mathcal{A}_{\bar{z}}$.

In the reduction to two dimensions having fixed $\mathcal{A}_{\bar{z}}$ at the boundary of I implies having fixed a holomorphic bundle on C, leaving the $(1,0)$ part of the complex gauge field unconstrained. The moduli space $\mathcal{M}_{dR}(G)$ of pairs $(\mathcal{E}, \nabla'_\zeta)$ is diffeomorphic to the Hitchin moduli space $\mathcal{M}_H(G)$ via the non-abelian Hodge (NAH) correspondence[(g)]. Fixing $\mathcal{E}$ therefore defines a submanifold in $\mathcal{M}_H(G)$ which is Lagrangian with respect to the holomorphic symplectic form Ω_ζ, and holomorphic w.r.t. to the complex structure I_ζ. For $\zeta = i$ one has $\Omega_\zeta = \Omega_J$, $I_\zeta = J$, leading to the identification of the brane coming from the reduction of the zero Nahm pole boundary condition as an (A,B,A)-brane in the A-model with the symplectic structure ω_I used in [**41**].

We are proposing that the zero Nahm pole boundary condition represents the presence of a surface operator of co-dimension two. Indeed, as was argued in [**17**], the presence of a co-dimension two surface operator naturally introduces additional background data which can be geometrically represented as the choice of a holomorphic bundle on C.

3.3. Relation to conformal blocks. The Lagrangian submanifolds $\mathsf{L}^{(2)}_{\mathcal{E},\zeta}$ of $\mathcal{M}_H(G)$ defined by considering pairs $(\mathcal{E}, \nabla')$ with fixed $\mathcal{E}$ are sometimes called the fibers of Hitchin's second fibration. Picking a system x of coordinates for Bun_G we will use the notation $\mathsf{L}^{(2)}_x$ for $\mathsf{L}^{(2)}_{\mathcal{E}_x,i}$ with $\mathcal{E}_x$ being a bundle representing the point in Bun_G specified by the coordinates x.

The results on the generalisation of the AGT-correspondence in the presence of the surface operators of co-dimension two mentioned above suggest that the space

$\mathcal{H}_x^{(2)} = \mathrm{Hom}_{\mathcal{M}_H(G)}(\mathcal{B}_{\mathrm{cc}}, \mathsf{L}_x^{(2)})$ can be identified with the space of conformal blocks of the affine Lie algebra $\hat{\mathfrak{g}}_k$ at level $k = -h^\vee - \frac{\epsilon_2}{\epsilon_1}$ on C.

We need to note, however, that finding the proper definition of both $\mathcal{H}_x^{(2)}$ and the relevant spaces of conformal blocks is nontrivial in the infinite-dimensional situation at hand. The usual algebraic definition of conformal blocks defines a space $\mathcal{CB}_{\mathrm{al}}(C, \hat{\mathfrak{g}}_k, \mathcal{E}_x)$ that is too large to be relevant for us. We'll need to consider a subspace denoted $\mathcal{CB}_{\mathrm{te}}(C, \hat{\mathfrak{g}}_k, \mathcal{E}_x)$ of "tempered" conformal blocks. The relation between these two spaces is in some respects similar to the relation between spaces $\mathcal{K}_{\mathrm{al}}$ of *formal* power series $\sum_{n\in\mathbb{Z}} a_n z^n$ to the spaces $\mathcal{K}_{\mathrm{te}}$ of tempered distributions on the unit circle. The latter can be represented by Fourier series $\sum_{n\in\mathbb{Z}} a_n e^{in\sigma}$ in the distributional sense, leading to an embedding of $\mathcal{K}_{\mathrm{te}}$ into $\mathcal{K}_{\mathrm{al}}$. However, being tempered imposes growth conditions on the coefficients a_n, making $\mathcal{K}_{\mathrm{te}}$ strictly smaller than $\mathcal{K}_{\mathrm{al}}$.

An algebraic counterpart for $\mathcal{H}_x^{(2)}$, here denoted $\mathcal{H}_{\mathrm{al},x}^{(2)}$, was proposed in the work [**16**] of E. Frenkel, where it was proven that $\mathcal{H}_{\mathrm{al},x}^{(2)} \simeq \mathcal{CB}_{\mathrm{al}}(C, \hat{\mathfrak{g}}_k, \mathcal{E}_x)$. We will argue that a different definition for $\mathcal{H}_x^{(2)}$ is appropriate in this context, leading to an isomorphism $\mathcal{H}_x^{(2)} \simeq \mathcal{CB}_{\mathrm{te}}(C, \hat{\mathfrak{g}}_k, \mathcal{E}_x)$ with a subspace $\mathcal{CB}_{\mathrm{te}}(C, \hat{\mathfrak{g}}_k, \mathcal{E}_x)$ of "tempered" conformal blocks within $\mathcal{CB}_{\mathrm{al}}(C, \hat{\mathfrak{g}}_k, \mathcal{E}_x)$. We can not discuss the definition of $\mathcal{CB}_{\mathrm{te}}(C, \hat{\mathfrak{g}}_k, \mathcal{E}_x)$ fully in this short note, we plan to return to this point elsewhere. Instead we will in the following discuss evidence for $\mathcal{H}_x^{(2)} \simeq \mathcal{CB}_{\mathrm{te}}(C, \hat{\mathfrak{g}}_k, \mathcal{E}_x)$ coming from the $2d$ sigma model.

3.3.1. *Module structures.* To begin with, let us note that $\mathcal{H}_x^{(2)}$ has *two* natural module structures coming from the vertex operators associated to the elements of the algebra $\mathcal{A}_{\mathrm{cc}}^{\hbar} = \mathrm{Hom}_{\mathcal{M}_G(C)}(\mathsf{B}_{\mathrm{cc}}, \mathsf{B}_{cc})$. For the case of interest one should represent $\mathcal{M}_H(G)$ as moduli space of flat complex connections on C using the NAH correspondence, allowing us to consider *two* algebraic structures coming from the representation of $\mathcal{M}_H(G)$ as character variety$^{(g)}$ $\mathcal{M}_B(G)$ and as moduli space $\mathcal{M}_{dR}(G)$ of pairs $(\mathcal{E}, \nabla')$, respectively. Different algebraic structures determine different sets of basic field variables to be used in the definition of the vertex operators in $\mathcal{A}_{\mathrm{cc}}$. One may accordingly distinguish $\mathcal{A}_B^{\hbar} = \mathrm{Hom}_{\mathcal{M}_B(G)}(\mathsf{B}_{\mathrm{cc}}, \mathsf{B}_{cc})$ and $\mathcal{A}_{dR}^{\hbar} = \mathrm{Hom}_{\mathcal{M}_{dR}(G)}(\mathsf{B}_{\mathrm{cc}}, \mathsf{B}_{cc})$, where $\hbar = \epsilon_1/\epsilon_2$.

On the one hand it was argued in [**41**] that the algebra $\mathcal{A}_B^{\hbar}$ can be identified with the quantised algebra of functions on $\mathcal{M}_B(G)$. The arguments in [**33**] may, on the other hand, be applied to the situation studied in [**41**] leading to the conclusion that $\mathcal{A}_{dR}^{\hbar} \simeq \mathcal{D}_{\hbar}$, the quantised algebra of functions on $\mathcal{M}_{dR}(G)$, which may be identified with the algebra of differential operators on a certain line bundle $\mathcal{L}^{-h^\vee - \epsilon_2/\epsilon_1}$ over Bun_G.

It well-known that the space $\mathcal{CB}_{\mathrm{al}}(C, \hat{\mathfrak{g}}_k, \mathcal{E}_x)$ of conformal blocks of the affine Lie algebra $\hat{\mathfrak{g}}_k$ at level $k = -h^\vee - \frac{\epsilon_2}{\epsilon_1}$ has a module structure with respect to $\mathcal{A}_{dR}^{\hbar}$. The $\mathcal{A}_{dR}^{\hbar}$-module structure is a direct consequence of the defining Ward identities.

A $\mathcal{A}_B^{\hbar}$-module structure can defined on spaces of conformal blocks by using degenerate representations of $\hat{\mathfrak{g}}_k$ to define analogs of the Verlinde line operators [**2, 11**] in this case. It will be shown elsewhere that the algebra of Verlinde line operators on affine Lie algebra conformal blocks may indeed be identified with $\mathcal{A}_B^{\hbar}$. The known definitions work with conformal blocks constructed using the gluing construction. Such conformal have particularly nice properties, one of which being that the canonical connection may be integrated over all of $\mathrm{Bun}_G^{\mathrm{vs}}$, the subset of Bun_G containing the "very stable" bundles not admitting a nilpotent Higgs field which is not the case for generic elements of $\mathcal{CB}_{\rm al}(C,\hat{\mathfrak{g}}_k,\mathcal{E}_x)$. This suggests that the $\mathcal{A}_B^\hbar$-module structure can only be defined on suitable subspaces $\mathcal{CB}_{\rm te}(C,\hat{\mathfrak{g}}_k,\mathcal{E}_x)$ of "tempered" conformal blocks.

The existence of two module structures on $\mathcal{H}_x^{(2)}$ can be regarded as a first piece of evidence for the conjectured isomorphism $\mathcal{H}_x^{(2)} \simeq \mathcal{CB}_{\rm te}(C,\hat{\mathfrak{g}}_k,\mathcal{E}_x)$. Further evidence will be given below.

3.3.2. *Relation to $\bar{\partial}$-cohomology.* We note that the space $\mathrm{Hom}_{\mathcal{M}_H(G)}(\mathsf{B}_{\rm cc},\mathsf{B}_{\rm op})$ has a realisation in the A-model with symplectic structure $\omega=\omega_I$. The boundary conditions on the strip are given as the canonical coisotropic A-branes $\mathsf{B}_{\rm cc}$ and the Lagrangian A-brane $\mathsf{B}_{\rm op}$. The Chan-Paton curvature F is given by the symplectic form ω_J on Hitchin moduli space. It follows that the complex structure determined by the canonical coisotropic A-brane $\mathsf{B}_{\rm cc}$ is $\omega^{-1}F=K$. It is argued in [**41**] that in this case the space $\mathrm{Hom}_{\mathcal{M}_H(G)}(\mathsf{B}_{\rm cc},\mathsf{B}_{\rm op})$ can be identified with the space of holomorphic sections of the line bundle $K_{\rm op}^{1/2}$ on the subspace $\mathrm{Op}_{\mathfrak{sl}_2}(C)$ of opers in $\mathcal{M}_{dR}(G)$. Indeed, the space $\mathrm{Hom}_{\mathcal{M}_H(G)}(\mathsf{B}_{\rm cc},\mathsf{B}_{\rm op})$ will have a realisation as space of functions of the zero modes of the A-model on the strip. We have Neumann-type boundary conditions for the sigma model fields representing coordinates on $\mathrm{Op}_{\mathfrak{sl}_2}(C)$ on both ends of the strip, while the remaining fields have Dirichlet-type boundary conditions on one end only. The zero modes of the A-model may therefore be represented in terms of coordinates on $\mathrm{Op}_{\mathfrak{sl}_2}(C)$. Standard arguments briefly reviewed in [**41**] may then be used to identify the space of supersymmetric ground states of the sigma model on the strip with the $\bar{\partial}$-cohomology with values in $K_{\rm op}^{1/2}$.

3.3.3. *Relation to conformal blocks: Virasoro case.* Note that $\mathrm{Op}_{\mathfrak{sl}_2}(C)$ is (non-canonically) isomorphic to the complex vector space $H^0(C,K_C^2)$ which is topologically trivial. The space $H^0(C,K_C^2)$ is canonically isomorphic to the cotangent fiber $T^*\mathcal{T}(C)|_C$ to the Teichmüller space $\mathcal{T}(C)$. It will be important for us to observe that there is a map from the cotangent space $T^*\mathcal{T}(C)|_C$ to the Teichmüller space $\mathcal{T}(C)$ relating the natural complex structures on these spaces.

In order to introduce the complex structure on $\mathcal{T}(C)$ one may describe the Teichmüller variations in terms of the harmonic Beltrami differentials, which are $(-1,1)$-forms μ satisfying $\partial_z(\eta\mu)=0$, where η is the hyperbolic metric uniquely determined by the complex structure on C. The complex structure on the vector space of harmonic Beltrami differentials thereby defines the complex structure on $\mathcal{T}(C)$. To a holomorphic $(2,0)$-form θ one may assign the corresponding harmonic Beltrami differential $\eta^{-1}\bar{\theta}$. This defines a complex anti-linear map from $H^0(C,K_C^2)$ to the space of harmonic Beltrami differentials expressing the relation between the natural complex structures on $\mathcal{T}(C)$ and on $H^0(C,K_C^2)$, respectively. It allows us to identify the space of holomorphic functions on $\mathrm{Op}_{\mathfrak{sl}_2}(C)$ with the space $\mathrm{Fun}_{\rm hol}(\mathcal{T}(C))$ of (anti-) holomorphic functions on $\mathcal{T}(C)$.

The spaces $\mathrm{Fun}_{\rm hol}(\mathcal{T}(C))$, on the other hand, are known to be isomorphic with (sub-)spaces of Virasoro conformal blocks. Conformal blocks on closed[4] surfaces C are defined as linear functionals f on the vacuum representation V_0 of the Virasoro algebra satisfying the conformal Ward identities describing invariance of f under the natural action of the algebra of vector fields holomorphic away from a point. The vector space of conformal blocks carries a canonical projectively flat connection defined by means to the Virasoro action on V_0. The curvature of the

[4]We temporarily restrict to this case to simplify the exposition.

canonical connection may be trivialised locally on the moduli space $\mathcal{M}_g$ of complex structures. On suitable subspaces of the space of all algebraically defined conformal blocks one may integrate the resulting flat connection to define horizontal sections f_τ on open subsets of $\mathcal{M}_g$ with local coordinates τ. The values of the conformal blocks $\mathcal{Z}_f(\tau) = f_\tau(v_0)$ on the highest weight vector v_0 are called chiral partition functions. The Virasoro uniformization theorem and conformal Ward identities relate the derivatives of $\mathcal{Z}_f(\tau)$ to the data characterising the conformal blocks f. One thereby gets a one-to-one correspondence between (locally defined) functions $\mathcal{Z}(\tau)$ and "integrable" conformal blocks f. This correspondence is essentially canonical: The Virasoro uniformisation describes the local structure of $\mathcal{M}_g$, and this is encoded in the definition of the conformal blocks.

3.3.4. *Relation to conformal blocks: Kac-Moody case.* It is possible to argue that $\mathcal{H}_x^{(2)}$ is isomorphic to the space $\mathcal{CB}(C, \hat{\mathfrak{g}}_k, \mathcal{E}_x)$ of conformal blocks for $\hat{\mathfrak{g}}_k$ twisted by the bundle $\mathcal{E}_x$ in a similar way as was argued in Section 3.3.3. Following the same reasoning as described there one would arrive at the conclusion that $\mathcal{H}_\mathcal{E}$ is isomorphic to the space of holomorphic functions on $H^0(C, \mathrm{End}(\mathcal{E}) \otimes K_C)$. The space $H^0(C, \mathrm{End}(\mathcal{E}) \otimes K_C)$ is the holomorphic cotangent space $T^*\mathrm{Bun}_G$. There exists a canonical hermitian metric on $T\mathrm{Bun}_G$ which is analogous to the Weil-Petersson metric in Teichmüller theory, defined using the Narasimhan-Seshadri theorem as analog of the uniformisation theorem. This metric identifies the holomorphic cotangent space with the anti-holomorphic tangent space, relating the natural complex structures on $H^0(C, \mathrm{End}(\mathcal{E}) \otimes K_C)$ and $T\mathrm{Bun}_G$.

The space of holomorphic functions on $T\mathrm{Bun}_G$ embeds naturally into $\mathrm{Jet}_\mathcal{E}$, the space of functions on a formal neighbourhood of $\mathcal{E}$ in Bun_G. It remains to notice that $\mathrm{Jet}_\mathcal{E}$ is isomorphic to the space of conformal blocks, $\mathrm{Jet}_\mathcal{E} \simeq \mathcal{CB}_{\mathrm{al}}(C, \hat{\mathfrak{g}}_k, \mathcal{E}_x)$ [**14**, Section 18.2]. This observation was used by E. Frenkel in [**16**] to prove the algebraic counterpart $\mathcal{H}_{\mathrm{al},x}^{(2)} \simeq \mathcal{CB}_{\mathrm{al}}(C, \hat{\mathfrak{g}}_k, \mathcal{E}_x)$ of our claim. It follows that spaces of holomorphic functions on $T\mathrm{Bun}_G$ can be identified with subspaces of "well-behaved" conformal blocks within $\mathcal{CB}(C, \hat{\mathfrak{g}}_k, \mathcal{E}_x)$, further supporting the conjecture $\mathcal{H}_x^{(2)} \simeq \mathcal{CB}_{\mathrm{te}}(C, \hat{\mathfrak{g}}_k, \mathcal{E}_x)$ proposed above.

4. Partition functions versus conformal blocks

Partition functions play a central role in the AGT-correspondence and its generalisations, while they do not appear in the approach of Kapustin and Witten to the geometric Langlands correspondence. In order to develop a unifying framework we will now discuss how to define the relevant partition functions in the effective description furnished by the two-dimensional sigma models.

4.1. TQFT set-up in four and two dimensions. Using the reduction of class $\mathcal{S}$-theories to two-dimensions we will in the following motivate a description of the partition functions within the two-dimensional sigma model with target $\mathcal{M}_H(G)$. It will be based on yet another type of boundary condition denoted $\mathsf{L}_a^{(1)}$. The notation $\mathsf{L}_a^{(1)}$ is motivated by a link to Hitchin's *first* fibration which will be disccussed later.

Following [**41**] we will consider topologically twisted class $\mathcal{S}$-theories on hemispheres $B^4_{\epsilon_1\epsilon_2}$ with Omega-deformation. The topologically twisted class $\mathcal{S}$-theory associates a vector space $\mathcal{H}_{\mathrm{top}} = Z(M^3_{\epsilon_1\epsilon_2})$ to $M^3_{\epsilon_1\epsilon_2} = \partial B^4_{\epsilon_1\epsilon_2}$, here identified with the cohomology of Q, the supercharge that can be preserved on $B^4_{\epsilon_1\epsilon_2}$. One may

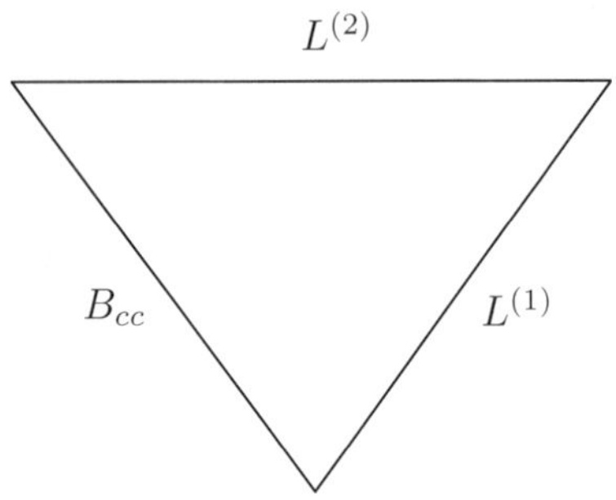

FIGURE 2. The 2d TQFT setup to calculate chiral partition functions $\mathcal{Z}(x, \check{a}, \check{\epsilon})$.

use the path integral over the 4d hemisphere $B^4_{\epsilon_1\epsilon_2}$ to define a vector $\Psi \in \mathcal{H}_{\rm top}$. Wave-functions $\Psi(a)$ of the vector Ψ may be identified with the partition functions $Z(B^4_{\epsilon_1,\epsilon_2}; \mathcal{B}_a)$ defined by imposing suitable Q-invariant boundary conditions $\mathcal{B}_a$ labelled by parameters a at $M^3_{\epsilon_1\epsilon_2}$. Such boundary conditions can be identified with the boundary conditions at the infinity of $\mathbb{R}^4_{\epsilon_1\epsilon_2}$ used to define the Nekrasov partition functions, fixing in particular the zero modes of the scalars in the vector multiplets to have values collected in the vector $a = (a_1, \dots, a_{3g-3+n})$. The boundary conditions $\mathcal{B}_a$ define a family of boundary states β_a, allowing us to represent $Z(B^4_{\epsilon_1,\epsilon_2}; \mathcal{B}_a)$ as an overlap $\langle \beta_a, \Psi \rangle$.

In the reduction of the class $\mathcal{S}$-theory to a $2d$ topological sigma model one should get the following representation of the 4d TQFT data introduced above:

- The vector space $\mathcal{H}_{\rm top} \simeq Z(S^3_{\epsilon_1,\epsilon_2}) \to Z(I)$.
- The vector $\Psi = Z(B^4_{\epsilon_1,\epsilon_2}) \to Z(T_{\epsilon_1,\epsilon_2}) \in \mathcal{H}_{\rm top}$, where $T_{\epsilon_1,\epsilon_2}$ is the open triangle with "upper" side removed, topologically equivalent to $\mathbb{R}_- \times I$, partially compactified by adding a point at the infinity of $\mathbb{R}_-$. The boundary of $T_{\epsilon_1,\epsilon_2}$ is $\{0\} \times I$.
- The partition function $Z(B^4_{\epsilon_1,\epsilon_2}; \mathcal{B}_a) \to Z(T_{\epsilon_1,\epsilon_2}; \mathsf{B}_a)$ gets associated to a triangle $T_{\epsilon_1,\epsilon_2}$ with a boundary condition $\mathsf{L}^{(1)}_a$ assigned to the upper side $\{0\} \times I$. $\mathsf{L}^{(1)}_a$ is defined from the boundary condition $\mathcal{B}_a$ assigned to $M^3_{\epsilon_1\epsilon_2}$ by the reduction to one dimension.

This means that the instanton partition functions $\mathcal{Z}(a; x; \tau; \epsilon_1, \epsilon_2)$ get represented by partition functions of the sigma model on a triangle which has sides coloured by $(\mathsf{B}_{\rm cc}, \mathsf{L}^{(1)}_a, \mathsf{L}^{(2)}_x)$.

4.2. $\mathcal{D}$-modules versus partition functions. States ψ in quantum theory are abstractly represented by vectors in a Hilbert space $\mathcal{H}$. A concrete representation as space of wave-functions $\psi(x)$ is obtained by introducing a family of elements δ_x of the (hermitian) dual $\mathcal{H}^\dagger$ of $\mathcal{H}$ allowing us to represented $\psi(x)$ as $\langle \delta_x, \psi \rangle$, with $\langle ., . \rangle : \mathcal{H}^\dagger \times \mathcal{H} \to \mathbb{C}$ being the natural pairing. In the present context we may interpret the partition functions $\mathcal{Z} = \mathcal{Z}(a; x, \tau; \epsilon_1, \epsilon_2)$ as wave-functions $\phi_x(a) = \langle \delta^{(1)}_a, \phi_x \rangle$ of an element $\phi_x \in \mathcal{H}^{(2)}_x := \mathrm{Hom}_{\mathcal{M}_H(G)}(\mathsf{B}_{\rm cc}, \mathsf{L}^{(2)}_x)$ created by the sigma model path integral over the semi-infinite strip $\mathbb{R}_- \times I$. The "boundary state" $\delta^{(1)}_a$ represents the boundary condition $\mathsf{L}^{(1)}_a$ at $\{0\} \times I$.

The overlap $\langle \delta_a^{(2)}, \phi_x \rangle$ has a representation as a partition function in the topological sigma model on a triangle with three different types of boundary conditions $(\mathsf{B}_{\rm cc}, \mathsf{L}_x^{(2)}, \mathsf{L}_a^{(1)})$ assigned to the three sides, see Figure 2. Time-reversal symmetry identifies the complex conjugate $\mathcal{Z}^*$ of $\mathcal{Z}$ with a partition function in the topological sigma model on a triangle with boundary conditions $(\mathsf{B}_{\rm cc}, \mathsf{L}_a^{(1)}, \mathsf{L}_x^{(2)})$ appearing in a different order. $\mathcal{Z}^*$ admits a Hamiltonian interpretation as a wave-function $\varphi_a(x) = \langle \delta_x^{(2)}, \varphi_a \rangle$ of an element $\varphi_a \in \mathcal{H}_x^{(1)} := \mathrm{Hom}_{\mathcal{M}_H(G)}(\mathsf{B}_{\rm cc}, \mathsf{L}_a^{(1)})$ created by the sigma model path integral over the semi-infinite strip $\mathbb{R}_- \times I$.

Both spaces $\mathcal{H}_a^{(1)}$ and $\mathcal{H}_x^{(2)}$ have a $\mathcal{A}_{dR}^{\hbar}$-module structure coming from the natural action of the algebra $\mathcal{A}_{dR}^{\hbar}$ on $\mathrm{Hom}_{\mathcal{M}_H(G)}(\mathsf{B}_{\rm cc}, \mathsf{B})$. We have argued above that this module structure coincides with the one on spaces of conformal blocks following from the affine algebra Ward identities. The resulting structure as twisted $\mathcal{D}$-module coincides with the one defined by the KZB equations satisfied by the chiral partition functions corresponding to the conformal blocks of affine Lie algebras. The partition functions $\mathcal{Z}$ are solutions to the KZB equations. The $\mathcal{D}$-modules $\mathcal{H}_a^{(1)}$ and $\mathcal{H}_x^{(2)}$ can be fully characterised by the corresponding spaces of wave-functions, physically represented as the partition functions $\mathcal{Z}$.

In the limit $\epsilon_2 \to 0$ discussed below we will propose that $\mathcal{H}_a^{(1)}$ appears in the Kapustin-Witten approach to the geometric Langlands correspondence while $\mathcal{H}_x^{(2)}$ appears to be a more natural space to consider in the context of generalisations of the AGT correspondence. The relation between the two Hamiltonian interpretations of the partition function $\mathcal{Z}(x, a, \tau; \epsilon_1, \epsilon_2)$ following from an open TQFT version of modular invariance represents a crucial link between these two stories.

4.3. Sigma model description of the brane $\mathsf{L}_a^{(1)}$. Let us now propose a description of the boundary condition $\mathsf{L}_a^{(1)}$ as a Lagrangian submanifold in $\mathcal{M}_H(G)$. For future use we will include the label σ referring to the Lagrangian representation of class $\mathcal{S}$ theories used to define the branes $\mathsf{L}_a^{(1)}$ into the notation, changing it into $\mathsf{L}_{\sigma,a}^{(1)}$. We propose that the relevant branes $\mathsf{L}_{\sigma,a}^{(1)}$ can be defined as follows. To any pants decomposition σ one may assign a collection of complex Fenchel-Nielsen coordinates[9)] used in a related context in [**39**]. Half of these coordinates parameterise the traces of holonomies along the curves γ_i, $i = 1, \dots, 3g - 3 + n$, defining σ as $L_i = 2\cosh(a_i/2)$. The functions a_i are mutually Poisson-commuting with respect to the natural symplectic form. Fixing the values of the functions a_i, $r = 1, \dots, 3g - 3 + n$, defines a Lagrangian submanifold, the orbit swept out by the Hamiltonian flows generated by the length functions $a_1, \dots, a_{3g-3+n}$. These orbits are mapped via holonomy map and the NAH correspondence to the Hitchin moduli space $\mathcal{M}_H(G)$. The branes $\mathsf{L}_{\sigma,a}^{(1)}$ are natural objects from the point of view of the integrable structure of the character variety $\mathcal{M}_B(G)$.

Our proposal is based on the known relations between Wilson loop observables in class $\mathcal{S}$ theories and co-dimension four defects in the six-dimensional $(2, 0)$ theory supported on products of two circles $S^1_{\mathcal{S}}$ and S^1_C in $\mathcal{M}^4$ and C, respectively. These relations carry over to the case where surface operators are present. Restricting attention to Wilson loops not sharing the support of the surface operators in $\mathcal{M}^4$ one may argue as before that the eigenvalues of Wilson loop observables can be expressed in terms of the scalar zero modes a appearing in the arguments of the instanton partition functions $\mathcal{Z}(a; x, \tau; \epsilon_1, \epsilon_2)$.

One may alternatively adapt the arguments in [**2**] to the case with additional surface operators of codimension two, which can be based on the results described in Section 6.

5. Another type of surface operators

One of the key ingredients in the geometric Langlands correspondence is the Hecke eigenvalue property. In order to prepare for the discussion of the Hecke eigenvalue property in the following section we will now discuss another type of surface operators giving us a physical realisation of the Hecke functors.

5.1. String-theoretical motivation. From the perspective of the six-dimensional $(2,0)$ theory it seems natural to investigate the effects of modifying the theory by defects of co-dimension four. This defines natural probes of the six-dimensional $(2,0)$ theories and their dimensional reductions which have already been investigated extensively in the context of the AGT correspondence [**2**]. We will now discuss briefly how the picture gets modified if one considers surface operators of co-dimensions two and four at the same time. The existence of such systems is suggested by string theory. Following the discussion in [**17**] one may, in particular, consider a configuration of M5-M5'-M2-branes in M-theory having support indicated in the following table:

Brane	0	1	2	3	4	5	6	7	8	9	10
M5	x	x	x	x			x				x
M5'	x	x					x		x	x	x
M2	x	x						x			

Having support of this type means that the co-dimension four surface operators on $\mathcal{M}^4 \times C$, with $\mathcal{M}^4$ a fibered product locally of the form[5] $\mathbb{R} \times I \times S_1^1 \times S_2^1$ will be represented as

- surface operators on $\mathcal{M}^4$ supported on $\mathcal{M}^2 \simeq \mathbb{R} \times S_1^1$,
- line operators in $N = 4$ SYM on $\mathbb{R} \times I \times C$, supported on $\mathbb{R} \times \{\pi\} \times \{P\}$,
- or line operators in CS theory on $\mathbb{R} \times C$, supported on $\mathbb{R} \times \{P\}$.

It easily follows from these observations that the presence of such co-dimension four surface operators modifies the boundary conditions in the two-dimensional sigma model describing the reduction to $\mathbb{R} \times I$. We will now briefly describe the modifications that result from the presence of surface operators of co-dimension four in the effective description as two-dimensional CFT.

5.2. Resulting modification of the (generalised) AGT-correspondence. It has been argued in [**2**] that the presence of co-dimension four surface operators in the six-dimensional $(2,0)$ theory is described after compactification on C as a system where a $4d$ theory of class $\mathcal{S}$ is coupled to certain two-dimensional gauged linear sigma models (GLSM) on a two-dimensional surface $\mathcal{M}^2$ in $\mathcal{M}^4$. The characteristic feature of the relevant GLSM is the infrared description in terms of nonlinear sigma model with Grassmannians targets. In the simplest case, $\mathfrak{g} = \mathfrak{sl}_2$, one gets, for example, the GLSM flowing to the $\mathbb{CP}^1$ sigma model.

[5]The lower subscripts in the notation S_i^1 refer to the rotation symmetry used to define the Omega-deformation with parameter ϵ_i.

Our main conjecture concerning the realisation of co-dimension four surface operators in the presence of co-dimension two surface operators within CFT is the following:

(5.1)
$$\boxed{\begin{array}{c}\text{Addition of a surface operator of co-dimension four at } P\in C\\ \text{gets represented by an insertion of the } \widehat{\mathfrak{sl}}_{2,k}\text{-module } W_{1/2\hbar} \text{ at } P.\end{array}}$$

To list the available support for this conjecture, let us start by noticing that it is suggested by a slight generalisation of the picture described in [**17**] . It was observed in this reference that a system composed out of M5-M5'-branes displays an interesting IR duality relating it to a system of M5-M2-branes, with a collection of M2-branes located at various points on C. This IR duality follows from a variant of the Hanany-Witten brane creation effect describing a family of IR-equivalent effective theories obtained by separating the M5-M5'-branes present in the configuration above along the x_7-axis. In order to get an IR-equivalent effective description one needs to introduce additional $M2$-branes suspended between the branes getting separated, and extended along the x_7-axis. The end-points of the $M2$-branes define a divisor $\sum_k P_k$ on C. It was argued in [**17**] that this IR duality is expressed in relations between supersymmetric partition functions which get by generalisations of the AGT-correspondence related to known relations between the conformal blocks of the affine algebra $\widehat{\mathfrak{sl}}_{2,k}$ and $\mathfrak{Vir}_c$. The relevant conformal blocks of $\mathfrak{Vir}_c$ are defined using insertions of additional degenerate representations $V_{-1/2b}$ with $b^2 = \hbar$ at the points $P_1, \dots, P_h$, $h = 4g - 4 + n$. The relations between these two types of conformal blocks follow from a generalisation of Sklyanin's separation of variables (SOV) method.

A simple variant of the set-up considered in [**17**] is to modify the initial system of M5-M5'-branes by having an additional "primordial" M2-brane localised at a point $P_0 \in C$ in the background, and extended along the x_7-axis. In the IR dual description one would then get an additional degenerate representation $V_{-1/2b}$ inserted at P_0 on top of the insertions at $P_1, \dots, P_h$. As explained in [**45**], this is nothing but the representation for conformal blocks of the affine algebra $\widehat{\mathfrak{sl}}_{2,k}$ obtained by the SOV when there is an additional insertion of the $\widehat{\mathfrak{sl}}_{2,k}$-module $W_{1/2\hbar}$ at P.

5.3. Further support. An alternative argument can be obtained by a suitable generalisation of the arguments leading [**2**] to describe the effect of additional M2-branes by the insertion of degenerate fields. Addition of such a vertex operator modifies the space of instanton partition functions by tensoring it with the two-dimensional vector space of vacua of the $\mathbb{CP}^1$ sigma model. Variation of the position of the co-dimension four surface operator generates a monodromy representation on this two-dimensional vector space. Taking the limit $\epsilon_2 \to 0$ results in a partial decoupling of the degrees of freedom on the surface operator and in the bulk, leading to a factorisation of the instanton partition functions into a four-dimensional singular, and a two-dimensional regular part. The $\widehat{\mathfrak{sl}}_{2,k}$-module $W_{1/2\hbar}$ is the unique candidate for an insertion in affine Lie algebra conformal blocks that has the properties (i) to generate a two-dimensional monodromy representation, and (ii) insertions of $W_{1/2\hbar}$ factor off the conformal blocks in the critical level limit. In the following section we will describe some elements of this line of arguments in more detail.

It would be nice to check our conjecture (5.1) by localisation calculations, generalising the results for codimension 2 surface operators cited above on the one hand, and the known results for codimension 4 surface operators (see [**28**] and references therein) on the other hand.

In the reduction to $N = 4$ SYM one would represent the co-dimension four surface operators in terms of the Wilson and 't Hooft line operators in $N = 4$ SYM [**48**]. The relation of these line operators to the Hecke functors of the geometric Langlands correspondence was extensively discussed in [**34**]. In the reduction scheme studied by Cordova and Jafferis in [**10**] one will find the Wilson lines of complex Chern-Simons theory, which should give an alternative view on the conjecture formulated in (5.1).

One may note, finally, that the Hecke functors in the work of Beilinson and Drinfeld are described by the critical level limit of insertions of the modules $W_{-(k+h^\vee)\lambda}$ of $\mathfrak{g}_k$ into conformal blocks, where λ is the weight of a finite-dimensional representation of ${}^LG$, see [**14**, Section 20.5] for a review. The modules appearing in (5.1) are associated to the case where $\lambda = 1/2$ corresponding to the fundamental representation of ${}^LG = \mathrm{PSL}(2, \mathbb{C})$.

6. Recovering the geometric Langlands correspondence

Our goal is to elaborate on the relations between generalised versions of the AGT-correspondence in the presence of surface operators to the geometric Langlands correspondence. We will argue that a direct relation emerges in the Nekrasov-Shatshavili limit $\epsilon_2 \to 0$. To this aim we will first observe that the asymptotics of the partition functions $\mathcal{Z}_\sigma(a; x, \tau; \epsilon_1, \epsilon_2)$ contains solutions of the Hitchin eigenvalue equations describing the $\mathcal{D}$-module structures in the geometric Langlands correspondence. We will then discuss how the crucial Hecke eigenvalue property of the relevant $\mathcal{D}$-modules can be understood from the point of view of the Nekrasov-Shatshavili limit of class $\mathcal{S}$ theories with surface operators. Possible interpretations of these phenomena in term of the $2d$ sigma model with target $\mathcal{M}_H(G)$ will be discussed in the next section.

6.1. Nekrasov-Shatashvili limit. The limit $\epsilon_2 \to 0$ is called Nekrasov-Shatashvili limit following [**42**]. The relation $k = -h^\vee - \epsilon_2/\epsilon_1$ identifies it with the critical level limit in CFT. In CFT is is known [**17**, **45**] that the chiral partition functions of affine Lie algebra conformal blocks behave in the critical level limit as

$$\mathcal{Z}_\sigma(a; x, \tau; \epsilon_1, \epsilon_2) \sim e^{-\frac{1}{\epsilon_2}\mathcal{Y}_\sigma(a;\tau;\epsilon_1)}\Psi(a; x, \tau; \epsilon_1)\big(1 + \mathcal{O}(\epsilon_2)\big)\,, \tag{6.1}$$

where

- $\mathcal{Y}_\sigma(a; \tau; \epsilon_1)$ is the generating function for the Lagrangian submanifold $\mathrm{Op}_{\mathfrak{sl}_2}(C_\tau)$ of opers, considered in a related context in [**39**], with a being the complex Fenchel-Nielsen coordinates for $\mathrm{Op}_{\mathfrak{sl}_2}(C_\tau)$ defined from the traces of holonomies along a set of curves on C defining the pants decomposition σ.
- $\Psi(a; x, \tau; \epsilon_1)$ is an eigenfunction of the Hitchin Hamiltonians, $D_r\Psi = E_r\Psi$, with eigenvalues $E_r = E_r(a; \tau; \epsilon_1)$ obtained from $\mathcal{Y}_\sigma(a; \tau; \epsilon_1)$ as $E_r = \partial_{\tau_r}\mathcal{Y}_\sigma(a; \tau; \epsilon_1)$.

The behaviour (6.1) can be seen as a concrete manifestation of the geometric Langlands correspondence: The points in the character variety having complex Fenchel-Nielsen coordinates $(a, \partial_a \mathcal{Y}_\sigma(a;\tau;\epsilon_1))$ may be represented as the holonomies of the ϵ_1-opers $(\mathcal{E}_o, \nabla'_{\epsilon_1,u}) = \big(\mathcal{E}_o, \big(\epsilon_1\partial_z + \big(\begin{smallmatrix}0 & u\\ 1 & 0\end{smallmatrix}\big)\big)dz\big)$. To an oper with $u = u^\sharp + \sum_{r=1}^{3g-3+n} E_r \vartheta_r(x)$, where $u^\sharp$ defines a fixed reference oper, and $\vartheta_r(x)(dz)^2$, $r = 1,\dots,3g-3+n$ is a basis for $H^0(C,K^2)$, we may associate the system of equations $D_r\Psi = E_r\Psi$, $r = 1,\dots,3g-3+n$, defining the $\mathcal{D}$-module associated to the oper $(\mathcal{E}_o, \nabla'_{\epsilon_1,u})$ by the geometric Langlands correspondence.

A behaviour of the type (6.1) is easily understood from the point of view of class $\mathcal{S}$-theories with surface operators. In the limit $\epsilon_2 \to 0$ one would expect that a modification of the boundary conditions for the gauge fields along a submanifold $\mathcal{M}^2$ that stays effectively compact when $\epsilon_2 \to 0$ can not affect the leading singular behaviour in this limit.

6.2. Hecke eigenvalue property in CFT. In the presence of both types of surface operators one may define further generalisations of the partition functions $\mathcal{Z}_v(a;x,\tau;t;\epsilon_1,\epsilon_2)$ depending in addition to all the variables considered previously on a variable t which may be identified with the coordinate of a point on C, the end-point of an "primordial" M2-brane. We have argued above that such partition functions get represented in CFT as chiral partition functions of conformal blocks with an additional insertion of a degenerate field at t. The extra label v refers to the elements of a basis for the space of vacua of the GLSM providing a $4d+2d$ Lagrangian description. Considering a fixed choice of pants decomposition we will temporarily suppress the label σ.

From CFT [**17**, **45**] we know that the partition functions $\mathcal{Z}_v(a;x,\tau;t;\epsilon_1,\epsilon_2)$ behave in the limit $\epsilon_2 \to 0$ as

$$(6.2)\qquad \mathcal{Z}(a;x,\tau;t;\epsilon_1,\epsilon_2) \sim e^{-\frac{1}{\epsilon_2}\mathcal{Y}(a;\tau;\epsilon_1)}\Psi(x;a,\tau;\epsilon_1)\,\psi_v(t;a,\tau;\epsilon_1)\big(1+\mathcal{O}(\epsilon_2)\big)\,,$$

where $\mathcal{Y}(a;\tau;\epsilon_1)$ and $\Psi(x;a,\tau;\epsilon_1)$ coincide with the functions appearing in (6.1), and $\psi_v(t) = \psi_v(t;a,\tau;\epsilon_1)$ generate a basis for the space of solutions of the differential equation $(\epsilon_1^2\partial_t^2 - u(t))\psi_v(t) = 0$ where $u = u(t)$ represents the oper connection as $\epsilon_1\partial_t + \big(\begin{smallmatrix}0 & u\\ 1 & 0\end{smallmatrix}\big)$. The index $v = +,-$ labels a basis for the space of solutions.

The fact that $\psi_v(t)$ factors off is related to the Hecke eigenvalue property, as has been pointed out in [**45**]. As before we may interpret $\mathcal{Z}_v^*$ as overlap $\langle \delta_x^{(2)}, \xi_a\rangle$, where $\xi_a \in \mathcal{H}_a^{(2)}$. If (6.2) holds for all x, it implies a similar factorisation for all $\xi_a \in \mathcal{H}_a^{(2)}$. The factor $\psi_v(t)$ does furthermore not depend on the choice of vector $\xi_a \in \mathcal{H}_a^{(2)}$ used to represent the partition function as $\langle \delta_x^{(2)}, \xi_a\rangle$. The monodromy generated by analytic continuation of $\psi_v(t)$ defines an oper local system χ. Being independent of $\xi_a \in \mathcal{H}_a^{(2)}$, one may attribute this modification to a modification of the space $\mathcal{H}_a^{(2)}$ itself, satisfying the Hecke eigenvalue property.

6.3. Hecke eigenvalue property from the tt^*-equations. Our goal in this section will be to elaborate on the (4d+2d)-description of the Hecke functors. Due to the fact that the Ω-deformation localises fluctuations to the fixed points of the circle actions, one may think of the limit $\epsilon_2 \to 0$ as a decompactification of the hemisphere $B^4_{\epsilon_1\epsilon_2}$ into $B^2_{\epsilon_1} \times \mathbb{R}^2$, where $B^2_{\epsilon_1}$ is a two-dimensional hemisphere. The surface operators of interest will be supported on $B^2_{\epsilon_1}$.

The goal is to offer an interpretation for the behaviour (6.2) in four-dimensional terms. Recall that the partition function $\mathcal{Z}_v(a;x,\tau,t;\epsilon_1,\epsilon_2)$ can be interpreted as a partition function of the gauge theory on a 4d hemisphere $B^4_{\epsilon_1\epsilon_2}$ with Ω-deformation, defined using boundary conditions projecting the state created by the gauge theory path integral on $B^4_{\epsilon_1\epsilon_2}$ to an eigenstate of the Wilson-loops [**41**, **46**]. The variables a parameterise the Wilson loop eigenvalues, and the collection of UV gauge couplings of the 4d SUSY gauge theory is denoted by τ. The variables x parameterise the boundary conditions defining a codimension-2 surface operator [**17**], while the variable t represents the FI parameter t of the GLSM representing the additional surface operator of co-dimension four [**2**].

6.3.1. *Factorisation.* The degrees of freedom on the surface operator can not contribute to the divergence of the free energy when $\epsilon_2 \to 0$ as the two-ellipsoid appearing in this limit ends up having finite volume. We therefore expect that $\mathcal{F} = \log \mathcal{Z}$ behaves as

$$\mathcal{F} \sim -\frac{1}{\epsilon_2}\mathcal{Y}^{\rm 4d}(a;\tau;\epsilon_1) + \mathcal{W}^{\rm 2d}(a,\tau;x,t;\epsilon_1) + \dots, \tag{6.3}$$

where $\mathcal{W}^{\rm 4d}(a;\tau;\epsilon_1)$ is independent of the 2d couplings x and t. Comparison with (6.1) gives

$$\mathcal{W}^{\rm 2d}(a,\tau;x,t;\epsilon_1) = \mathcal{W}^{\rm M2}(a,\tau;t;\epsilon_1) + \mathcal{W}^{\rm M5'}(a,\tau;x;\epsilon_1),$$

where $\mathcal{W}^{\rm M2}(a,\tau;t;\epsilon_1) = \log\psi_v(t;a,\tau;\epsilon_1)$ and $\mathcal{W}^{\rm M5'}(a,\tau;x;\epsilon_1) = \log\Psi(x;a,\tau;\epsilon_1)$. This decoupling can probably be understood as a consequence of the fact that the boundary condition defining a codimension 2 surface operator only involves the bulk fields and not the degrees of freedom used to describe the codimension 4 surface operator.

It is then natural to interpret $\mathcal{Z}^{\rm 2d}_v(a,\tau;t;\epsilon_1) := \psi_v(t)$ as the partition function of the GLSM on the lower 2d hemisphere $B^2_{\epsilon_1}$ with Ω-deformation, coupled to the class $\mathcal{S}$ theory in the bulk, and subject to boundary conditions denoted as $\mathfrak{b}_v$. Such partition functions have recently been studied in [**18**], where a close relation was found to the hemisphere partition functions studied in [**30**, **32**]. It was observed that the dependence of $\mathcal{Z}^{\rm 2d}_v(a,\tau;t;\epsilon_1)$ on the FI parameter t is holomorphic. The partition function is furthermore related to an overlap in the topological twisted version of the GLSM as [**30**, **32**]

$$\mathcal{Z}^{\rm 2d}_v(a,\tau;t;\epsilon_1) = \langle\, \mathfrak{b}_v \,|\, 0\,\rangle, \tag{6.4}$$

where $|\,0\,\rangle$ is the state created by the path integral on the hemisphere with no operator insertion, and $\langle\, \mathfrak{b}_v\,|$ is the boundary state associated to the boundary condition $\mathfrak{b}_v$.

6.3.2. *tt^*-equations.* The overlap $\langle\, \mathfrak{b}_v \,|\, 0\,\rangle$ has a natural generalisation obtained by replacing $|\,0\,\rangle$ by other supersymmetric ground states $|\,i\,\rangle$ created from $|\,0\,\rangle$ by means of the chiral ring generators Φ_i, $|\,i\,\rangle = \Phi_i|\,0\,\rangle$. It has been observed in [**31**] that $\Pi^v_i := \langle\, \mathfrak{b}_v\,|\,i\,\rangle$ represents a horizontal section of the tt^*-connection of Cecotti and Vafa [**9**]. In our case we have a single coupling t and a two-dimensional space of SUSY vacua. The tt^*-connection therefore has the form of a flat $SL(2,\mathbb{C})$-connection $\nabla' = (\partial_t + \mathcal{A}_t)dt$, $\nabla'' = (\bar\partial_{\bar t} + \mathcal{A}_{\bar t})d\bar t$ [**19**]. The holomorphicity observed in [**18**, **30**, **32**] implies that $\mathcal{A}_{\bar t} = 0$ for the case at hand. The horizontality condition $\nabla'\Pi^v = 0 = \bar\partial_{\bar t}\Pi^v$ then implies a second order differential equation for the components Π^v_i.

It follows from the arguments presented in [**2**] that the couplings t can be identified with local coordinates on the Riemann surface C_τ. On physical grounds one would expect that the ordinary differential equation following from the horizontality condition should be non-singular on C_τ. This would imply that the tt^*-connection must be gauge equivalent to an oper connection. The results of [**18**, **30**, **32**] provide support for this claim.

6.3.3. *Fixing the monodromies.* A weakly coupled Lagrangian description of the 4d theories of class $\mathcal{S}$ exists when the complex structure of C_τ is near a boundary of the moduli space of Riemann surfaces where C_τ can be represented as a collection of pairs of pants connected by long tubes $T_1, \ldots, T_{3g-3+n}$. The coupled 4d-2d-system will be weakly coupled if t is located on a particular tube T_r. The Lagrangian description involves a coupling of the GLSM to the 4d bulk theory of the form ta_r/ϵ_1, where a_r is the restriction of the scalar field in the vector multiplet associated to T_r to the support of the codimension-4 surface operator [**2**]. The monodromy of the partition function $\mathcal{Z}_v^{2\mathrm{d}}(a;\tau;t;\epsilon_1)$ corresponding to $t \to t + 2\pi$ is therefore diagonal, and can be represented by multiplication with $e^{2\pi v a_r/\epsilon_1}$. This fixes half of the monodromy of the tt^*-connection which is enough to fix it completely, being gauge equivalent to an oper connection.

The tt^*-connection thereby gets identified with the oper local system ρ appearing on one side of the geometric Langlands correspondence. Taking the quotient of the algebra of global differential operators on Bun_G by the ideal generated from $D_r - E_r$, $r = 1, \ldots, 3g-3+n$ defines the corresponding D-module Δ_ρ on the other side of the Langlands correspondence [**15**]. The factorisation (6.2) can be seen as a manifestation of the Hecke eigenvalue property on the level of the solutions of differential equations associated to the D-module Δ_ρ.

7. Sigma model interpretation?

The discussion in the previous section has revealed close connections between generalisations of the AGT-correspondence in the presence of surface operators and the Beilinson-Drinfeld approach to the geometric Langlands correspondence. We had identified the geometric Langlands correspondence as the limit $\epsilon_2 \to 0$ of a description for the spaces of conformal blocks in terms of the quantised character varieties. In this final section we are going to return to the question how the emerging picture is related to the Kapustin-Witten approach to the geometric Langlands correspondence. In Section 7.1 we will raise the question how to describe the limit $\epsilon_2 \to 0$ in terms of the $2d$ sigma model. Sections 7.2 and 7.3 outline a speculative answer to this question. In the rest of this Section we will propose a more explicit comparison between the approaches of Beilinson-Drinfeld and Kapustin-Witten based on similar arguments.

7.1. Comparison I: Relations between AGT- and geometric Langlands correspondences. We had argued that the $\mathcal{D}$-modules appearing in this context can be characterised in terms of the spaces of solutions to the corresponding differential equations, physically realised as partition functions. In the context of the AGT-correspondence with surface operators we had argued that the spaces $\mathcal{H}_x^{(2)} := \mathrm{Hom}_{\mathcal{M}_H(G)}(\mathsf{B}_{\mathrm{cc}}, \mathsf{L}_x^{(2)})$ can be identified with the fibers of affine algebra conformal blocks over the bundles $\mathcal{E}_x$. In CFT we can view the chiral partition functions $\mathcal{Z}(a;x,\tau;k)$ solving the KZB equations as representatives of elements φ_a of subspaces $\mathcal{CB}_a^\sigma(C,\hat{\mathfrak{g}}_k,\mathcal{E}_x)$ of the spaces $\mathcal{CB}(C,\hat{\mathfrak{g}}_k,\mathcal{E}_x)$ conformal blocks having fixed eigenvalues parameterised by a under a maximal commuting subset of the Verlinde line operators.

Modular invariance of the open sigma model TQFT suggests that we can identify $\mathcal{CB}^{\sigma}_{a}(C,\hat{\mathfrak{g}}_k) \simeq \mathrm{Hom}_{\mathcal{M}_H(G)}(\mathsf{B}_{\rm cc},\mathsf{L}^{(1)}_{\sigma,a})$. This identification can be supported more directly by observing that the boundary condition $\mathsf{L}^{(1)}_{\sigma,a}$ fixes the values of some zero modes in the sigma model on the strip, which should lead to an eigenvalue property w.r.t. to the subalgebra of $\mathcal{A}^{\hbar}_{B}$ generated by the corresponding quantised trace functions, associated to the curved defining the pants decomposition σ.

In the last section we have discussed how the subspaces $\mathcal{CB}^{\sigma}_{a}(C,\hat{\mathfrak{g}}_k,\mathcal{E}_x)$ get related to the Hecke-eigen $\mathcal{D}$-modules appearing the the geometric Langlands correspondence in the limit $\epsilon_2\to 0$, with local systems representing the "eigenvalue" being represented by families of opers parameterised by the variables a. What is not clear, however, is how our observations concerning the appearance of the Hecke eigenvalue property in the limit $\epsilon_2\to 0$ can be understood from the perspective of the sigma model. There appears to be an immediate obstacle: The sigma model considered in [**34**] in the context of the geometric Langlands correspondence are A-models with symplectic structure ω_K, and the cc-brane has Chan-Paton curvature ω_J. A different choice appears in the Nekrasov-Witten approach to the AGT-correspondence, where the A-model with symplectic structure $\frac{\epsilon_2}{\epsilon_1}\omega_I$ is used. There is no obvious parameter allowing us to move continuously between these two cases.

In view of the fact that within CFT one can obtain the geometric Langlands correspondence in the critical level limit it seems very natural to ask if this can be understood within the $2d$ sigma model with target $\mathcal{M}_H(G)$. Starting from Subsection 7.2 we'll speculate about a possible way to see this.

7.2. Hyperkähler rotation. It is tempting to modify the 2d TQFT set-up of Nekrasov and Witten using the hyperkähler$^{(g)}$ rotations coming from the circle action $\varphi\to e^{i\theta}\varphi$ on Hitchin's moduli spaces. It is shown in [**29**, Section 9] that this circle action can be complexified into a $\mathbb{C}^*$-action relating all complex and symplectic structures in the family (I_ζ,Ω_ζ) apart from $\zeta=0,\infty$. The hyperkähler rotations allow us to identify the key geometric structure on $\mathcal{M}_H(G)$ for $\zeta\neq 0,\infty$ in the sense that the rotations can be represented by diffeomorphisms of $\mathcal{M}_H(G)$ [**29**, Proposition (9.1)]. This amounts to reparametrizations of the sigma model fields, which may not affect the physical content.

Invariance of the sigma model path integral under field redefinitions may be a subtle issue. The circle action $\varphi\to e^{i\theta}\varphi$ should be easy to understand, but it is not clear to us if the complexification to a $\mathbb{C}^*$-action is easy to understand as a symmetry of the sigma model TQFT in a suitable sense. What we'd like to point out is that validity of a certain form of invariance of the topological sigma models with target $\mathcal{M}_H(G)$ under the $\mathbb{C}^*$-action would give us a simple way to interpolate between the sigma models considered in [**34**] and [**41**], respectively. In this way one could get an attractive explanation of the observation made in the previous section.

We are interested in open A-models on surfaces with boundaries with boundary conditions being either of the canonical coisotropic or Lagrangian type. We may use the family of holomorphic symplectic forms Ω_ζ to define a family of A-models with symplectic structure $\mathrm{Im}(\Omega_\zeta)$ having coisotropic branes with Chan-Paton curvature $\mathrm{Re}(\Omega_\zeta)$. Submanifolds of $\mathcal{M}_H(G)$ which are Lagrangian with respect to Ω_ζ define branes in this family of A-models. If an appropriate form of $\mathbb{C}^*$-invariance of the sigma model TQFT holds, one would expect that the partition functions $\mathcal{Z}_\zeta$ defined in this family of A-models are in fact ζ-independent.

One should note, however, that all Lagrangian submanifolds we use to define boundary conditions will have to be varied consistently to keep $\mathcal{Z}_\zeta$ unchanged. The submanifolds of our interest are the orbits of the complex Fenchel-Nielsen$^{(g)}$ twist flows in the character variety, considered as submanifolds of $\mathcal{M}_H(G)$ via NAH-correspondence and holonomy map. Considering a fixed orbit in the character variety, one gets a one-parameter family of submanifolds of $\mathcal{M}_H(G)$ upon varying the hyperkähler parameter ζ.

Given we have invariance under the $\mathbb{C}^*$-action, we could combine variations of the coupling parameter ϵ_1/ϵ_2 with a suitable hyperkähler rotation in such a way that we obtain a one-parameter family of topological A-models with target $\mathcal{M}_H(G)$ that interpolates between the ones considered in [**41**] and [**34**], respectively. This could be done by setting $\zeta = \epsilon_2\xi$. The expression for $\hat{\omega}_\zeta = \frac{\epsilon_2}{\epsilon_1}\mathrm{Im}(\Omega_\zeta)$ reduces in the limit $\epsilon_2 \to 0$ to $\hat{\omega}_\zeta = \frac{1}{2\epsilon_1|\xi|^2}(\mathrm{Re}(\zeta)\omega_K - \mathrm{Im}(\zeta)\omega_J)$, reproducing for real ζ the symplectic form used to define the A-model studied in [**34**].

7.3. Nekrasov-Shatashvili limit of the branes $\mathsf{L}^{(1)}_{\sigma,a}$. In order to see that the resulting scenario may indeed resolve the puzzle stated above, let us first note that the (A,B,A) branes $\mathsf{L}^{(1)}_{\sigma,a}$ considered in Section (4.3) can easily be generalised into $(B,A,A)_\zeta$-branes $\mathsf{L}^{(1)\zeta}_{\sigma,a}$ by using the ζ-dependent NAH-correspondence in their definition. Choosing ζ in an $\hbar$-dependent way, as suggested above, would turn them into branes $\mathsf{L}^{(1)\hbar}_{\sigma,a}$ in the family of A-models with hyperkähler parameter $\hbar = \epsilon_1/\epsilon_2$.

We will observe that something interesting happens in the Nekrasov-Shatashvili limit of the family of branes $\mathsf{L}^{(1)\hbar}_{\sigma,a}$: The branes $\mathsf{L}^{(1)\hbar}_{\sigma,a}$ will have a well-defined limit $\hbar \to \infty$ if the parameters a are scaled in this limit as $a_r = \frac{\epsilon_1}{\epsilon_2}\breve{a}_r$ with $\breve{a}_r$ finite. In order to understand the limit $\epsilon_2 \to 0$ one mainly needs to study the WKB approximation for the holonomy of $\partial_u + \mathcal{A}_u$ where $\mathcal{A}_u = A_u + \frac{1}{\epsilon_2}\varphi$. By gauge transformations one can always locally reach the form $\mathcal{A}_u = \left(\begin{smallmatrix} 0 & u_{\epsilon_2} \\ 1 & 0 \end{smallmatrix}\right)$. The function $u_{\epsilon_2}(t)$ appearing in the upper right matrix element of $\mathcal{A}_u$ will in general only be meromorphic in u, but the residues are of order ϵ_2. We note that A_u contributes only in subleading orders of the expansion. It follows easily from these facts that $\vartheta_{\epsilon_2}(t) = \frac{1}{2}\mathrm{tr}(\varphi^2) + \mathcal{O}(\epsilon_2)$. To leading order in ϵ_2 one may therefore represent the solutions s to $(\partial_u + \mathcal{A}_u)s_{\epsilon_2} = 0$ in the form

$$s_{\epsilon_2}(t) \sim e^{\pm\frac{1}{\epsilon_2}\int_{C_t}\lambda}, \tag{7.1}$$

where C_t is a path on the spectral curve Σ ending at a lift $\hat{t} \in \Sigma$ of $t \in \mathbb{C}$, and λ is the canonical differential on Σ. The conclusion is that the rescaled Fenchel-Nielsen length coordinates $\breve{a}_r$ behave in the limit $\epsilon_2 \to 0$ as $\breve{a}_r = \frac{1}{\epsilon_1}\mathsf{a}_r + \mathcal{O}(\epsilon_2)$, where a_r are the periods of the differential λ defined along a Lagrangian subspace of a canonical homology basis determined by σ.

One may in this sense view the "quantum periods" $\breve{a}_r$ as deformations of the action variables a_r for the Hitchin integrable system. It is interesting to note that the dependence on σ disappears in the limit $\epsilon \to 0$: Even if the definition of the coordinates a_r carries a residual dependence on σ, this is not the case for the fiber of Hitchin's fibration determined by the values of the a_r. In this way we are led to the conclusion that the scaling limits of the branes $\mathsf{L}^{(1)\hbar}_{\sigma,a}$ get represented by the (B,A,A)-branes supported on the fibers $\mathsf{F}_{(u_\mathsf{a},0)}$ of Hitchin's (first) fibration with trivial Chan-Paton bundle over a point u_a on the Hitchin base $\mathcal{B}$ determined by the coordinates a_r.

Let us finally note that one could discuss the branes $\mathsf{L}_x^{(2)\hbar}$ in a similar way. It is easy to see that one obtains the branes $\mathsf{L}_x^{(2)}$ supported on fibers of Hitchin's second fibration when $\hbar \to 0$.

7.4. Comparison II: Relation between Kapustin-Witten approach and CFT?. Any comparison between the approaches of Beilinson-Drinfeld and Kapustin-Witten will need to address the following point. Hecke-eigenbranes are described in [**34**] as skyscraper sheaves on $\mathcal{M}_H(G)$. However, in order to use SYZ mirror symmetry on the Hitchin fibration, Kapustin and Witten use the representation of $\mathcal{M}_H(G)$ and $\mathcal{M}_H(^LG)$ adapted to the complex structure I, whereas the Beilinson-Drinfeld approach considers ^LG-local systems $(\mathcal{E}, \nabla')$ on one side of the correspondence. In order to relate the two, one needs to use the NAH-correspondence. If μ represents a point of $\mathcal{M}_H(G)$, represented as a torus fibration, let $\check{\mathsf{F}}_\mu$ be the corresponding skyscraper sheaf in the B-model, and F_μ be the Hitchin fiber in the SYZ-dual A-model. In order to formulate a conjectural relation between the two approaches we need to use the (inverse of the) NAH-correspondence to find the point $\mu(\chi) \in \mathcal{M}_H(G)$ associated to the local system χ.

A natural guess for a possible relation between the approaches of Beilinson-Drinfeld and Kapustin-Witten could then be the validity of the isomorphism of $\mathcal{D}$-modules

$$\begin{aligned} \mathrm{Hom}_{\mathcal{M}_H(G)}(\mathsf{B}_{\mathrm{cc}}, \mathsf{F}_{\mu(\chi)}) &\simeq \mathcal{CB}(C, \hat{\mathfrak{g}}_{-h^\vee}, \mathcal{E}) \;\; \text{for} \;\; \mathcal{E} \in \mathrm{Bun}_G^{\mathrm{vs}} \\ &\simeq \big[\text{Fiber of } \Delta_\chi \;\; \text{over} \;\; \mathcal{E} \in \mathrm{Bun}_G^{\mathrm{vs}}\big] \end{aligned} \tag{7.2}$$

where $\mathrm{Bun}_G^{\mathrm{vs}}$ is the space of "very stable" bundles $\mathcal{E}$, bundles that do not admit a nilpotent Higgs field, and Δ_χ is the $\mathcal{D}$-module represented by $\mathcal{CB}(C, \hat{\mathfrak{g}}_{-h^\vee}, \mathcal{E})$. The right hand side does not depend on the choice of $\mathcal{E}$ due to the existence of a canonical flat connection identifying fibers associated to different $\mathcal{E} \in \mathrm{Bun}_G^{\mathrm{vs}}$. The dimension of $\mathcal{CB}(C, \hat{\mathfrak{g}}_{-h^\vee}, \mathcal{E})$ may jump away from $\mathrm{Bun}_G^{\mathrm{vs}}$, see [**15**, Section 9.5] for a discussion.

At this point we may note another puzzle arising in the comparison of the approaches of Beilinson-Drinfeld and Kapustin-Witten. As noted above, and illustrated by the examples studied in [**45**], there is a somewhat discontinuous behaviour of the $\mathcal{D}$-modules appearing in the Beilinson-Drinfeld approach to the geometric Langlands correspondence away from the sub-variety of opers within $\mathcal{M}_{dR}(^LG)$, described by the appearance of a number of additional degenerate representations in the representation of the $\mathcal{D}$-modules as conformal blocks. No such discontinuous behaviour is seen in the approach of Kapustin and Witten.

7.5. Conformal limit. We would also like to suggest a way which might lead to an answer for the questions raised in Subsection 7.4 above. It is based on the observation made in [**12, 20**] that the NAH correspondence may simplify drastically in the conformal limit where the parameter R introduced into the NAH correspondence by scaling $\varphi \to R\varphi$ is sent to zero together with the hyperkähler parameter ζ such that ζ/R stays finite. As a preparation we'd here like to discuss the possible relevance of this limit. The issues are similar, but non-identical to the ones concerning the possible relevance of hyperkähler rotations discussed in Section 7.2, motivating a separate discussion.

Replacing $\varphi \to R\varphi$ defines a one-parameter family of deformations of the NAH correspondence, leading to an apparent modification of the hyperkähler metric on

$\mathcal{M}_H(G)$. However, the parameter R is inessential for the geometry of Hitchin's moduli spaces in the sense that hyperkähler metrics associated to different values of R are related by diffeomorphisms. One may define one-parameter families of sigma model actions S_R using the hyperkähler metrics on $\mathcal{M}_H(G)$ obtained from the modified NAH correspondence. However, being constructed out of hyperkähler metrics related by diffeomorphisms, one may be tempted to identify two actions S_{R_1} and S_{R_2} differing only in the choice of the parameter R as physically equivalent Lagrangian representations for the *same* sigma model QFT. In suitable coordinates like $u = \frac{1}{2}\mathrm{tr}(\varphi^2)$ one finds that the relevant diffeomorphism is realised as a simple scaling $u \to ur^2$. The corresponding field redefinition should indeed lead to a rescaling of partition functions by inessential overall factors only.

If the partition functions $Z_R(u)$ defined using the actions S_R depend on a boundary parameter u associated to a coordinate for $\mathcal{M}_H(G)$ scaling under $\varphi \to R\varphi$ as $u \to ur^2$, we find that the R-independence of the sigma model metric modulo diffeomorphisms is expressed in the fact that $Z_{rR}(u) \propto Z_R(ur^2)$, again possibly up to inessential overall factors.

A partition function $Z'(\chi)$ in the boundary B-model with a boundary condition defined by a point χ on $\mathcal{M}_{dR}({}^LG)$ may be represented in terms of a partition function $Z''_R(\mu_{R,\zeta}(\chi))$ depending on a point $\mu_{R,\zeta}(\chi) \in \mathcal{M}_H({}^LG)$ since the R-dependence resulting from the modification of the NAH correspondence is compensated by the corresponding modification of the sigma model action. This observation may be useful if there is a limit where $\mu_{R,\zeta}(\chi)$ simplifies considerably. We will see that such a limit is the so-called conformal limit $R \to 0$, $\zeta \to 0$ keeping ζ/R finite. This will lead to an interesting reformulation of the proposed relation (7.2), as we shall now discuss.

7.6. Relation between Kapustin-Witten approach and CFT, more explicitly. We will mostly restrict attention to the case where the local systems χ are opers $\rho_u = \big(\mathcal{E}_o, \big(\partial_t + \big(\begin{smallmatrix} 0 & u \\ 1 & 0 \end{smallmatrix}\big)\big)dt\big)$, playing a basic role in the approach of Beilinson and Drinfeld. It is interesting to note that the conformal limit of the NAH correspondence becomes particularly simple in this case [**12**], relating opers ρ_u to Higgs pairs of a very particular form. For ${}^LG = \mathrm{PSL}(2,\mathbb{C})$ one finds Higgs pairs $(\mathcal{E},\varphi)$ of the form $(K^{1/2} \oplus K^{-1/2}, \big(\begin{smallmatrix} 0 & u \\ 1 & 0 \end{smallmatrix}\big)dz)$, where u is the quadratic differential representing ρ_u in the Fuchsian uniformisation of C. Passing to the description of $\mathcal{M}_H({}^LG)$ as a torus fibration represented by pairs $(\Sigma_u, \mathcal{L})$, where Σ_u is the spectral curve associated to a point $u \in \mathcal{B}$ of the Hitchin base, and $\mathcal{L}$ is a line bundle on Σ_u, one gets the line bundle $\mathcal{L}_0 = \pi^*(K^{1/2})$, with $\pi : \Sigma_u \to C$ being the covering projection. This line bundle represents a canonical "origin" of the Jacobian/Prym parametrising the choices of $\mathcal{L}$ [**49**].

Let $\check{\mathsf{F}}_{(u,0)}$ be the skyscraper sheaf on $\mathcal{M}_H({}^LG)$ supported at the point $(\Sigma_u, \mathcal{L}_0)$ and let $\mathsf{F}_{(u,0)}$ be the fiber of the Hitchin fibration which is SYZ dual to $\check{\mathsf{F}}_{(u,0)}$. The possible relation (7.2) between the approaches of Beilinson-Drinfeld and Kapustin-Witten may then be formulated more explicitly as

$$\mathrm{Hom}_{\mathcal{M}_H(G)}(\mathsf{B}_{\mathrm{cc}}, \mathsf{F}_{(u,0)}) \simeq \big[\text{Fiber of } \Delta_{\rho_u} \text{ over } \mathcal{E} \in \mathrm{Bun}_G^{\mathrm{vs}}\big]. \tag{7.3}$$

We remark that the image of generic points $(u, 0) \in \mathcal{M}_H({}^LG)$ under the NAH correspondence will be represented by an oper if and only if the map from $\mathcal{M}_H({}^LG)$

to $\mathcal{M}_{dR}({}^LG)$ is defined using the conformal limit of the non-abelian Hodge correspondence.[6] This indicates that this limit is well-suited for formulating the relation between the approaches of Beilinson-Drinfeld and Kapustin-Witten.

Concerning the generalisation of (7.3) to more general local systems χ we conjecture that there exist natural stratifications of $\mathcal{M}_{dR}({}^LG)$ and $\mathcal{M}_H({}^LG)$, having strata related to each other by the conformal limit of the NAH correspondence. This would allow us to extend the relation (7.3) to generic irreducible local systems, linking the discontinuous behaviour of the $\mathcal{D}$-modules appearing in the geometric Langlands correspondence to the passage from one stratum to another. We plan to return to this point in a forthcoming publication.

7.7. Outlook. We will elsewhere discuss available evidence for the existence of relations of the form

$$\mathcal{H}_x^{(2)} \simeq \int^{\oplus} d\mu_\sigma(a)\, \mathcal{H}_{\sigma,a}^{(1)}, \tag{7.4}$$

and for the existence of linear relations between the spaces $\mathcal{H}_{\sigma,a}^{(1)}$ associated to different pants decompositions σ. This restores a weaker version of σ-independence within the story associated to nonzero ϵ_2.

The geometric Langlands correspondence is sometimes presented as an analog of the spectral decompositions of spaces of automorphic forms appearing in the classical Langlands program. We view the contents of this note as hints that it may not be outrageous to dream of a variant of the geometric Langlands program extending it by transcendental and analytic aspects. The transcendental aspects involve the partition functions representing solutions to the systems of differential equations defined by the $\mathcal{D}$-modules, and the analytic aspects concern spectral decompositions as proposed in (7.4). Identifying the partition functions as analogs of the automorphic forms would strengthen the analogies to the classical Langlands program even further. The partition functions represent the bridge between the algebraic structures of $\mathcal{M}_H(G)$ associated to the representation as moduli space $\mathcal{M}_{dR}(G)$ of local systems, and as character variety $\mathcal{M}_B(G)$, respectively. In this way one may expect to get a larger picture unifying topological and complex structure dependent aspects of the geometric Langlands program.

We plan to discuss these matters, the interpretation as "quantum geometric Langlands duality", and the relation to another incarnation of Langlands duality patterns referred to as modular duality in [**45**] in forthcoming publications.

Let us finally note that recent progress on the geometric Langlands program from the gauge theory side has been described in [**21, 22**]. It should be interesting to analyse the relations to our work.

Acknowledgements: A.B would like to thank D. Ben-Zvi, M. Mulase, A. Neitzke and R. Wentworth for discussions and the organizers of String-Math 2016 for putting together a stimulating conference. A.B would also like to thank L'Institut Henri Poincaré (Paris) and ICTS-TIFR (Bangalore) for hospitality during visits when this work was in progress.

---

[6]The direction "if" was shown in [**12**]. The image of points $(u, 0) \in \mathcal{M}_H({}^LG)$ under NAH consists of connections with real holonomy, intersecting the variety of opers only discretely. We thank A. Neitzke for this remark.

J.T. would like to thank M. Mulase, A. Neitzke for discussions. Special thanks to E. Frenkel for various discussions, for communicating the content of his unpublished work [**16**], and for critical remarks on the draft.

Appendix A. Hitchin's moduli spaces

We assume that $G = SL(2)$, and that C is a Riemann surface with genus g and n punctures.

Hitchin moduli space $\mathcal{M}_H(G)$ [**29**]. Moduli space of pairs $(\mathcal{E}, \varphi)$, where $\mathcal{E} = (E, \bar{\partial}_{\mathcal{E}})$ is a holomorphic structure on a smooth vector bundle E, and $\varphi \in H^0(C, \mathrm{End}(\mathcal{E}) \otimes K)$. The moduli space of such pairs modulo natural gauge transformations is denoted by $\mathcal{M}_H(G)$.

Hitchin's integrable system [**29**]. Given $(\mathcal{E}, \varphi)$ one constructs the spectral curve $\Sigma = \{(u, v); v^2 = \frac{1}{2}\mathrm{tr}(\varphi^2)\} \subset T^*C$, and the line bundle $\mathcal{L}$ representing the cokernel of $\varphi - v$. One may reconstruct $(\mathcal{E}, \varphi)$ from $(\Sigma, \mathcal{L})$ as $\mathcal{E} = \pi_*(\mathcal{L})$ and $\varphi = \pi_*(v)$. This describes $\mathcal{M}_H(G, C)$ as a torus fibration over the base $\mathcal{B} \simeq H^0(C, K^2)$, with fibres representing the choices of $\mathcal{L}$ identified with the Jacobian of Σ if $G = GL(2)$, and with the Prym variety if $G = SL(2)$. Natural coordinates for the base $\mathcal{B}$ are provided by Hitchin's Hamiltonians, defined by expanding $\frac{1}{2}\mathrm{tr}(\varphi^2) = \sum_{r=1}^{3g-3+n} \vartheta_r H_r$, with $\{\vartheta_r, r = 1, \ldots, 3g - 3 + n\}$ being a basis for $H^0(C, K^2)$.

Local systems. Pairs $(\mathcal{E}, \nabla'_\epsilon)$, where $\mathcal{E}$ is a holomorphic vector bundle as above, and ∇'_ϵ is a holomorphic ϵ-connection, satisfying $\nabla'_\epsilon(fs) = \epsilon(\partial f)s + f\nabla'_\epsilon s$ for functions f and smooth sections s of E. The moduli space of such pairs is denoted $\mathcal{M}_{dR}(G)$. Local systems are here often identified with the corresponding *flat bundles*, systems of local trivialisations with constant transitions functions, or the representations of the fundamental group (modulo conjugation) obtained as holonomy of $(\mathcal{F}, \nabla'_\epsilon)$, leading to the isomorphism between $\mathcal{M}_{dR}(G)$ and the

Character variety $\mathcal{M}_B(G)$: The space of representations of $\pi_1(C)$ into G, modulo overall conjugation, as algebraic variety described as a GIT quotient $\mathbb{C}[\mathrm{Hom}(\pi_1(C), G)]^G$.

Opers. Special local systems, where $\mathcal{E} = \mathcal{E}_{\mathrm{op}}$, the unique extension $0 \to K^{1/2} \to \mathcal{E}_{\mathrm{op}} \to K^{-1/2} \to 0$ allowing a holomorphic connection ∇'_ϵ of the form $\nabla'_\epsilon = dz\big(\epsilon\partial_z + \left(\begin{smallmatrix} 0 & u \\ 1 & 0 \end{smallmatrix}\right)\big)$.

Non-Abelian Hodge (NAH) correspondence [**29, 44**]. Given a Higgs pair $(\mathcal{E}, \varphi)$, there exists a unique harmonic metric h on E satisfying $F_{\mathcal{E},h} + R^2[\varphi, \varphi^{\dagger_h}] = 0$ where $F_{\mathcal{E},h}$ is the curvature of the unique h-unitary connection $D_{\mathcal{E},h}$ having $(0,1)$-part $\bar{\partial}_{\mathcal{E}}$. One may then form the corresponding two-parameter family of flat connections $\nabla_{\zeta,R} = \zeta^{-1}R\varphi + D_{\mathcal{E},h} + R\zeta\,\varphi^{\dagger_h}$. Decomposing $\nabla_{\zeta,R}$ into the $(1,0)$ and $(0,1)$-parts defines a pair $(\mathcal{F}, \nabla'_\epsilon)$ consisting of $\mathcal{F} = (E, \bar{\partial}_{\mathcal{F}})$ and the ϵ-connection $\nabla'_\epsilon = \epsilon\nabla' = \epsilon\partial_{\mathcal{E},h} + \varphi$, with $\epsilon = \zeta/R$, holomorphic in the complex structure defined by $\bar{\partial}_{\mathcal{F}}$.

Hyperkähler structure [**29**]. There exists a $\mathbb{P}^1$ worth of complex structures I_ζ and holomorphic symplectic structures Ω_ζ. The latter are defined as $\Omega_\zeta = \frac{1}{2}\int_C \mathrm{tr}(\delta\mathcal{A}_\zeta \wedge \delta\mathcal{A}_\zeta)$. A triplet of symplectic forms $(\omega_I, \omega_J, \omega_K)$ can be defined by expanding Ω_ζ as $\Omega_\zeta = \frac{1}{2\zeta}(\omega_J + i\omega_K) + i\omega_I + \frac{1}{2}\zeta(\omega_J - i\omega_K)$. The corresponding complex structures are $I_\zeta = \frac{1}{1+|\zeta|^2}((1 - |\zeta|^2)I - i(\zeta - \bar{\zeta})J - (\zeta + \bar{\zeta})K)$.

Complex Fenchel-Nielsen coordinates [**39**]. Darboux coordinates for $\mathcal{M}_B(G)$ associated to pants decompositions σ of C obtained by cutting along closed curves γ_i, $i = 1, \ldots, 3g-3+n$. The complex length coordinates parameterise the trace functions $L_i = \mathrm{tr}(\rho(\gamma_i))$ as $L_i = 2\cosh(a_i/2)$. One may define canonically conjugate coordinates κ_r such that the natural Poisson structure gets represented as $\{a_r, \kappa_s\} = \delta_{r,s}$, $\{a_r, a_s\} = 0 = \{\kappa_r, \kappa_s\}$.

References

[1] Mina Aganagic, Edward Frenkel, and Andrei Okounkov, *Quantum q-Langlands Correspondence*, (2017).

[2] Luis F. Alday, Davide Gaiotto, Sergei Gukov, Yuji Tachikawa, and Herman Verlinde, *Loop and surface operators in $\mathcal{N}=2$ gauge theory and Liouville modular geometry*, J. High Energy Phys. **1** (2010), 113, 50. MR2660780

[3] Luis F. Alday, Davide Gaiotto, and Yuji Tachikawa, *Liouville correlation functions from four-dimensional gauge theories*, Lett. Math. Phys. **91** (2010), no. 2, 167–197. MR2586871

[4] Luis F. Alday and Yuji Tachikawa, *Affine* SL(2) *conformal blocks from 4d gauge theories*, Lett. Math. Phys. **94** (2010), no. 1, 87–114. MR2720257

[5] D. Arinkin, *Irreducible connections admit generic oper structures*, arXiv:1602.08989 (2016).

[6] D. Arinkin and D. Gaitsgory, *Singular support of coherent sheaves and the geometric Langlands conjecture*, Selecta Math. (N.S.) **21** (2015), no. 1, 1–199. MR3300415

[7] D. Ben-Zvi and D. Nadler, *Betti Geometric Langlands*, arXiv:1606.08523 (2016).

[8] Alexander Braverman, *Instanton counting via affine Lie algebras. I. Equivariant J-functions of (affine) flag manifolds and Whittaker vectors*, Algebraic structures and moduli spaces, CRM Proc. Lecture Notes, vol. 38, Amer. Math. Soc., Providence, RI, 2004, pp. 113–132. MR2095901

[9] Sergio Cecotti and Cumrun Vafa, *Topological–anti-topological fusion*, Nuclear Phys. B **367** (1991), no. 2, 359–461. MR1139739

[10] Clay Cordova and Daniel L. Jafferis, *Toda Theory From Six Dimensions*, (2016).

[11] Nadav Drukker, Jaume Gomis, Takuya Okuda, and Jörg Teschner, *Gauge theory loop operators and Liouville theory*, J. High Energy Phys. **2** (2010), 057, i, 61. MR2672742

[12] O. Dumitrescu, L. Fredrickson, G. Kydonakis, R. Mazzeo, M. Mulase, and A. Neitzke, *Opers versus nonabelian Hodge*, arXiv:1607.02172 (2016).

[13] Boris Feigin and Edward Frenkel, *Affine Kac-Moody algebras at the critical level and Gel′fand-Dikiĭ algebras*, Infinite analysis, Part A, B (Kyoto, 1991), Adv. Ser. Math. Phys., vol. 16, World Sci. Publ., River Edge, NJ, 1992, pp. 197–215. MR1187549

[14] Edward Frenkel and David Ben-Zvi, *Vertex algebras and algebraic curves*, 2nd ed., Mathematical Surveys and Monographs, vol. 88, American Mathematical Society, Providence, RI, 2004. MR2082709

[15] Pierre Cartier, Bernard Julia, Pierre Moussa, and Pierre Vanhove (eds.), *Frontiers in number theory, physics, and geometry. I*, Springer-Verlag, Berlin, 2006. On random matrices, zeta functions, and dynamical systems; Papers from the meeting held in Les Houches, March 9–21, 2003; Corrected second printing of the 2006 original. MR2251029

[16] ———, *On the brane interpretation of conformal blocks*, available at http://edwardfrenkel.com/confblocks.pdf (2010).

[17] Edward Frenkel, Sergei Gukov, and Jörg Teschner, *Surface operators and separation of variables*, J. High Energy Phys. **1** (2016), 179, front matter+52. MR3471401

[18] Toshiaki Fujimori, Taro Kimura, Muneto Nitta, and Keisuke Ohashi, *2d partition function in Ω-background and vortex/instanton correspondence*, J. High Energy Phys. **12** (2015), 110, front matter+40. MR3464692

[19] Davide Gaiotto, *Surface operators in $\mathcal{N}=2$ 4d gauge theories*, J. High Energy Phys. **11** (2012), 090, front matter + 24. MR3036474

[20] Davide Gaiotto, *Opers and TBA*, arXiv:1403.6137 (2014).

[21] Davide Gaiotto *S-duality of boundary conditions and the Geometric Langlands program*, arXiv:1609.09030 (2016).

[22] Davide Gaiotto *Twisted compactifications of 3d $N=4$ theories and conformal blocks*, arXiv:1611.01528 (2016).

[23] Davide Gaiotto, Gregory W. Moore, and Andrew Neitzke, *Wall-crossing, Hitchin systems, and the WKB approximation*, Adv. Math. **234** (2013), 239–403. MR3003931

[24] Davide Gaiotto and Edward Witten, *S-duality of boundary conditions in $\mathcal{N}=4$ super Yang-Mills theory*, Adv. Theor. Math. Phys. **13** (2009), no. 3, 721–896. MR2610576

[25] Davide Gaiotto and Edward Witten, *Supersymmetric boundary conditions in $\mathcal{N}=4$ super Yang-Mills theory*, J. Stat. Phys. **135** (2009), no. 5-6, 789–855. MR2548595

[26] Davide Gaiotto and Edward Witten, *Knot invariants from four-dimensional gauge theory*, Adv. Theor. Math. Phys. **16** (2012), no. 3, 935–1086. MR3024278

[27] Dennis Gaitsgory, *Outline of the proof of the geometric Langlands conjecture for GL_2* (English, with English and French summaries), Astérisque **370** (2015), 1–112. MR3364744

[28] Jaume Gomis, Bruno Le Floch, Yiwen Pan, and Wolfger Peelaers, *Intersecting Surface Defects and Two-Dimensional CFT*, (2016).

[29] N. J. Hitchin, *The self-duality equations on a Riemann surface*, Proc. London Math. Soc. (3) **55** (1987), no. 1, 59–126. MR887284

[30] Daigo Honda and Takuya Okuda, *Exact results for boundaries and domain walls in 2d supersymmetric theories*, J. High Energy Phys. **9** (2015), 140, front matter+63. MR3430625

[31] Kentaro Hori, Amer Iqbal, and Cumrun Vafa, *D-branes and mirror symmetry*, (2000).

[32] Kentaro Hori and Mauricio Romo, *Exact Results In Two-Dimensional (2,2) Supersymmetric Gauge Theories With Boundary*, (2013).

[33] Anton Kapustin, *A Note on Quantum Geometric Langlands Duality, Gauge Theory, and Quantization of the Moduli Space of Flat Connections*, (2008).

[34] Anton Kapustin and Edward Witten, *Electric-magnetic duality and the geometric Langlands program*, Commun. Number Theory Phys. **1** (2007), no. 1, 1–236. MR2306566

[35] Taro Kimura and Vasily Pestun, *Quiver W-algebras*, arXiv:1512.08533 (2015).

[36] Can Kozçaz, Sara Pasquetti, Filippo Passerini, and Niclas Wyllard, *Affine* $\mathrm{sl}(N)$ *conformal blocks from $\mathcal{N}=2\,\mathrm{SU}(N)$ gauge theories*, J. High Energy Phys. **1** (2011), 045, 40. MR2792288

[37] Satoshi Nawata, *Givental J-functions, quantum integrable systems, AGT relation with surface operator*, Adv. Theor. Math. Phys. **19** (2015), no. 6, 1277–1338. MR3501061

[38] Andrei Negut, *Affine Laumon Spaces and the Calogero-Moser Integrable System*, arXiv:1112.1756 (2011).

[39] N. Nekrasov, A. Rosly, and S. Shatashvili, *Darboux coordinates, Yang-Yang functional, and gauge theory*, Nuclear Phys. B Proc. Suppl. **216** (2011), 69–93. MR2851597

[40] Nikita Nekrasov, *BPS/CFT correspondence: non-perturbative Dyson-Schwinger equations and qq-characters*, JHEP **03** (2016), 181.

[41] Nikita Nekrasov and Edward Witten, *The omega deformation, branes, integrability and Liouville theory*, J. High Energy Phys. **9** (2010), 092, i, 82. MR2776942

[42] Nikita A. Nekrasov and Samson L. Shatashvili, *Quantization of integrable systems and four dimensional gauge theories*, XVIth International Congress on Mathematical Physics, World Sci. Publ., Hackensack, NJ, 2010, pp. 265–289. MR2730782

[43] Vasily Pestun, *Localization of gauge theory on a four-sphere and supersymmetric Wilson loops*, Comm. Math. Phys. **313** (2012), no. 1, 71–129. MR2928219

[44] Carlos T. Simpson, *Higgs bundles and local systems*, Inst. Hautes Études Sci. Publ. Math. **75** (1992), 5–95. MR1179076

[45] J. Teschner, *Quantization of the Hitchin moduli spaces, Liouville theory and the geometric Langlands correspondence I*, Adv. Theor. Math. Phys. **15** (2011), no. 2, 471–564. MR2924236

[46] J. Teschner and G. S. Vartanov, *Supersymmetric gauge theories, quantization of $\mathcal{M}_{\mathrm{flat}}$, and conformal field theory*, Adv. Theor. Math. Phys. **19** (2015), no. 1, 1–135. MR3418499

[47] Edward Witten, *Geometric Langlands From Six Dimensions*, arXiv:0905.2720 (2009).

[48] Edward Witten, *Fivebranes and knots*, Quantum Topol. **3** (2012), no. 1, 1–137. MR2852941

[49] Edward Witten, *More On Gauge Theory And Geometric Langlands*, arXiv:1506.04293 (2015).

Department Mathematik, Universität Hamburg, Bundesstrasse 55, 20146 Hamburg, Germany; and DESY theory, Notkestrasse 85, 22607 Hamburg, Germany
Email address: aswin@physics.rutgers.edu

Department Mathematik, Universität Hamburg, Bundesstrasse 55, 20146 Hamburg, Germany; and DESY theory, Notkestrasse 85, 22607 Hamburg, Germany
Email address: joerg.teschner@desy.de

Proceedings of Symposia in Pure Mathematics
Volume **98**, 2018
http://dx.doi.org/10.1090/pspum/098/01726

A journey from the Hitchin section to the oper moduli

Olivia Dumitrescu

To Anthony

ABSTRACT. This paper provides an introduction to the mathematical notion of *quantum curves.* We start with a concrete example arising from a graph enumeration problem. We then develop a theory of quantum curves associated with Hitchin spectral curves. A conjecture of Gaiotto, which predicts a new construction of opers from a Hitchin spectral curve, is explained. We give a step-by-step detailed description of the proof of the conjecture for the case of rank 2 Higgs bundles. Finally, we identify the two concepts of *quantum curve* arising from the topological recursion formalism with the limit oper of Gaiotto's conjecture.

CONTENTS

1. Introduction

Mathematical research is a journey. We start from one place, often a remote place nobody cares. Guided by mysteries one after another, we arrive at a place we have never imagined. We then suddenly realize that many people have come to the same place, starting from totally different origins.

2010 *Mathematics Subject Classification.* Primary 58E15, 53C07; Secondary 14D21, 81T13.
Key words and phrases. Hitchin's equations, moduli space of Higgs bundles, opers, Non-abelian Hodge correspondence, quantum curves.

The author is a member of the Simion Stoilow Institute of Mathematics of the Romanian Academy.

These are the lectures that the author has delivered in the last few years in many places of the world. They are meant to be an introduction to the notion of *quantum curves*. Yet the honest feeling that the author has now is that these are more a record of how her understanding of quantum curves has evolved. The mathematics of quantum curves itself has been changing over the years. We have started from one place, based on what is known as *topological recursion*. When we have arrived at the current position, we find ourselves dealing with *opers*.

The notion of *quantum curves* was conceived in string theory by Aganagic, Dijkgraaf, Gukov, Hollands, Klemm, Marino, Sulkowski, Vafa, and others [**1**, **5**, **6**, **22**]. We are far from establishing a complete theory at this moment. Yet we hope these lectures give a snapshot of what is understood in the mathematics community now, at least one of the many sides of the story of quantum curves.

This paper is organized as follows. In Section 2, we start from a simple question in enumerative geometry, and obtain the Dijkgraaf-Verlinde-Verlinde formula [**7**] for intersection numbers of ψ-classes on moduli space of stable curves $\overline{\mathcal{M}}_{g,n}$. More precisely, in Section 2.3, we use the *edge contraction* operations of ribbon graphs to generalize a count of graphs, and establish a recursion of *Catalan numbers* of arbitrary genera. Then in Section 2.5, we present how the Laplace transform of the recursion of Catalan numbers surprisingly gives the DVV formula for intersection numbers on $\overline{\mathcal{M}}_{g,n}$. By the WKB analysis the recursion relation becomes equivalent to the *quantization of the spectral curve* of the Catalan numbers. In Section 2.8, we present how the same set of edge contraction operations on ribbon graphs give the *cut-and-join equations* for orbifold Hurwitz numbers.

We start with presenting an introduction to the geometry of the Hitchin moduli spaces of holomorphic Higgs bundles and connections in Section 4. We then generalize the quantization theorem of Catalan recursion 2.5, replacing the concept of spectral curves of Section 2.4 by the framework of Hitchin spectral curves. More precisely, following [**10**, **11**, **13**], we present the quantization results of spectral curves for holomorphic and meromorphic Higgs bundles of rank 2. Here, the algebro-geometric technique presented in Section 4.1 was indispensable in quantizing *singular* Hitchin spectral curves [**11**].

In Section 5, using the work of Gunning [**23**], we propose to identify the two concepts: *quantum curves* and *opers*. The new idea of quantization in these notes is based on a recent solution [**9**] of a conjecture due to the physicist Gaiotto [**18**], presented in Sections 6 and 7.

2. Enumeration of ribbon graphs

2.1. A combinatorial model for the moduli space of curves $\mathcal{M}_{g,n}$. *Ribbon graphs* are combinatorial objects first used by G. 't Hooft [**41**] in quantum gauge theory, and later by Kontsevich [**27**] in random matrix theory as the first approach to Gromov-Witten theory. They appeared independently in the work of Grothendieck [**21**] and are also known as *dessins d'enfants*. A ribbon graph as a graph has a *cyclic ordering* of the set of incident half-edges at each vertex and labeled faces. A ribbon graph embeds into an oriented compact topological surface of type (g, n), where g represents the genus of the surface and n the number of marked points corresponding to the faces of the ribbon graph.

Decorating ribbon graphs by a positive real number on each edge fixes a topological surface of type (g, n) together with *a complex structure* on it. Introduce

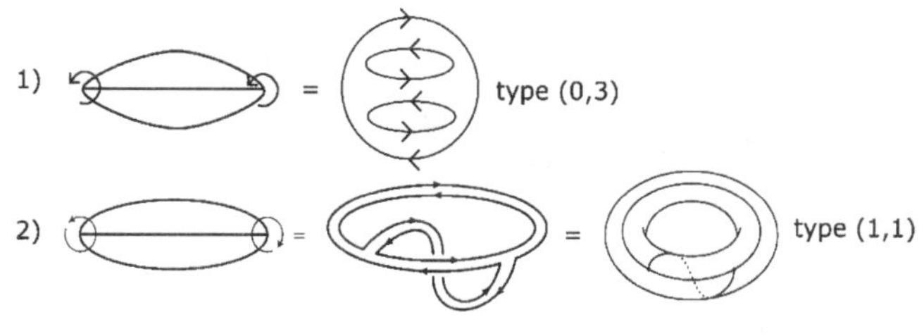

FIGURE 1

the total space of graphs as an orbifold parametrizing decorated ribbon graphs of a given topological type (g, n):

$$\mathcal{R}_{g,n} = \coprod_{\substack{\Gamma \text{ boundary labeled} \\ \text{ribbon graph} \\ \text{of type } (g,n) \\ \text{and } e(\Gamma) \text{ edges}}} \frac{\mathbb{R}_+^{e(\Gamma)}}{\mathrm{Aut}(\Gamma)}$$

The space $\mathcal{R}_{g,n}$ is a smooth orbifold (see [**30**, Section 3] and [**39**]). The combinatorial model of moduli space was constructed by Thurston [**39**], Harer [**24, 25**], Mumford [**33**], and Strebel [**40**] (cf. [**30**]). There exists an orbifold isomorphism between the total space of graphs of type (g, n) and the product of $\mathbb{R}_+^n$ and the moduli space $\mathcal{M}_{g,n}$ of smooth algebraic curves of genus g with n marked points:

$$\mathcal{R}_{g,n} \cong \mathcal{M}_{g,n} \times \mathbb{R}_+^n. \tag{2.1.1}$$

The isomorphism (2.1.1) gives a cell-decompositions of the moduli space $\mathcal{M}_{g,n}$ for each choice of $p \in \mathbb{R}_+^n$, and generalized Catalan numbers are related to a count of *lattice points* in each cell-decomposition for $p \in \mathbb{Z}_+^n$. The isomorphism (2.1.1) enables us to use the combinatorial model for the study of topology of $\mathcal{M}_{g,n}$ via ribbon graphs and their geometry. Starting from a count of graphs, or the number of orbi-cells in $\mathcal{R}_{g,n}$, the corresponding enumerative problem on $\mathcal{M}_{g,n}$ surprisingly becomes the intersection numbers of the ψ classes on $\overline{\mathcal{M}}_{g,n}$.

2.2. The Combinatorics of Catalan recursion. In combinatorics, the Catalan numbers form a sequence of natural numbers that occur in various counting problems for recursively defined objects. They also appear in nature, and have more than twenty alternative definitions. To extend one of these interpretations we define the generalized Catalan number to count a number of graphs on a Riemann surface of genus g with n marked points. We define a *cell graph* to be a ribbon graph with labeled vertices. We introduce the generalized Catalan numbers, $C_{g,n}(\mu_1, \ldots, \mu_n)$, as the *number of cell graphs* of type (g, n) with an outgoing arrow and degree μ_i at each vertex i. In Figure (2(1)), we give an example of a vertex of degree 7.

EXAMPLE 2.1. For $(g, n) = (0, 1)$, the Catalan numbers C_m count graphs on a Riemann sphere with one vertex (so all edges are loops). We recall the classical definition of Catalan numbers counting the number of expressions containing m pairs of parentheses which are correctly matched. It is easy to see that $C_{0,1}(2m) = C_m$. In Figure (2(2)) we provide an example explaining why each cell graph corresponds to a pair of parenthesis for $m = 3$.

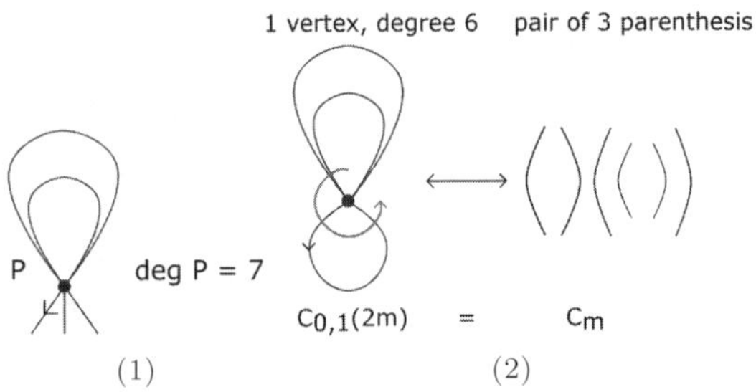

FIGURE 2

Since for the classical Catalan numbers there is the recursion

$$C_m = \sum_{a+b=m-1} C_a C_b,$$

we expect to find a similar recursion for the generalized Catalan number $C_{g,n}(\mu_1, \ldots, \mu_n)$.

2.3. Edge contraction operation and Catalan recursion.

THEOREM 2.1 (Theorem 3.2, [**16**], [**42**]). *For* $2g-2+n \geq 0$, $n \geq 1$, *the generalized Catalan numbers satisfy the following recursion*

$$C_{g,n}(\mu_1, \ldots, \mu_n) = \sum_{j=2}^{n} \mu_j \cdot C_{g,n-1}(\mu_1 + \mu_j - 2, \mu_2, \ldots, \widehat{\mu_j}, \ldots, \mu_n)$$

$$+ \sum_{\alpha+\beta=\mu_1-2} \Bigg[C_{g-1,n+1}(\alpha, \beta, \mu_2, \ldots, \mu_n)$$

$$+ \sum_{g_1+g_2=g \, I \sqcup J=\{2,\ldots,n\}} C_{g_1,|I|+1}(\alpha, \mu_I) \cdot C_{g_2,|J|+1}(\beta, \mu_J) \Bigg].$$

PROOF. Starting from a cell graph with an arrowed edge at vertex 1 we will contract this edge to a point, and we call this an **Edge Contraction Operation**. We distinguish two cases.

Case 1. The arrowed edge connects vertex 1 of degree μ_1 with the vertex $j > 1$ of degree μ_j. We contract the edge and we join the two vertices 1 and j together as shown in figure below. The resulting graph has one less vertex, but the genus is the same, the degree of the newly created vertex is $\mu_1 + \mu_j - 2$; we mark the edge that was immediately counterclockwise of the contracted edge, as indicated in Figure 3(1).

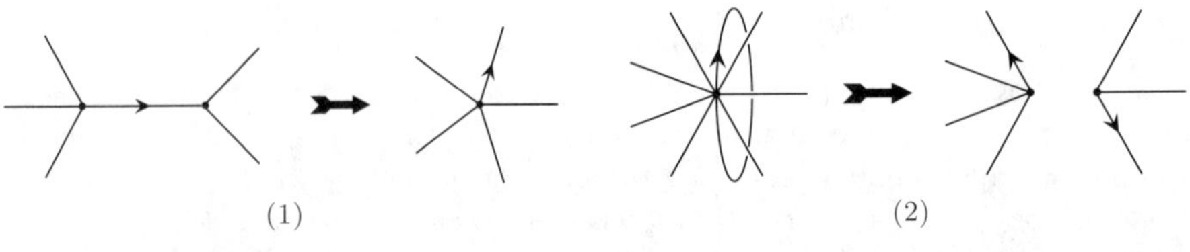

FIGURE 3

Case 2. The arrowed edge is a loop attached to vertex 1, then we remove this loop from the cell graph, and separate the vertex into two vertices. The loop classifies all other incident edges into two groups with α edges in one group, and β the rest. Then $\alpha + \beta = \mu_1 - 2$, and we create two vertices of degrees α and β as in the Figure 3(2).

□

2.4. Spectral curve for the Catalan recursion.

QUESTION 2.2. *What is the mirror dual of Catalan numbers?*

We will construct the *spectral curve* of Catalan numbers from the $(g, n) = (0, 1)$ unstable geometry. We observe first that the recursion in Theorem 2.1 for the unstable range gives the well-known Catalan recursion of Example 2.1, i.e.,

$$C_m = \sum_{a+b=m-1} C_a \cdot C_b. \tag{2.4.1}$$

We define the generating function of Catalan numbers

$$z(x) := \sum_{m=0}^{\infty} C_m \cdot x^{-(2m+1)}.$$

The Catalan recursion (2.4.1) is equivalent to the series $x - z(x)$ being the inverse of the $z(x)$. We thus discover the *spectral curve of the Catalan recursion*:

$$z^2 - z \cdot x + 1 = 0. \tag{2.4.2}$$

This is the *mirror dual* of Catalan numbers.

2.5. Genesis of Enumerative Geometry: Gromov-Witten invariants of a point. By the orbifold isomorphism (2.1.1) we will deduce that the count of graphs is equivalent to an enumerative question on $\overline{\mathcal{M}}_{g,n}$. In the stable range we will consider first the generating function of generalized Catalan numbers, or *free energies*:

$$F_{g,n}(x_1, \ldots, x_n) := \sum_{\mu_1, \ldots, \mu_n > 0} \frac{C_{g,n}(\mu_1, \ldots, \mu_n)}{\mu_1 \cdots \mu_n} \prod_{i=1}^{n} x_i^{-\mu_i}.$$

Surprisingly, the generating function $F_{g,n}$ knows $\chi(\mathcal{M}_{g,n})$ and intersection numbers of $\overline{\mathcal{M}}_{g,n}$!

We now perform a change of coordinates. For each of the variables x_i we introduce a variable t_i by

$$x_i := 2 \cdot \frac{t_i^2 + 1}{t_i^2 - 1}, \quad z_i := \frac{t_i + 1}{t_i - 1}. \tag{2.5.1}$$

With this change of variables $F_{g,n}(t_1, \ldots, t_n)$ becomes a *Laurent polynomial* of degree
$3 \cdot (2n - 2 + n)$ with beautiful geometric properties

(1) $F_{g,n}(1, \ldots, 1) = (-1)^n \chi(\mathcal{M}_{g,n})$
(2) $F_{g,n}(t_1, \ldots, t_n) = F_{g,n}(\frac{1}{t_1}, \ldots, \frac{1}{t_n})$

In Section 4.1 we will answer the following question:

QUESTION 2.3. *Why do we have to perform the change of variables* (2.5.1) *in order to see the topological information encoded by* $F_{g,n}$*?*

The leading terms of $F_{g,n}(t_1,\dots,t_n)$ form a homogeneous polynomial of degree $3(2n-2+n)$

$$F_{g,n}(t_1,\dots,t_n)^{top} = \frac{(-1)^n}{2^{2g-2+n}} \cdot \sum_{\substack{d_1+\cdots+d_n\\=3g-3+n}} \langle \tau_{d_1}\cdots\tau_{d_n}\rangle \cdot \prod_{i=1}^{n} (|2d_i-1|)!! \cdot \left(\frac{t_i}{2}\right)^{2d_i+1},$$

where the symbol

$$\langle \tau_{d_1}\cdots\tau_{d_n}\rangle := \int_{\overline{\mathcal{M}}_{g,n}} c_1(\mathbb{L}_1)^{d_1}\cdots c_1(\mathbb{L}_n)^{d_n}$$

denotes cotangent class intersection numbers on $\overline{\mathcal{M}}_{g,n}$. Furthermore, the Catalan recursion obtained via Edge Contraction Operations in Theorem 2.1 translates into an infinite system of differential equations known as the *Dijkgraaf-Verlinde-Verlinde equation* [**7**] of the intersection numbers. This implies the celebrated theorem of Kontsevich, Mirzakhani, Okounkov-Pandharipande and others, on the Witten conjecture.

THEOREM 2.4 (Theorem 6.1, [**16**]). *The Witten-Kontsevich intersection numbers satisfy the DVV equation.*

Surprisingly, the recursion relations of Theorem 2.4 can be encoded compactly into a single ordinary differential equation. Namely, there exist a differential operator, what we call a *quantum curve*, that annihilates the generating function of the free energies. More precisely, the first quantization result was proved by Mulase and Sułkowski following a conjecture of Gukov and Sułkowski:

THEOREM 2.5 ([**31**]). *Let $\hbar$ be a formal parameter. Then we have*

$$\left(\hbar^2\cdot\frac{d^2}{dx^2}+\hbar\cdot x\cdot\frac{d}{dx}+1\right)\exp\left(\sum_{2g-2+n\geq -1}\frac{1}{n!}\cdot\hbar^{2g-2+n}\cdot F_{g,n}(t,\dots,t)\right)=0.$$

Here, the scalar $\hbar$ is a deformation parameter. In Section 5, we will see that mathematically $\hbar$ is an extension class of line bundles on an algebraic curve.

QUESTION 2.6. (1) *Why does this complicated function satisfy such a simple differential equation?*
(2) *Where does this differential equation come from?*

Let us replace z by $-\hbar\frac{d}{dx}$ in the *the spectral curve equation of the Catalan recursion* (2.4.2)

$$z^2 - x\cdot z + 1 = 0.$$

We obtain precisely the differential operator known as the *the quantum curve of the Catalan spectral curve* in Theorem 2.5:

$$\hbar^2\cdot\frac{d^2}{dx^2}+\hbar\cdot x\cdot\frac{d}{dx}+1.$$

The following questions are natural.

QUESTION 2.7. (1) *From the shape of the above equations, it looks like a canonical quantization of the spectral curve. Is it really the case?*
(2) *If so, then what is the mathematical framework that explains this surprising phenomenon?*

2.6. Cut-and-Join equation for orbifold Hurwitz numbers. Another example of the use of edge contraction operations on cell graphs gives a surprising enumerative problem of Hurwitz numbers. Let $H^r_{g,n}(\mu_1,\dots,\mu_n)$ denote the number of topological types of regular maps from a smooth curve of genus g to $\mathbb{P}^1$ with profile $(\overbrace{r,\dots,r}^{m})$ over $0\in\mathbb{P}^1$, labeled profile $(\mu_1,\dots,\mu_n)$ over $\infty\in\mathbb{P}^1$, and simple ramification at any other ramification points, weighted with automorphisms of such maps. These numbers are referred to as *orbifold Hurwitz numbers*. For $r=1$, they count simple Hurwitz numbers. In [**12**] we generalized the notion of *branching graph* of Okounkov and Pandharipande [**36**] ($r=1$) to *r-Hurwitz graphs* (arbitrary r), via combinatorics. An *r-Hurwitz graphs* is a cell graph with a collection of dots associated to each vertex ([Definition 6.1, [**12**]]). Counting the number of *r-Hurwitz graphs* via *edge contraction operations* we recover the Cut-and-Join equation for orbifold Hurwitz numbers. Denoting $\mathcal{H}^r_{g,n}(\mu_1\dots,\mu_n)=\mu_1\dots\mu_n H^r_{g,n}(\mu_1\dots,\mu_n)$, we obtain

THEOREM 2.8 (Theorem 6.6, [**12**], Cut-and-Join equation for orbifold Hurwitz numbers). *The number of arrowed Hurwitz graphs satisfy the following edge-contraction formula.*

$$
\begin{aligned}
&\left(2g-2+\frac{d}{r}+n\right)\mathcal{H}^r_{g,n}(\mu_1\dots,\mu_n)\\
&\quad=\sum_{i<j}\mu_i\mu_j\mathcal{H}^r_{g,n-1}(\mu_1,\dots,\mu_{i-1},\mu_i+\mu_j,\mu_{i+1},\dots,\widehat{\mu_j},\dots,\mu_n)\\
&\quad+\frac{1}{2}\sum_{i=1}^{n}\mu_i\sum_{\substack{\alpha+\beta=\mu_i\\ \alpha,\beta\geq 1}}\Bigg[\mathcal{H}^r_{g-1,n+1}(\alpha,\beta,\mu_1,\dots,\widehat{\mu_i},\dots,\mu_n)\\
&\quad+\sum_{\substack{g_1+g_2=g\\ I\sqcup J=\{1,\dots,\hat{i},\dots,n\}}}\mathcal{H}^r_{g_1,|I|+1}(\alpha,\mu_I)\mathcal{H}^r_{g_2,|J|+1}(\beta,\mu_J)\Bigg].
\end{aligned}
\tag{2.6.1}
$$

Here, $\widehat{\ }$ *indicates the omission of the index, and* $\mu_I=(\mu_i)_{i\in I}$ *for any subset* $I\subset\{1,2,\dots,n\}$.

The restriction to the $(0,1)$ unstable cases of Theorem 2.8 recovers the *spectral curve of the orbifold Hurwitz numbers*, that is known as the r-Lambert curve. The computations are similar to the ones explained in Section 2.2, leading to the *mirror curve of Hurwitz numbers*

$$x^r=ye^{-ry}.$$

Edge contraction operations are graphical manifestation of a Frobenius algebra structure and it was shown in [Corollary 4.8, [**12**])] that they give alternative axiomatic definition of 2 dimensional *topological quantum field theory* (2d TQFT). We further emphasize the importance of these operations by relating the 2d TQFT formula of [Corollary 4.8, [**12**])] with the count of points of a character variety for a finite group and Hodge-Deligne polynomial of a character variety in [**14, 15**].

While Catalan numbers have an algebraic spectral curve, Hurwitz numbers have an analytic spectral curve. Later on we will focus on rank 2 Higgs bundles, whose spectral curves are algebraic. In Section 4.1, we will encounter with the familiar Catalan example.

3. A walk into the woods of Higgs bundles and connections

Let C be a smooth projective curve, and K_C the canonical bundle of C whose sections are holomorphic 1-forms.

3.1. Moduli spaces of vector bundles. We recall the somewhat anachronistic definition of a holomorphic vector bundle E over C. For an open cover of affine sets $C = \cup_\alpha U_\alpha$, we denote by $f_{\alpha\beta} : U_\alpha \cap U_\beta \to GL_r(\mathbb{C})$ the holomorphic transition functions that satisfy the 1-cocyle condition $f_{\alpha\beta} = f_{\alpha\gamma} \cdot f_{\gamma\beta}$ on $U_\alpha \cap U_\beta \cap U_\gamma$. Two points $(x_\alpha, \zeta_\alpha(x_\alpha)) \in U_\alpha \times \mathbb{C}^r$ and $(x_\beta, \zeta_\beta(x_\beta)) \in U_\beta \times \mathbb{C}^r$ are glued if $\zeta_\alpha(x) = f_{\alpha\beta}(x) \cdot \zeta_\beta(x)$ for $x \in U_\alpha \cap U_\beta$. Two transition functions $f_{\alpha\beta}$ and $f'_{\alpha\beta}$ subordinating the same open covering U_α define the isomorphic vector bundle if and only if there exists a family $u_\alpha : U_\alpha \to GL_r(\mathbb{C})$ of holomorphic maps, called *gauge transformations*, satisfying

$$f'_{\alpha\beta} = u_\alpha \cdot f_{\alpha\beta} \cdot u_\beta.$$

A global holomorphic section of the vector bundle E, $s \in H^0(C, E)$, is given by a collection of holomorphic maps $s_\alpha : U_\alpha \to \mathbb{C}^r$ compatible with the transition functions: $s_\alpha(x) = f_{\alpha\beta}(x) \cdot s_\beta(x)$ for $x \in U_\alpha \cap U_\beta$.

The *degree* of a vector bundle is the first Chern class, $\deg(E) := c_1(\Lambda^r(E))$. Over a compact connected Riemann surface C, topologically vector bundles are completely classified by the discrete invariants, rank and degree. However, introducing a holomorphic structure, classification results of vector bundles on a smooth, irreducible, complex projective curve become more elaborate. Define the slope of E to be the rational number $\mu(E) = \deg(E)/\text{rank}(E)$ - this is a topological quantity with important implications on holomorphic structures. A holomorphic vector bundle E is called *stable* (resp. *semi-stable*) if for any non-trivial holomorphic subbundle F, $\mu(F) < \mu(E)$ (resp. $\mu(F) \leq \mu(E)$) holds. There are complete classification results for holomorphic vector bundles for rational and elliptic curves, due to Grothendieck for the case of the Riemann sphere [**20**], and Atiyah for the case of elliptic curves [**2**]. However, for genus higher than one there are no such classification results available, therefore such question is replaced by the construction of *moduli space of stable holomorphic vector bundles* of rank r and degree e, denoted by $\mathcal{U}_C(r, e)$, whose geometry has been intensely studied. Over a smooth projective curve C of genus $g > 1$, the moduli space $\mathcal{U}_C(r, e)$ is a quasi-projective complex variety of dimension $r^2(g-1)+1$ (Narasimhan-Seshadri [**35**], Seshadri [**37**], see also Atiyah-Bott [**3**] and Mumford-Fogarty-Kirwan [**34**] for more information on the moduli theory of stable vector bundles over Riemann surfaces).

By reducing the Yang-Mills self-duality equations from dimension 4 to dimension 2 on a compact Riemann surface, Hitchin introduced in [**26**] *the moduli space of solutions to Hitchin's equations*, having a $\mathbb{CP}^1$ of complex structures, parametrized by $\zeta \in \mathbb{CP}^1$. If ζ is zero, then this space is identified with the *Dolbeault moduli space* $\mathcal{M}_{\text{Dol}}$ *of holomorphic stable Higgs bundles* consisting of (E, ϕ), where E is a vector bundle of rank r and fixed degree e, and $\phi \in H^0(C, \text{End}(E) \otimes K_C)$ is a *Higgs field.* If r and e are coprime, then $\mathcal{M}_{\text{Dol}}$ becomes quasi-projective variety. If ζ is non-zero and $e = 0$, then the the moduli space of solutions to Hitchin's equations can be identified, as a complex analytic variety, with *the de Rham moduli space*

$\mathcal{M}_{\mathrm{deR}}$ consisting of irreducible flat connections ∇ in holomorphic vector bundles V of rank r and degree 0.

The cotangent bundle $T^*\mathcal{U}_C(r,e)$ is an open dense subset of $\mathcal{M}_{\mathrm{Dol}}$ whose complement has codimension 2 or higher. Therefore, $\dim \mathcal{M}_{\mathrm{Dol}} = \dim\, T^*\mathcal{U}_C(r,e) = 2\dim\, \mathcal{U}_C(r,e) = 2 \cdot r^2 \cdot (g-1) + 2$, and it acquires a holomorphic symplectic structure.

These two moduli spaces, $\mathcal{M}_{\mathrm{Dol}}$ and $\mathcal{M}_{\mathrm{deR}}$, will play a key role in our discussion.

DEFINITION 3.1. Let C be a smooth projective curve of genus at least two, E and V two holomorphic rank r vector bundles on C and d the exterior differential on C.

(1) A *holomorphic Higgs bundle* is a pair (E,ϕ), where E is a holomorphic vector bundle and $\phi : E \to E \otimes K_C$ is a $\mathcal{O}_C$-module homomorphism, i.e., $\phi(f\cdot s) = f \cdot \phi(s)$, $\forall f \in \mathcal{O}_C, s \in E$.
(2) A *stable* (resp. *semi-stable*) Higgs bundle is a Higgs pair (E,ϕ) such that for any ϕ-invariant sub-bundle F of E, $\phi : F \to F \otimes K_C$, $\mu(F) < \mu(E)$ (resp. $\mu(F) \leq \mu(E)$) holds.
(3) A *holomorphic connection* is a $\mathbb{C}$-linear homomorphism $\nabla : V \to V \otimes K_C$ of a holomorphic vector bundle V such that $\nabla(f\cdot s) = df \otimes s + f \cdot \nabla(s)$, $\forall f \in \mathcal{O}_C$ and holomorphic sections $s \in V$. A *differentiable connection* is defined in the same way, replacing f by a differentiable function on C and $s \in V$ by a differentiable section of V.
(4) An *irreducible connection* is a connection ∇ in V for which no sub-bundle of V is ∇-invariant.
(5) A *hermitian metric* on the complex vector bundle E is a positive definite Hermitian form h on each fiber E_p, $p \in C$. It is a smooth section of $\Gamma(E \otimes \bar{E}^*)$ such that for all $\eta, \zeta \in E_p$,
$$\langle \eta, \zeta\rangle := h_p(\eta, \bar{\zeta}) = h_p(\zeta, \bar{\eta}) \text{ and } h_p(\zeta, \bar{\zeta}) > 0.$$
(6) Let h be a hermitian metric in a vector bundle V, and $\langle\cdot,\cdot\rangle$ the hermitian inner product. An *h-unitary connection* on a vector bundle V is a differentiable connection ∇ such that for any differentiable sections s and t of E, $\langle s,t\rangle = \langle \nabla(s), \nabla(t)\rangle$.

REMARK 3.2. If ∇_1 and ∇_2 are holomorphic connections in a holomorphic vector bundle V, then the difference $\nabla_1 - \nabla_2$ is an $\mathcal{O}_C$-module homomorphism. Therefore, $(V, \nabla_1 - \nabla_2)$ is a holomorphic Higgs pair. This proves that the Dolbeault and the de Rham moduli spaces, $\mathcal{M}_{\mathrm{Dol}}$ and $\mathcal{M}_{\mathrm{deR}}$, have the same dimensions.

3.2. Spectral curve of the Hitchin fibration. The Higgs field $\phi : E \to E \otimes K_C$ induces a map $\wedge^i \phi : \wedge^i E \to \wedge^i E \otimes K_C^i$ for every $i \geq 0$, locally given by the alternative sum of the i-th minors of ϕ, with trace $\mathrm{tr}(\wedge^i\phi) \in H^0(C, K_C^i)$. If $i = r$ then $\wedge^r\phi = \mathrm{tr}(\wedge^r\phi) = \det\phi \in H^0(C, \mathrm{End}(\wedge^r E) \otimes K_C^r)$. This defines a holomorphic map, called the *Hitchin map* H, which induces an algebraically completely integrable Hamiltonian system

$$\begin{array}{c} \mathcal{M}_{\mathrm{Dol}} \\ \big\downarrow H \\ B := \bigoplus_{i=1}^r H^0(C, K_C^i). \end{array} \tag{3.2.1}$$

For a Higgs pair (E,ϕ) let s denote the *spectral data*

$$s := \left((-1)^i \mathrm{tr}(\wedge^i\phi)\right)_{i=1,\dots,r} \in B \cong \mathbb{C}^{r^2(g-1)+1}.$$

The Hitchin map H sends a Higgs bundle to its spectral data

$$(E,\phi) \overset{H}{\to} s.$$

Obviously, the fiber of H over zero contains all Higgs bundles of the form $(E,\phi=0)$, where E is a stable vector bundle, so $\mathcal{U}_C(r,e) \subset H^{-1}(0)$.

The total space of the canonical bundle K_C is the cotangent bundle T^*C. There is a tautological 1-form $\eta \in H^0(T^*C, \pi^*K_C)$ on the symplectic variety T^*C defined by

$$\begin{array}{ccc} T^*C & \longleftarrow & \pi^*K_C \\ {\scriptstyle\pi}\downarrow & & \\ C & \longleftarrow & K_C. \end{array}$$

Locally η is defined by ydx for $x \in C$ and y the fiber coordinate, while $-d\eta$ is the natural holomorphic symplectic form on T^*C.

The characteristic polynomial of a Higgs bundle (E,ϕ) defines the *spectral curve* denoted by Σ_s in T^*C as the divisor of the zero of the following global section

$$\det(\eta\cdot I_r - \pi^*\phi) := \sum_{i=0}^{r}(-1)^i \mathrm{tr}(\wedge^i\phi)\otimes\eta^{\otimes(r-i)} \in H^0(T^*C, \pi^*K_C^r). \tag{3.2.2}$$

The coefficients of the defining equation of Σ_s are given by the spectral data s of the Higgs bundle (E,ϕ). Observe that η induces a 1-form on the spectral curve Σ_s by pulling back via the inclusion map i of Σ_s in T^*C.

REMARK 3.3. General Properties of Σ_s.

(1) Σ_s is non-singular for generic s.

(2) Σ_s is a curve inside the cotangent bundle T^*C of genus

$$g(\Sigma) = r^2(g-1)+1.$$

(3) The fiber over a generic point is the Jacobian:

$$H^{-1}(s) = \mathrm{Jac}(\Sigma_s).$$

(4) There is a degree r cover

$$\begin{array}{c} \Sigma_s \\ \downarrow {\scriptstyle r:1} \\ C. \end{array}$$

From now on we will consider the Hitchin theory for the Lie group $G = SL_r(\mathbb{C})$. In other words, $\mathcal{M}_{\mathrm{Dol}}$ denotes the moduli space of holomorphic Higgs bundles (E,ϕ) with $\mathrm{tr}(\phi) = 0$ such that E has the fixed trivial determinant. The fiber of the Hitchin map $H^{-1}(s)$ at a generic point

$$s \in B := \bigoplus_{i=2}^{r} H^0(C, K_C^i)$$

becomes the Prym variety

$$H^{-1}(s) = \mathrm{Prym}(\Sigma_s \to C) := \mathrm{Ker}(Nm)$$

of the spectral covering $\pi : \Sigma_s \longrightarrow C$ with the spectral data s, i.e., the kernel of the norm map

$$Nm : \mathrm{Jac}(\Sigma_s) \ni \sum_{p\in\Sigma_s} m_p \cdot p \longmapsto \sum_{p\in\Sigma_s} m_p \cdot \pi(p) \in \mathrm{Jac}(C).$$

The moduli space $\mathcal{M}_{\mathrm{Dol}}$ is a generically Abelian fibration over B.

3.2.1. *Rank* 2 *simplification.* We will focus next on rank 2 and degree 0.

(1) The $SL_2(\mathbb{C})$-Higgs bundle (E,ϕ), has $\mathrm{tr}(\phi) = 0$ and trivial determinant $\det(E) = \wedge^2 E = \mathcal{O}_C$.
(2) The Hitchin map

$$\begin{array}{c}\mathcal{M}_{\mathrm{Dol}}\\ \big\downarrow H\\ B := H^0(C, K_C^2) \ni s\end{array}$$

sends $(E,\phi) \overset{H}{\to} \det(\phi) = s$.

EXAMPLE 3.1 (rank two $SL_2(\mathbb{C})$ stable Higgs bundles). Choose a spin structure on a curve C of genus $g \geq 2$, i.e., a choice of the line bundle $K_C^{\frac{1}{2}}$, (5.1.2). For any quadratic differential $q \in H^0(C, K_C^2)$ on C, $\left(K_C^{\frac{1}{2}} \oplus K_C^{-\frac{1}{2}}, \begin{pmatrix} 0 & q \\ 1 & 0\end{pmatrix}\right)$ is a stable Higgs bundle on C. This example will play a key role in our later analysis (see Definition (5.2)).

Another example is on the trivial vector bundle. For any non-zero holomorphic 1-form $p \in H^0(C, K_C)$, $\left(\mathcal{O}_C \oplus \mathcal{O}_C, \begin{pmatrix} 0 & p \\ p & 0\end{pmatrix}\right)$ is again a stable Higgs bundle.

REMARK 3.4. Let $\{f_{\alpha\beta}\}$ be a transition function for E.

(1) Locally on C, $\nabla|_U = d + A|_U$, where $A : E \to E \otimes K_C$.
(2) Any Higgs field is compatible with the transition functions of E. Indeed, if $\phi_\alpha = \phi|_{U_\alpha}$, then

$$\phi_\alpha = f_{\alpha\beta} \cdot \phi_\beta \cdot f_{\alpha\beta}^{-1}. \tag{3.2.3}$$

(3) If $s \in E$ is a holomorphic section, then $s_\alpha(x) = f_{\alpha\beta}(x) \cdot s_\beta(x)$ for $x \in U_\alpha \cap U_\beta$. We note that derivatives of sections do not make sense as sections: $ds_\alpha(x) \neq f_{\alpha\beta}(x) \cdot ds_\beta(x)$. Nevertheless, if $A_\alpha = A|_{U_\alpha}$, then $(d + A_\alpha)\cdot s_\alpha = f_{\alpha\beta}\cdot[(d+A_\beta)\cdot s_\beta]$ on $U_\alpha \cap U_\beta$. Therefore, for any connection $\nabla = \{d + A_\alpha\}_\alpha$ on E, the Gauge transformation holds

$$A_\alpha = f_{\alpha\beta} \cdot A_\beta \cdot f_{\alpha\beta}^{-1} - f_{\alpha\beta}^{-1} \cdot df_{\alpha\beta}. \tag{3.2.4}$$

Since $f_{\alpha\beta} \cdot f_{\alpha\beta}^{-1} = 1$, by applying the exterior differential d, it is obvious that (3.2.4) is equivalent to $A_\alpha = f_{\alpha\beta} \cdot A_\beta \cdot f_{\alpha\beta}^{-1} + f_{\alpha\beta} \cdot df_{\alpha\beta}^{-1}$, or simply $df_{\alpha\beta} = f_{\alpha\beta} \cdot A_\beta - A_\alpha \cdot f_{\alpha\beta}$.

Locally, ϕ and A are both r by r matrices of 1-forms, but satisfying different rules with respect to the transition function of the vector bundle. We emphasize here that given a Higgs bundle (E,ϕ) in $\mathcal{M}_{\mathrm{Dol}}$, it is not obvious how to obtain a connection (V,∇) in $\mathcal{M}_{\mathrm{deR}}$. The goal of these lecture notes is to reveal the holomorphic *path* that a Higgs bundle on a Hitchin section travels to become a connection, called an *oper*.

4. From Higgs bundles to quantum curves

4.1. Higgs bundles for Catalan numbers. We wish to consider the curve $z^2 - z \cdot x + 1 = 0$ of (2.4.1) as a local expression of a singular spectral curve (divisor) inside the Hirzebruch surface $\mathbb{F}_2$ associated with a meromorphic Higgs bundle.

EXAMPLE 4.1. The **Spectral curve** of Catalan numbers as the **spectral curve** of a Higgs bundle.

(1) The curve is $C = \mathbb{P}^1$, and the vector bundle $E = K_C^{\frac{1}{2}} \oplus K_C^{-\frac{1}{2}} = \mathcal{O}_{\mathbb{P}^1}(-1) \oplus \mathcal{O}_{\mathbb{P}^1}(1)$.

(2) The **meromorphic Higgs field** $\phi : E \to E \otimes K_C(*)$ is given by

$$\phi = \begin{pmatrix} 0 & -(dx)^2 \\ 1 & x \cdot dx \end{pmatrix}$$

on the affine line $\mathbb{A}^1 \subset \mathbb{P}^1$.

(3) The *spectral curve of a Higgs bundle*, denoted by

$$\Sigma \subset \overline{T^*\mathbb{P}^1} = \mathbb{F}_2 := \mathbb{P}\big(\mathcal{O}_{\mathbb{P}^1}(-2) \oplus \mathcal{O}_{\mathbb{P}^1}\big),$$

is given by the characteristic polynomial of the Higgs field ϕ

$$\det(\eta \cdot I_2 - \phi) = \det(z \cdot dx \cdot I_2 - \phi) = (z^2 - x \cdot z + 1) \cdot (dx)^2 = 0$$

in $T^*\mathbb{A}^1$.

- We further consider a **resolution of singularity** of curve Σ by blowing up $\mathbb{F}_2$.
- Σ is smooth in $T^*\mathbb{A}^1$ near $(0,0)$ (Figure 4(1)), and has a *double point* in $\mathbb{F}_2$ at (∞, ∞) (Figure 4(2)).

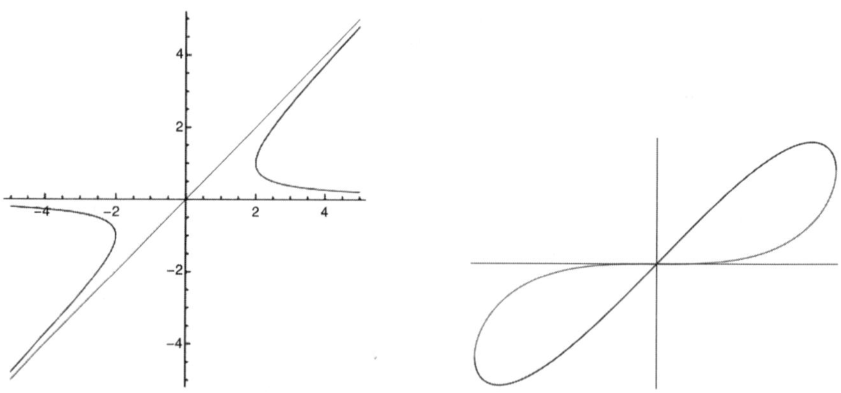

(1): Σ around $(0,0)$: $z^2 - x \cdot z + 1 = 0$

(2): Σ at (∞, ∞): $u^4 - u \cdot w + w^2 = 0$

FIGURE 4

- Blow up the surface $\mathbb{F}_2$ and denote by $\widetilde{\Sigma}$ the strict transform of Σ. Introduce a new local parameter w_1 such that $w = w_1 \cdot u$. The strict transform $\widetilde{\Sigma}$ becomes a conic

$$\widetilde{\Sigma} : u^2 + (w_1 - \frac{1}{2})^2 = \frac{1}{4}.$$

- $\widetilde{\Sigma}$ is a rational curve, and t of (2.5.1) **is the normalization coordinate** that parametrizes the affine part of the spectral curve by

$$u = \frac{t^2 - 1}{2 \cdot (t^2 + 1)}$$
$$w_1 = \frac{1}{2} - \frac{t}{t^2 + 1}.$$

- Denote by F the class of a fiber of $\mathbb{F}_2 \to \mathbb{P}^1$, by B the negative section, i.e., the zero section of $T^*\mathbb{P}^1$, and by E the exceptional divisor created on $\mathbb{F}_2$ after the blow-up at the double point of Σ at (∞, ∞). Then the proper transform $\widetilde{\Sigma}$ on the blown-up of $\mathbb{F}_2$ is written as the divisor

$$\widetilde{\Sigma} = 4F + 2B - 2E.$$

4.2. Higgs bundles and quantum curves. We are now ready to generalize Theorem 2.5 and results in Section 4.1 to any meromorphic Higgs bundle (E, ϕ) of rank two.

In [**10**] we have established a **new connection** between the Hitchin theory/Higgs bundles and topological recursion/quantum curve theory. These are two apparently different broad theories that share the notion of **spectral curves**. To establish the notations, let C be a smooth projective curve of *arbitrary genus*, and K_C the canonical bundle. We denote by E a holomorphic *rank two* vector bundle on C, and by $\phi : E \to E \otimes K_C(*)$ a Higgs field.

- In [**10**] we considered a *holomorphic* Higgs pair (E, ϕ). Hitchin constructed the spectral curve Σ of ϕ by the characteristic polynomial of ϕ (3.2.2), $\Sigma \hookrightarrow T^*C$.
- In [**11**] we considered a *meromorphic* Higgs pair. We construct the spectral curve Σ as the zero divisor of the characteristic polynomial of ϕ inside the compactified cotangent bundle of C that is a ruled surface over C:

$$\Sigma := (\det(\eta \cdot I_r - \pi^*\phi))_0 \hookrightarrow \overline{T^*C},$$

where $\eta \in H^0(T^*C, \pi^*K_C)$ is the tautological 1-form on T^*C extended as a meromorphic 1-form on the compactification $\overline{T^*C}$. We consider a **resolution of singularities** of Σ by blowing up the ruled surface $\overline{T^*C}$ over C, along the base locus of Σ.

$$\begin{array}{ccc} \widetilde{\Sigma} & \xrightarrow{\ i\ } & Bl(\overline{T^*C}) \\ \downarrow & & \downarrow{\scriptstyle blow-up} \\ \Sigma & \xrightarrow[i]{} & \overline{T^*C} \end{array}$$

In [**10**], [**11**] (see also [**13**]), we extended the framework of topological recursion [**17**] to singular Hitchin spectral curves, utilizing the birational geometry of ruled surfaces. As a consequence, this extension has led to the discovery of the relation between **Hitchin spectral curves** and **Gromov-Witten invariants** in few examples (as the one in Section 2.1 and Section 4.1). More precisely, the novelty of this approach is the discovery of the **PDE differential recursions of free energies** $F_{g,n}$ in [Definition 6.6, [**11**]] (as well as [Equation 6.5, [**10**]]) that implies the WKB analysis of the quantization Theorem 4.1. Moreover, the **PDE differential recursions of free energies** $F_{g,n}$ also implies the well-known integral **topological**

recursion of Eynard-Orantin for a spectral curve of genus 0. The PDE recursion relates the Hitchin spectral curve with enumerative geometry.

THEOREM 4.1 (Quantization Theorem [**10**], [**11**]). *For a rank 2 Higgs bundle and $x \in C$, we construct locally a second order differential operator $P(x, \hbar \cdot d/dx)$ whose semi-classical limit recovers the spectral curve Σ. We also construct a solution $\psi(x, \hbar)$ of equation $P(x, \hbar \cdot d/dx)\psi(x, \hbar) = 0$ in terms of principal specialization of the PDE recursion.*

The enumerative geometry example of the Catalan numbers emphasized by equation (2.4.2) is locally encaptured in the framework of Hitchin systems by Example in the Section 4.1. Following this approach, assume that the spectral curve of the Higgs bundle has the local expression

$$y^2 - \operatorname{tr}\phi(x) \cdot y + \det \phi(x) = 0.$$

The quantum curve associated to this spectral curve is a Rees D-module, locally given by the second order differential operator obtained by replacing the y variable by $\hbar \frac{d}{dx}$ (as in Theorem 2.5)

$$P(x, \hbar \cdot d/dx) = \left(\hbar \cdot \frac{d}{dx}\right)^2 - \operatorname{tr}\phi(x) \cdot \hbar \cdot \frac{d}{dx} + \det \phi(x).$$

The generating function of free energies is

$$\psi(x, \hbar) = \exp\left(\sum_{2g-2+n \geq -1} \frac{1}{n!} \cdot \hbar^{2g-2+n} \cdot F_{g,n}(x, \ldots, x) \right) = 0,$$

where $F_{g,n}(x_1, \ldots, x_n)$ are the free energies defined by the PDE recursion of [Definition 6.6, [**11**]]. If the spectral curve Σ is a singular curve, then the differential operator $P(x, \hbar \cdot d/dx)$ has irregular singularities and ψ has essential singularities. The asymptotic expansion of ψ (see e.g. [Definition 1.1, [**13**]]) as in the Catalan example (2.4.2) around its singularity has coefficients that encode information of Gromov-Witten invariants (Section 2.5, see also the Airy example of [Section 1, [**13**]]).

5. The metamorphosis of quantum curves into opers

From now on we will focus on holomorphic Higgs bundles (E, ϕ) on a Riemann surface C of genus at least two.

5.1. Projective coordinate system. We recall that a *universal covering* is a covering space that is simply connected. By Riemann uniformization theorem, every simply connected Riemann surface is biholomorphic to $\mathbb{P}^1$, $\mathbb{C}$, or to the upper half-plane $\mathbb{H} := \{z \in \mathbb{C} | Im(z) > 0\}$ with a global coordinate z. Therefore for a Riemann surface of genus at least two, the universal covering is the upper half-plane.

Notice that the global coordinate on $\mathbb{H}$ induces, by the quotient map $\pi : \mathbb{H} \to C$, a particular coordinate system on the Riemann surface C. Indeed, there is a faithful representation

$$\rho : \pi_1(C) \longrightarrow SL(2, \mathbb{R})$$

such that $C \cong \mathbb{H}/\rho(\pi_1(C))$, where $SL(2, \mathbb{R})$ acts on $\mathbb{H}$ through the projection

$$0 \longrightarrow \mathbb{Z}/2\mathbb{Z} \longrightarrow SL(2, \mathbb{R}) \longrightarrow PSL(2, \mathbb{R}) = \operatorname{Aut}(\mathbb{H}) \longrightarrow 0.$$

We can give a particular coordinate system on C using the universal covering map $\pi : \mathbb{H} \longrightarrow C$. Let

$$C = \bigcup_\alpha U_\alpha$$

be an open finite cover of C. For each coordinate neighborhood U_α, choose a contractible open subset $\widetilde{U}_\alpha \subset \mathbb{H}$ for which the map

$$\pi : \widetilde{U}_\alpha \xrightarrow{\sim} U_\alpha \subset C$$

is a biholomorphic map. Let us denote by z_α the local coordinate defined on U_α that corresponds to the global coordinate z restricted on $\widetilde{U}_\alpha$. Then on each $U_\alpha \cap U_\beta$, we have a *Möbius* coordinate transformation

$$z_\alpha = \frac{a_{\alpha\beta} \cdot z_\beta + b_{\alpha\beta}}{c_{\alpha\beta} \cdot z_\beta + d_{\alpha\beta}}, \qquad \begin{bmatrix} a_{\alpha\beta} & b_{\alpha\beta} \\ c_{\alpha\beta} & d_{\alpha\beta} \end{bmatrix} \in SL(2, \mathbb{R}). \tag{5.1.1}$$

In what follows, we choose and fix a Möbius coordinate system on C.

Since

$$dz_\alpha = \frac{1}{(c_{\alpha\beta} \cdot z_\beta + d_{\alpha\beta})^2} \cdot dz_\beta,$$

the transition function for the canonical line bundle K_C of C is given by the cocycle

$$\left\{ \xi^2_{\alpha\beta} = \frac{dz_\beta}{dz_\alpha} = (c_{\alpha\beta} \cdot z_\beta + d_{\alpha\beta})^2 \right\} \quad \text{on} \quad U_\alpha \cap U_\beta.$$

(1) We choose and fix a theta characteristic, or **a spin structure for** C, i.e. a line bundle $K_C^{\frac{1}{2}}$ such that $(K_C^{\frac{1}{2}})^{\otimes 2} \cong K_C$.
(2) Let $\{\xi_{\alpha\beta}\}$ denote the 1-cocycle corresponding to $K_C^{\frac{1}{2}}$ with respect to the Möbius coordinate system.
(3) The transition functions for $K_C^{\frac{1}{2}}$ are given by

$$\xi_{\alpha\beta} = \pm(c_{\alpha\beta} \cdot z_\beta + d_{\alpha\beta}). \tag{5.1.2}$$

The choice of the $\pm$ sign here is exactly an element of $H^1(C, \mathbb{Z}/2\mathbb{Z}) = (\mathbb{Z}/2\mathbb{Z})^{2g}$, which classifies the spin structure of C.

DEFINITION 5.1 (Gunning 1967 [**23**]). A *projective coordinate system* on C is a coordinate system on which transition function is given by a *Möbius* transformation.

$$C = \bigcup_\alpha U_\alpha, \quad z_\alpha \in U_\alpha, \quad z_\alpha = \frac{a_{\alpha\beta} \cdot z_\beta + b_{\alpha\beta}}{c_{\alpha\beta} \cdot z_\beta + d_{\alpha\beta}}, \quad \begin{bmatrix} a_{\alpha\beta} & b_{\alpha\beta} \\ c_{\alpha\beta} & d_{\alpha\beta} \end{bmatrix} \in SL(2, \mathbb{C}).$$

5.2. Hitchin section in rank two. Equipped with the choice of a spin structure for C and the transition functions $\xi_{\alpha\beta}$ for the line bundle $K_C^{\frac{1}{2}}$, we define the *Hitchin section in rank two*.

Recall the Hitchin map for the $SL_2(\mathbb{C})$-Higgs bundles sends $\mathcal{M}_{\text{Dol}} \ni (E, \phi) \xrightarrow{H} \det(\phi) \in B$.

DEFINITION 5.2. For each choice of $q \in H^0(C, K_C^2) = B$, the **Hitchin section** is the holomorphic Lagrangian inside the Dolbeault moduli space $\mathcal{M}_{\text{Dol}}$, given by

$$s(q) = \left(E_0 := K_C^{\frac{1}{2}} \oplus K_C^{-\frac{1}{2}}, \phi(q) := \begin{bmatrix} 0 & q \\ 1 & 0 \end{bmatrix} \right).$$

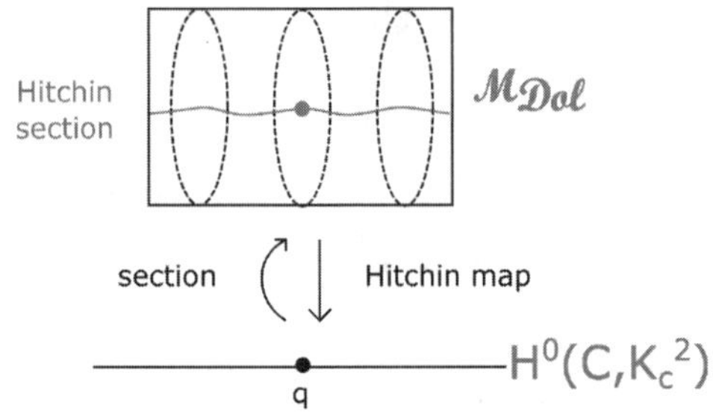

Let $f_{\alpha\beta} = \begin{pmatrix} \xi_{\alpha\beta} & 0 \\ 0 & \xi_{\alpha\beta}^{-1} \end{pmatrix}$ be the transition functions of the vector bundle $K_C^{\frac{1}{2}} \oplus K_C^{-\frac{1}{2}}$. If the quadratic differential q has a local form $q(z)|_{U_\alpha} = q_\alpha(z_\alpha) \cdot dz_\alpha^2$, then the Higgs field on the Hitchin section $\phi(q)$, which is a matrix valued 1-form, is given by

$$\phi_\alpha = \begin{pmatrix} 0 & q_\alpha(z_\alpha) \cdot dz_\alpha \\ dz_\alpha & 0 \end{pmatrix}.$$

Notice that the Higgs field $\phi(q)$ satisfies the compatibility condition (3.2.3):

$$\begin{aligned} f_{\alpha\beta} \cdot \phi_\beta \cdot f_{\alpha\beta}^{-1} &= \begin{pmatrix} \xi_{\alpha\beta} & 0 \\ 0 & \xi_{\alpha\beta}^{-1} \end{pmatrix} \cdot \begin{pmatrix} 0 & q_\beta(z_\beta) \cdot dz_\beta \\ dz_\beta & 0 \end{pmatrix} \cdot \begin{pmatrix} \xi_{\alpha\beta}^{-1} & 0 \\ 0 & \xi_{\alpha\beta} \end{pmatrix} \\ &= \begin{pmatrix} 0 & \xi_{\alpha\beta}^2 \cdot q_\beta(z_\beta) \cdot dz_\beta \\ \xi_{\alpha\beta}^{-2} \cdot dz_\beta & 0 \end{pmatrix} = \phi_\alpha. \end{aligned}$$

It follows from noticing $q_\beta(z_\beta)dz_\beta^2 = q_\alpha(z_\alpha)dz_\alpha^2$ and $\xi_{\alpha\beta}^2 = \frac{dz_\beta}{dz_\alpha}$, concluding that

$$\xi_{\alpha\beta}^2 \cdot q_\beta(z_\beta) \cdot dz_\beta = q_\beta(z_\beta) \cdot \frac{dz_\beta^2}{dz_\alpha} = q_\alpha(z_\alpha) \cdot dz_\alpha.$$

The stability of the Higgs bundle $s(q)$ with

$$\phi(q) = \begin{bmatrix} 0 & q \\ 1 & 0 \end{bmatrix} : K_C^{\frac{1}{2}} \oplus K_C^{-\frac{1}{2}} \to K_C^{\frac{3}{2}} \oplus K_C^{\frac{1}{2}}$$

is seen as follows. First observe that if $q \neq 0$, then any vector sub-bundle of E_0, either $0 \oplus K_C^{-\frac{1}{2}}$ or $K_C^{\frac{1}{2}} \oplus 0$, is not ϕ invariant. If $q = 0$, then the vector sub-bundle $0 \oplus K_C^{-\frac{1}{2}}$ is ϕ-invariant since it is mapped to zero by ϕ. However, its slope $1 - g$ is negative, since we assume $g \geq 2$.

REMARK 5.3. The Hitchin section (sometimes called the Hitchin component) is a *section* of the Hitchin fibration 3.2.1 in the sense that it intersects with each fiber of H exactly once. For the case of $SL_2(\mathbb{C})$-Higgs bundles (rank 2), it is also a section with respect to the Hitchin map H, since $H \circ s = Id_B$. However, in general, the Hitchin sections we construct in Section 6 are not the section with respect to the Hitchin map H, because $H \circ s \neq Id_B$.

For $q \in H^0(C, K_C^2)$, the differential operator for a Higgs pair on a Hitchin section (5.2) in Theorem 4.1, i.e., the *quantum curve* $P(x, \hbar \cdot d/dx)|_{\hbar=1} = \frac{d^2}{dx^2} - q(x)$, is not globally defined. This is because unlike the exterior differentiation d which is globally defined, the second order differentiation d^2/dx^2 has no global meaning. However, in the *projective coordinate system*, the differential equation

$\left(\frac{d^2}{dx^2} - q(x)\right)\psi(x) = 0$ makes sense globally on the curve C, provided that ψ is a (multi-valued) section of $K_C^{-\frac{1}{2}}$. More precisely, with respect to a coordinate change $x = x(u)$, we have

$$\psi\big(x(u)\big)\frac{1}{\sqrt{dx}} = \psi(u)\frac{1}{\sqrt{du}} \qquad \Longleftrightarrow \qquad e^{-\frac{1}{2}\log x'(u)}\psi\big(x(u)\big) = \psi(u),$$

where $x'(u) = \frac{dx}{du}$, and $q(u)du^2 = q(x)dx^2$. Then

$$\begin{aligned}
0 &= du^2 \cdot e^{\frac{1}{2}\log x'(u)}\left[\left(\frac{d}{du}\right)^2 - q(u)\right]\psi(u)\\
&= du^2 \cdot e^{\frac{1}{2}\log x'(u)}\left[\left(\frac{d}{du}\right)^2 - q(u)\right]e^{-\frac{1}{2}\log x'(u)}\psi\big(x(u)\big)\\
&= du^2 \cdot \left(e^{\frac{1}{2}\log x'(u)}\frac{d}{du}e^{-\frac{1}{2}\log x'(u)}\right)^2\psi\big(x(u)\big) - dx^2 q(x)\psi(x)\\
&= du^2 \cdot \left(\frac{d}{du} - \frac{1}{2}\frac{x''}{x'}\right)^2\psi\big(x(u)\big) - dx^2 q(x)\psi(x)\\
&= du^2 \cdot \left[\left(\frac{d}{du}\right)^2 - \frac{x''}{x'}\frac{d}{du} - \frac{1}{2}\left(\left(\frac{x''}{x'}\right)' - \frac{1}{2}\left(\frac{x''}{x'}\right)^2\right)\right]\psi\big(x(u)\big) - dx^2 q(x)\psi(x)\\
&= du^2 \cdot \left[\left(\psi_x(x)x'\right)_u - \frac{x''}{x'}\psi_x(x)x'\right] - dx^2 q(x)\psi(x)\\
&= \psi_{xx}(x)\left(\frac{dx}{du}\right)^2 du^2 + du^2 \cdot \left(\psi_x(x)x'' - x''\psi_x(x)\right) - dx^2 q(x)\psi(x)\\
&= dx^2 \cdot \left[\left(\frac{d}{dx}\right)^2 - q(x)\right]\psi(x).
\end{aligned}$$

Here, we have used the fact that the Schwarzian derivative

$$s_u(x) := \left(\frac{x''(u)}{x'(u)}\right)' - \frac{1}{2}\left(\frac{x''(u)}{x'(u)}\right)^2$$

is identically 0 if $x = x(u) = \frac{a_{\alpha\beta}\cdot u + b_{\alpha\beta}}{c_{\alpha\beta}\cdot u + d_{\alpha\beta}}$, $\begin{bmatrix} a_{\alpha\beta} & b_{\alpha\beta}\\ c_{\alpha\beta} & d_{\alpha\beta}\end{bmatrix} \in SL(2,\mathbb{C})$. is a Möbius transformation of Definition (5.1). We thus conclude that the quantum curve $P(x, \hbar\cdot d/dx)|_{\hbar=1}$ in Theorem 4.1 is globally defined as a *twisted* D-module acting on the sheaf $K_C^{-\frac{1}{2}}$. We leave to the interested reader to check the details of this computation. The details of this consideration will be provided in [**14, 15**].

Let us now present an *intuitive definition* of opers.

DEFINITION 5.4. An *oper* on an algebraic curve C is a *globally defined* differential operator of order r acting on $K_C^{-\frac{r-1}{2}}$.

REMARK 5.5. **Importance of Gunning's definition.** In a projective coordinate system of C, the quantum curve in Theorem 4.1 $P(x, \hbar\cdot d/dx)|_{\hbar=1}$ is an oper!

5.3. A family of Deligne's $\hbar$-connections. For a Higgs bundle in Definition 5.2, we interpret $P(x,\hbar\cdot d/dx)\psi=0$ as $\nabla^\hbar\begin{bmatrix}-\hbar\psi'\\ \psi\end{bmatrix}=0$. Indeed, the quantum curve of the Higgs field

$$\left(E_0=K_C^{\frac{1}{2}}\oplus K_C^{-\frac{1}{2}},\phi=\begin{pmatrix}0 & q(x)\cdot dx\\ dx & 0\end{pmatrix}\right)$$

is $\left(\hbar^2\frac{d^2}{dx^2}-q(x)\right)\cdot\psi(x,\hbar)=0$. This second order differential equation corresponds to the linear system of ODE, $\left(\hbar\cdot\nabla^\hbar\right)\begin{bmatrix}-\hbar\cdot\psi\\ \psi\end{bmatrix}=0$, where $\nabla^\hbar$ is an $\hbar$-deformation family of opers

$$\nabla^\hbar=d+\frac{1}{\hbar}\cdot\begin{pmatrix}0 & q(x)\cdot dx\\ dx & 0\end{pmatrix}. \tag{5.3.1}$$

QUESTION 5.6. *What is the corresponding vector bundle in which this family of connections $\nabla^\hbar$ is defined as in Remark* 3.4, (3.2.4)*?*

To answer this question we interpret the complex number $\hbar$ of Theorem 4.1 as an extension class of line bundles $\hbar\in\mathbb{C}=Ext^1(K_C^{-\frac{1}{2}},K_C^{\frac{1}{2}})\cong H^1(C,K_C)\cong H^0(C,\mathcal{O})$.

THEOREM 5.7 ([**23**]). *For every $\hbar\in\mathbb{C}$, there exists a unique extension*

$$0\to K_C^{\frac{1}{2}}\to V_\hbar\to K_C^{-\frac{1}{2}}\to 0 \tag{5.3.2}$$

such that

(1) *the rank 2 vector bundle $V_\hbar$ is given by transition functions $\{g^\hbar_{\alpha\beta}\}$, $g^\hbar_{\alpha\beta}:=$ $\begin{pmatrix}\xi_{\alpha\beta} & \hbar\cdot\frac{d\xi_{\alpha\beta}}{dz_\beta}\\ 0 & \xi_{\alpha\beta}^{-1}\end{pmatrix}$,*
(2) $V_0\cong K_C^{\frac{1}{2}}\oplus K_C^{-\frac{1}{2}}$, *and*
(3) *for $\hbar\neq 0$, all the vector bundles $V_\hbar$ are isomorphic.*

We denote by $V:=V_\hbar|_{\hbar=1}$ the unique non-trivial extension of $K_C^{-\frac{1}{2}}$ by $K_C^{\frac{1}{2}}$. We will give a more detail of higher-rank cases in [**14, 15**].

PROOF. We recall that $\xi_{\alpha\beta}$ of (5.1.2) are transition functions of $K_C^{\frac{1}{2}}$.

(1) It is an easy computation to check that $g^\hbar_{\alpha\beta}$ are transition functions of $V_\hbar$ satisfying the 1-cocycle condition. Denote by $\sigma_{\alpha\beta}=\frac{d\xi_{\alpha\beta}}{dz_\beta}$; according to (5.1.2), it is a constant. First, we have

(5.3.3)
$$g^\hbar_{\alpha\beta}\cdot g^\hbar_{\beta\gamma}=\begin{pmatrix}\xi_{\alpha\beta} & \hbar\cdot\sigma_{\alpha\beta}\\ 0 & \xi_{\alpha\beta}^{-1}\end{pmatrix}\cdot\begin{pmatrix}\xi_{\beta\gamma} & \hbar\cdot\sigma_{\beta\gamma}\\ 0 & \xi_{\beta\gamma}^{-1}\end{pmatrix}=\begin{pmatrix}\xi_{\alpha\beta}\cdot\xi_{\beta\gamma} & \hbar\cdot(\xi_{\alpha\beta}\sigma_{\beta\gamma}+\xi_{\beta\gamma}^{-1}\sigma_{\alpha\beta})\\ 0 & (\xi_{\alpha\beta}\cdot\xi_{\beta\gamma})^{-1}\end{pmatrix}.$$

From the 1-cocycle condition $\xi_{\alpha\beta}\cdot\xi_{\beta\gamma}=\xi_{\alpha\gamma}$ and $\xi^2_{\alpha\beta}=\frac{dz_\beta}{dz_\alpha}$, we claim

$$\sigma_{\alpha\gamma}=\xi_{\alpha\beta}\sigma_{\beta\gamma}+\xi_{\beta\gamma}^{-1}\sigma_{\alpha\beta}. \tag{5.3.4}$$

To see this, apply the logarithmic differentiation to $\xi_{\alpha\gamma}=\xi_{\alpha\beta}\cdot\xi_{\beta\gamma}$. We obtain

$$dz_\gamma\cdot\xi_{\alpha\gamma}^{-1}\cdot\frac{d\xi_{\alpha\gamma}}{dz_\gamma}=dz_\beta\cdot\xi_{\alpha\beta}^{-1}\cdot\frac{d\xi_{\alpha\beta}}{dz_\beta}+dz_\gamma\cdot\xi_{\beta\gamma}^{-1}\cdot\frac{d\xi_{\beta\gamma}}{dz_\gamma}.$$

Hence

$$\xi_{\alpha\gamma}^{-1} \cdot \frac{d\xi_{\alpha\gamma}}{dz_\gamma} = \frac{dz_\beta}{dz_\gamma} \cdot \xi_{\alpha\beta}^{-1} \cdot \sigma_{\alpha\beta} + \xi_{\beta\gamma}^{-1} \cdot \sigma_{\beta\gamma},$$
$$\begin{aligned}\sigma_{\alpha\gamma} &= \xi_{\alpha\gamma} \cdot \xi_{\gamma\beta}^2 \cdot \xi_{\alpha\beta}^{-1} \cdot \sigma_{\alpha\beta} + \xi_{\alpha\gamma} \cdot \xi_{\beta\gamma}^{-1} \cdot \sigma_{\beta\gamma} \\ &= (\xi_{\alpha\beta} \cdot \xi_{\beta\gamma}) \cdot \xi_{\beta\gamma}^{-2} \cdot \xi_{\alpha\beta}^{-1} \cdot \sigma_{\alpha\beta} + (\xi_{\alpha\beta} \cdot \xi_{\beta\gamma}) \cdot \xi_{\beta\gamma}^{-1} \cdot \sigma_{\beta\gamma} \\ &= \xi_{\beta\gamma}^{-1} \cdot \sigma_{\alpha\beta} + \xi_{\alpha\beta} \cdot \sigma_{\beta\gamma}.\end{aligned}$$

Therefore, $g_{\alpha\beta}^\hbar \cdot g_{\beta\gamma}^\hbar = g_{\alpha\gamma}^\hbar$ in (5.3.3).

(2) Since the matrix $g_{\alpha\beta}^\hbar|_{\hbar=0}$ is diagonal, the vector bundle it defines splits

$$V_0 \cong K_C^{\frac{1}{2}} \oplus K_C^{-\frac{1}{2}}.$$

(3) For every $\hbar \neq 0$, we show that vector bundle $V_\hbar$ is isomorphic to V. Indeed, the transition functions $g_{\alpha\beta}^\hbar$ and $g_{\alpha\beta}^\hbar|_{\hbar=1}$ are compatible with the change of trivialization $u_\alpha : U_\alpha \to GL_r(\mathbb{C})$ (see Section 3.1)

$$\begin{aligned}u_\alpha \cdot g_{\alpha\beta}^\hbar \cdot u_\beta^{-1} &= \begin{pmatrix} \sqrt{\hbar}^{-1} & 0 \\ 0 & \sqrt{\hbar} \end{pmatrix} \cdot \begin{pmatrix} \xi_{\alpha\beta} & \hbar \cdot \sigma_{\alpha\beta} \\ 0 & \xi_{\alpha\beta}^{-1} \end{pmatrix} \cdot \begin{pmatrix} \sqrt{\hbar} & 0 \\ 0 & \sqrt{\hbar}^{-1} \end{pmatrix} \\ &= \begin{pmatrix} \xi_{\alpha\beta} & \sigma_{\alpha\beta} \\ 0 & \xi_{\alpha\beta}^{-1} \end{pmatrix} = g_{\alpha\beta}^\hbar|_{\hbar=1}. \end{aligned} \tag{5.3.5}$$

□

LEMMA 5.8. *$\nabla^\hbar$ in (5.3.1) is a connection on $V_\hbar$.*

PROOF. We first recall equation (3.2.4) for a connection $\nabla^\hbar = \{d + A_\alpha^\hbar\}$ (5.3.1) on a vector bundle $V_\hbar$ given by transition functions $g_{\alpha\beta}^\hbar$, where

$$A_\alpha^\hbar = \frac{1}{\hbar}\begin{pmatrix} 0 & q_\alpha(z_\alpha) \cdot dz_\alpha \\ dz_\alpha & 0 \end{pmatrix}.$$

$$A_\alpha^\hbar = g_{\alpha\beta}^\hbar \cdot A_\beta^\hbar \cdot (g_{\alpha\beta}^\hbar)^{-1} + g_{\alpha\beta}^\hbar \cdot d(g_{\alpha\beta}^\hbar)^{-1} \tag{5.3.6}$$

$$\begin{aligned}g_{\alpha\beta}^\hbar \cdot A_\beta^\hbar \cdot (g_{\alpha\beta}^\hbar)^{-1} &= \frac{1}{\hbar}\begin{pmatrix} \xi_{\alpha\beta} & \hbar \cdot \sigma_{\alpha\beta} \\ 0 & \xi_{\alpha\beta}^{-1} \end{pmatrix} \cdot \begin{pmatrix} 0 & q_\beta(z_\beta) \cdot dz_\beta \\ dz_\beta & 0 \end{pmatrix} \cdot \begin{pmatrix} \xi_{\alpha\beta}^{-1} & -\hbar \cdot \sigma_{\alpha\beta} \\ 0 & \xi_{\alpha\beta} \end{pmatrix}. \\ &= \begin{pmatrix} \sigma_{\alpha\beta} \cdot \xi_{\alpha\beta}^{-1} & -\hbar \cdot \sigma_{\alpha\beta}^2 + \frac{1}{\hbar}\xi_{\alpha\beta}^2 \cdot q_\beta(z_\beta) \\ \frac{1}{\hbar} \cdot \xi_{\alpha\beta}^{-2} & -\xi_{\alpha\beta}^{-1} \cdot \sigma_{\alpha\beta} \end{pmatrix} \cdot dz_\beta.\end{aligned}$$

$$\begin{aligned}g_{\alpha\beta}^\hbar \cdot d(g_{\alpha\beta}^\hbar)^{-1} &= -\frac{dg_{\alpha\beta}^\hbar}{dz_\beta} \cdot (g_{\alpha\beta}^\hbar)^{-1} \cdot dz_\beta \\ &= -\begin{pmatrix} \sigma_{\alpha\beta} & \hbar \cdot \frac{d\sigma_{\alpha\beta}}{dz_\beta} \\ 0 & \frac{d\xi_{\alpha\beta}^{-1}}{dz_\beta} \end{pmatrix} \cdot \begin{pmatrix} \xi_{\alpha\beta}^{-1} & -\hbar \cdot \sigma_{\alpha\beta} \\ 0 & \xi_{\alpha\beta} \end{pmatrix} \cdot dz_\beta \\ &= -\begin{pmatrix} \sigma_{\alpha\beta} \cdot \xi_{\alpha\beta}^{-1} & -\hbar \cdot \sigma_{\alpha\beta}^2 + \hbar \cdot \xi_{\alpha\beta} \cdot \frac{d^2\xi_{\alpha\beta}}{dz_\beta^2} \\ 0 & -\xi_{\alpha\beta}^{-1} \cdot \sigma_{\alpha\beta} \end{pmatrix} \cdot dz_\beta.\end{aligned}$$

Relation (5.1.2) implies $\frac{d^2\xi_{\alpha\beta}}{dz_b^2}=0$. We conclude that

$$g^\hbar_{\alpha\beta}\cdot A^\hbar_\beta\cdot(g^\hbar_{\alpha\beta})^{-1}+g^\hbar_{\alpha\beta}\cdot d(g^\hbar_{\alpha\beta})^{-1}$$
$$=\frac{1}{\hbar}\cdot\left(0q_\beta(z_\beta)\cdot\frac{dz_\beta^2}{dz_a}dz_\alpha 0\right)=\frac{1}{\hbar}\cdot\left(0q_\alpha(z_\alpha)\cdot dz_a dz_\alpha 0\right)=A^\hbar_\alpha.$$

□

By fixing a complex structure of the curve C Gunning proved the following isomorphism as affine spaces in [**23**]

$$\begin{aligned}H^0(C,K_C^2)&\cong\text{moduli space of } SL_2(\mathbb{C})\text{-opers on } C\\&\cong\text{moduli space of \textit{projective coordinate systems} on } C\\&\quad\text{that subordinate the complex structure of } C.\end{aligned}\tag{5.3.7}$$

Since the space of quadratic differentials $H^0(C,K_C^2)$ is a vector space, it seems to imply that the holomorphic Lagrangian of opers also inherits an origin, corresponding to $q=0$. Indeed, $\nabla_{unif}=d+\begin{pmatrix}0&0\\dx&0\end{pmatrix}$ that we call the *uniformizing oper*, will play an important role in the next two sections. However, we note that such a choice does not come from algebraic geometry, as we see below.

The computations performed in Definition 5.2 and Lemma 5.8 show that the family $\hbar\cdot\nabla^\hbar$, as well as the *quantum curve* of Theorem 4.1, is a **$\hbar$-connection of Deligne**. This is a family of deformations that interpolates a Higgs field $\hbar\cdot\nabla^\hbar|_{\hbar=0}$ and a genuine connection $\hbar\cdot\nabla^\hbar|_{\hbar=1}$. We thus conclude that the Dumitrescu-Mulase quantum curve of Theorem 4.1, $\hbar\nabla^\hbar$, is an $\hbar$-deformation family of connections constructing a *holomorphic passage* form a Higgs field on the Hitchin section $\hbar\cdot\nabla^\hbar|_{\hbar=0}$ to an oper $\hbar\cdot\nabla^\hbar|_{\hbar=1}$, once we choose a Möbius coordinate system:

$$\left(K_C^{\frac{1}{2}}\oplus K_C^{-\frac{1}{2}},\begin{bmatrix}0&q\\1&0\end{bmatrix}\right)\overset{DM}{\rightarrow}\left(V,d+\begin{bmatrix}0&q(x)dx\\dx&0\end{bmatrix}\right).$$

6. Hitchin moduli spaces for the Lie group $G=SL_r(\mathbb{C})$

To introduce Gaiotto's correspondence, we need to consider Hitchin moduli spaces for simple complex Lie group G. In this paper, we restrict ourselves to the case of $G=SL_r(\mathbb{C})$. An $SL_r(\mathbb{C})$-Higgs bundle is a pair (E,ϕ) consisting of a holomorphic vector bundle E over a smooth projective curve C with a fixed determinant $\det(E)=\wedge^2E=\mathcal{O}_C$, and a traceless Higgs field ϕ. We use the same notations in Definition (3.1) in Section 3.1.

- E,V denote holomorphic vector bundles of rank r and degree 0 with trivial determinant.
- $\phi:E\to E\otimes K_C$ is a traceless holomorphic Higgs field.
- $\nabla:V\to V\otimes K_C$ is an irreducible holomorphic connection.

Let $E^{top}:=E$ denote the underlying topological structure of the rank r vector bundle E, obtained by forgetting its complex structure. Topological complex vector bundles over a compact topological surface are classified by their degrees and ranks, while the complete topological classification of complex vector bundles over a higher dimensional smooth topological manifold is given by their Chern classes. Since E^{top} has rank r and degree zero, it is topologically isomorphic to the direct sum $\mathcal{O}_C^{\oplus r}$ of r copies of the trivial line bundle $\mathcal{O}_C$ over C.

As mentioned earlier, a classical result of Narasimhan-Seshadri [**35**] shows that the moduli space $\mathcal{U}_C(r,d)$ of stable holomorphic vector bundles of rank r and degree d defined on a smooth projective algebraic curve C is *diffeomorphic* to the space of projectively flat irreducible unitary connections on C. A connection is said to be *projectively flat* if its curvature takes values in the center of the Lie algebra of the structure group of the vector bundle. For the case of degree 0, there is a one-to-one correspondence between stable holomorphic vector bundles and flat irreducible unitary connections. Through the Riemann-Hilbert correspondence, these flat irreducible connections correspond to representations of the fundamental group $\pi_1(C)$ into the unitary group modulo conjugation [**3**, **34**]. The equivalence classes of representations form a *character variety*

$$\mathrm{Hom}^{\mathrm{irr}}\left(\pi_1(C), U_n\right)/U_n.$$

The work of Hitchin [**26**], Donaldson [**8**] and Simpson [**38**] generalizes the above result to the moduli theory of Higgs bundles, holomorphic connections, and complex character varieties. According to this generalization, a stable holomorphic Higgs bundle (E,ϕ) of degree 0 corresponds to (D,ϕ,h) consisting of the following data that satisfy *Hitchin's equations*:

- h is a *hermitian metric* on E^{top}.
- D is a *unitary connection* on E^{top} with respect to the hermitian metric h. The connection D decomposes into the holomorphic and antiholomorphic part
$$D = D^{1,0} + D^{0,1}.$$
In terms of a local coordinate z of C, D can locally be given by $D = d + A$ with the exterior differential $d = \partial + \bar{\partial}$, where $\partial = \frac{\partial}{\partial z}\cdot dz$ and $\bar{\partial} = \frac{\partial}{\partial \bar{z}}\cdot d\bar{z}$, and an $r\times r$ skew-hermitian matrix A of 1-forms on C.
- $\phi : E^{top} \longrightarrow E^{top}\otimes\Omega_C^1$ is a traceless $r\times r$ matrix of differentiable 1-forms on C.

We note that the Cauchy-Riemann part $D^{0,1}$ of the connection D induces a holomorphic structure in E^{top}, which we denote simply by E. The unitarity condition means that the connection D is determined by $D^{0,1}$.

The great discovery of Donaldson [**8**] is that *the stability condition for a Higgs bundle (E,ϕ) is the system of non-linear PDEs that Hitchin discovered through the reduction of* 4*D Yang-Mills self-duality equations.* Denote by F_D the curvature of the connection D,

$$F_D = \frac{1}{2}\cdot[D,D] = [D^{1,0}, D^{0,1}],$$

and by $\phi^{\dagger_h}$ the adjoint of ϕ with respect to the hermitian metric h. The following system of non-linear PDEs is known as Hitchin's equations:

$$\begin{cases} F_D + [\phi, \phi^{\dagger_h}] = 0 \\ D^{0,1}\phi = 0. \end{cases} \tag{6.0.1}$$

For our purpose, it is important that Hitchin's equations (6.0.1) are equivalent to the flatness of the family of connections

$$D(\zeta) := \frac{1}{\zeta}\cdot\phi + D + \zeta\cdot\phi^{\dagger_h} \tag{6.0.2}$$

for all $\zeta \in \mathbb{C}^*$. We can see this equivalence as follows. A straightforward calculation shows

$$[D(\zeta), D(\zeta)] = \frac{1}{\zeta^2} \cdot [\phi, \phi] + \zeta^2 \cdot [\phi^{\dagger_h}, \phi^{\dagger_h}] + 2 \cdot (F_D + [\phi, \phi^{\dagger_h}])$$
$$+ \frac{1}{\zeta} \cdot ([\phi, D] + [D, \phi]) + \zeta \cdot \left([\phi^{\dagger_h}, D] + [D, \phi^{\dagger_h}]\right).$$

Clearly (6.0.1) implies the flatness of $D(\zeta)$, because the second equation makes ϕ holomorphic with respect to the complex structure of C and the holomorphic structure of E. Conversely, from the flatness of $D(\zeta)$, the first equation of (6.0.1) follows from the constant terms with respect to ζ. From the $1/\zeta^2$ terms, we see that ϕ contains only dz or $d\bar{z}$ term, and from the $1/\zeta$ and ζ terms we see that either ϕ or $\phi^{\dagger_h}$ is holomorphic. We rename the holomorphic one ϕ to obtain (6.0.1).

We thus have the following correspondences

A stable Higgs bundle (E, ϕ) $\longleftrightarrow$ (D, ϕ, h) satisfying (6.0.1) $\longleftrightarrow$ $[D(\zeta), D(\zeta)] = 0$ of (6.0.2).

To deal with three different appearances of complex moduli spaces in the Hitchin theory, we use the terminology *gauge theoretical moduli space*, denoted by $\mathcal{M}_{\text{Gauge}}$, to describe the differential geometric moduli space of solutions (D, ϕ, h) satisfying Hitchin's equations (6.0.1). It is a hyperKähler manifold with $\mathbb{P}^1$-worth of complex structures. Customary, we assign the complex structure of $\mathcal{M}_{\text{Dol}}$, the moduli space of stable Higgs bundles, to the origin of $\mathbb{P}^1$, and the algebraic structure of $\mathcal{M}_{\text{deR}}$, the moduli space of irreducible holomorphic connections, to $1 \in \mathbb{P}^1$. They are both *diffeomorphic* to $\mathcal{M}_{\text{Gauge}}$.

A particular diffeomorphism, known as the *nonabelian Hodge correspondence*, between $\mathcal{M}_{Dol}$ and $\mathcal{M}_{\text{deR}}$ is given as follows [**8, 26, 38**]. Firstly, we assign the flat connection $D(\zeta)$ to $(E, \phi) \in \mathcal{M}_{\text{Dol}}$. Secondly, we define a new holomorphic vector bundle $V = (E^{top}, D(\zeta = 1)^{0,1})$ by using the Cauchy-Riemann part of the flat connection $D(\zeta)$ at $\zeta = 1$. With respect to this complex structure, the $(1, 0)$-part of the connection $\nabla := D(\zeta = 1)^{1,0}$ is automatically a holomorphic connection in V, since $D(\zeta)$ is flat. Thus we obtain $(V, \nabla) \in \mathcal{M}_{\text{deR}}$.

$$\boxed{\mathcal{M}_{\text{Dol}} \ni (E, \phi) \overset{NAH}{\longrightarrow} (V, \nabla := D(1)^{1,0}) \in \mathcal{M}_{\text{deR}}.}$$

This is a generalization of the classical results of Narasimhan-Seshadri to Higgs bundles. A character variety also comes in to the picture, as the *Betti moduli space*

$$\mathcal{M}_{\text{Betti}} := \text{Hom}^{\text{irr}}(\pi_1(C), SL_r(\mathbb{C})) /\!/ SL_r(\mathbb{C}).$$

The classical unitary group is now replaced by a complex Lie group $G = SL_r(\mathbb{C})$. The complex structure of $\mathcal{M}_{\text{Betti}}$ comes from that of the group $SL_r(\mathbb{C})$. The Riemann-Hilbert correspondence gives a highly transcendental biholomorphic map between $\mathcal{M}_{\text{deR}}$ and $\mathcal{M}_{\text{Betti}}$. We thus have

$$\begin{aligned}
\mathcal{M}_{\text{Dol}} &= \text{moduli space of stable holomorphic Higgs bundles } (E, \phi) \\
&\quad \text{on } C \text{ of rank } r \\
&\wr \ \text{[Diffeomorphic NAH, Donaldson-Hitchin-Simpson]} \\
\mathcal{M}_{\text{deR}} &= \text{moduli space of rank r irreducible connections } (V, \nabla) \text{ on } C \\
&\cong \ \text{[Biholomorphic Riemann-Hilbert]} \\
\mathcal{M}_{\text{Betti}} &= \text{Hom}^{\text{irr}}(\pi_1(C), SL_r(\mathbb{C})) /\!/ SL_r(\mathbb{C}).
\end{aligned}$$

6.1. Hitchin section for $SL_r(\mathbb{C})$-Higgs bundles (principal $sl_2(\mathbb{C})$). We fix a spin structure $K_C^{\frac{1}{2}}$ on C given by transition functions $\{\xi_{\alpha\beta}\}$. To define a Hitchin section of G-Higgs bundles for a simple complex Lie group G, we need the notion of Konstant's *principal three-dimensional subgroup* (TDS) of [**29**]. For the case of $G = SL_r(\mathbb{C})$, it simply comes from the unique r-dimensional irreducible representation of $SL_2(\mathbb{C})$. The Lie algebra of principal TDS is the linear span $\langle X_+, X_-, H\rangle$, where

- $X_+ := \begin{bmatrix} 0 & \sqrt{p_1} & 0 & \cdots & 0 \\ 0 & 0 & \sqrt{p_2} & \cdots & 0 \\ \vdots & \vdots & \vdots & \ddots & \vdots \\ 0 & 0 & 0 & \cdots & \sqrt{p_{r-1}} \\ 0 & 0 & 0 & \cdots & 0 \end{bmatrix}, \qquad p_i := i(r-i),$
- $X_- := X_+^t$,
- $H := [X_+, X_-] = \begin{bmatrix} r-1 & 0 & \cdots & 0 & 0 \\ 0 & r-3 & \cdots & 0 & 0 \\ \vdots & \vdots & \ddots & \vdots & \vdots \\ 0 & 0 & \cdots & -(r-3) & 0 \\ 0 & 0 & \cdots & 0 & -(r-1) \end{bmatrix}.$

Define the split vector bundle $E_0 := K_C^{\frac{r-1}{2}} \oplus K_C^{\frac{r-1}{2}-1} \oplus \ldots \oplus K_C^{-\frac{r-1}{2}}$, whose transition function is given by $\{\xi^H_{\alpha\beta} = \exp(H \cdot \log \xi_{\alpha\beta})\}$. We note that every $q_i \in H^0(C, K_C^{i+1})$ satisfies that $q_i|_{U_\alpha} = q_i|_{U_\beta} \cdot \xi_{\alpha\beta}^{2(i+1)}$.

Now we can generalize Definition 5.2 of Section 5.2 as follows.

DEFINITION 6.1. The **Hitchin section** is a holomorphic Lagrangian inside $\mathcal{M}_{\mathrm{Dol}}$ consisting of stable Higgs pairs $(E_0, \phi(q))$ for every $q = (q_1, \ldots, q_{r-1}) \in B = \bigoplus_{i=1}^{r-1} H^0(C, K_C^{i+1})$, where

$$\phi(q) := X_- + \sum_{i=1}^{r-1} q_i \cdot X_+^i.$$

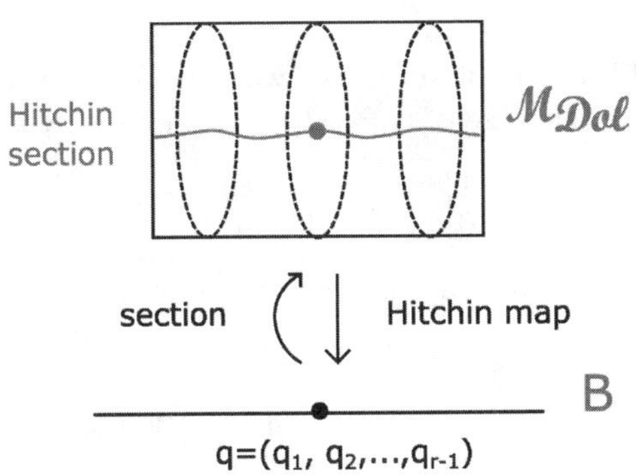

6.2. On a conjecture of Gaiotto. In 2014, Gaiotto [**18**] conjectured the following.

CONJECTURE 6.2. *Let (E_0, ϕ) be a stable Higgs pair on a Hitchin section in $\mathcal{M}_{\mathrm{Dol}}$, and $D(\zeta)$, $\zeta \in \mathbb{C}^*$, the corresponding one-parameter family of flat connections. Define a* ***two-parameter*** *family of connections by*

$$D(\zeta, R) := \zeta^{-1} \cdot R \cdot \phi + D + \zeta \cdot R \cdot \phi^\dagger, \tag{6.2.1}$$

where $R \in \mathbb{R}_+$ is a positive real number. Then the scaling limit

$$\lim_{\substack{R\to 0,\ \zeta\to 0\\ \zeta/R=\hbar}} D(\zeta, R)$$

exists, and defines an $SL_r(\mathbb{C})$-oper for every $\hbar \in \mathbb{C}^*$.

The data (D, ϕ, h) corresponding to any point $(E_0, \phi(q))$ on the Hitchin section satisfies Hitchin's equations (6.0.1). Scaling the Higgs field $\phi(q)$ by any real parameter $R \in \mathbb{R}_+$ does not affect the stability condition. Therefore, the scaled data $(D, R\phi, h)$ corresponding to the pair $(E_0, R\phi(q)) \in \mathcal{M}_{\mathrm{Dol}}$ also satisfies Hitchin's equations. This time, the equation is R-twisted:

$$\begin{cases} F_D + R^2 \cdot [\phi, \phi^{\dagger_h}] = 0 \\ D^{0,1}\phi = 0. \end{cases} \tag{6.2.2}$$

By the same argument as before, the R-twisted Hitchin equations are equivalent to the flatness of the two-parameter family of connections $D(\zeta, R)$.

A surprising fact of this conjecture is that the scaling limit of the differential geometric object $D(\zeta, R)$ is automatically an $\hbar$-family of holomorphic connections defined on an algebraic $\hbar$-deformation family of filtered vector bungles, generalizing the extension

$$0 \to K_C^{\frac{1}{2}} \to V_\hbar \to K_C^{-\frac{1}{2}} \to 0.$$

Naïvely, it looks that the scaling limit of (6.2.1) is simply $D + \frac{1}{\hbar}\phi$. There is a problem here, because the hermitian metric h that solves (6.2.2) explodes as $R \to 0$. Since the h-unitary connection D also depends on h, the limit of D does not make sense as R tends to 0.

THEOREM 6.3 (Dumitrescu, Fredrickson, Kydonakis, Mazzeo, Mulase, Neitzke, [**9**]). *Conjecture* 6.2 *holds for an arbitrary simple and simply connected complex Lie group G.*

6.3. Sketch of the proof in rank two. We present here the main steps to prove Theorem 6.3 for the case of $G = SL_2(\mathbb{C})$. We use the basis

$$\left\langle X_+ = \begin{bmatrix} 0 & 1 \\ 0 & 0 \end{bmatrix}, X_- = \begin{bmatrix} 0 & 0 \\ 1 & 0 \end{bmatrix}, H = \begin{bmatrix} 1 & 0 \\ 0 & -1 \end{bmatrix} \right\rangle$$

for $sl_2(\mathbb{C})$ to simplify our calculations. Their commutation relations are

$$[X_+, X_-] = H, \qquad [H, X_\pm] = \pm 2X_\pm.$$

STEP 1. We first notice that a hermitian metric on the canonical bundle K_C naturally comes from a hermitian metric on the curve C itself. Since we start from a Higgs pair on a Hitchin section with vector bundle $E_0 = K_C^{\frac{1}{2}} \oplus K_C^{-\frac{1}{2}}$, a fiber metric is determined by a hermitian metric on C. Recall that

$$\boxed{\text{A Higgs bundle } (E_0, \phi) \text{ on the Hitchin section} \longrightarrow (D, \phi, h) \text{ satisfying (6.0.1).}}$$

A choice of a hermitian metric on C determines the fiber metric h, and hence the unitary connection D. Thus we wish to see how it translates into Hitchin's equations.

STEP 2. We start from a complex structure on C with a holomorphic local parameter z, together with a hermitian metric

$$g = \lambda^2 \cdot dz \cdot d\bar{z}$$

on C, where λ is a positive real function depending on R. The hermitian metric on C is the same as the fiber metric of the tangent bundle of C, which is K_C^{-1}. Hence λ naturally gives a fiber metric of $K_C^{-\frac{1}{2}}$. Therefore, the hermitian metric on the split vector bundle $E_0 = K_C^{\frac{1}{2}} \oplus K_C^{-\frac{1}{2}}$ is given by the matrix $h = \begin{bmatrix} \lambda^{-1} & 0 \\ 0 & \lambda. \end{bmatrix}$. Then D becomes the Chern connection $D = D^{1,0} + D^{0,1}$, where $D^{0,1} = \bar{\partial}$ and $D^{1,0} = \partial + h^{-1} \cdot \partial h$. In terms of λ, we have

$$D = d + h^{-1} \cdot \partial h = d - \partial \log \lambda \cdot \begin{bmatrix} 1 & 0 \\ 0 & -1 \end{bmatrix} = d - \partial \log \lambda \cdot H.$$

STEP 3. Let us introduce a Higgs field

$$\phi = \phi(q) = \begin{bmatrix} 0 & q \\ 1 & 0 \end{bmatrix} \cdot dz = (X_- + q \cdot X_+) \cdot dz$$

so that $(E_0, \phi(q))$ is on the Hitchin section, where $q \in H^0\left(C, K_C^2\right)$. The hermitian conjugate of the Higgs field is calculated by

$$\begin{aligned} \phi^{\dagger_h} = h^{-1} \cdot \overline{\phi^t} \cdot h &= \begin{bmatrix} \lambda & 0 \\ 0 & \lambda^{-1} \end{bmatrix} \cdot \overline{\begin{bmatrix} 0 & 1 \\ q & 0 \end{bmatrix} \cdot dz} \cdot \begin{bmatrix} \lambda^{-1} & 0 \\ 0 & \lambda \end{bmatrix} \\ &= \begin{bmatrix} 0 & \lambda^2 \\ \lambda^{-2} \cdot \bar{q} & 0 \end{bmatrix} \cdot d\bar{z} = \left(\lambda^{-2} \cdot \bar{q} \cdot X_- + \lambda^2 \cdot X_+\right) \cdot d\bar{z}. \end{aligned}$$

STEP 4. Since we have identified all the ingredients, we can now write the two-parameter family of connections in this local coordinate as

$$\begin{aligned} D(\zeta, R) &= \frac{R}{\zeta} \cdot \begin{bmatrix} 0 & q \\ 1 & 0 \end{bmatrix} \cdot dz + d - \partial \log \lambda \cdot \begin{bmatrix} 1 & 0 \\ 0 & -1 \end{bmatrix} \cdot dz + R \cdot \zeta \cdot \begin{bmatrix} 0 & \lambda^2 \\ \bar{q} \cdot \lambda^{-2} & 0 \end{bmatrix} \cdot d\bar{z} \\ &= d + \frac{R}{\zeta} \cdot dz \cdot (X_- + q \cdot X_+) - \partial \log \lambda \cdot dz \cdot H \\ &\quad + R \cdot \zeta \cdot d\bar{z} \cdot \left(\bar{q} \cdot \lambda^{-2} \cdot X_- + \lambda^2 \cdot X_+\right). \end{aligned}$$

STEP 5. A simple calculation shows that the coefficient of H in the flatness condition

$$[D(\zeta, R), D(\zeta, R)] = 0$$

of $D(\zeta, R)$ yields

$$\begin{aligned} 0 &= [d, -\partial \log \lambda \cdot dz \cdot H] + R^2 \cdot dz \wedge d\bar{z} \cdot \left(\lambda^2 \cdot [X_-, X_+] + q \cdot \bar{q} \cdot \lambda^{-2} \cdot [X_+, X_-]\right) \\ &= -d\partial \log \lambda \cdot dz \cdot H + R^2 \cdot dz \wedge d\bar{z} \cdot \left(-\lambda^2 \cdot H + q \cdot \bar{q} \cdot \lambda^{-2} \cdot H\right) \\ &= \left(\bar{\partial}\partial \log \lambda + R^2 \cdot (\lambda^{-2} \cdot q \cdot \bar{q} - \lambda^2)\right) \cdot dz \wedge d\bar{z}. \end{aligned}$$

Therefore, we obtain

$$\bar{\partial}\partial \log \lambda + R^2 \cdot (\lambda^{-2} \cdot q \cdot \bar{q} - \lambda^2) = 0. \tag{6.3.1}$$

We thus conclude that the flatness condition for the two-parameter family of connections $D(\zeta, R)$ gives the harmonicity condition (6.3.1) for the hermitian metric λ.

STEP 6. For $q = 0$, i.e., $\phi = X_-$, the harmonicity equation (6.3.1) becomes

$$\bar{\partial}\partial \log \lambda - R^2 \cdot \lambda^2 = 0, \tag{6.3.2}$$

which can be solved explicitly. We obtain

$$\lambda_0 = \frac{1}{R} \cdot \frac{i}{z - \bar{z}}. \tag{6.3.3}$$

Let us denote by

$$\lambda_\natural = \frac{i}{z - \bar{z}} = \frac{1}{2 \cdot y}, \tag{6.3.4}$$

where $z = x + iy$. The corresponding hermitian metric is then

$$g_\natural = \frac{dz \cdot d\bar{z}}{4 \cdot y^2},$$

whose Gaussian curvature is

$$K := -\frac{4}{\lambda_\natural^2} \cdot \partial\bar{\partial} \log \lambda_\natural = -4. \tag{6.3.5}$$

Indeed, $g_\natural$ is the globally defined constant curvature metric on the upper half plane $\mathbb{H}$, which is invariant under the action of $PSL_2(\mathbb{R}) = \mathrm{Aut}(\mathbb{H})$. Since we are dealing with a Riemann surface C of genus $g \geq 2$, its universal covering is $\mathbb{H}$, and we have a non-canonical isomorphism $C \cong \mathbb{H}/\pi_1(C)$, where $\pi_1(C)$ acts on $\mathbb{H}$ through a representation $\rho : \pi_1(C) \longrightarrow SL_2(\mathbb{R})$. By inducing a metric by the push-forward of the covering map $\mathbb{H} \longrightarrow C$, we conclude that the harmonicity equation (6.3.2) can be solved globally on C with the hyperbolic metric on C of constant curvature $-4R^2$.

Since $\frac{\zeta}{R} = \hbar$, we obtain

$$\begin{aligned} D(\zeta, R) &= d + \frac{1}{\hbar} \cdot \begin{bmatrix} 0 & 0 \\ 1 & 0 \end{bmatrix} \cdot dz - \partial \log \lambda_\natural \cdot \begin{bmatrix} 1 & 0 \\ 0 & -1 \end{bmatrix} dz + \hbar \cdot \begin{bmatrix} 0 & \lambda_\natural^2 \\ 0 & 0 \end{bmatrix} d\bar{z} \\ &= d + \frac{1}{\hbar} \cdot dz \cdot X_1 - \partial \log \lambda_\natural \cdot dz \cdot H + \hbar \cdot \lambda_\natural^2 \cdot d\bar{z} \cdot X_+, \end{aligned} \tag{6.3.6}$$

which *does not* depend on R.

STEP 7. The case when $q \neq 0$ in general. We remark that any hermitian metric compatible with the complex structure of C is conformal to the constant curvature metric $g_\natural$. Therefore, we can write

$$\lambda = \lambda_0 \cdot e^{f(R)} = \frac{\lambda_\natural}{R} \cdot e^{f(R)}$$

with a conformal factor $e^{f(R)}$ depending on a real valued function $f(R)$ on C. We plug this expression into (6.3.1) and apply the implicit function theorem to yield that f is real analytic, and more significantly, that $f(R) = f_4 \cdot R^4 +$ *higher order terms.* This implies

$$\lambda^{-2} = \frac{R^2}{\lambda_\natural^2} \cdot e^{-2f(R)} = \frac{R^2}{\lambda_\natural^2} + \mathcal{O}(R^6), \quad \lambda = \frac{\lambda_\natural}{R} + \mathcal{O}(R^3), \quad \partial \log \lambda = \partial \log \lambda_\natural + \mathcal{O}(R^4).$$

Therefore, we obtain the scaling limit as

(6.3.7)

$$
\begin{aligned}
D(\zeta, R) &= d + \frac{1}{\hbar}\cdot\begin{bmatrix}0 & q\\ 1 & 0\end{bmatrix}\cdot dz - \partial\log\lambda\cdot\begin{bmatrix}1 & 0\\ 0 & -1\end{bmatrix}\cdot dz + R^2\cdot\hbar\begin{bmatrix}0 & \lambda^2\\ \overline{q}\cdot\lambda^{-2} & 0\end{bmatrix}\cdot d\bar{z}\\
&= d + \frac{1}{\hbar}\cdot\begin{bmatrix}0 & q\\ 1 & 0\end{bmatrix}\cdot dz - \partial\log\lambda_\natural\cdot\begin{bmatrix}1 & 0\\ 0 & -1\end{bmatrix}\cdot dz + \mathcal{O}(R^4)\cdot\begin{bmatrix}1 & 0\\ 0 & -1\end{bmatrix}\cdot dz\\
&\quad + \hbar\cdot\begin{bmatrix}0 & \lambda_\natural^2 + \mathcal{O}(R^4)\\ \overline{q}\cdot\frac{R^4}{\lambda_\natural^2} + \mathcal{O}(R^8) & 0\end{bmatrix}\cdot d\bar{z}\\
&\xrightarrow[\zeta/R=\hbar]{R\to 0,\zeta\to 0} d + \frac{1}{\hbar}\cdot\begin{bmatrix}0 & q\\ 1 & 0\end{bmatrix}\cdot dz - \partial\log\lambda_\natural\cdot\begin{bmatrix}1 & 0\\ 0 & -1\end{bmatrix}\cdot dz + \hbar\cdot\begin{bmatrix}0 & \lambda_\natural^2\\ 0 & 0\end{bmatrix}\cdot d\bar{z}\\
&= d + \frac{1}{\hbar}\cdot dz\cdot(X_- + q\cdot X_+) - \partial\log\lambda_\natural\cdot dz\cdot H + \hbar\cdot\lambda_\natural^2\cdot d\bar{z}\cdot X_+.
\end{aligned}
$$

We can see that the only dependence on the quadratic differential q is in the form of $\phi(q)$.

7. The limit oper of Gaiotto's correspondence and the quantum curve

In this section we will prove that the scaling limit $\lim_{R,\zeta\to 0,\frac{\zeta}{R}=\hbar} D(\zeta, R)$ is an oper.

A surprising fact is that the limit oper of Theorem 6.3 [**9**] is gauge equivalent to $\nabla^\hbar = d + \frac{1}{\hbar}\cdot\phi$ in the Möbius coordinate system obtained by the uniformization of the base curve C. In other words, the limit oper of Gaiotto correspondence in rank two is the quantum curve of Theorem 4.1 ([**10**]). We use Theorem 5.7 and Lemma 5.8 of Section 5.3 to imply that the scaling limit of Gaiotto's correspondence is actually an oper. For the sake of completeness, we include detailed computations here.

PROPOSITION 7.1 (Gauge transform of the scaling Limit). *The limit expression of* (6.3.7),

$$D(\hbar) := d + \frac{1}{\hbar}\cdot dz\cdot(X_- + q\cdot X_+) - \partial\log\lambda_\natural\cdot dz\cdot H + \hbar\cdot\lambda_\natural^2\cdot d\bar{z}\cdot X_+,$$

is gauge equivalent to an oper

$$\nabla^\hbar := d + \frac{1}{\hbar}\cdot\begin{bmatrix}0 & q\\ 1 & 0\end{bmatrix}\cdot dz = d + \frac{1}{\hbar}\cdot\phi(q). \tag{7.0.1}$$

Here, $\lambda_\natural = \frac{i}{z-\bar{z}}$ *is the local expression of the hermitian metric of constant curvature* -4 *on* C. *If we introduce a Möbius coordinate system on* C *induced by the uniformization covering* $\mathbb{H}\longrightarrow C$, *then* (7.0.1) *determines a globally defined oper on* C.

PROOF. The claim is that $\nabla^\hbar = g\cdot D(\hbar)\cdot g^{-1}$ with the gauge transformation

$$g = \begin{bmatrix}1 & \hbar\cdot\partial\log\lambda_\natural\\ 0 & 1\end{bmatrix} = e^{\hbar\cdot\partial\log\lambda_\natural\cdot X_+}.$$

For brevity, let us denote $a = \partial\log\lambda_\natural$ and $b = \lambda_\natural^2$. Then

$$D(\hbar) = d + \frac{1}{\hbar}\cdot dz\cdot(X_- + q\cdot X_+) - a\cdot dz\cdot H + \hbar\cdot b\cdot d\bar{z}\cdot X_+, \qquad g = e^{\hbar\cdot a\cdot X_+}.$$

Recall that for any elements A and B of a Lie algebra and a central parameter ϵ, we have the adjoint formula

$$e^{\epsilon A} \cdot B \cdot e^{-\epsilon A} = \sum_{n=0}^{\infty} \frac{1}{n!} \epsilon^n \cdot \mathrm{ad}_A^n(B),$$

where

$$\mathrm{ad}_A^n(B) := \overbrace{[A, [A, [\cdots, [A}^{n}, B] \cdots]]].$$

From the commutation relations of the basis for $sl_2(\mathbb{C})$, we see that

$$g \cdot X_- \cdot g^{-1} = X_- + \hbar \cdot a \cdot [X_+, X_-] + \frac{1}{2}(\hbar \cdot a)^2 \cdot [X_+, [X_+, X_-]] = X_- + \hbar \cdot a \cdot H - (\hbar \cdot a)^2 \cdot X_+.$$

Therefore,

$$g \cdot \frac{1}{\hbar} \cdot dz \cdot (X_- + q \cdot X_+) \cdot g^{-1} = \left(\frac{1}{\hbar} \cdot (X_- + q \cdot X_+) + a \cdot H - \hbar \cdot a^2 \cdot X_+ \right) \cdot dz.$$

Similarly,

$$g \cdot H \cdot g^{-1} = H + \hbar \cdot a \cdot [X_+, H] = H - 2\hbar \cdot a \cdot X_+,$$

which yields

$$-g \cdot a \cdot dz \cdot H \cdot g^{-1} = -a \cdot dz \cdot H + 2\hbar \cdot a^2 \cdot dz \cdot X_+.$$

It is obvious that

$$g \cdot \hbar \cdot b \cdot d\bar{z} \cdot X_+ \cdot g^{-1} = \hbar \cdot b \cdot d\bar{z} \cdot X_+.$$

Finally, we calculate

$$\begin{aligned} g \cdot d \cdot g^{-1} &= d - \hbar \cdot da \cdot X_+ \\ &= d - \hbar \cdot (\partial a \cdot dz + \bar{\partial} \cdot a \cdot d\bar{z}) \cdot X_+ \\ &= d - \hbar \cdot dz \cdot \partial^2 \log \lambda_\natural \cdot X_+ - \hbar \cdot d\bar{z} \cdot \bar{\partial}\partial \log \lambda_\natural \cdot X_+ \\ &= d + \hbar \cdot dz \cdot \frac{(\partial \lambda_\natural)^2 - \lambda_\natural \cdot \partial^2 \lambda_\natural}{\lambda_\natural^2} \cdot X_+ - \hbar \cdot d\bar{z} \cdot \lambda_\natural^2 \cdot X_+, \end{aligned}$$

where we used the constant curvature property (6.3.5) in the last step.

Adding all together, we obtain the gauge transformation formula

$$\begin{aligned} g \cdot D(\hbar) \cdot g^{-1} = d &+ \left(\frac{1}{\hbar} \cdot (X_- + q \cdot X_+) + \partial \log \lambda_\natural \cdot H - \hbar \cdot (\partial \log \lambda_\natural)^2 \cdot X_+ \right) \cdot dz + \\ &+ \hbar \cdot dz \cdot \frac{(\partial \lambda_\natural)^2 - \lambda_\natural \cdot \partial^2 \lambda_\natural}{\lambda_\natural^2} \cdot X_+ + (\hbar \cdot b \cdot d\bar{z} - \hbar \cdot d\bar{z} \cdot \lambda_\natural^2) \cdot X_+ - \\ &- \partial \log \lambda_\natural \cdot dz \cdot H + 2\hbar \cdot (\partial \log \lambda_\natural)^2 \cdot dz \cdot X_+ \\ = d &+ \frac{1}{\hbar} \cdot (X_- + q \cdot X_+) \cdot dz + \hbar \cdot dz \cdot \frac{2 \cdot (\partial \lambda_\natural)^2 - \lambda_\natural \cdot \partial^2 \lambda_\natural}{\lambda_\natural^2} \cdot X_+ \\ = d &+ \frac{1}{\hbar} \cdot \phi(q). \end{aligned}$$

Here, we use the fact that $2 \cdot (\partial \lambda_\natural)^2 - \lambda_\natural \cdot \partial^2 \lambda_\natural = 0$, which follows from the expression (6.3.4). ◻

The computation we have shown above is only for $SL_2(\mathbb{C})$. Yet it is valid for proving vastly general Theorem 6.3. The key idea is to use Kostant's principal TDS [**29**], replacing the basis $\langle H, X_\pm \rangle$ by the one for TDS. Then almost exactly the same formulas hold for the general situation. Here, we only indicate the oper we obtain through Gaiotto's scaling limit for the case of $SL_r(\mathbb{C})$-Higgs bundles. We will give a geometric definition of *oper* that generalizes Gunning's Definition encaptured by Theorem 5.7 from rank two to arbitrary rank r.

DEFINITION 7.2 (Beilinson-Drinfeld, 1993). Let V be a holomorphic vector bundle of rank r and degree 0. An $SL_r(\mathbb{C})$-oper is a pair $(V, \nabla) \in \mathcal{M}_{\text{deR}}$ satisfying the following conditions:

(1) There is a global filtration $0 = F_r \hookrightarrow F_{r-1} \hookrightarrow \ldots \hookrightarrow F_0 = V$ in V.
(2) Griffiths transversality. The connection ∇ induces a map $\nabla|_{F_{i+1}} : F_{i+1} \to F_i \otimes K_C$ for every $i = 0, \ldots, r-1$.
(3) $\nabla|_{F_{i+1}}$ induces an $\mathcal{O}_C$-linear isomorphism $F_{i+1}/F_{i+2} \cong F_i/F_{i+1} \otimes K_C$ for every i.

Let $(E_0, \phi(q))$ be a point on the Hitchin section of Section 6.1. Then Gaiotto's scaling limit produces an $\hbar$-family of opers $(V_\hbar, \nabla^\hbar)$ defined as follows. First, we choose once and for all the Möbius coordinate system associated with the uniformization mentioned above.

- $V_\hbar$ is given by the transition function $\{f^\hbar_{\alpha\beta}\}$, where

$$f^\hbar_{\alpha\beta} = \exp(H \cdot \log \xi_{\alpha\beta}) \cdot \exp\left(\hbar \cdot \frac{d \log \xi_{\alpha\beta}}{dz_\beta} \cdot X_+\right).$$

- The connection is defined by

$$\nabla^\hbar := d + \frac{1}{\hbar} \cdot \phi(q).$$

  Note that this definition is globally valid with respect to a Möbius coordinate system.
- $V_0 = K_C^{\frac{r-1}{2}} \oplus \ldots \oplus K_C^{-\frac{r-1}{2}} = E_0$, since $f^{\hbar=0}_{\alpha\beta} = \exp(H \cdot \log \xi_{\alpha\beta}) = \xi^H_{\alpha\beta}$.
- There is a unique filtration in the vector bundle $V_\hbar$ with $F_{r-1} = K_C^{\frac{r-1}{2}}$ that satisfies the conditions of Definition 7.2.
- The vector bundles $V_\hbar$ are isomorphic for all $\hbar \neq 0$.

We refer to [**14, 15**] for more detail.

8. Conclusion

We emphasize again that the nonabelian Hodge correspondence is a diffeomorphism between $\mathcal{M}_{\text{Dol}}$ and $\mathcal{M}_{\text{deR}}$. ([**8, 26, 38**]). Conjecture 6.2 of Gaiotto realizes a **holomorphic point by point correspondence** between two holomorphic Lagrangians, the *Hitchin section* in $\mathcal{M}_{\text{Dol}}$ and the *moduli space of opers* in $\mathcal{M}_{\text{deR}}$. Since the quantum curve should depend holomorphically on the spectral curve, we consider the Gaiotto correspondence as the desired construction of quantum curves.

- Donaldson, Hitchin, Simpson Nonabelian-Hodge correspondence

$$\mathcal{M}_{\text{Dol}} \overset{\text{diffeomorphism}}{\longrightarrow} \mathcal{M}_{\text{deR}}$$

- Gaiotto's conjecture gives a holomorphic correspondence between Lagrangians

$$\text{Hitchin Section} \overset{\text{holomorphic}}{\longrightarrow} \text{moduli space of opers}$$

Quantization is never unique. Yet the Catalan example we presented earlier clearly shows why we are interested in the unique process of quantization. A quantum curve quantizes the B-model geometry, which provides a generating function of genus g A-models for all g. Thus we wish a unique quantization result.

Starting from a Hitchin spectral curve, we identify the Higgs bundle on the Hitchin section. This is unique, once the spin structure of the curve C is chosen. Then the correspondence given by Gaiotto's scaling limit constructs, again, a unique oper in the moduli space of holomorphic connections on C. Thus the process from the spectral curve to the quantum curve (oper) is unique, and depends holomorphically on the moduli of spectral curves, when the complex structure of C is fixed.

We present below a local picture of the two Lagrangians inside the Dolbeault moduli space together with their images under the nonabelian-Hodge and Gaiotto's correspondences. In the figure, $V_1 = V_\hbar|_{\hbar=1}$. The picture does not show the global relations between various Lagrangians in the moduli spaces. For example, the $SL_r(\mathbb{R})$-Hitchin component in $\mathcal{M}_{\text{deR}}$ and the oper moduli space intersects infinitely many times. Only locally they intersect at a point, here at $(V_1, d + X_-)$.

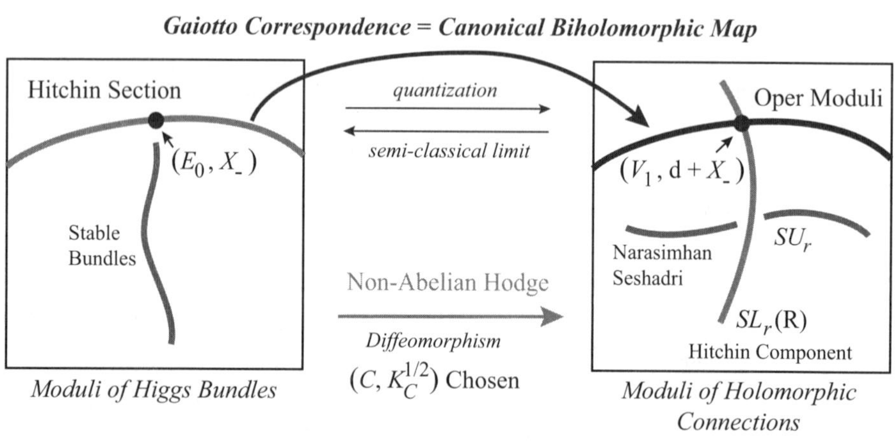

Acknowledgments

The author would like to express her gratitude to the organizers of *String-Math 2016* held in Collège de France, Paris, and to the Institute Henri Poincaré, for their hospitality. These lecture notes grew out from the author's paper [**9**] in collaboration with L. Fredrickson, G. Kydonakis, R. Mazzeo, M. Mulase, and A. Neiztke, that solves a conjecture of Davide Gaiotto [**18**]. This work was initiated at the AIM workshop, "New perspectives on spectral data for Higgs bundles." The author also thanks the organizers of the workshop, in particular Philip Boalch and Laura Schaposnik, for motivating interest in this problem by posing the question which led to this analysis.

The author is deeply indebted to Motohico Mulase for his generosity in mathematical discussions, enthusiasm, passion and encouragement that stimulated our collaboration throughout the years. This work could not have been produced without his support, for which the author would like to express all her gratitude.

The research of the author was supported by a grant from the Max-Planck Institute for Mathematics, Bonn. These lectures are based on a collaboration and discussions of the author with Motohico Mulase that took place in 2016 at the Max-Planck Institute for Mathematics, Bonn, and the Institute of Mathematics "Simion Stoilow" in Bucharest. These lectures will be continued in [**14**, **15**].

References

[1] M. Aganagic, R. Dijkgraaf, A. Klemm, M. Mariño, and C. Vafa, *Topological strings and integrable hierarchies*, Comm. Math. Phys. **261** (2006), no. 2, 451–516. MR2191887

[2] M. F. Atiyah, *Vector bundles over an elliptic curve*, Proc. London Math. Soc. (3) **7** (1957), 414–452. MR0131423

[3] M. F. Atiyah and R. Bott, *The Yang-Mills equations over Riemann surfaces*, Philos. Trans. Roy. Soc. London Ser. A **308** (1983), no. 1505, 523–615. MR702806

[4] A. Beilinson and V. Drinfeld, *Opers*, arXiv:math/0501398v1 [math.AG] (2005).

[5] R. Dijkgraaf, L. Hollands, and P. Sułkowski, *Quantum curves and $\mathcal{D}$-modules*, J. High Energy Phys. **11** (2009), 047, 59. MR2628886

[6] R. Dijkgraaf, L. Hollands, P. Sułkowski, and C. Vafa, *Supersymmetric gauge theories, intersecting branes and free fermions*, J. High Energy Phys. **2** (2008), 106, 57. MR2385939

[7] R. Dijkgraaf, H. Verlinde, and E. Verlinde, *Loop equations and Virasoro constraints in nonperturbative two-dimensional quantum gravity*, Nuclear Phys. B **348** (1991), no. 3, 435–456. MR1083914

[8] S. K. Donaldson, *A new proof of a theorem of Narasimhan and Seshadri*, J. Differential Geom. **18** (1983), no. 2, 269–277. MR710055

[9] O. Dumitrescu L. Fredrickson, G. Kydonakis, R. Mazzeo, M. Mulase and A. Neitzke, *Opers versus nonabelian Hodge* http://arxiv.org/pdf/1607.02172v1.pdf, under review.

[10] O. Dumitrescu and M. Mulase, *Quantum curves for Hitchin fibrations and the Eynard-Orantin theory*, Lett. Math. Phys. **104** (2014), no. 6, 635–671. MR3200933

[11] O. Dumitrescu and M. Mulase, *Quantization of spectral curves for meromorphic Higgs bundles through topological recursion*, http://arxiv.org/pdf/1411.1023v1.pdf, under review.

[12] O. Dumitrescu and M. Mulase, *Edge-contraction on dual ribbon graphs, 2D TQFT, and the mirror of orbifold Hurwitz numbers*, http://arxiv.org/pdf/1508.05922v1.pdf, under review in *Journal of Algebra.*

[13] O. Dumitrescu and M. Mulase, *Lectures on the topological recursion for Hitchin spectral curves and quantization*, http://arxiv.org/pdf/1509.09007v1.pdf to appear in Lecture Notes Series, Institute for Mathematical Sciences, National University of Singapore.

[14] O. Dumitrescu and M. Mulase, *Lectures on topological quantum field theory and character varieties*, Lecture Notes.

[15] O. Dumitrescu and M. Mulase, *Weyl quantization of Hitchin spectral curves and opers*, "Proceedings of the 2016 AMS von Neumann Symposium," Proceedings of Symposia in Pure Mathematics, American Mathematical Society.

[16] O. Dumitrescu, M. Mulase, B. Safnuk, and A. Sorkin, *The spectral curve of the Eynard-Orantin recursion via the Laplace transform*, Algebraic and geometric aspects of integrable systems and random matrices, Contemp. Math., vol. 593, Amer. Math. Soc., Providence, RI, 2013, pp. 263–315. MR3087960

[17] B. Eynard and N. Orantin, *Invariants of algebraic curves and topological expansion*, Commun. Number Theory Phys. **1** (2007), no. 2, 347–452. MR2346575

[18] D. Gaiotto,*Opers and TBA*, arXiv:1403.6137 [hep-th], (2014).

[19] D. Gaiotto, G. W. Moore, and A. Neitzke, *Wall-crossing, Hitchin systems, and the WKB approximation*, Adv. Math. **234** (2013), 239–403. MR3003931

[20] A. Grothendieck, *Sur la classification des fibrés holomorphes sur la sphère de Riemann* (French), Amer. J. Math. **79** (1957), 121–138. MR0087176

[21] A. Grothendieck, *Esquisse d'un programme*, (1984).

[22] S. Gukov and P. Sułkowski, *A-polynomial, B-model, and quantization*, J. High Energy Phys. **2** (2012), 070, front matter+56. MR2996110
[23] R. C. Gunning, *Special coordinate coverings of Riemann surfaces*, Math. Ann. **170** (1967), 67–86. MR0207978
[24] J. L. Harer, *The cohomology of the moduli space of curves*, Theory of moduli (Montecatini Terme, 1985), Lecture Notes in Math., vol. 1337, Springer, Berlin, 1988, pp. 138–221. MR963064
[25] J. Harer and D. Zagier, *The Euler characteristic of the moduli space of curves*, Invent. Math. **85** (1986), no. 3, 457–485. MR848681
[26] N. J. Hitchin, *The self-duality equations on a Riemann surface*, Proc. London Math. Soc. (3) **55** (1987), no. 1, 59–126. MR887284
[27] M. Kontsevich, *Intersection theory on the moduli space of curves and the matrix Airy function*, Comm. Math. Phys. **147** (1992), no. 1, 1–23. MR1171758
[28] M. Kontsevich and Yu. Manin, *Gromov-Witten classes, quantum cohomology, and enumerative geometry*, Comm. Math. Phys. **164** (1994), no. 3, 525–562. MR1291244
[29] B. Kostant, *The principal three-dimensional subgroup and the Betti numbers of a complex simple Lie group*, Amer. J. Math. **81** (1959), 973–1032. MR0114875
[30] M. Mulase and M. Penkava, *Ribbon graphs, quadratic differentials on Riemann surfaces, and algebraic curves defined over* $\overline{\mathbf{Q}}$, Asian J. Math. **2** (1998), no. 4, 875–919. Mikio Sato: a great Japanese mathematician of the twentieth century. MR1734132
[31] M. Mulase and P. Sułkowski, *Spectral curves and the Schrödinger equations for the Eynard-Orantin recursion*, Adv. Theor. Math. Phys. **19** (2015), no. 5, 955–1015. MR3487649
[32] D. Mumford, *Projective invariants of projective structures and applications*, Proc. Internat. Congr. Mathematicians (Stockholm, 1962), Inst. Mittag-Leffler, Djursholm, 1963, pp. 526–530. MR0175899
[33] D. Mumford, *Towards an enumerative geometry of the moduli space of curves*, Arithmetic and geometry, Vol. II, Progr. Math., vol. 36, Birkhäuser Boston, Boston, MA, 1983, pp. 271–328. MR717614
[34] D. Mumford, J. Fogarty, and F. Kirwan, *Geometric invariant theory*, 3rd ed., Ergebnisse der Mathematik und ihrer Grenzgebiete (2) [Results in Mathematics and Related Areas (2)], vol. 34, Springer-Verlag, Berlin, 1994. MR1304906
[35] M. S. Narasimhan and C. S. Seshadri, *Stable and unitary vector bundles on a compact Riemann surface*, Ann. of Math. (2) **82** (1965), 540–567. MR0184252
[36] A. Okounkov and R. Pandharipande, *Gromov-Witten theory, Hurwitz numbers, and matrix models*, Algebraic geometry—Seattle 2005. Part 1, Proc. Sympos. Pure Math., vol. 80, Amer. Math. Soc., Providence, RI, 2009, pp. 325–414. MR2483941
[37] C. S. Seshadri, *Space of unitary vector bundles on a compact Riemann surface*, Ann. of Math. (2) **85** (1967), 303–336. MR0233371
[38] C. T. Simpson, *Higgs bundles and local systems*, Inst. Hautes Études Sci. Publ. Math. **75** (1992), 5–95. MR1179076
[39] D. D. Sleator, R. E. Tarjan, and W. P. Thurston, *Rotation distance, triangulations, and hyperbolic geometry*, J. Amer. Math. Soc. **1** (1988), no. 3, 647–681. MR928904
[40] K. Strebel, *Quadratic differentials*, Ergebnisse der Mathematik und ihrer Grenzgebiete (3) [Results in Mathematics and Related Areas (3)], vol. 5, Springer-Verlag, Berlin, 1984. MR743423
[41] G. 't Hooft, *A planer diagram theory for strong interactions*, Nuclear Physics **B72**, 461–473 (1974).
[42] T. R. S. Walsh and A. B. Lehman, *Counting rooted maps by genus. I*, J. Combinatorial Theory Ser. B **13** (1972), 192–218. MR0314686

Department of Mathematics, Pierce Hall 209, Central Michigan University, Mount Pleasant Michigan 48859; and Simion Stoilow Institute of Mathematics, Romanian Academy, 21 Calea Grivitei Street, 010702 Bucharest, Romania
Email address: dumit1om@cmich.edu

Proceedings of Symposia in Pure Mathematics
Volume **98**, 2018
http://dx.doi.org/10.1090/pspum/098/01721

S-duality of boundary conditions and the Geometric Langlands program

Davide Gaiotto

Abstract. Maximally supersymmetric gauge theory in four dimensions admits local boundary conditions which preserve half of the bulk supersymmetries. The S-duality of the bulk gauge theory can be extended in a natural fashion to act on such half-BPS boundary conditions. The purpose of this note is to explain the role these boundary conditions can play in the Geometric Langlands program. In particular, we describe how to obtain pairs of Geometric Langland dual objects from S-dual pairs of half-BPS boundary conditions.

1. Introduction and conclusions

The beautiful work of [**1**] established a rich dictionary between S-duality in four-dimensional $\mathcal{N} = 4$ gauge theory and the mathematical subject of Geometric Langlands duality. We refer to that paper and to the more recent work [**2**] for an extensive discussion of the relationship and references to the mathematical literature on the subject.

The Geometric Langlands duality is a relation between two categories associated to a Riemann surface C and Langlands dual Lie groups G and ${}^\vee G$. These two categories are interpreted as an abstract description of two categories of boundary conditions ("branes") for a two-dimensional physical system, the twisted compactification of $\mathcal{N} = 4$ gauge theory on C. In turn, the duality between the two categories is a form of two-dimensional mirror symmetry which follows from the S-duality between the G and ${}^\vee G$ gauge theories [**3**].

A natural way to define boundary conditions in two-dimensions is to start from a local boundary condition in four-dimensional $\mathcal{N} = 4$ gauge theory, compactified along C. This was already crucial in the original work on the subject [**1**]: half-BPS Dirichlet boundary conditions in the four-dimensional gauge theory map to the most basic objects on one side of the GL duality, labelled by a G local system on C. The S-duality image ${}^\vee\mathfrak{D}$ of Dirichlet boundary conditions should then map to the GL dual image of these objects, the so-called "Hecke eigensheaves".

This relation is explained in [**1**] with the help of half-BPS line defects in the four-dimensional gauge theory. Half-BPS line defects inserted at points on C map to interfaces in the two dimensional system and thus to interesting functors acting on the categories of branes. Wilson lines act on the Chan-Paton bundle of branes

2010 *Mathematics Subject Classification.* Primary 81T60.

in a simple manner, by tensor product with a certain reference bundle. 'T Hooft operators act by Hecke correspondences, functors which play a key role in Geometric Langlands.

Dirichlet boundary conditions $\mathfrak{D}$ produce "eigenbranes" for the action of Wilson lines, simply because they fix the value at the boundary of the connection and scalar fields which enter the Wilson line path-ordered exponential. S-duality exchanges Wilson lines and 't Hooft lines. As a consequence, the S-dual boundary condition ${}^\vee\mathfrak{D}$ will produce "eigenbranes" for the action of 't Hooft lines, i.e. Hecke eigensheaves.

The analysis of [**1**] did not employ directly the S-dual boundary condition ${}^\vee\mathfrak{D}$, as the action of S-duality on boundary conditions for non-Abelian gauge theory was not understood at the time. This provided an important motivation for the work of [**4**, **5**], which described general half-BPS boundary conditions four-dimensional $\mathcal{N} = 4$ gauge theory and their behaviour under the action of S-duality.

One conclusion of that work was that ${}^\vee\mathfrak{D}$ can be defined with the help of a very special three-dimensional $\mathcal{N} = 4$ SCFT $T[G]$, which unfortunately lacks a simple UV Lagrangian description. For some classical groups a UV Lagrangian description is available, but breaks symmetries which are crucial for GL applications. On the other hand, the analysis also provided us with a large list of S-dual pairs of boundary conditions and interfaces which do admit an effective UV Lagrangian description.

The purpose of this paper is to make explicit the map from half-BPS boundary conditions which admit an effective Lagrangian description to objects in the categories relevant for the Geometric Langlands duality (see [**6**] for some early work on the subject). The expected payoff is a long list of Geometric Langlands dual pairs of mathematical objects which can be defined uniformly on any Riemann surface.

The obvious limitation of the computational techniques employed in this paper is the requirement of an effective UV description of the boundary conditions. This prevents us from, say, constructing generic Hecke eigensheaves directly from ${}^\vee\mathfrak{D}$. In a separate paper, [**7**], we will discuss alternative computational approaches based on Vertex Operator Algebras, which appear to have a broader range of applicability.

1.1. The twisted compactification of half-BPS boundary conditions. We employ a standard twisted compactification of maximally supersymmetric four-dimensional gauge theory with some gauge group G on a Riemann surface C. The R-symmetry of the theory is $SO(6)_R$, which has an obvious $SO(2)_V \times SO(4)_A$ block-diagonal subgroup. The twist on C involves the $SO(2)_V$ factor in the subgroup. The $SO(4)_A$ factor survives the twisting and becomes an R-symmetry group of the compactified theory.

The twisted compactification preserves $(4,4)$ two-dimensional supersymmetry. After the twist, the fields of the four-dimensional gauge theory can be organized into a two-dimensional $(4,4)$ gauged linear sigma model with infinite-dimensional target space and gauge group: the target space is the cotangent bundle to the space of G connections on C and the gauge group consists of maps from C to G.

Concretely, we denote the connection on C as A and the $(1,0)$ and $(0,1)$ components of the corresponding covariant derivative as ∂_A and $\bar{\partial}_A$. We denote the $(1,0)$ and $(0,1)$ components of the Higgs field (the scalar superpartner parameterizing the cotangent directions) as Φ, $\bar{\Phi}$.

1.1.1. *Hitchin equations.* An important branch of the moduli space of 2d supersymmetric vacua of the system is given by the space of solutions $\mathcal{M}_H(C, G)$ of G

Hitchin equations on C modulo gauge transformations. Recall that Hitchin equations are two-dimensional differential equations for a connection and Higgs field on C [**8**]:

$$\bar{\partial}_A \Phi = 0 \qquad F_A + [\Phi, \bar{\Phi}] = 0 \tag{1.1}$$

The moduli space $\mathcal{M}_H(C, G)$ is an hyper-Kähler manifold.

For most purposes, the result of the twisted compactification can be described as a two-dimensional non-linear sigma model with target $\mathcal{M}_H(C, G)$. The underlying gauge theory description may become important at special loci in $\mathcal{M}_H(C, G)$ where some gauge symmetry is unbroken [**9**]. Whenever possible, we will focus on situations for which the $\mathcal{M}_H(C, G)$ sigma model description is adequate.

Recall that a hyper-Kähler manifold can be endowed with a $\mathbb{P}^1$ worth of different complex structures, the "twistor sphere", parameterized by a complex variable ζ.

In complex structure "I", i.e. $\zeta = 0$, $\mathcal{M}_H(C, G)$ can be identified with the moduli space of Higgs bundles, i.e. pairs $(\bar{\partial}_A, \Phi)$ with

$$\bar{\partial}_A \Phi = 0 \tag{1.2}$$

modulo complexified gauge transformations. A conjugate description holds at the opposite complex structure $\bar{I}$, i.e. $\zeta = \infty$.

In all other complex structures, $\mathcal{M}_H(C, G)$ can be identified with a space $\mathrm{Loc}(C, G)$ of complexified G flat connections modulo complexified gauge transformations, sending the pair (A, Φ) into the Lax connection

$$D^\zeta = \partial_A + \zeta^{-1}\Phi \qquad \bar{D}^\zeta = \bar{\partial}_A + \zeta\bar{\Phi} \tag{1.3}$$

The S-duality relation between G and ${}^\vee G$ four-dimensional gauge theories maps to a mirror symmetry relation between $\mathcal{M}_H(C, G)$ and $\mathcal{M}_H(C, {}^\vee G)$. This is a mirror symmetry in the sense of SYZ [**10**]: both spaces are complex integrable systems in complex structure I with the same base, parameterized by gauge-invariant polynomials built from the Higgs fields Φ and ${}^\vee\Phi$. The non-singular fibers are dual Abelian varieties.

Mirror symmetry relates BPS boundary conditions in the two sigma models. The Geometric Langlands program is expected to encode mathematical aspects of the mirror relationship between two specific classes of half-BPS boundary conditions: BBB and BAA branes.[1]

1.1.2. *BBB branes.* BBB branes are boundary conditions which are compatible with the B-model twist in every complex structure of $\mathcal{M}_H(C, G)$. A prototypical BBB brane is defined by a Chan-Paton bundle on $\mathcal{M}_H(C, G)$ equipped with a hyper-holomorphic connection, i.e. a connection whose curvature is of type $(1, 1)$ in all complex structures.

More general BBB branes should be described by some hyper-holomorphic version of complexes of coherent sheaves, as a BBB brane will map to a standard B-brane in every complex structure. We do not know a necessary set of mathematical conditions for such a family of B-branes over the twistor sphere to give a BBB brane, but in Appendices B and C we will discuss important examples of BBB branes and some tentative characterization of a general story.

[1] Mirror symmetry and the Geometric Langlands program can also be thought of as matching the larger categories of B and A branes in a specific complex structure "K". The sub-categories of BBB and BAA branes, though, have useful extra structures and occur most naturally in our setup.

The "classical" discussion of the mirror symmetry transformation of BBB branes is easiest in complex structure I, but in standard Geometric Langlands applications the description in other complex structures becomes important.

The basic BBB objects in Geometric Langlands are Dirichlet boundary conditions D_μ, which constrain the fields at the boundary to lie at a specific point μ in the Hitchin moduli space. In particular, they are labelled by a point in $\mathrm{Loc}(C, G)$.

1.1.3. *BAA branes.* BAA branes are boundary conditions which are compatible with B twist in complex structure I and A-twist in complex structures orthogonal to I, such as J and K. A prototypical BAA brane is supported on a complex Lagrangian submanifold in the moduli space of Higgs bundles, equipped with a Chan-Paton bundle with a connection which is both flat and holomorphic in complex structure I.

In standard Geometric Langlands applications, one roughly identifies $\mathcal{M}_H(C, G)$ with the cotangent bundle to the space $\mathrm{Bun}(C, G)$ of G bundles on C and "quantizes" the BAA brane to a (twisted) D-module on $\mathrm{Bun}(C, G)$. The quantization is explained in [**1**] through the notion of a "canonical coisotropic brane" (see also [**11**, **12**]).

Another useful perspective based on $\mathcal{N} = 4$ supersymmetric quantum mechanics (see Appendix B) can be given with the help of auxiliary BAA branes F_p supported on the fiber of the cotangent bundle over a point p of $\mathrm{Bun}(C, G)$: the stalks of the sheaf underlying the D-module consist of the spaces $\mathrm{Hom}_A(F_p, B)$ of A-model morphisms from F_p to the BAA brane B in consideration.[2]

1.1.4. *Twisted compactification.* A natural way to produce BBB and BAA branes in the two-dimensional theory is to adapt the twisted compactification along C to standard half-BPS boundary conditions for the four-dimensional gauge theory.[3] These boundary conditions naturally preserve an $SO(3)_H \times SO(3)_C$ subgroup of $SO(6)_R$. Embedding $SO(2)_V$ as a Cartan subgroup in $SO(3)_H$ or $SO(3)_C$ respectively, we obtain either a BAA or a BBB brane in the two-dimensional theory.

For example, starting from four-dimensional Dirichlet boundary conditions $\mathfrak{D}$, the BAA twist produces the F_p branes mentioned above, while the BBB twist produces Dirichlet boundary conditions D_μ.

S-duality acts on boundary conditions in a non-trivial manner, while also exchanging the two $SO(3)$ R-symmetry factors. Each pair $(\mathfrak{B}, {}^\vee\mathfrak{B})$ of S-dual four-dimensional boundary conditions will map to two pairs of mirror 2d boundary conditions: the BAA twist of $\mathfrak{B}$ is dual to the BBB twist of ${}^\vee\mathfrak{B}$ and vice-versa.

In particular, the S-dual of Dirichlet boundary conditions ${}^\vee\mathfrak{D}$ would map under BAA twist to the Hecke eigensheaves ${}^\vee D_\mu$ dual to the BBB Dirichlet boundary conditions, while it would map under the BBB twist to some BBB branes ${}^\vee F_p$. The latter branes should have the property that B-model morphisms $\mathrm{Hom}_B({}^\vee F_p, E)$

---

[2]Yet another perspective on this relation (again discussed in Appendix B) is that general BAA branes in a neighbourhood of $\mathrm{Bun}(C, G)$ in $\mathcal{M}_H(C, G)$ can be parameterized by a solution $(\mathcal{A}, \Phi)$ of generalized Hitchin equations on $\mathrm{Bun}(C, G)$. The connection $\mathcal{A}$ is interpreted as the connection of a Chan-Paton bundle for a brane wrapping $\mathrm{Bun}(C, G)$ and Φ as a matrix-valued transverse deformation of the shape of the BAA brane. In complex structure I the spectral data of the pair $(\mathcal{A}, \Phi)$ gives a complex Lagrangian submanifold in $T^* \mathrm{Bun}(C, G)$ equipped with a line bundle. In general complex structure, the flat Lax connection $D_{\mathcal{A}} + \zeta^{-1}\Phi$ and $\bar{D}_{\mathcal{A}} + \zeta\bar{\Phi}$ can be interpreted as a holomorphic D-module on $\mathrm{Bun}(C, G)$.

[3]Another natural way is to consider surface defects in class S theories [**13**].

are the stalks of the D-module dual to any sheaf E on $\mathrm{Loc}(C, G)$, i.e. should be "kernels" for the Geometric Langlands duality.

In most of the paper we will discuss the "classical" descriptions of the BAA and BBB branes associated to $\mathfrak{B}$ as complex Lagrangian manifolds $\mathcal{L}(C, G, \mathfrak{B})$ and sheaves $\mathcal{V}(C, G, \mathfrak{B})$ in the moduli space of Higgs bundles. In particular, we will test the expectation that $\mathcal{L}(C, G, \mathfrak{B})$ and $\mathcal{V}(C, {}^\vee G, {}^\vee\mathfrak{B})$ should be related by fiberwise mirror-symmetry at generic points in the base of the complex integrable system.

At the end of the paper we will discuss more briefly the corresponding D-modules on $\mathrm{Bun}(C, G)$ and the sheaves on $\mathrm{Loc}(C, G)$, leaving further discussion to the companion paper [**7**].

1.1.5. *Symmetries and background fields.* In some situations, a single half-BPS boundary condition in four dimensions will map to a *family* of BAA and BBB boundary conditions in two dimensions. This will occur if the four-dimensional boundary condition is equipped with extra global symmetries, which can be coupled to background gauge multiplets on C in the twisted compactification process.

On general grounds, an half-BPS boundary condition may admit two types of global symmetry groups, exchanged by S-duality. The group of "Higgs branch" symmetries G_H can be coupled to background vectormultiplets, meaning a G_H bundle for a BAA twist or a G_H Hitchin system for a BBB twist. The group of "Coulomb branch" symmetries G_C can be coupled to background twisted vector-multiplets, meaning a G_C bundle for a BBB twist or a G_C Hitchin system for a BAA twist.

As a result, a single four-dimensional boundary condition will give us a family of two-dimensional BAA boundary conditions parameterized by $\mathrm{Bun}(G_H, C) \times \mathcal{M}_H(G_C, C)$ or a family of two-dimensional BBB boundary conditions parameterized by $\mathrm{Bun}(G_C, C) \times \mathcal{M}_H(G_H, C)$.

S-duality will then relate the two families associated to a pair of S-dual four-dimensional boundary conditions.

1.2. General strategy. Maximally supersymmetric four-dimensional gauge theory admits a large variety of half-BPS boundary conditions. We adopt a divide-and-conquer strategy, focussing our analysis on simple boundary conditions and interfaces, which can be concatenated to produce more general examples. In particular, all boundary conditions which admit an explicit Lagrangian description can be decomposed in such a manner. Concatenation of interfaces in the four-dimensional gauge theory will map to concatenation of interfaces in the two-dimensional setup and then to appropriate convolution operations in the categories of branes.

The most important and novel part of our analysis concerns boundary conditions and interfaces where the four-dimensional gauge fields are coupled to three-dimensional hypermultiplets in some symplectic representation M of G.

The BAA twist of such boundary conditions will give us a class of complex Lagrangian submanifolds $\mathcal{L}(C, G, M)$ of the space of Higgs bundles labelled by M. The mathematical treatment of such Lagrangians has been initiated recently by [**14**]. These Lagrangian submanifolds admit a natural quantization to D-modules given by conformal blocks for a free symplectic boson vertex algebra.

The BBB twist of such boundary conditions gives us a hyper-holomorphic bundle $\mathcal{V}(C, G, M)$ built as the "quantization" of the Dirac-Higgs bundle, itself defined as an hyper-holomorphic bundle of zeromodes of a Dirac operator couples to a solution of the Hitchin system (See [**15**] for a review of the Dirac-Higgs bundle).

The corresponding sheaf in any complex structure can be computed rather explicitly as the cohomology of a complex of forms on C valued in the bundle associated to M.

Other useful ingredients are various generalization of Dirichlet boundary conditions, which we review in Appendix D.

1.3. Open problems and future directions. Even in the simplest examples discussed in this paper, our analysis barely scratches the surface of very rich mathematical structures, which should be explored in further detail. Many interesting calculations are left for future work.

Line defects play a prominent role in the analysis of Kapustin and Witten and in the comparison with Geometric Langlands. Boundary conditions on four-dimensional gauge theory may support Wilson-like and vortex-like BPS line defects related by S-duality, which can be added at points in C during the twisted compactification in order to modify the resulting BAA and BBB branes. The S-duality map should be a generalization of the mirror map discussed in detail in [**16**].

Bulk line defects brought to the boundary will map to boundary line defects. This may allow one to systematically discuss the effect of Hecke modifications on the D-modules defined in this paper. Boundary line defects will also arise as the endpoint of bulk surface defects, which occur in the study of Geometric-Langlands duality for ramified Hitchin systems [**17**]. Work is currently in progress on the Geometric Langland applications of boundary line defects [**18**].

Finally, in this paper we only consider boundary conditions which preserve a full three-dimensional $\mathcal{N} = 4$ supersymmetry algebra, including the full three-dimensional Lorentz group. More general half-BPS local boundary conditions which break Lorentz symmetry do exist and can play an important role for Geometric Langlands applications.

2. Neumann-like boundary conditions and matter interfaces

Neumann-like half-BPS boundary conditions preserve the full gauge symmetry at the boundary. The four-dimensional G gauge fields can be coupled at the boundary to any three-dimensional $\mathcal{N} = 4$ SQFT with G (Higgs branch) global symmetry [**4**].

The boundary conditions for the six scalar fields of SYM gauge theory follow from the boundary conditions on the gauge fields. The three scalar fields rotated by $SO(3)_H$ receive Dirichlet boundary conditions: they are set to equal the three hyper-Kähler moment map operators of the $\mathcal{N} = 4$ SQFT. The three scalar fields rotated by $SO(3)_C$ receive Neumann boundary conditions and act as mass parameters for the three-dimensional $\mathcal{N} = 4$ SQFT.

As the system is compactified on C, the $\mathcal{N} = 4$ SQFT is twisted as well. The BBB and BAA twists are respectively analogous to the Rozansky-Witten (RW) twist [**19**] and its mirror (mRW). More precisely, the B or A supercharges in complex structure J literally coincide with the three-dimensional topological RM or mRW supercharges. This should allow for some interesting tests of our claims using three-dimensional TFT tools.

Another simple system one can associate to a $\mathcal{N} = 4$ SQFT with G global symmetry is a "matter interface": the four-dimensional gauge fields are defined on the whole of space-time and the thee-dimensional $\mathcal{N} = 4$ SQFT is coupled to the restriction of the four-dimensional gauge multiplet to a hyper-surface. The three bulk scalar fields rotated by $SO(3)_C$ still act as mass parameters for the $\mathcal{N}=4$ SQFT. The three scalar fields rotated by $SO(3)_H$ are discontinuous at the interface, with discontinuity given by the three hyper-Kähler moment map operators of the $\mathcal{N}=4$ SQFT.[4]

2.1. Pure Neumann boundary conditions. As a warm-up for the general case, we can review a simple example: pure Neumann boundary conditions without any extra boundary degrees of freedom.

The main subtlety we need to keep track of occurs if the gauge group contains Abelian factors: the Neumann boundary condition has a dual Abelian boundary Coulomb branch global symmetry G_C whose currents are simply the boundary values of the Abelian bulk field strengths. Concretely, if the gauge group has n_C Abelian factors then $G_C = U(1)^{n_C}$.

In particular, upon compactification on C, configurations involving a gauge bundle of degree d give rise to states or Chan-Paton bundles which carry G_C charge d. Furthermore, the system can be coupled to three-dimensional background twisted G_C vectormultiplets: a G_C Higgs bundle for a BAA twist and a G_C bundle for a BBB twist.

The BAA image of pure Neumann boundary condition sets the Higgs field to zero or, more generally, to equal some reference one-form t valued in the Abelian part $\mathfrak{g}_A$ of the gauge Lie algebra. The gauge connection on C is unconstrained and thus the BAA brane is supported on the full fiber $\mathcal{L}_t$ of the complex integrable system over t in the Hitchin fibration. The definition of an BAA brane also involves a flat connection over $\mathcal{L}_t$. We take that to be a $U(1)$ connection depending only on the Abelian part of the gauge connection, built from a G_C connection on C. This connection combines with t to give the data of a G_C Higgs bundle.

As the brane passes through the singular locus of the Hitchin moduli space, this sigma model description of the BAA image of pure Neumann boundary conditions may be incomplete.

The BBB twist of pure Neumann boundary conditions leaves both the connection and the Higgs field free to fluctuate. The corresponding BBB brane is thus supported on the whole of $\mathcal{M}_H(C,G)$. It should be essentially the same as the structure sheaf on $\mathcal{M}_H(C,G)$, up to the possible twist by a point in $\mathrm{Bun}(C,G_C)$, interpreted as a choice of line bundle on the torus fibers of the Abelian factors of the Hitchin moduli space.

The S-dual image of pure Neumann boundary conditions is known in general: it is a modification of Dirichlet boundary conditions known as a regular Nahm pole boundary condition. We will review in Appendix D the BAA and BBB twists of such boundary conditions. The BAA twist of the regular Nahm pole boundary conditions leads to the Hitchin section of the complex integrable system, with is the natural mirror to the structure sheaf on $\mathcal{M}_H(C,G)$. The BBB twist gives a brane whose description lies outside the scope of the sigma-model description.

Notice that a Neumann-like boundary condition can be realized as the concatenation of a matter interface and a pure Neumann boundary condition.

2.2. Boundary hypermultiplets and interface hypermultiplets. We now come to the main ingredient of our analysis: boundary hypermultiplets valued

[4]In the language of [**4**], a matter interface is an interface between G and G gauge theories with the gauge group restricted to the diagonal $G \subset G \times G$ at the interface.

in some symplectic representation M of G. The scalar fields in the hypermultiplet transform as a doublet of $SO(3)_H$, while the fermions transform as a doublet of $SO(3)_C$.

We will denote the complex scalar fields in the hypermultiplets as Z, the symplectic pairing on M as $\langle Z, Z' \rangle$ and the complex moment map for the G action on M as $\mu(Z)$.

In general, the representation M may carry an action of an extra flavor group G_H commuting with the G action. This allows one to couple the system to three-dimensional background G_H vectormultiplets: a G_H bundle for a BAA twist and a G_H Higgs bundle for a BBB twist.

The $G_C = U(1)^{n_C}$ symmetry of Neumann boundary conditions is unaffected by the additional matter fields and thus we will still be able to couple the system to a G_C Higgs bundle for a BAA twist and a G_C bundle for a BBB twist if the gauge group has Abelian factors.

2.2.1. *BAA twist.* The BAA twist makes the hypermultiplet scalars into spinors on C. Thus Z is a spinor section of the symplectic bundle M_E associated to E, i.e. a section of $M_E \otimes K_C^{1/2}$. Supersymmetry requires the section to be holomorphic. We can call $\mathcal{Z}(C, G, M)$ the space of pairs (E, Z).

The boundary conditions for the gauge theory scalars require furthermore

$$\Phi = \mu(Z) \tag{2.1}$$

Thus the BAA brane should be supported on the image $\mathcal{L}_M(C, G)$ of $\mathcal{Z}_M(C, G)$ under the map

$$(E, Z) \to (E, \Phi = \mu(Z)) \tag{2.2}$$

It is far from obvious that this image should be a Lagrangian submanifold. It becomes clear if we observe that the submanifold manifold $\mathcal{L}_M$ belongs to a general class of Lagrangian submanifolds associated to generating functions, discussed in Appendix A. The generating function for $\mathcal{L}_M(C, G)$ is the functional

$$W[Z, \bar{A}] \equiv \int_C \langle Z, \bar{\partial}_A Z \rangle \tag{2.3}$$

Indeed the equations defining $\mathcal{L}_M(C, G)$ can be re-cast in the form

$$\frac{\delta W}{\delta Z} = \bar{\partial}_A Z = 0 \qquad \Phi = \frac{\delta W}{\delta \bar{A}} = \mu(Z) \tag{2.4}$$

In order to specify a BAA brane, we need both the support $\mathcal{L}_M$ and a choice of flat Chan-Paton bundle. The Chan-Paton bundle is given by the bundle of ground states for the system of hypermultiplets on C coupled to E and constrained by Equation 2.1. The system is analogous to a supersymmetric quantum mechanics with Kähler target space and a super-potential W (See Appendix B for details).

If the pre-image of a point under the projection $\mathcal{Z}_M(C, G) \to \mathcal{L}_M(C, G)$ is a point or a collection of points, they will give a basis for the bundle of ground states. If the pre-image of a point p is a manifold $\mathcal{F}_p$, the flat bundle should be defined by the cohomology of $\mathcal{F}_p$.

If there is a non-trivial flavor group G_H, we can couple the system to a G_H bundle on C simply by modifying accordingly the $\bar{\partial}_A$ operator.

Similarly, if G has Abelian factors, we can couple the system to a G_C Higgs bundle by deforming Equation 2.1 to $\Phi = \mu(Z) + t$ and twisting the Chan-Paton bundle by the appropriate line bundle.

Similar considerations apply for a matter interface consisting of 3d hypermultiplets coupled to the gauge theory at an interface. The interface will give a Lagrangian correspondence between two copies of $\mathcal{M}(C, G)$:

$$E = E' \qquad \Phi = \Phi' + \mu(Z) \tag{2.5}$$

2.2.2. *BBB twist.* The BBB twist does not affect the hypermultiplet scalars, but twists the fermions into Grassmann-odd scalars which are 1-forms on C.

The setup is rather analogous to a Nahm transform. The analogy is rather precise, in the sense that both the compactification of hypermultiplets on C and all examples of Nahm transforms can be interpreted in the language of $N = 4$ supersymmetric quantum mechanics, with super-charges described by appropriate Dirac operators. See Appendix B for details.

Generic enough Higgs field and connection make the bosonic fields and appropriate fermionic partners massive, removing them from the calculation. Fermionic zeromodes in the kernel of the Dirac operator survive and are quantized to give a fermionic Fock space of supersymmetric ground states.[5]

The Dirac operator is acting on the space of M-valued forms on C. It is known in the mathematical literature as the Dirac-Higgs operator [**15**]. The space of zeromodes of the Dirac-Higgs operator is a hyper-holomorphic sheaf $\mathcal{D}(C, G, M)$ called the Dirac-Higgs bundle. The bundle can be described in complex structure J as the deformed de-Rahm complex of forms valued in M with differential $d + \Phi + \bar{\Phi}$ and in complex structure I as the deformation of the Dolbeault complex by $\Phi\wedge$ ("hyper-cohomology"). Generically, the cohomology lies in degree 1 and gives the Dirac zeromodes.

The phase space has a (graded) symplectic pairing inherited from the pairing of M, which induces a symplectic pairing on the Dirac-Higgs bundle. When the bundle lives in degree 1 we can promote $\mathcal{D}(C, G, M)$ to a sheaf of Clifford algebras and the sheaf of ground states as the corresponding Clifford module.

If we are dealing with full hyper-multiplets, i.e. $M = N + N^*$, we can build the corresponding Fock space by using creation operators valued in N. Thus we can take the Dirac-Higgs sheaf $D_N \equiv \mathcal{D}(C, G, N)$ associated to N and build the Fock space

$$\mathcal{V}(C, G, M) \equiv (\det D_N)^{-1/2} \Lambda^* D_N. \tag{2.6}$$

This is our proposal for the hyper-holomorphic sheaf $\mathcal{V}(C, G, M)$ defining our BBB brane.

If the Higgs field and connection are not generic enough, the hypermultiplet scalars and their fermionic partners will have zeromodes. If the hypermultiplet scalars acquire a supersymmetric vev, they will induce a vev for the bulk scalar fields which lie outside the non-linear sigma-model description of the 2d system. We leave a proper understanding of that situation to future work.

The Dirac-Higgs bundle has an intuitive description over loci in the Hitchin moduli space where the determinant of the Higgs field in representation N has isolated simple zeroes on C: the hypercohomology lies in degree 1 and decomposes as a direct sum of line bundles, the co-kernels for the action of Φ. The corresponding Fock space similarly decomposes into a direct product of two-dimensional Fock

[5]We are indebted to Kevin Costello for explaining this point to us

spaces. If the zeroes of Φ are isolated but not simple, each multiple zero will contribute an appropriate vector bundle to the hypercohomology.

This description is physically reasonable. The Higgs field appears as a mass term in the hypermultiplet Lagrangian. We can integrate out the hypermultiplets except at the loci on C where the Higgs field does not have full rank, where zero-modes may be localized. We can think about the contributions of each zero to the space of ground states as the Hilbert space of a vortex.

If the Higgs field in representation N vanishes or has non-maximal rank over the whole of C, these simplifications do not occur and the hypercohomology must be computed more carefully.

If M includes half-hypermultiplets, i.e. has no Lagrangian splitting, the situation is more subtle. One route may be to double-up the representation and look for a square root of the Fock space built from D_M. Notice that half-hypermultiplets have a potential anomaly, which may be reflected into the absence of such a square root.

For a matter interface, we would place the same sheaf upon the diagonal in $\mathcal{M}_H(C,G) \times \mathcal{M}_H(C,G)$. Thus acting with a matter interface onto a boundary condition simply tensors the corresponding BBB brane by the above sheaf $\mathcal{B}_M$.

If the hypermultiplet representation M also carries a G_H action, we can deform the BBB brane by including a solution of G_H Hitchin equations into the Dirac-Higgs operator. In particular, the hypercohomology will be supported on loci in C where $\Phi + \Phi_H$ drops in rank.

In Appendix B we will propose an alternative description of the Dirac-Higgs sheaf involving BAA branes in T^*C.

2.3. Boundary gauge theories. A more general class of boundary conditions may be obtained by coupling the boundary hypermultiplets to some G_B boundary gauge fields.

We can describe this system as the composition of an interface with boundary hypermultiplets coupled to G and G_B four-dimensional gauge fields and a G_B pure Neumann boundary condition.

The BAA and BBB twisted descriptions of the system should analogously follow from the composition of the appropriate 2d interfaces and boundary conditions.

2.3.1. *BAA twist.* We will now consider an auxiliary space $\mathcal{Z}(C,G,G_B,M)$ consisting of triples (E, E_B, Z), with Z coupled to both the G bundle E and the G_B bundle E_B, subject to the further constraint

$$\mu_B(Z) = 0 \tag{2.7}$$

where $\mu_B(Z)$ is the moment map for the G_B action. If G_B has Abelian factors, the right hand side can be shifted by a 1-form.

The BAA brane should be again supported on the image $\mathcal{L}(C,G,G_B,M)$ of $\mathcal{Z}_M(C,G,G_B,M)$ under the map

$$(E, E_B, Z) \to (E, \Phi = \mu(Z)) \tag{2.8}$$

2.3.2. *BBB twist.* The analysis of the previous section gave us a hyper-holomorphic bundle, or sheaf, $\mathcal{V}(C,G,M)$ on $\mathcal{M}_H(C,G) \times \mathcal{M}_H(C,G_B)$. In order to understand the boundary conditions involving a G_B gauge theory, we need to convolve that sheaf with the structure sheaf on $\mathcal{M}_H(C,G_B)$.

In a given complex structure, the B-model prescription for the convolution is that the stalks of the resulting sheaf at a point on $\mathcal{M}_H(C,G)$ are the global sections or, better, the cohomology of $\mathcal{V}(C,G,M)$ restricted to p.

Hyper-holomorphically, the resulting sheaf should be the sheaf of supersymmetric ground states of an $\mathcal{N}=4$ SQM with $\mathcal{M}_H(C,G_B)$ target space, associated to the hyper-holomorphic connection on $\mathcal{B}_M$. See Appendix B for details. This is roughy a Nahm transform of $\mathcal{V}(C,G,M)$ over $\mathcal{M}_H(C,G_B)$.

It is reasonable to worry, though, that this naive description may miss contributions from the singular loci of $\mathcal{M}_H(C,G_B)$, where unbroken G_B gauge fields are present. A description as an $\mathcal{N}=4$ gauged SQM may be more suitable. We leave this problem to future work.

2.4. Generalization to smooth sigma models. Our discussion of the free hypermultiplet case does generalize in a natural way to boundary conditions involving three-dimensional non-linear sigma model. The sigma models must have a smooth hyper-Kähler target space $\mathcal{X}$ equipped with a tri-holomorhic G action.

The Higgs branch global symmetry G_H of such a model consists of extra tri-holomorhic isometries commuting with G. If $\mathcal{X}$ is endowed with suitable $(1,1)$ forms, such as these associated with resolution/deformation parameters, they can be pulled back to space-time to define currents for Abelian G_C symmetries. The mass parameters for such isometries are identified with the resolution/deformation parameters. Notice that the sigma model description will break down if these parameters are tuned to a value which makes $\mathcal{X}$ singular.

The BAA twist of such a boundary condition requires $\mathcal{X}$ to admit a $SO(2)_H$ Kähler isometry in complex structure I which rotates the complex symplectic form, realizing the Cartan of $SO(3)_H$. The $SO(2)_H \times G$ action can be used to make $\mathcal{X}$ into the fiber of a bundle $\mathcal{X}_C$ on C, in such a way that the complex symplectic form on $\mathcal{X}$ is a section of K_C and the complex G moment map a section of $\mathrm{End}(E)\otimes K_C$. Then it makes sense to consider holomorphic sections Z of $\mathcal{X}_C$ and write $\Phi=\mu(Z)$ to define a BAA brane.

Notice that the complex deformation parameters $t_{\mathbb{C}}$ of $\mathcal{X}$ would break $SO(2)_H$. They can be introduced in the story if they are twisted to a 1-form $t(z)$ on C. This is the background Higgs field for the G_C twisted vectormultiplet.

In the BBB twist, we can consider a generalization of the $(0,4)$ supersymmetric quantum mechanics of free hypermultiplets on C discussed in Appendix B. We can take $(0,4)$ hypermultiplets valued in $\mathcal{X}$ together with a bundle of $(0,4)$ Fermi multiplets modelled on 1-forms on C valued in the pull-back of the tangent bundle to $\mathcal{X}$.

An example of a situation where the non-linear sigma model construction is fully applicable involves a family of hyper-Kähler manifolds defined as moduli spaces of solutions of Nahm equations on a segment with structure group G. When the boundary conditions are selected to be Dirichlet at one end of the segment, a Nahm pole on the other end, the result is a smooth hyper-Kähler manifold with $G\times Z_\rho$ hyperholomorphic isometries, where Z_ρ is the stabilizer of the Nahm pole (see **[4]** for a review and several mathematical references). Modified Neumann boundary conditions involving these manifolds are equivalent to the Dirichlet and Nahm pole boundary conditions reviewed in Appendix D. They provide an alternative way to understand the image of these boundary conditions upon compactification on C.

An important potential application of non-linear sigma model construction would be to approximate the theory $T[G]$ by its resolved Higgs branch, the cotangent bundle to the flag variety for G. This approximation would not be compatible with the most general possible background for the ${}^\vee G$ valued twisted vectormultiplets, but should be sufficient to deal with backgrounds in an Abelian subgroup of ${}^\vee G$. It would be interesting to match this construction with the mathematical construction of the Geometric Langland dual to local systems in an Abelian subgroup of ${}^\vee G$. See e.g. [**20**].

3. A rich example: Particle-vortex duality in $U(1)$ gauge theory

The $U(1)$ Hitchin equations are linear and thus the corresponding moduli space is rather simple: it is the product of the space of 1-forms Φ on C and of the space $\mathrm{Jac}(C)$ of degree 0 line bundles E on C.

There are two mirror subtleties which are related to the existence of an unbroken four-dimensional gauge symmetry. Boundary conditions can carry magnetic and electric charges d and q which are roughly exchanged by S-duality.

- The magnetic charge modifies the Hitchin moduli space by shifting the degree of the line bundle from 0 to d so that the fiber of the Hitchin fibration is $\mathrm{Pic}_d(C)$ We obtain an infinite family of moduli spaces $\mathcal{M}_d$, labelled by the degree d of the line bundle E. Giving a boundary condition means giving a brane B_d in each of these moduli spaces.
- There is an unbroken 2d gauge symmetry and thus the Chan-Paton bundles should really be thought of as a choice of boundary condition for that 2d gauge theory. Essentially, they should carry an action of the unbroken gauge group. Boundary conditions can be decomposed into branes labelled by such extra representation data q. Geometrically, this is recognizable as an ambiguity in the Chan-Paton bundle which can be fixed non-canonically by picking a specific representative for E.

These subtleties are discussed at length in [**2**] as they affect the treatment of Wilson and 't Hooft lines. A 't Hooft line maps to a BAA interface which identifies the Higgs fields Φ and Φ' on the two sides, but relates the line bundles E and E' by a Hecke modification at a point, which shifts the degree of the bundle. A Wilson line maps to a BBB brane on the diagonal of the two Hitchin moduli spaces, whose Chan-Paton bundle E_z is given by the line bundle restricted to a point z on C. Different choices representative for E over the space of line bundles give a different meaning to the Chan-Paton bundle.

At Neumann boundary conditions, there is a "topological" $U(1)_t$ factor in G_C which is generated by the current $*F$. The degree d can be identified with the $U(1)_t$ charge of the state. In the presence of matter fields, there is a slightly spurious, but still useful, $U(1)_b$ factor in G_H which acts on the boundary matter fields in the same way as the bulk gauge group. The charge q can be identified with the $U(1)_b$ charge of the state.

The simplest non-trivial example of the action of S-duality on boundary conditions arises for a $U(1)$ gauge theory coupled at the boundary to a single hypermultiplet of charge 1. This boundary condition is self-mirror. It has both G_H and G_C equal to $U(1)$.

3.1. BAA analysis. We decompose the hypermultiplet scalars into two complex fields X and Y of gauge charge 1 and -1 respectively.

Given a $U(1)$ bundle E, X is a section of $E \otimes K_C^{1/2}$ and Y is a section of $E^{-1} \otimes K_C^{1/2}$. The Higgs field is then $\Phi = XY$. Notice that the difference in the dimension of the spaces of X and Y global sections is controlled by an index theorem and equals the degree d of E.

At first, we can take the degree of E to be 0. Generically there are no holomorphic X and Y and we have $\Phi = 0$. At a special co-dimension 1 locus on $\mathrm{Bun}_0(C, U(1))$, the theta divisor Θ_0, holomorphic solutions for X and Y will occur simultaneously.

At generic points on Θ_0, there is a one-dimensional space of solutions for X and Y and thus a 1-dimensional space of possible Φ, which parameterizes the co-normal bundle to Θ_0. Above a point in that Lagrangian image we have a $\mathbb{C}^*$ worth of possible pairs (X, Y), acted upon by the unbroken 2d gauge symmetry. At special loci on the Θ_0 divisor the dimension of the space of solutions for X and Y may grow larger and the local geometry of the Lagrangian manifold will become more intricate and possibly singular.

In positive degree d, there will generically be a d-dimensional space of solutions for X and none for Y. We have again $\Phi = 0$ but a non-trivial $\mathbb{C}^d$ fiber acted upon by the unbroken gauge transformations. At a co-dimension $d+1$ locus Θ_d in the space of bundles a solution for Y will appear and there will be a $(d+1)$-dimensional space of solutions for X. Now Φ will also live in a $(d+1)$-dimensional space. More complicated geometry will again occur in higher co-dimension.

For $d < 0$ the roles of X and Y are exchanged.

In order to study fiberwise mirror symmetry, we need to understand the intersection of the Lagrangian manifold with the fibres of the Hitchin fibration. A non-zero Higgs field Φ has $2g-2$ zeroes. For simplicity, we can assume that they are distinct. The zeroes of Φ must be either zeroes of X or Y: $g-1+d$ zeroes of X and $g-1-d$ zeroes of Y.

Conversely, we can find a solution for any of the $\binom{2g-2}{g-1+d}$ possible ways of splitting the zeroes of Φ among X and Y: given the zeroes, the bundle E exists and is unique. We can roughy represent it by a divisor $\frac{\sum_i x_i - \sum_j y_j}{2}$, where x_i and y_j are the zeroes in X and Y, by looking at the meromorphic section X/Y of E^2.

Thus the Lagrangian manifold is a nice multi-section of the Hitchin fibration, with $\binom{2g-2}{g-1+d}$ sheets, ramified above the locus where Φ has higher order zeroes. Above a generic point in the manifold there is a $\mathbb{C}^*$ worth of possible values of X and Y, acted upon by the unbroken 2d gauge symmetry.

Above the singular point $\Phi = 0$, instead, we have the whole fiber of the Hitchin fibration.

The mirror of this BAA brane should be a collection of sheaves which at generic Φ have rank $\binom{2g-2}{g-1+d}$ and when restricted to the fiber at that point is a sum of line bundles of the form $\left(\otimes_i L_{x_i}^{\frac{1}{2}}\right) \otimes \left(\otimes_j L_{y_j}^{-\frac{1}{2}}\right)$, where L_z is the line bundle on the fiber which associates to an element L of Pic_d the restriction of L to the point z. We refer to [**2**] for details of the Abelian fiberwise mirror map.

3.2. BBB analysis. If the zeroes of Φ are distinct and E has degree 0, the hyper-cohomology of (E, Φ) in a charge 1 representation consists of a direct sum

$$\mathbf{H}^1(E, \Phi) = \oplus_{i=1}^{2g-2} (E \otimes K)_{z_i} \tag{3.1}$$

where z_i are the zeroes of Φ. and the summands on the right hand side are the restriction of $E \otimes K$ to these points.

Correspondingly, the Fock space is

$$F\mathbf{H}^1(E, \Phi) = \otimes_{i=1}^{2g-2} ((E \otimes K)_{z_i}^{1/2} \oplus (E \otimes K)_{z_i}^{-1/2}) \tag{3.2}$$

The two summands in each product have G_H charges $\pm 1/2$ respectively.

Thus the BBB brane, for non-trivial Φ with distinct zeroes, looks like a direct sum of rank $\binom{2g-2}{g-1+q}$ vector bundles. The dimension of this space, including the G_H grading, matches the number of points in the fiber above Φ in the conjectural mirror BAA brane, graded by degree.

Each summand of the Fock space can be written as $\otimes_i (E \otimes K)_{x_i}^{1/2} \otimes_j (E \otimes K)_{y_j}^{-1/2}$, where we split the zeroes of Φ into the two groups x_i and y_j. This line bundle is naturally mirror to the divisor $\frac{\sum x_i - \sum y_i}{2}$, matching the BAA analysis. Indeed L there is the same as E here and the extra factor of K does not change the resulting bundle on the fiber of the Hitchin fibration.

We expect a similar analysis to hold for degree d bundles, giving again a sum of terms of the form $\otimes_i (E \otimes K)_{x_i}^{1/2} \otimes_j (E \otimes K)_{y_j}^{-1/2}$. The existence of this infinite collection of branes of the same form for various d should be mirror to the presence of the $\mathbb{C}^*$ boundary degrees of freedom in the BAA analysis.

In conclusion, we verified that at least above a generic point in the base of the integrable system the BBB and BAA images of the four-dimensional boundary condition are mirror to each other.

4. Bifundamental and fundamental interfaces

A simple modification of the setup in the last section is to take a matter interface for a $U(1)$ gauge theory consisting of a single charge 1 hypermultiplet. Applying S-duality to both sides of the interface, we end up with a "bi-fundalental" interface: Neumann boundary conditions for both $U(1)$ gauge groups, with a hypermultiplet of charge $(1, -1)$ at the interface.

Although this system is closely related to the one in the previous section, it has a nice advantage: it can be generalized to $U(N)$ gauge theory. S-duality relates a matter interface with hypers in a fundamental representation for a $U(N)$ gauge theory ("D5 interface") to an interface involving bi-fundamental hypermultiplets coupled to two $U(N)$ gauge theories with Neumann b.c. ("NS5 interface"). We will now consider such a generalization.

4.1. BAA analysis of a D5 interface. As this is a matter interface and the gauge fields continue across the interface, the gauge bundles on the two sides of the interface coincide, i.e. we have $E = E'$. On the other hand, the Higgs fields satisfy

$$\Phi' = \Phi + XY \tag{4.1}$$

where X and Y are sections of $E \otimes K_C^{1/2}$ and $E^* \otimes K_C^{1/2}$ respectively.

Consider a generic location on the bases of the Hitchin moduli spaces where we can encode the Higgs bundles into spectral curves Σ and Σ' in T^*C. In the notation of [**2**], the spectral data of the Higgs bundle (Φ, E) is the spectral curve

$\det(y-\Phi)=0$ in T^*C equipped with a line bundle L such that E is the pushforward of L along the projection to C.

Over the spectral curve Σ we have both the 1-form λ of Φ eigenvalues and the corresponding eigenvector v, which can be thought of as a section of the pull-back of E to the spectral curve Σ tensored with the line bundle $\tilde{L}$ whose pushforward to C is E^*. Over Σ' we have both the 1-form λ' of Φ' eigenvalues and the corresponding left eigenvector v', section of the pull-back of E^* to Σ' tensored with the line bundle L' whose pushforward to C is E' .

Pulling back the above equation to the combined curve parameterized by the pair (λ, λ') over C and contracting with the eigenvectors, we find

$$(\lambda' - \lambda)(v', v) = (v', X)(Y, v) \tag{4.2}$$

The zeroes of $\lambda - \lambda'$ are zeroes of a polynomial in the traces of Φ and Φ', the *resultant* of their characteristic polynomials. It is a differential of degree N^2 on C. These zeroes will have to match either a zero of (v', X) on the point of Σ' labelled by λ' or of (Y, v) on the point of Σ labelled by λ.

These two inner products are sections of the pull-back of $K^{\frac{1}{2}}$ to Σ' or Σ tensored respectively with L' and $\tilde{L}$. The degree of these bundles is $N^2(g-1) \pm d$ and thus the total number of zeroes they have is $n = N^2(2g-2)$, which is the same as the total number of zeroes of the resultant.

Thus if the resultant has simple zeroes, in order to find all solutions of 4.1 we can pick any of the 2^n ways to split the zeroes of the resultant among (v', X) and (Y, v) and then characterize $\tilde{L}$ and L' by the corresponding divisors. Each of these possible 2^n choices will fix the line bundle on the spectral curves and determine an intersection point of our Lagrangian manifold with the fibers of the complex integrable systems.

Upon fiberwise mirror symmetry, this BAA brane will map to a sheaf with rank 2^n over the product of the two Hitchin systems, which above generic points on the base of the integrable systems will be described as a sum of line bundles which are mirror dual to the product of the L and L' associated to the possible splittings of the zeroes of the resultant.

4.2. BBB analysis of a D5 interface. The BBB twist of the interface will provide a sheaf supported on the diagonal of the product of the two $U(N)$ Hitchin systems associated to the gauge theories on either sides of the matter interface.

The sheaf is the Fock space built from the fundamental Dirac-Higgs bundle for the Hitchin system. It has rank $2^{2N(g-1)}$.

At least over loci where $\det \Phi$ has simple zeroes, the Dirac-Higgs bundle is again a direct sum of sub-bundles, the co-kernels of the map $E_{z_i} \to (E \otimes K)_{z_i}$ given by the action of $\Phi(z_i)$, where z_i are the zeroes of $\det \Phi$. The Fock space is built as before.

The co-kernels can be recognized as the restriction of the line bundle $L \otimes K_C$ on the spectral curve to the points where the spectral curve intersects the zero section of T^*C. The Fock space will be a direct sum of terms of the rough form

$$\left(\otimes_i (L \otimes K_C)^{\frac{1}{2}}_{x_i}\right) \otimes \left(\otimes_j (L \otimes K_C)^{-\frac{1}{2}}_{y_j}\right) \tag{4.3}$$

where x_i and y_j are a partition of the zeroes of $\det \Phi$ in two groups. See Appendix B for an A-model version of this statement.

4.3. BAA analysis of an NS5 interface. The Higgs fields for the two gauge theories satisfy now

$$\Phi' = XY \qquad \Phi = YX \tag{4.4}$$

where X and Y are sections of $E' \otimes E^* \otimes K_C^{\text{-}1/2}$ and $E \otimes (E')^* \otimes K_C^{\text{-}1/2}$ respectively.

At least as long as Φ and Φ' have full rank, the two Higgs fields have the same characteristic polynomial. Taking the determinant of the relations, we get that over loci where $\det \Phi$ has simple zeroes, the $2N(g-1)$ zeroes of $\det \Phi$ must be distributed among $\det X$ and $\det Y$.

Furthermore X and Y map eigenvectors of Φ into eigenvectors of Φ' with the same eigenvalues and viceversa. They provide an identification of the pair of bundles L and $K^{\frac{1}{2}} \otimes L'$ or the pair of bundles L' and $K^{\frac{1}{2}} \otimes L$ on the spectral curve away from the zeroes of λ. Thus L and $K_C^{\frac{1}{2}} \otimes L'$ or L' and $K_C^{\frac{1}{2}} \otimes L$ can only differ by a modification at these zeroes. As λ has $2N(g-1)$ zeroes and $K_C^{\frac{1}{2}}$ has degree $N(g-1)$, it is reasonable to conclude that

$$\begin{aligned} K_C^{\frac{1}{2}} \otimes L' &= L \otimes (\otimes_i O(x_i)) \\ K_C^{\frac{1}{2}} \otimes L &= L' \otimes (\otimes_j O(y_j)) \end{aligned} \tag{4.5}$$

where x_i and y_j are a partition of the zeroes of $\det \Phi$ in two groups.

We can confirm the linear relation between L and L' by observing that

$$\Phi' X = X \Phi \qquad Y \Phi' = \Phi Y \tag{4.6}$$

These relations imply directly that the Lagrangian submanifold is fixed by the simultaneous Hamiltonian flow along the fibers of the two complex integrable systems. Indeed, the flow is generated by shifting $\bar{\partial}_E$ by an amount proportional to a polynomial the Higgs field. Then equations such as $\bar{\partial}_{E \times (E')^*} X = 0$ change along the flow by amounts such as $\Phi' X - X\Phi$, which vanish on the Lagrangian submanifold.

Thus these intertwining relationships show that the Lagrangian submanifold is invariant under differences of flows in pairs of copies, generated by corresponding pairs $u - u'$ of Hamiltonians in the two systems. In terms of angular coordinates θ and θ' on generic fibers, the Lagrangian should be defined locally by equations of the form $u = u'$ and $\theta' - \theta = f(u)$.

This is a rather reasonable property for a Lagrangian correspondence which sits over the diagonal on the bases of the complex integrable system: the fiber consists of $2^{2N(g-1)}$ shifted images of the diagonal, labelled by the splitting of zeroes of $\det \Phi$ among x_i and y_j.

Upon mirror symmetry, we should get a BBB brane associated to a rank $2^{2N(g-1)}$ sheaf on the diagonal. This agrees with the prediction of S-duality between NS5 and D5 interfaces. Indeed, above general points in the base we can match the difference between L and L' for each summand of the sheaf to the result of the BBB analysis for the D5 interface. This agrees with the prediction of S-duality between NS5 and D5 interfaces.

4.4. BBB analysis of an NS5 interface. The BBB twist of the interface will provide a sheaf on the product of the two rank N Hitchin systems. The sheaf is the Fock space built from the Dirac-Higgs bundle in the bi-fundamental representation over the product of Hitchin system.

If the rank of the linear map $f \to \Phi f - f\Phi'$ is maximal away from isolated points on C. the Dirac-Higgs bundle should decompose again into sub-bundles, the co-kernels of that linear map. These n points are the zeroes of the resultant of the characteristic polynomials of Φ and Φ'.

The co-kernel consists essentially of the tensor product of the Φ eigenvector and Φ' left eigenvector with matching eigenvalues. Thus the Dirac-Higgs bundle can be expressed in terms of the product of the line bundles on the spectral curves at the points where they intersect. See Appendix B for an A-model version of this statement.

At such generic loci, the Fock bundle has thus rank 2^n and is a sum of line bundles which can be matched under fiberwise mirror symmetry with the results of the BAA analysis for the D5 interface. This agrees with the prediction of S-duality between NS5 and D5 interfaces.

5. General NS5 and D5 interfaces for unitary groups

The D5 and NS5 interfaces have a generalization to interfaces between $U(N)$ and $U(M)$ gauge theories with $M < N$. On the NS5 side the generalization is obvious: the interface supports $N \times M$ hypermultiplets in a bi-fundamental representation of $U(N) \times U(M)$. On the D5 side, one has no matter, but an interface with reduced gauge symmetry. At the interface, $U(M)$ is identified with a block-diagonal subgroup of $U(N)$ and a Nahm pole is inserted in the remaining $U(N-M)$ block of $U(N)$.

Due to the Nahm pole, the BBB analysis of the D5 interface goes outside of the sigma model description: the BBB brane is supported on the singular locus in the product of Hitchin systems where the solution of $U(N)$ Hitchin system reduces to a solution of the $U(M)$ Hitchin system times a trivial solution in of the $U(N-M)$ Hitchin system.

Within limits, we can reproduce this from the BAA twist of the NS5 interface. The equations

$$\Phi' = XY \qquad \Phi = YX \tag{5.1}$$

now tell us that the $U(N)$ Higgs field Φ has rank at most M, with X and Y providing the inter-twiners between the $U(N)$ and $U(M)$ Higgs bundles.

More precisely, if Φ' has full rank, then Φ must have rank M and the characteristic polynomial of Φ should be $\det(x - \Phi)x^{N-M} \det(x - \Phi')$.

The relations

$$\Phi' X = X\Phi \qquad Y\Phi' = \Phi Y \tag{5.2}$$

still hold, showing that the Lagrangian manifold is fixed by the simultaneous Hamiltonian flow along the fibers of the two complex integrable systems

The other twists are more interesting.

5.1. BBB twist of NS5 interface. The Dirac Higgs bundle is again built from the cokernels of the map $f \to \Phi f - f\Phi'$. The rank of that map drops at the $n = NM(2g-2)$ points which are the zeroes of the resultant of the characteristic polynomials of Φ and Φ'.

Again, if the zeroes are simple, the cokernels can be identified with matrices which are the tensor product of the eigenvectors of Φ and Φ' with a common eigenvalue.

At such generic loci on the bases of the Hitchin systems, the Fock bundle has thus rank 2^n and is a sum of line bundles built from the cokernels in the usual fashion.

5.2. BAA twist of D5 interface, $M = N - 1$. It is instructive to specialize at first to $N - M = 1$. The boundary condition imposes (see Appendix D)

$$E = E' \oplus L \tag{5.3}$$

for some fixed line bundle L. Here L gives the coupling to a background vector-multiplet for the $U(1)$ global symmetry of the interface.[6]

We also have the constraint

$$\Phi|_{\mathrm{End}(E')} = \Phi' \tag{5.4}$$

Pulling back this relation to the (fiberwise product of the) spectral curves and sandwiching between eigenvectors $v = (v_{E'}, v_L)$ for Φ and $(v', 0)$ with v' eigenvector for Φ' we get

$$(\lambda' - \lambda)(v_{E'}, v') = v_L(\Phi|_{\mathrm{Hom}(E',L)} v') \tag{5.5}$$

The zeroes of $\lambda' - \lambda$ are zeroes of the resultant of the two characteristic polynomials of the Higgs fields, which is a differential of degree NM. Thus it has $n = N(N-1)(2g-2)$ zeroes.

Again, the n zeroes of $\lambda' - \lambda$ will be distributed among the sections v_L and $(\Phi|_{\mathrm{Hom}(E',L)} v')$ on the two spectral curves. We expect each of these possible 2^n choices to determine the line bundle on the spectral curve and thus an intersection point of the Lagrangian manifold with the fibers of the complex integrable systems.

With a bit more work, it should be straigthforward to fully verify the fiberwise mirror symmetry at generic fibers which follows from the prediction of S-duality between NS5 and D5 interfaces.

5.3. BAA twist of D5 interface, general $N - M$. Because of the Nahm pole (see Appendix D), the bundles on the two sides are related as

$$E = E' \oplus \left(L \otimes K^{\frac{N-M-1}{2}}\right) \oplus \cdots \oplus \left(L \otimes K^{\frac{-N+M+1}{2}}\right) \tag{5.6}$$

with $\Phi|_{\mathrm{End}(E')} = \Phi'$.

We can use gauge transformations to restrict the form of Φ further, analogously to what happens for the Hitchin section at $M = 0$. This can be encoded in terms of the regular $su(2)$ embedding in $U(N - M)$. The canonical form for Φ is that of a Slodowy slice, the sum of a raising operator t^+ and a piece which commutes with the lowering operator t^-. In particular, the off-diagonal blocks of Φ reduce to a single column or row with M elements each.

Now we can sandwich Φ between eigenvectors for Φ and Φ' as before, to get the usual equation

$$(\lambda' - \lambda)(v_{E'}, v') = v_{L\otimes K^{\frac{N-M-1}{2}}} (\Phi|_{\mathrm{Hom}(E', L\otimes K^{\frac{N-M-1}{2}})} v') \tag{5.7}$$

and proceed as before to match the zeroes of $\lambda' - \lambda$ with the zeroes of the two factors on the right hand side.

[6]This could also have been included in the $N = M$ case by twisting the bundle X and Y transformed in.

5.4. Hanany-Witten moves. Hanany-Witten moves play a crucial role in understand the S-duality properties of boundary conditions in $U(N)$ gauge theory.

The basic idea is that boundary conditions defined by linear quiver gauge theories can be built as a sequence of D5 and NS5 interfaces. The NS5 interfaces provide the bi-fundamental hypermultiplets and the D5 interfaces can provide extra fundamental hypermultiplets.

S-duality is implemented by mapping D5 to NS5 and viceversa. After this map, it is often useful to change the relative order of D5 and NS5s in such a way to eliminate D5 interfaces involving Nahm poles or symmetry breaking. This can be done with the help of the Hanany-Witten moves. It would be very useful to prove directly the relations between convolutions of D5 and NS5 interfaces which follow from the Hanany-Witten rules: it would allow one to repeat the brane manipulations directly in the language of Geometric Langlands. We leave this problem for future work.

6. Ortho-symplectic examples

Brane constructions extend to orthogonal and symplectic gauge groups, albeit with several non-trivial complications.

A simple generalization of NS5 interfaces for unitary groups are the half-NS5 interfaces between an orthogonal group $SO(2N)$ and a symplectic group $USp(2M)$. They consist simply of a bi-fundamental hypermultiplet, i.e. a hypermultiplet in the product of the fundamental representations, which is obviously symplectic.

The S-dual of these interfaces are half-D5 interfaces between the orthogonal group $SO(2N)$ and the orthogonal group $SO(2M+1)$. They are defined by a Nahm pole of dimension $|2N-2M-1|$ together with an embedding of the smaller group into the larger.

It is also possible to consider half-NS5 interfaces between an orthogonal group $SO(2N+1)$ and a symplectic group $USp(2M)$, defined in an analogous manner. Notice that this interface involves an odd number of half-hypermultiplets for $USp(2M)$, which have an anomaly. The anomaly is compensated by a discrete theta angle in the $USp(2M)$ gauge theory, which we can thus denote as $USp(2M)'$.

The theta angle affects the S-duality properties of the $USp(2M)'$ gauge theory, which is mapped back to $USp(2M)'$. The S-dual interface is expected to be a half-D5 interface between $USp(2N)$ and $USp(2M)'$. If $N = M$, this involves a fundamental half-hypermultiplet. Otherwise, it involves the embedding of the smaller gauge group into the larger and a Nahm pole of dimension $2|N-M|$.

It would be very interesting to work out these general examples in full detail. Instead, in the next sections we will limit ourselves to discussing some simple examples involving $SU(2)$ gauge groups. For example, the half-NS5 interface between $SO(2)$ and and $USp(2)$ will be discussed as an interface between $U(1)$ and $SU(2)$. This is dual to the half-D5 interface between $SO(2)$ and $SO(3)$. Similarly, the half-NS5 interface between $SO(1)$ and $USp(2)'$ consists of a single half-hypermultiplet for a single $SU(2)$ gauge theory and is dual to the half-D5 interface between $USp(0)$ and $USp(2)'$, i.e. the maximal Nahm pole for $SU(2)'$. Finally, the half-NS5 interface between $SO(4)$ and $USp(2)$ is really a tri-valent interface between three $SU(2)$ gauge groups with very nice properties, S-dual to the half-D5 interface between $SO(4)$ and $SO(3)$, which identifies the three $SU(2)$ gauge groups with each other.

7. Tri-fundamental $SU(2) \times SU(2) \times SU(2)$ interface

Until now we have considered either boundary conditions for a single four-dimensional gauge theory, leading to branes in a single Hitchin moduli space, or interfaces between two four-dimensional gauge theories, leading to branes in the product of two Hitchin moduli spaces, possibly interpreted as correspondences or functors.

In this section we look at an example of interface between three gauge theories, leading to branes in the product of three Hitchin moduli spaces. This could also be used to produce a functorial operation on the category of branes. The particular example we consider here actually gives an associative operation. It would be nice to explore it further.

The simplest possible tri-valent interface is a "tri-transparent" interface between three G gauge theories, where the three gauge groups are simply identified at the interface. In other words, we preserve at the interface the diagonal combination of the three G gauge groups, without adding any extra matter fields.

For $G = SU(2)$, the interface is S-dual to a Neumann boundary condition for all three gauge groups coupled to a single non-anomalous half-hypermultiplet transforming in the product of the fundamental representations of $SU(2) \times SU(2) \times SU(2)$.

This fact follows from the properties of half-NS5 and half-D5 interfaces for orthogonal and symplectic gauge theories, but can also be derived with a bit of help of the 6d $(2,0)$ theory. The generalization to other groups G involves a non-trivial interface SCFT which brings the problem out of the scope of this paper.

We will thus analyze this three-way interface for $SU(2) \times SU(2) \times SU(2)$ and compare it to the behaviour of the "tri-transparent" interface.

7.1. BAA analysis of tri-fundamental. A mathematical discussion of this system was recently given in [**14**].

The hypermultiplet scalar field Z is a section of $E \otimes E' \otimes E'' \otimes K^{1/2}$, where E, E', E'' are the bundles associated to the irreducible doublet representations of the three $SU(2)$ groups. We will explicitly denote the 8 components as $Z_{\alpha\beta\gamma}$, where the indices transform under the three $SU(2)$.

The corresponding Lagrangian manifold in the product of three copies of the $SU(2)$ Hitchin system is defined by the equations

$$\begin{aligned} &D_{\bar z} Z = 0 \\ &\Phi^{(a)} = \mu^{(a)}(Z) \end{aligned} \tag{7.1}$$

there $\mu^{(a)}(Z)$ are the moment maps for the three $SU(2)$ groups and $\Phi^{(a)}$ the corresponding Higgs fields. We will sometimes denote the three Higgs fields as Φ, Φ', Φ''.

In components, we can write the equations as

$$\begin{aligned} \Phi^{(1)}_{\alpha\alpha'} &= \frac{1}{2} Z_{\alpha\beta\gamma} Z_{\alpha'\beta'\gamma'} \epsilon^{\beta\beta'} \epsilon^{\gamma\gamma'} \\ \Phi^{(2)}_{\beta\beta'} &= \frac{1}{2} Z_{\alpha\beta\gamma} Z_{\alpha'\beta'\gamma'} \epsilon^{\alpha\alpha'} \epsilon^{\gamma\gamma'} \\ \Phi^{(3)}_{\gamma\gamma'} &= \frac{1}{2} Z_{\alpha\beta\gamma} Z_{\alpha'\beta'\gamma'} \epsilon^{\alpha\alpha'} \epsilon^{\beta\beta'} \end{aligned} \tag{7.2}$$

where the ϵ symbols denote the symplectic pairing in each doublet ($\epsilon^{+-} = 1$, $\epsilon^{-+} = -1$) and the Higgs fields are represented as symmetric matrices.

It is also often useful to represent Z as a pair of 2×2 matrices X and Y and the Higgs fields as traceless matrices by raising some indices:

$$\begin{aligned} X_\alpha^{\beta'} &= Z_{\alpha\beta+}\epsilon^{\beta\beta'} \\ Y_\beta^{\alpha'} &= Z_{\alpha\beta-}\epsilon^{\alpha\alpha'} \end{aligned} \tag{7.3}$$

which allows us to write

$$\begin{aligned} \Phi &= XY - \frac{1}{2}\mathrm{Tr}XY \\ \Phi' &= -YX + \frac{1}{2}\mathrm{Tr}XY \\ \Phi'' &= \begin{pmatrix} \frac{1}{2}\mathrm{Tr}XY & \det X \\ -\det Y & -\frac{1}{2}\mathrm{Tr}XY \end{pmatrix} \end{aligned} \tag{7.4}$$

The first consequence of these equations is that the three Higgs fields have the same characteristic polynomial, i.e. the Lagrangian submanifold projects down to the tri-diagonal in the bases of the three Hitchin systems:

$$\det \Phi = \det \Phi' = \det \Phi'' \tag{7.5}$$

In terms of the Hamiltonians $u_i^{(a)}$ of the integrable system, that means $u_i^{(a)} = u_i^{(b)}$ for all pairs a, b.

Furthermore, we have intertwining relations such as

$$\Phi X + X\Phi' = 0 \qquad\qquad Y\Phi + \Phi' Y = 0 \tag{7.6}$$

and cyclic rotations thereof. As the Hamiltonian flows along the fibres of the integrable system can be understood as shifts of $D_{\bar{z}}$ by monomials in the $\Phi^{(a)}$, the intertwining relationships show that the Lagrangian submanifold is invariant under differences of flows in pairs of copies, generated by Hamiltonians $u_i^{(a)} - u_i^{(b)}$. In terms of angular coordinates $\theta^{(a)}$ on a generic fiber, the Lagrangian should be defined locally by equations of the form $\theta^{(1)} + \theta^{(2)} + \theta^{(3)} = f(u)$.

It is easy to see that the Lagrangian includes the tri-diagonal of the Hitchin section of the integrable system, associated to the same spin bundle $K^{1/2}$ used in the definition of Z: if the $SU(2)$ bundles are all $K^{1/2} \oplus K^{-1/2}$, then the 8 components of Z are sections of $K^{1/2 \pm 1/2 \pm 1/2 \pm 1/2}$. We can take as the only non-zero components

$$Z_{+++} = \phi(z) \qquad\qquad Z_{+--} = Z_{-+-} = Z_{--+} = 1 \tag{7.7}$$

i.e.

$$\begin{aligned} X &= \begin{pmatrix} 0 & -1 \\ \phi(z) & 0 \end{pmatrix} \\ Y &= \begin{pmatrix} -1 & 0 \\ 0 & 1 \end{pmatrix} \end{aligned} \tag{7.8}$$

to get the canonical Hitchin section for each Higgs field:

$$\Phi = \Phi' = \Phi'' = \begin{pmatrix} 0 & 1 \\ \phi(z) & 0 \end{pmatrix} \tag{7.9}$$

Following the analysis of [**14**] or proceeding analogously to our previous examples, one may conclude that at above a generic point in the bases of the Hitchin system, the complex Lagrangian is defined by the equations

$$u_i^{(1)} = u_i^{(2)} = u_i^{(3)} \qquad \theta^{(1)} + \theta^{(2)} + \theta^{(3)} = 0 \tag{7.10}$$

where "0" is defined in terms of the Hitchin section for a given choice of $K^{1/2}$.

7.2. BBB analysis of tri-fundamental. In order to complete the BBB analysis, we need to understand how to define the Fock space for a situation where the representation M does not split. As the system should not have an anomaly, the problem should not be too hard.

The Dirac-Higgs bundle for the tri-fundamental representation is generically a sum of lines defined by the cokernel of the map $\Phi \oplus \Phi' \oplus \Phi''$ on the tri-fundamental bundle. The map has rank 8 away from points on C where $P = P_1^2 + P_2^2 + P_3^2 - 2P_1P_2 - 2P_1P_3 - 2P_2P_3$ vanishes, where P_i are the determinants of the Higgs fields. It is not obvious, but at these loci the rank drops by 2. These are the loci where the Higgs field eigenvalues satisfy $\lambda \pm \lambda' \pm \lambda'' = 0$.

As the rank drops by 2 at the special loci, each will contribute a two-dimensional sub-bundle. Thus the rank of the Fock space should be 2^n, where $n = 8g - 8$ is the number of zeroes of P.

In order to describe the sheaf of ground-states we need to compute the symplectic form on the Dirac-Higgs bundle for the tri-fundamental representation and seek a Lagrangian splitting. We will not attempt to do so, although we may observe that the images of holomorphic sections of $E \otimes E' \otimes E'' \otimes K$ modulo the image of $\Phi \oplus \Phi' \oplus \Phi''$ should give a Lagrangian subbundle of the Dirac-Higgs bundle.

7.3. BAA analysis of tri-transparent interface. As the gauge fields for the three theories are identified, we must have

$$E = E' = E'' \tag{7.11}$$

Dually, the Higgs fields must satisfy

$$\Phi + \Phi' + \Phi'' = 0 \tag{7.12}$$

We can visualize the Higgs fields on each point of C as three vectors in $\mathbb{C}^3$ forming a triangle. The square lengths P_i of the vectors determine the geometry of the triangle, which collapses at the points on C where $P = P_1^2 + P_2^2 + P_3^2 - 2P_1P_2 - 2P_1P_3 - 2P_2P_3$ vanishes.

It should be possible to show that there are 2^n possible choices of bundle E given the P_i and match them with the S-dual description. We leave this for future work.

7.4. BBB analysis of tri-transparent interface. The interface should be the structure sheaf of the tri-diagonal in the product of the three Hitchin systems. This is compatible with the BAA analysis of the dual interface.

8. More examples with gauge group reductions

There is a large class of examples of simple interfaces whose S-dual description involves some kind of reduction of the gauge group to a subgroup. We review the properties of interfaces with gauge group reduction in Appendix D.

8.1. Abelianization interface between $SU(2)$ and $U(1)$ gauge theories. Consider an interface between $SU(2)$ and $U(1)$ gauge theories, both with Neumann boundary conditions, coupled to a set of hypermultiplets in the fundamental representation of $SU(2)$, of charge 1 under $U(1)$.

This interface is expected to be S-dual to an interface where the $SU(2)$ gauge group is reduced to $U(1)$ and identified with the $U(1)$ gauge group on the other side.

8.1.1. *BAA twist.* The equations

$$\Phi_{SU(2)} = XY - \frac{1}{2}Y \cdot X \qquad \Phi_{U(1)} = Y \cdot X \tag{8.1}$$

imply that $\mathrm{Tr}\Phi^2_{SU(2)} = \frac{1}{4}\Phi^2_{U(1)}$. Hence the BAA brane is supported on the locus where the eigenvalues of the $SU(2)$ Higgs field are globally defined.

This agrees with the proposed mirror BBB brane, which is a sheaf supported on the graph of the diagonal embedding of $U(1)$ Hitchin systems into $SU(2)$ Hitchin systems, but makes the analysis more complicated and brings us outside the scope of a sigma model analysis.

Notice that the sections X and Y are globally defined left and right eigenvectors of $\Phi_{SU(2)}$ with eigenvalue $\frac{1}{2}\Phi_{U(1)}$. As discussed in previous examples, these relations imply that the Lagrangian is invariant under flows generated by the Hamiltonian $\mathrm{Tr}\Phi^2_{SU(2)} - \frac{1}{4}\Phi^2_{U(1)}$, as it should.

8.1.2. *BBB twist.* The Dirac-Higgs bundle over the product of Hitchin moduli spaces, in the doublet representation of $SU(2)$, charge 1 representation of $U(1)$, can be described most easily at loci where $\Phi_{SU(2)} + \Phi_{U(1)}$ has generically full rank, dropping to rank 1 at $4g-4$ isolated points in C where $\mathrm{Tr}\Phi^2_{SU(2)} = \frac{1}{4}\Phi^2_{U(1)}$. Then the bundle is a sum of the co-kernels of $\Phi_{SU(2)} + \Phi_{U(1)}$.

The associated Fock bundle will thus have rank 2^{4g-4}. This agrees with the expected mirror: the BAA twist of the S-dual interface is supported on the locus where $E_{SU(2)} = E_{U(1)}^{\frac{1}{2}} \oplus E_{U(1)}^{-\frac{1}{2}}$ and

$$\Phi_{SU(2)} = \begin{pmatrix} \frac{1}{2}\Phi_{U(1)} & U \\ -V & -\frac{1}{2}\Phi_{U(1)} \end{pmatrix} \tag{8.2}$$

where U and V are sections of $KE_{U(1)}$ and $KE^{-1}_{U(1)}$. Thus we have

$$\mathrm{Tr}\Phi^2_{SU(2)} - \frac{1}{4}\Phi^2_{U(1)} = UV \tag{8.3}$$

and we can solve the problem by distributing the zeroes of $\mathrm{Tr}\Phi^2_{SU(2)} - \frac{1}{4}\Phi^2_{U(1)}$ among U and V, thus fixing $E_{U(1)}$.

Thus the mirror BAA brane intersects the fiber of the complex integrable systems generically at 2^{4g-4} points. The d and q charges match as well, as for the $U(1)$ example. We leave as an exercise to confirm the fiberwise mirror identification of the points in the fiber with the line bundles of the original BBB brane.

8.2. Half-fundamental for $SU(2)$. As the fundamental representation of $SU(2)$ is pseudo-real, one can consider fundamental half-hypermultiplets. These cannot be coupled to three-dimensional gauge fields due to a Z_2 anomaly, but can be coupled at a Neumann boundary condition or matter interface to four-dimensional gauge fields. The anomaly, though, still has to be cancelled by anomaly inflow from a non-trivial discrete θ angle.

In other words, the four-dimensional gauge theory compatible with coupling to a single fundamental half-hypermultiplet at a Neumann boundary condition must be the theory indicated sometimes as $Sp(1)'$, which is mapped to itself under S-duality, rather than being mapped to $SO(3)$ gauge theory.

Furthermore, the S-duality map between boundary conditions is also affected by this. The interface we are discussing here is S-dual to a maximal Nahm pole for the $Sp(1)'$ gauge theory.

This setup generalizes to $Sp(n)'$ gauge theory for all n: Neumann boundary conditions with half-fundamental matter are dual to a maximal Nahm pole for $Sp(n)'$ gauge theory.

The discrete theta angle must modify the Hitchin moduli space, by coupling the connection on C to an appropriate gerbe. Roughly, this affects the way the Abelian fibers of the integrable system are glued together globally. The BAA and BBB branes must only make sense in the modified moduli space.

An accurate analysis goes beyond the scope of this paper. In this section we mostly outline some general perspective on the problem.

8.2.1. *BAA twist.* A mathematical discussion of this system can be found in [**14**].

The BAA brane is supported on the origin of the base of the complex integrable system, $\mathrm{Tr}\Phi^2 = 0$, as we have

$$\Phi = X\, X\wedge \tag{8.4}$$

This is compatible with the expected mirror, but brings us outside the scope of a sigma model analysis.

8.2.2. *BBB twist.* The Dirac-Higgs bundle in the fundamental representation for general Φ receives contributions from the $4g-4$ points on C where $\mathrm{Tr}\Phi^2$ vanishes. Generically, the rank of Φ only drops by 1 at these loci. Thus the rank of the Dirac-Higgs bundle is generically $4g-4$.

The associated Fock space bundle should have rank 2^{2g-2}. As we are dealing with an half-hypermultiplet, in order to describe the Fock space bundle we need to figure out the symplectic form on the Dirac-Higgs bundle and pick a Lagrangian splitting. We will not attempt to do so here.

The mirror BAA brane should be some version of the Hitchin section appropriate for the twisted Hitchin system.

8.3. Real forms. There is a neat class of boundary conditions with gauge group reductions, which reduce the gauge group G to a subgroup H which is the maximally compact subgroup of some non-compact real form G' of G. In particular, the BAA twist of such boundary conditions has a neat description in the complex structure where the holomorphic data is the complexified flat connection $d_A+\Phi+\bar{\Phi}$: the connection lies in G'. The S-dual boundary conditions are known in many examples, see Table 3 in [**5**].

In particular, if $G = SU(2N)$ and $H = S(U(N)\times U(N))$, so that the real form is $G' = SU(N,N)$, the S-dual boundary condition involves a gauge group reduction to $Sp(2N)$ coupled to a set of fundamental boundary hypermultiplets.

Thus the BAA brane associated to $SU(N,N)$ should be mirror of the BBB brane supported on the $Sp(2N)$ Hitchin moduli space inside $SU(2N)$, given by the Fock space bundle built from the fundamental Dirac Higgs bundle for $Sp(2N)$. This is precisely the proposal of section 7 in [**21**].

9. D-modules

9.1. Hypermultiplets and symplectic bosons. Following the finite-dimensional analogy in Appendix A, the quantization of the Lagrangian submanifold $\mathcal{L}(C, G, M)$ should be associated to the path integral

$$\int DZ e^{\frac{1}{\hbar}\int_\Sigma \langle Z, \bar{\partial}_A Z\rangle} \tag{9.1}$$

This is a well known object: it defines the partition function of a set of chiral symplectic bosons in two dimensions.

In the absence of zeromodes, conformal blocks for these chiral symplectic bosons provides an interesting (twisted) rank 1 D-module, where the vector field $\hbar \frac{\delta}{\delta A_{\bar{z}}}$ acts as the insertion of a current

$$J =: \mu(Z) : \tag{9.2}$$

Concretely, the result of the path integral, which is a flat section for that D-module, is the inverse of the determinant of the $\bar{\partial}_A$ operator valued in N (or square root of the determinant of the Dirac operator valued in M), which is a certain theta function $\Theta_M(A)$. In particular, the D-module has regular singularities on the locus where the $\bar{\partial}_A$ operator admits global sections, which is the locus where the classical description of the BAA brane has components with $\Phi \neq 0$.

The space of conformal blocks in the neighbourhood of that singular locus is somewhat intricate. We refer the reader to [**7**] for a more in-depth discussion.

9.2. Boundary gauge theories. Following the finite-dimensional analogy in Appendix A, the quantization of the Lagrangian submanifold $\mathcal{L}(C, G, G_B, M)$ should be associated to the path integral

$$\int DZ D\bar{A}_B e^{\frac{1}{\hbar}\int_\Sigma \langle Z, \bar{\partial}_{A+A_B} Z\rangle} \tag{9.3}$$

The meaning of this path integral is not immediately obvious, but a natural conjecture is that it should produce the conformal blocks of a coset

$$\frac{\{Z_M\}}{\hat{G}_B} \tag{9.4}$$

of the symplectic boson theory by the $\hat{G}_B$ WZW current algebra generated by the moment maps $\mu_B(Z)$.

We will pursue this idea at length in [**7**]. It passes some rather non-trivial tests.

10. Sheafs on the moduli space of local systems

10.1. Hypermultiplets. The analysis of the Dirac-Higgs bundle in complex structure J is rather straightforward: it gives us the sheaf $H^*_{\mathrm{de}Rahm}(C, G, M)$ over $\mathrm{Loc}(C, G)$ of de Rahm cohomology for forms on C valued in M (seen as a local system over C).

The cohomology is generically located at degree 1 and the corresponding sheaf $\mathcal{V}(C, G, M)$ over $\mathrm{Loc}(C, G)$ is given by the Fock space built from that degree 1 cohomology. If $M = N \oplus N^*$, we can write

$$\mathcal{V}(C, G, M) = (\det H^1_{\mathrm{de}Rahm}(C, G, N))^{-\frac{1}{2}} \Lambda^* H^1_{\mathrm{de}Rahm}(C, G, N) \tag{10.1}$$

10.2. Boundary gauge theories. We expect that in the presence of boundary gauge fields, the sheaf on $\mathrm{Loc}(C, G)$ will be the convolution of the sheaf $\mathcal{V}(C, G \times G_B, M)$ on $\mathrm{Loc}(C, G) \times \mathrm{Loc}(C, G_B)$ built from the boundary matter fields and the structure sheaf on $\mathrm{Loc}(C, G_B)$.

Again, we stress that this picture may miss some important contributions from the singular locus of $\mathrm{Loc}(C, G_B)$.

Appendix A. Lagrangian submanifolds and generating functions

Consider the complex symplectic manifold T^*B for some complex manifold B, with local coordinates q along B and p along the fibre. The complex symplectic form is $dpdx$ and the simplest complex Lagrangian submanifold $\mathcal{L}_0$ is defined by $p = 0$.

Given an holomorphic superpotential $W(q)$ defined on B, we can define a complex Lagrangian submanifold $\mathcal{L}_W$ as

$$p = \frac{\partial W}{\partial q} \tag{A.1}$$

This is Lagrangian because

$$dpdq|_{\mathcal{L}_W} = \frac{\partial^2 W}{\partial q^2} dqdq = 0 \tag{A.2}$$

We can extend such construction by looking at a superpotential which depends on some auxiliary variables. We could take a superpotential function on $U \times B$, for some auxiliary complex manifold U with coordinates u, or perhaps even defined on some complex fibration $F \to B$. We can consider the equations

$$p = \frac{\partial W}{\partial q} \qquad \frac{\partial W}{\partial u} = 0 \tag{A.3}$$

defining a submanifold F_W in F. The projection of such submanifold onto B is a complex Lagrangian L_W:

$$dpdq|_{L_W} = \frac{\partial^2 W}{\partial q^2} dqdq + \frac{\partial^2 W}{\partial q \partial u} dqdu = -\frac{\partial^2 W}{\partial^2 u} dudu = 0 \tag{A.4}$$

The fibration F_W can be used to produce certain locally constant sheaves over L_W. It is particularly natural to consider the cohomology of the fiber.

This construction is compatible with symplectic quotient operations as long as W is invariant under the group one quotients T^*M by.

A.1. Quantization. There is a natural quantization of the Lagrangian submanifolds defined above. We refer to the reader to [**7**] for a more in-depth discussion.

In the case of an holomorphic superpotential $W(q)$ defined on B, we can define a D-module on B whose sections are holomorphic functions on B, with

$$\hat{p} f(q) = \hbar \partial_q f(q) + \frac{\partial W}{\partial q} f(q) \tag{A.5}$$

This D-module encodes the behaviour of functions of the form $f(q) \exp \frac{W}{\hbar}$.

More generally, we can define a D-module on B whose sections are holomorphic functions on F, modulo the subspace of functions of the form

$$\hbar \partial_u g(q, u) + \frac{\partial W}{\partial u} g(q, u) \tag{A.6}$$

with vector fields on B acting by

$$\hat{p}f(q,u) = \hbar\partial_q f(q,u) + \frac{\partial W}{\partial q} f(q,u) \tag{A.7}$$

This D-module encodes the behaviour of functions of the form

$$\oint_U f(q,u)\exp\frac{W}{\hbar} \tag{A.8}$$

Appendix B. Supersymmetric Berry connections for $\mathcal{N}=4$ SQM

B.1. Supersymmetric Berry connection. Consider a Supersymmetric Quantum Mechanics, defined by two supercharges Q and $Q^\dagger$, with $Q^2=0$ and Hamiltonian

$$H = \{Q, Q^\dagger\} \tag{B.1}$$

Supersymmetric ground states are annihilated by both super-charges. If the system has a gap, supersymmetric ground states have an equivalent description as the Q-cohomology of the Hilbert space.

Next, consider a complex family of $\mathcal{N}=2$ SQM systems, where the supercharge $Q(u)$ depends holomorphically on some parameters u^a and $Q^\dagger(\bar{u}^a)$ depends anti-holomorphically on them. Then the ground states form a holomorphic sheaf on the parameter space: the anti-holomorphic derivatives $\partial_{\bar{u}^a}$ commute with the supercharge and each other and descend to commuting anti-holomorphic derivatives $D_{\bar{u}^a}$ on the cohomology of $Q(u)$. We denote these parameters as "B-type" parameters, for reasons which will become apparent momentarily.

Concretely, if we have some local basis $|i\rangle$ of the space of ground states, with $\langle i||j\rangle = \delta_i^j$, the Berry connection is defined by

$$(A_a)_i^j = \langle i|\partial_{u^a}|j\rangle \qquad (A_{\bar{a}})_i^j = \langle i|\partial_{\bar{u}^a}|j\rangle \tag{B.2}$$

Thus along B-type directions, the curvature of the Berry connection is of type $(1,1)$.

Another interesting situation is a real family of $\mathcal{N}=2$ SQM systems, where the supercharge $Q(t)$ depends on parameters t^i in such a way that

$$\partial_{t^i}Q(t) = [Q(t), T_i] \qquad \partial_{t^i}Q^\dagger(t) = -[Q^\dagger(t), T_i] \tag{B.3}$$

for some Hermitean operators T_i, with $[\partial_{t^i}+T_i, \partial_{t^j}+T_j]$ being $Q(t)$-exact. Then the differential operators $\partial_t + T$ commute with the supercharge and descend to a complexified flat connection D_t on the sheaf of ground states. We denote these parameters as "A-type" parameters.

Thus along A-type directions, the Berry connection can be complexified to a flat supersymmetric Berry connection.

More generally, we can have some family $Q(u,t)$ depending on both types of operators, which will lead to a sheaf of ground states admitting two commuting structures: a holomorphic Berry connection $D_{\bar{u}}$ and a flat complexified Berry connection D_t.

B.1.1. *De Rahm SQM.* The simplest example of supersymmetric quantum mechanics takes the Hilbert space to consist of L^2 normalizable forms on a compact Riemannian manifold X and the supercharge to be the exterior derivative:

$$Q = d \qquad Q^\dagger = d^\dagger \tag{B.4}$$

so that the Hamiltonian is the Laplace operator on forms. The space of ground states coincides with the de Rahm cohomology of X.

In this gauge the metric dependence is all in $Q^\dagger$ and if we have some parameter space of metrics on X, the de Rahm cohomology is naturally a flat bundle over the parameter space. This is a rather trivial example of a real family of $\mathcal{N} = 2$ SQM systems, as the Berry connection is just flat, without need of complexification: T vanishes.

Morse-Witten quantum mechanics provides a somewhat more interesting example:

$$Q = d + dh \wedge \qquad Q^\dagger = d^\dagger + (dh\wedge)^\dagger \tag{B.5}$$

Here a variation of the Morse function leads to

$$\partial_t Q = d(\partial_t h)\wedge = [Q, \partial_t h] \tag{B.6}$$

and we get a complexified flat connection on spaces of Morse functions.

If we were to replace dh with a more general closed 1-form, the variation of Q along the parameter space would not be exact. We can get a holomorphic Berry connection if we enlarge the parameter space by including also a twist by a $U(1)$ connection on X:

$$Q = d_A \qquad Q^\dagger = d_A^\dagger \tag{B.7}$$

for some complex (i.e. $GL(1)$) flat connection A. The anti-Hermitean part of A is the $U(1)$ connection, while the Hermitean part of A is the generalization of the "$dh\wedge$" part of the Morse quantum mechanics.

Notice that unitary gauge transformations leave the quantum mechanics unchanged, while complexified gauge transformations are equivalent to adding a Morse function to the system. The parameter space will be a space of $GL(1)$ flat connections on X modulo unitary gauge transformations, to be thought of as a fibration over a parameter space of $GL(1)$ flat connections on X modulo $GL(1)$ gauge transformations, with fibres being spaces of Morse functions.

The space of $GL(1)$ flat connections on X modulo $GL(1)$ gauge transformations is naturally a complex manifold and Q varies holomorphically on it. Thus we obtain a supersymmetric Berry connection which is holomorphic along the space of $GL(1)$ flat connections on X modulo $GL(1)$ gauge transformations and flat along the fiber directions of the space of $GL(1)$ flat connections on X modulo unitary gauge transformations.

This system has obvious generalizations where X is equipped with vector bundle and a complexified flat connection.

More generally, one can consider some auxiliary quantum mechanics with X as a parameter space of A-type deformations and write the combined supercharge

$$\epsilon(d_x + T\wedge) + Q(x) \tag{B.8}$$

The A-type constraints on T and $\partial_x Q(x)$ guarantee that the above supercharge is nilpotent. In the Born-Oppenheimer approximation for small ϵ, the ground states of the system can be computed by the reduced super-charge d_A where A is the supersymmetric Berry connection for $Q(x)$.

B.1.2. *Dolbeault SQM.* Another natural example of supersymmetric quantum mechanics involves a complex manifold Y equipped with a holomorphic vector bundle E, both endowed with an Hermitean metric. We can define

$$Q = \bar{\partial}_E \qquad Q^\dagger = (\bar{\partial}_E)^\dagger \tag{B.9}$$

acting on $(0,q)$ forms valued in E. The Q cohomology is the Dolbeault cohomology of E.

The supercharge and cohomology depend holomorphically on the choice of vector bundle E and trivially on the choice of Hermitean metrics.

There is an useful extension to a complex of vector bundles $E^{\cdot}$, with some holomorphic differential δ and supercharge $Q = \bar{\partial}_{E^{\cdot}} + \delta$. The space of ground states is the *hypercohomology* of the complex.

It is also interesting to consider some auxiliary quantum mechanics with Y as a parameter space of B-type deformations and write the combined supercharge

$$\epsilon\bar{\partial}_y + Q(y) \tag{B.10}$$

In the Born-Oppenheimer approximation for small ϵ, the ground states of the system can be computed by the reduced super-charge $\bar{\partial}_E$ where E is the sheaf of ground states for $Q(y)$.

B.1.3. *From SQM to branes.* The $\mathcal{N}=2$ SQM systems discussed above can be used as boundary degrees of freedom in defining branes for two-dimensional sigma models. In particular, if the gap of the quantum mechanical system is much larger than the scale at which the sigma model becomes strongly coupled, the space of ground states of the quantum mechanical system becomes the Chan-Paton bundle for a traditional UV boundary condition for the sigma model.[7]

The natural way to define such coupling to a sigma model is to start from some reference boundary condition for the sigma model and then promote the boundary values of the bulk fields to parameters in the SQM.

B-type boundary conditions involve the promotion of B-type deformation parameters in the SQM. For example, we can start from some reference Neumann boundary condition associated to the structure sheaf of the sigma model target space U and couple it to a SQM with U as a parameter space of B-type deformations.

The result should be a standard B-brane based on the holomorphic sheaf of ground states of the SQM over U.

A-type boundary conditions involve the promotion of A-type deformation parameters in the SQM. It is instructive to first consider the example of a Morse-Witten SQM.

Consider a sigma model on T^*X, with the standard boundary condition given by the base X of the co-tangent bundle. In local coordinates (x,p) on T^*X we can denote it as $p=0$.

Consider a Morse-Witten SQM on some auxiliary manifold Y with a family of Morse functions $h(x,y)$ on Y parameterized by X. Classically, when h has isolated critical points as a function of y, the boundary coupling induced by $h(x,y)$ deforms the boundary condition to $p = \partial_x h(x,y^*)$, where the Morse function is evaluated at a critical point y^* solving $\partial_y(x,y^*) = 0$. This is a Lagrangian submanifold which is a multi-cover of X in T^*X. It supports a flat $U(1)$ bundle given by the cohomology class associated to the critical point.

A better way to think about this is an A-brane which is a deformation of multiple copies of X by the complexified Berry connection D_x: the unitary part of the connection is the connection of the CP bundle on X, while the Hermitean part

[7]It is also reasonable to consider situations where a finite number of non-ground states survive at the sigma model UV scale, giving rise to Chan-Paton complex equipped with a differential.

T should be understood as an Higgs field, a matrix-valued version of a deformation of the Lagrangian manifold along the normal direction to X. When T and A are approximately diagonal this reduces to the multi-cover of X mentioned above.

This perspective applies in general to any SQM with X as a space of A-type deformations.

B.1.4. *From branes so SQM.* Conversely, a nice example of families of $\mathcal{N}=2$ SQM arise when one considers the compactification of a $(2,2)$ sigma model on a segment, with families of supersymmetric boundary conditions at the endpoint.

As an example, consider families of A-branes. An A-brane described by a Lagrangian submanifold equipped with a flat connection has a moduli space of deformations given by deformations of the manifold or deformations of the flat connection.

A deformation of the Lagrangian submanifold is a section of the normal bundle, which can be mapped to a closed 1-form with the help of the symplectic form.

If the 1-form is exact, we obtain an Hamiltonian function which generates the deformation of the Lagrangian manifold. If the 1-form is not exact, then it gives us a flat bundle deformation.

Thus general deformations are complex, combining corresponding shape and flat connection deformations. They are actually B-type deformations of the SQM. On the other hand, exact shape deformations are A-type deformations, with the Hamiltonian playing the role of T.

B.2. Hyper-holomorphic Berry connections. Next we can consider quantum-mechanical systems with $\mathcal{N}=4$ supersymmetry. In the simplest situation, we have four Hermitean supercharges $\tilde{Q}_i$ which satisfy

$$\{\tilde{Q}_i, \tilde{Q}_j\} = \delta_{ij} H \tag{B.11}$$

The supersymmetric ground states of the system are annihilated by all supercharges.

It is also useful to consider complex supercharges $Q_1 = \tilde{Q}_1 + i\tilde{Q}_2$ and $Q_2 = \tilde{Q}_3 + i\tilde{Q}_4$, so that in particular

$$\{Q_\alpha, Q_\beta\} = 0 \tag{B.12}$$

and furthermore

$$\{Q_1, (Q_2)^\dagger\} = 0 \qquad \{Q_1, (Q_1)^\dagger\} = \{Q_2, (Q_2)^\dagger\} = H \tag{B.13}$$

Lowering the index of the second supercharge with an anti-symmetric tensor, we define $\bar{Q}_2 = (Q_1)^\dagger$ and $\bar{Q}_1 = -(Q_2)^\dagger$, so that

$$\{Q_\alpha, \bar{Q}_\beta\} = \epsilon_{\alpha\beta} H \tag{B.14}$$

In particular, the supersymmetric ground states can be identified with the cohomology of Q_1, or the cohomology of Q_2, or the cohomology of any other linear combination $Q_\zeta = Q_1 + \zeta Q_2 = Q_\alpha \zeta^\alpha$.

In a more "twistorial" language, the SUSy relations can be expressed in terms of Q_ζ and $\bar{Q}_\zeta = \bar{Q}_1 + \zeta\bar{Q}_2$ as

$$Q_\zeta^2 = 0 \qquad \{Q_\zeta, \bar{Q}_\zeta\} = 0 \qquad \bar{Q}_\zeta^2 = 0 \tag{B.15}$$

with a reality condition

$$(\bar{Q}_\zeta)^\dagger = \zeta^\dagger Q_{-1/\zeta^\dagger} \tag{B.16}$$

We will describe various ways the Berry connection for these ground states can be affected by the $\mathcal{N} = 4$ supersymmetry. We will be mostly interested in two possibilities. BBB-type deformations are analogous to Nahm transforms and will give us a Berry connection which is hyper-holomorphic. BAA deformations are analogous to the tt^* geometry of (2, 2) SQFTs [**22**] and will give us a Berry connection which combined to a Higgs field is a solution of generalized Hitchin's equations. There are other possibilities, which we will not need here, which give solutions of generalized Nahm or BPS equations [**23**].

A BAA-type deformation is in particular a B-type deformation for Q_1 and for $\bar{Q}_1 = Q_2^\dagger$ which is also an A-type deformation for all other Q_ζ. That means we have $\partial_{\bar{u}} Q_1 = 0$ and $\partial_u Q_2 = 0$, but

(B.17)
$$\partial_u Q_1 = \partial_u (Q_1 + \zeta Q_2) = [Q_1 + \zeta Q_2, T_u(\zeta)] \qquad \zeta \partial_{\bar{u}} Q_2 = [Q_1 + \zeta Q_2, T_{\bar{u}}(\zeta)]$$

which we solve by

$$\text{(B.18)} \quad \partial_u Q_1 = [Q_2, C_u] \qquad [Q_1, C_u] = 0 \qquad \partial_{\bar{u}} Q_2 = [Q_1, C_{\bar{u}}(\zeta)] \qquad [Q_2, C_{\bar{u}}] = 0$$

We also require $[C_u, C_{\bar{u}}] = 0$.

These relations mean that the Lax operators

$$\text{(B.19)} \qquad \partial_u + \frac{C_u}{\zeta} \qquad \partial_{\bar{u}} + \zeta C_{\bar{u}}$$

commute and descend to a supersymmetric Berry connection which is flat for all values of ζ, i.e. the Lax connection for a solution of Hitchin equations, with the Higgs field Φ being the projection of C on ground states.

For multiple deformation directions, we also require extra consistency conditions such as $[C_u, C_{u'}] = 0$, $\partial_u C_{u'} = \partial_{u'} C_u$, $[C_u, C_{\bar{u}'}] = 0$ in order to get a solution of generalized Hitchin equations on the parameter space.

A BBB deformation is in particular a B-type deformation for all Q_ζ, but in a different way for each ζ. That means the parameter space M is hyper-kähler and Q_ζ is holomorphic in complex structure ζ on M. If we model locally the deformation space on $\mathbb{C}^{2n}$, so that the anti-holomorphic derivatives take the form

$$\text{(B.20)} \qquad \begin{aligned} &\partial_{\bar{u}} + \zeta \partial_v \\ &\partial_{\bar{v}} - \zeta \partial_u \end{aligned}$$

we get relations of the form

$$\text{(B.21)} \qquad \partial_{\bar{u}} Q_1 = 0 \qquad \partial_v Q_1 + \partial_{\bar{u}} Q_2 = 0 \qquad \partial_v Q_2 = 0$$

and

$$\text{(B.22)} \qquad \partial_{\bar{v}} Q_1 = 0 \qquad \partial_u Q_1 - \partial_{\bar{v}} Q_2 = 0 \qquad \partial_u Q_2 = 0$$

More generally, if we decompose the tangent bundle of M as the tensor product of an $SU(2)$ and an $Sp(n)$ bundles, we would write

$$\text{(B.23)} \qquad \partial_{I(\alpha} Q_{\beta)} = 0$$

The Berry connection has curvature of type (1, 1) in all complex structures and thus the ground states form a hyper-holomorphic sheaf on M.

It is possible to have systems which admit both BAA and BBB deformations. Then the Lax connection along the BAA direction commutes with the anti-holomorphic derivatives in complex structure ζ along M.

Although we will not need it for the examples in this paper, the analysis can be generalized to situations with central charges. That means a more general supersymmetry algebra

$$\{Q_\alpha, Q_\beta\} = 0 \qquad \{Q_\alpha, \bar{Q}_\beta\} = \epsilon_{\alpha\beta} H + Z_{\alpha\beta} \tag{B.24}$$

where the triplet of central charges $Z_{\alpha\beta}$ commute with all other operators.

In sectors with non-zero $Z_{\alpha\beta}$ we can only find BPS ground states with $E = |Z|$, annihilated by half of the supercharges only. In particular, they are annihilated by Q_ζ only if ζ^α is an eigenvector of Z_α^β with eigenvalue $|Z|$. In the sector annihilated by the central charges $Z_{\alpha\beta}$, the supersymmetric ground states are annihilated by all supercharges and can be identified with the cohomology of any Q_ζ. These are the states to which our discussion would apply.

B.3. Dirac operators. Several constructions such as ADHM constructions and Nahm transforms, which are usually stated in terms of zeromodes of Dirac operators, arise naturally in the context of $\mathcal{N} = 4$ SQM.

Typically, one has some Dirac operator

$$\Delta = \begin{pmatrix} A_1 \\ A_2 \end{pmatrix} \tag{B.25}$$

a linear map or differential operator between two spaces V and $W \oplus W$, V and W being equipped with an Hermitean inner product, with the property that

$$\Delta\Delta^\dagger = \begin{pmatrix} A_1 \\ A_2 \end{pmatrix} \begin{pmatrix} A_1^\dagger & A_2^\dagger \end{pmatrix} = \begin{pmatrix} A_1 A_1^\dagger & A_1 A_2^\dagger \\ A_2 A_1^\dagger & A_2 A_2^\dagger \end{pmatrix} = \begin{pmatrix} h & 0 \\ 0 & h \end{pmatrix} \tag{B.26}$$

i.e. $A_\alpha \bar{A}_\beta = h\epsilon_{\alpha\beta}$. Here we lower indices with $\epsilon_{\alpha\beta}$ as before, so that $\bar{A}_2 = (A_1)^\dagger$ and $\bar{A}_1 = -(A_2)^\dagger$. Notice that $(\bar{A}_1)^\dagger = -A_2$ and $-(\bar{A}_2)^\dagger = -A_1$.

The zeromodes of Δ are then used produce the desired output, such as a hyper-holomorphic connection on parameter space, or a solution of Hitchin, Nahm or BPS equations. Concretely, zeromodes are elements of V annihilated by both A_α. In typical situations, there is some index theorem relating the number of zeromodes of D and $D^\dagger$. Furthermore, $D^\dagger$ has generically no zeromodes.

We can assemble four supercharges from the linear maps in the Dirac operator, acting on the Hilbert space $W \oplus V \oplus W$:

$$\begin{aligned} Q_\alpha &: W \xrightarrow{-\bar{A}_\alpha} V \xrightarrow{A_\alpha} W \\ \bar{Q}_\alpha &: W \xleftarrow{A_\alpha} V \xleftarrow{\bar{A}_\alpha} W \end{aligned} \tag{B.27}$$

Then $\{Q_\alpha, Q_\beta\} = 0$ and

$$\{Q_\alpha, \bar{Q}_\beta\} = \begin{pmatrix} h & 0 & 0 \\ 0 & A_1^\dagger A_1 + A_2^\dagger A_2 & 0 \\ 0 & 0 & h \end{pmatrix} \epsilon_{\alpha\beta} \tag{B.28}$$

The zeromodes of D coincide with the supersymmetric groundstates of the $\mathcal{N} = 4$ SQM which lie in V. On the other hand, supersymmetric ground states in either of the two W summands would in particular be zeromodes of $D^\dagger$ and will generically be absent.

The zeromodes or ground states can then be identified with the cohomology of Q_ζ for any ζ. The dimension of the cohomology is generically $\dim V - 2\dim W$.

B.3.1. *Quantization of Dirac zeromodes.* There are several physical situations, including one of the two main classes employed in this paper, where the complexes described in the previous section arise as a semi-classical phase space of a system. The actual $\mathcal{N} = 4$ quantum mechanical system involves a Fock space built from these complexes.

Consider a quantum mechanical system involving bosonic creation operators (x_+, y_+) valued in $W \oplus W$ and fermionic creation operators ξ_+ valued in V. We denote the corresponding destruction operators as (x_-, ξ_-, y_-). We can thus define supercharges

$$\begin{aligned} Q_\alpha &: -\xi_+ \bar{A}_\alpha x_- + y_+ A_\alpha \xi_- \\ \bar{Q}_\alpha &: x_+ A_\alpha \xi_- + \xi_+ \bar{A}_\alpha y_- \end{aligned} \tag{B.29}$$

so that the Hamiltonian becomes $H = x_+ h x_- + y_+ h y_- + \xi_+ (A_1^\dagger A_1 + A_2^\dagger A_2) \xi_-$.

This is a nice, free quantum mechanical system whose supersymmetric ground states are essentially the exterior algebra of the space of zeromodes in the previous section. More precisely, if we denote the space of zeromodes in V as Z, the supersymmetric ground states of the quantum mechanics become the fermionic Fock space

$$FZ \equiv (\det Z)^{-\frac{1}{2}} \Lambda^* Z \tag{B.30}$$

There is another important $\mathcal{N} = 4$ quantum mechanical system one may consider, which is the dimensional reduction of a $(0, 4)$ 2d theory. It involves a set of $(0, 4)$ hypermultiplets valued in $W \oplus W$ and $(0, 4)$ Fermi multiplets valued in V. These can be coupled supersymmetrically to background $(0, 4)$ twisted hypermultiplets A_α. We can write some supercharges:

$$\begin{aligned} Q_\alpha &= \bar{p}_y \eta_\alpha + \bar{\eta}_\alpha \bar{p}_x + \psi \bar{A}_\alpha x + y A_\alpha \bar{\psi} \\ \bar{Q}_\alpha &= p_y \bar{\eta}_\alpha - \eta_\alpha p_x - \bar{x} A_\alpha \bar{\psi} + \psi \bar{A}_\alpha \bar{y} \end{aligned} \tag{B.31}$$

with $\{\eta_\alpha, \bar{\eta}_\beta\} = \epsilon_{\alpha\beta}$ are fermions in W and $\{\psi, \bar{\psi}\} = 1$ are fermions in V, x and y are complex bosons. The Hamiltonian is roughly

$$|p_y|^2 + |p_x|^2 + h|x|^2 + h|y|^2 + \psi \bar{A}_\alpha \eta^\alpha + \bar{\eta}^\alpha A_\alpha \bar{\psi} \tag{B.32}$$

The gap is now determined by $\sqrt{h}$. As long as h is positive definite, the ground states are again the Fock space FZ built from the fermion zeromodes $\bar{\psi}$ annihilated by A_α. Intuitively, the mass terms pair up the η_α and some ψ modes and lift them.

B.3.2. *The ADHM system.* The ADHM constructions of $U(N)$ instantons in R^4 will provide our first example of BBB-type deformations, as hyper-holomorphic bundles in real dimension 4 are precisely instantons.

The ADHM data consists of $k \times k$ matrices B_1 and B_2 and $k \times N$ and $N \times k$ matrices I and J, satisfying appropriate moment map constraints for a $U(k)$ hyper-Kähler quotient.

The ADHM construction employs as a Dirac operator the $2k \times (N + 2k)$ matrix

$$\Delta = \begin{pmatrix} I & B_1 + z_1 & B_2 + z_2 \\ J^\dagger & B_2^\dagger + z_2^\dagger & -B_1^\dagger - z_1^\dagger \end{pmatrix} \tag{B.33}$$

with z_1 and z_2 being holomorphic coordinates on R^4.

The supercharge Q_1 thus takes the form

$$Q_1 : \mathbb{C}^k \xrightarrow{\begin{pmatrix} J \\ B_2 + z_2 \\ -B_1 - z_1 \end{pmatrix}} \mathbb{C}^{2k+N} \xrightarrow{\begin{pmatrix} I & B_1 + z_1 & B_2 + z_2 \end{pmatrix}} \mathbb{C}^k \tag{B.34}$$

and the remaining supercharges are defined as above. The B-type parameter space of the SQM consists both of R^4 and of the moduli space $\mathcal{M}_{\mathrm{ADHM}}$ of the ADHM data. Notice how the $U(k)$ hyper-Kähler quotient of the ADHM data happens here: the moment map constraints are required for the $\mathcal{N} = 4$ supersymmetry algebra to hold and the $U(k)$ action on the ADHM data gives equivalent quantum mechanics.

The supersymmetric ground states of the SQM give a hyper-holomorphic bundle on R^4 for every choice of ADHM data which is the instanton bundle itself. More generally, they give an hyper-holomorphic bundle on $R^4 \times \mathcal{M}_{\mathrm{ADHM}}$ which is the universal bundle on the instanton moduli space.

In any given complex structure, the corresponding holomorphic bundle is computed as the cohomology of the Q_ζ complex.

B.3.3. *Kähler SQM.* The standard $\mathcal{N} = 2$ SQM sigma model has enhanced supersymmetry when the target space X is Kähler: the exterior derivative splits as $d = \partial + \bar{\partial}$ and we can use ∂, $\bar{\partial}$, $\partial^\dagger$ and $\bar{\partial}^\dagger$ as four supercharges for an $\mathcal{N} = 4$ SQM.

If we set $Q_1 = \bar{\partial}$, we have two possible choices: we can take either $Q_2 = \partial$ or $Q_2 = \partial^\dagger$. Although the two choices give the same quantum mechanics, the two choices allow one to discuss two distinct and incompatible classes of supersymmetric deformations. In particular, the first choice includes complex structure and superpotential deformations while the second choice includes deformations of the Kähler metric and equivariant deformations. The first choice is the most natural to describe BBB and BAA spaces of deformations. The second would give Nahm- and BPS-type deformations.

At first, we can consider a moduli space U of complex structure deformations of a compact Kähler target manifold X. The super-charge at $\zeta = 1$ is simply the de Rahm differential. Thus the cohomology at $\zeta = 1$ is simply the de Rahm cohomology, which is independent of the complex structure and obviously gives a flat bundle on the space U of complex structure deformations. On the other hand, at $\zeta = 0$ we have the Dolbeault cohomology of X, which is a natural holomorphic bundle on U.

To explore the system in further detail, consider the variation of the $\bar{\partial}$ operator under a change in complex structure. Obviously, $\delta_{\bar{u}}\bar{\partial} = 0$. We can also write

$$\delta_u \bar{\partial} = [\partial, i_{\mu_u}] \tag{B.35}$$

where i_{μ_u} is an operation of contraction with the Beltrami differential corresponding to the variation of complex structure along u. The Beltrami differentials also satisfy the expected $[\bar{\partial}, i_{\mu_u}] = 0$ and the other relations we expect on C_u and $C_{\bar{u}}$, which follow from the definition as variations of the complex structure J.

As a result, we learn that variations of complex structure of X are BAA-type deformations and the Berry connection $(D_u, D_{\bar{u}})$ and the expectation values Φ_u and $\Phi_{\bar{u}}$ of the Beltrami differential operators i_{μ_u} and $i_{\mu_{\bar{u}}}$ solve the generalized Hitchin equations on U. In complex structure $\zeta = 0$, the data reduces to the Higgs bundle given by the holomorphic bundle of Dolbeault cohomology of X together with the operators of contraction with holomorphic Beltrami differentials. In complex structure $\zeta = 1$ the data reduces to the bundle of de Rahm comonology equipped with the flat Gauss-Manin connection. In general complex structure, we have the cohomology of the $\bar{\partial} + \zeta\partial$ operator and associated complexified flat connection.

Next, we can consider a non-compact Kähler target manifold X equipped with a holomorphic super-potential W. The relevant supercharges are

$$Q_1 = \bar{\partial} + \partial W \wedge \qquad Q_2 = \partial + \bar{\partial}\bar{W}\wedge \tag{B.36}$$

The moduli space of BAA deformations now include both complex structure deformations and changes in the super-potential. The corresponding C_u operators are simply multiplication by the holomorphic functions $\partial_u W$.

If we replace ∂W with a more general holomorphic $(1,0)$ form λ, only exact changes in λ will be associated to BAA deformations. General deformations can be promoted to BBB-type deformations by adding the extra data of a flat $U(1)$ connection a:

$$Q_1 = \bar{\partial}_a + \lambda \wedge \qquad Q_2 = \partial_a + \bar{\lambda}\wedge \tag{B.37}$$

so that the variation of Q_1 under a change in λ matches the variation of Q_2 under a change in a.

More generally, given any solution $(A, \Phi, \bar{\Phi})$ of generalized Hitchin's equations on X we can define a corresponding quantum mechanics:

$$Q_1 = \bar{\partial}_A + \Phi \wedge \qquad Q_2 = \partial_A + \bar{\Phi}\wedge \tag{B.38}$$

The deformations of $(A, \Phi, \bar{\Phi})$ are BBB-type deformations.

In complex structure $\zeta = 0$, the ground states are identified with the cohomology of $\bar{\partial}_A + \Phi\wedge$. In terms of the corresponding Higgs bundle (E, Φ), this is the *hypercohomology* of a complex of vector bundles on X: the bundles of forms valued in E with differential $\Phi\wedge$.

If X is a Riemann surface C and $(A, \Phi, \bar{\Phi})$ a solution of the standard Hitchin equations on C, the quantum mechanics above is associated to a Dirac operator

$$\Delta = \begin{pmatrix} \bar{D}_A & \Phi \\ \bar{\Phi} & -D_A \end{pmatrix} \tag{B.39}$$

The space of supersymmetric ground states is thus identified with the space of zeromodes of Δ, which in turn is known as the *Dirac-Higgs bundle*, a natural hyper-holomorphic bundle on the Hitchin moduli space.

This general type of quantum mechanics arises universally as the Born approximation of a system of the form

$$Q_1 = \bar{\partial}_x + C \wedge + Q_1^{\text{aux}}(x) \qquad Q_2 = \partial_x + \bar{C} \wedge + Q_2^{\text{aux}}(x) \tag{B.40}$$

where the "slow" degrees of freedom $x \in X$ act as BAA-type deformation parameters of the auxiliary degrees of freedom described by $Q_\alpha^{\text{aux}}(x)$. The constraint of BAA-type guarantee that the above supercharges define an $\mathcal{N} = 4$ quantum-mechanical system.

B.3.4. *Hyper-holomorphic SQM.* Our second example of quantum-mechanical system is based on a hyper-Kähler target space Y equipped with an hyper-holomorphic bundle E. It is precisely the same as the Dolbeault SQM, but the extra symmetries guarantee the enhancement of supersymmetry to $\mathcal{N} = 4$. In particular, besides $Q_1 = \bar{\partial}_E$ one finds a second $Q_2 = \tilde{\partial}_E^\dagger$ with the property that Q_ζ is equivalent to the Dolbeault differential $\bar{\partial}_E^\zeta$ in complex structure ζ. This relies on the identification between spaces of $(0,q)$ forms in different complex structure induced by the reduction of the structure group of the cotangent bundle to $SU(2) \times Sp(n)$ on hyperkähler manifolds.

For example, given an instanton bundle on R^4, we can consider the Dirac operator

$$\Delta = \begin{pmatrix} -\bar{D}_1 & \bar{D}_2 \\ D_2 & D_1 \end{pmatrix} \tag{B.41}$$

acting on L^2 normalizable functions on R^4.

The corresponding supercharge

$$Q_1 : L^2(R^4) \xrightarrow{\begin{pmatrix} \bar{D}_2 \\ \bar{D}_1 \end{pmatrix}} L^2(R^4) \oplus L^2(R^4) \xrightarrow{\begin{pmatrix} -\bar{D}_1 & \bar{D}_2 \end{pmatrix}} L^2(R^4) \tag{B.42}$$

can be identified with the $\bar{\partial}_E$ operator acting on $(0,q)$ forms. The general supercharge

$$Q_\zeta : L^2(R^4) \xrightarrow{\begin{pmatrix} \bar{D}_2+\zeta D_1 \\ \bar{D}_1-\zeta D_2 \end{pmatrix}} L^2(R^4) \oplus L^2(R^4) \xrightarrow{\begin{pmatrix} -\bar{D}_1+\zeta D_2 & \bar{D}_2+\zeta D_1 \end{pmatrix}} L^2(R^4) \tag{B.43}$$

can be identified with the $\bar{\partial}^\zeta_E$ operator acting on $(0,q)$ forms in complex structure ζ. Notice that the same middle term in the complex is identified with $(0,1)$ forms in different complex structures.

Hyper-holomorphic bundles can admit all sort of different deformation moduli, including BBB-, BAA-, BPS- and Nahm- type deformation parameters. Indeed, canonical examples are universal bundles for moduli spaces of solutions of instanton, Hitchin, BPS and Nahm equations.

This general type of quantum mechanics arises universally as the Born approximation of a system of the form

$$Q_1 = \bar{\partial}_y + Q_1^{\mathrm{aux}}(y) \qquad\qquad Q_2 = \tilde{\partial}_y + Q_2^{\mathrm{aux}}(y) \tag{B.44}$$

where the "slow" degrees of freedom $y \in Y$ act as BBB-type deformation parameters of the auxiliary degrees of freedom described by $Q_\alpha^{\mathrm{aux}}(y)$. The constraint of BBB-type guarantee that the above supercharges define an $\mathcal{N} = 4$ quantum-mechanical system.

B.3.5. *Hypermultiplets on a Riemann surface.* The BBB twist of three-dimensional hypermultiplets valued in M on a Riemann surface C can be thought of as the dimensional reduction of four-dimensional hypermultiplets twisted on C, which is a system with 2d $(0,4)$ supersymmetry.

The hypermultiplet scalars reduced on C together with half of the fermions give $(0,4)$ hypermultiplets, while the other half of the fermions give $(0,4)$ Fermi multiplets. In the notation used in B.3.1, W is the space of M-valued functions on C and V is the space of real M-valued 1-forms on C, where the reality conditions uses the symplectic pairing on M and the symplectic pairing on 1-forms.

The background twisted hypermultiplets A_α are identified with the differential operators $\bar{\partial}_A + \Phi\wedge$ and $\partial_A + \bar{\Phi}\wedge$, so that the Dirac operator Δ coincides with the Dirac-Higgs operator B.39. The Hamiltonian for the $(0,4)$ SQM can be readily matched to the Hamiltonian for the hypermultiplets.

As long as the h built from the connection and Higgs field is positive definite, the ground states are the Fock space built from the fermion zeromodes annihilated by A_α, i.e. the fermionic Fock space modelled on the Dirac-Higgs bundle.

If h has zeromodes, then the system has bosonic flat directions and a continuum of states touching the ground states.

B.3.6. *From $\mathcal{N}=4$ SQM to branes.* The $\mathcal{N}=4$ SQM systems discussed above can be used as boundary degrees of freedom in defining branes for two-dimensional sigma models with eight supercharges, generalizing the statement for $\mathcal{N}=2$ SQM systems.

Again, the natural way to define such coupling to a sigma model is to start from some reference boundary condition for the sigma model and then promote the boundary values of the bulk fields to parameters in the SQM.

BBB-type boundary conditions involve the promotion of BBB-type deformation parameters in the SQM. For example, we can start from some reference Neumann BBB boundary condition associated to the structure sheaf of the sigma model target space U and couple it to a $\mathcal{N}=4$ SQM with U as a parameter space of BBB-type deformations.

The result should be a standard BBB-brane based on the hyper-holomorphic sheaf of ground states of the SQM over U.

BAA-type boundary conditions involve the promotion of BAA-type deformation parameters valued in a complex manifold X, with T^*X being endowed with some local hyperkähler metric. We can consider a BAA-brane which is a deformation of multiple copies of the BAA brane X in T^*X by the Berry connection and Higgs field produced by the SQM: the connection is the connection of the CP bundle on X, while the Higgs field C should be understood as a matrix-valued version of a complex deformation of the Lagrangian manifold along the normal direction to X. When the Higgs field and connection are approximately diagonal this reduces to a complex Lagrangian multi-cover of X.

B.3.7. *From branes so SQM.* Conversely, a nice example of families of $\mathcal{N}=4$ SQM arise when one considers the compactification of a $(4,4)$ sigma model on a segment, with families of supersymmetric boundary conditions at the endpoint.

As a neat concrete example, consider families of BAA-branes in the cotangent bundle T^*C to a Riemann surface C [**24**] An obvious family of BAA branes consist of the fibers F_c of T^*C, parameterized by a point c in C. These are BAA-type deformations, as the Hamiltonian for the deformation is the fiber coordinate p.

A second family of BAA branes can be a collection of spectral curves $\Sigma_\mu^{(N)}$ in T^*C, a ramified N-sheeted cover of C equipped with a flat $U(1)$ bundle, which can be mapped to a line bundle. This family is parameterized by the spectral data μ of a solution of $U(N)$ Hitchin equations on C. This is a moduli space of BBB-type deformations, with shape deformations being paired up with bundle deformations. The space of ground states on a segment with boundary conditions F_c and $\Sigma_\mu^{(N)}$ clearly gives both the actual $U(N)$ Hitchin system on C and the universal bundle for the Hitchin moduli space as we vary c or μ [**24**]

We can also consider the segment with boundary conditions $\Sigma_\mu^{(N)}$ and C itself. This will give a hyper-holomorphic bundle on the $U(N)$ Hitchin moduli space, which we expect to coincide with the Dirac-Higgs bundle in the fundamental representation for $U(N)$.

Similarly, we can consider a segment with boundary conditions $\Sigma_\mu^{(N)}$ and $\Sigma_{\mu'}^{(M)}$ to get a hyper-holomorphic bundle on the product of $U(N)$ and $U(M)$ Hitchin moduli spaces, which we expect to coincide with the Dirac-Higgs bundle in the bi-fundamental representation for $U(N)\times U(M)$.

Similar considerations apply to branes in a complex integrable system $\mathcal{M}$. For example, the elliptic fibers F_μ equipped with a flat $U(1)$ bundle will have the Langlands dual complex integrable system as a moduli space of BBB-type deformations. Segments with F_p at one end and a generic BAA brane at the other end will produce the mirror BBB brane.

Conversely, there should be families of BBB branes on a Hitchin moduli space with a BAA-type deformation space consisting of bundles for the Langlands dual group, which implement the opposite Geometric Langlands duality. These should be the BBB branes arising as the BBB-twist of the $T[G]$ theory defined in [**5**].

Appendix C. The category of BBB branes

A BBB brane is a B-brane in each complex structure. A typical mathematical description of B-branes is a derived category of coherent sheaves on the complex manifold. We would like to understand better, through general considerations and examples, which general form should such complexes take for a BBB brane. Intuitively, we expect to find some sort of complexes of hyper-holomorphic coherent sheaves with ζ-dependent differentials which are holomorphic in every complex structure.

A simple way to map a B-brane B on a manifold X to an object in the derived category is to consider a segment with Dirichlet boundary D_x conditions at one end and B at the other end.[8] This is an $\mathcal{N} = 2$ SQM with x as a space of B-type deformations. If we consider a point x_0 with a finite gap between the space of supersymmetric ground states V_{x_0} and the next excited states, in a sufficiently small neighbourhood of x_0 we can describe the ground states as the cohomology of a complex $(V_0, \delta_{x_0,x})$ with a differential which depends holomorphically on x.

The result is a collection of open sets U_i and complexes of holomorphic vector bundles (V_i, δ_i) defined on U_i equipped with compatible quasi-isomorphisms on intersections of open sets. The mathematical term for this structure is a *perfect complex*. Perfect complexes are a nice sub-category of the derived category of coherent sheaves.

We can run this analysis equally well for a BBB brane. Now the segment with Dirichlet boundary D_x conditions at one end and B at the other end gives us an $\mathcal{N} = 4$ SQM with x as a space of BBB-type deformations. If we consider a point x_0 with a finite gap between the space of supersymmetric ground states V_{x_0} and the next excited states, in a sufficiently small neighbourhood of x_0 we can describe the ground states by an auxiliary finite-dimensional $\mathcal{N} = 4$ SQM with Hilbert space V_0 and supercharges δ_α, $\bar{\delta}_\alpha$.

The result is some sort of hyper-holomorphic perfect complex: a collection of open sets U_i and $\mathcal{N} = 4$ SQM $(V_i, \delta_{\alpha,i})$ with U_i as spaces of BBB-type deformations which have the same supersymmetric ground states on the intersection of open sets. It should be possible to improve this description further, possibly with the help of twistorial ideas.

[8]The following argument was developed in the course of an ongoing collaboration with M. Del Zotto and L. Bhardwaj

Appendix D. Gauge group reductions

We will begin with a discussion of Dirichlet-like boundary conditions and interfaces, which do not require adding extra degrees of freedom at the boundary and reduce the gauge group to a subgroup at the boundary.

This is somewhat orthogonal to the rest of the paper, which focuses on Neumann-like boundary conditions involving boundary degrees of freedom coupled to the full gauge group at the boundary.

Dirichlet-like boundary conditions can actually be recast as Neumann-like boundary conditions involving specific choices of boundary degrees of freedom. We prefer to review them separately as they are already rather well-understood (see e.g. [**25**]) and we will need the following results in order to understand some important S-duality examples.

D.1. Dirichlet boundary conditions. Half-BPS Dirichlet boundary conditions impose Dirichlet boundary conditions on both the connection and the three scalar fields rotated by $SO(3)_C$. The remaining three scalar fields rotated by $SO(3)_H$ receive Neumann boundary conditions. Dirichlet boundary conditions admit a global symmetry $G_H = G$.

In a BAA twist, the two scalar fields which give rise to the Higgs field on C receive Neumann boundary conditions and are thus free to fluctuate at the boundary. The connection on C, though, is fixed to some choice of background connection, which can be thought of as a connection for the global symmetry group G_H.

In the language of Higgs bundles (E, Φ), the boundary conditions fix the bundle E to some reference E_H and let Φ unconstrained. Thus the BAA brane is supported on the complex Lagrangian submanifold in $\mathcal{M}_H(C, G)$ given by the fiber over a point of the co-tangent bundle to $\mathrm{Bun}(C, G)$.

In the BBB twist, the two scalar fields which give rise to the Higgs field on C receive Dirichlet boundary conditions and are thus fixed at the boundary as well. The boundary value can be thought of as a G_H Higgs field. The background connection and Higgs field must satisfy Hitchin equations on C and give a point p in $\mathcal{M}_H(C, G)$. The BBB brane is a skyscraper sheaf at p.

Dirichlet boundary conditions can be generalized to interfaces between a G gauge theory and a $G' \subset G$ gauge theory, at which the gauge group is reduced from G to G'. Such interfaces have a global symmetry group $G_H = Z_{G'}(G)$, the commutant of G' in G.[9]

On the BAA side, we restrict the G bundle E to coincide with the G' bundle E' tensored with the reference G_H bundle E_H., We also require the G' Higgs field Φ' to coincide with the restriction of the G Higgs field Φ to $\mathrm{End}(E')$. Thus the BAA interface is supported on a simple Lagrangian correspondence.

On the BBB side, we obtain a sheaf supported on the graph of the embedding of $\mathcal{M}_H(C, G')$ into $\mathcal{M}_H(C, G)$ by tensor with a reference solution in $\mathcal{M}_H(C, G_F)$. We expect the sheaf to be trivial, so that convolution with a sheaf on $\mathcal{M}_H(C, G')$ would give the same sheaf on the image of $\mathcal{M}_H(C, G')$ in $\mathcal{M}_H(C, G)$.

D.2. Nahm pole boundary conditions and interfaces. Dirichlet boundary conditions admit a surprising modification, which imposes a singular boundary condition on the three scalar fields rotated by $SO(3)_H$. The modified boundary

[9] A better characterization would be $N_{G'}(G)/G'$, the quotient of the normalizer of G' in G by G'.

condition is labelled by an embedding ρ of $\mathfrak{su}(2)$ into the Lie algebra $\mathfrak{g}$ of the gauge group. The singular boundary condition breaks the boundary global symmetry to $G_H = Z_\rho(G)$, the commutant of ρ in G. For a previous discussion of the role of these boundary conditions in Geometric Langlands, see e.g. [**25**].

In the BAA twist, the singular boundary conditions force the Higgs field to lie in a Slodowy slice S_ρ. If $t^\pm$ and t^0 are the images of $\mathfrak{su}(2)$ raising, lowering and Cartan generators respectively, that means

$$\Phi = t^+ + \Phi^- \tag{D.1}$$

where $[t^-, \Phi^-] = 0$. The bundle E is fixed to have the form $E_H \otimes t^0(K_C^{1/2})$, where E_H is some bundle with G_H structure group and we used t^0 to promote $K_C^{1/2}$ to an $SU(2)$ bundle. This is compatible with the restriction on Φ, as $t^\pm$ are global sections of $\mathrm{End}(E) \otimes K_C$.[10]

These relations define the support for the BAA brane.

If ρ is a maximal embedding, then the support of the BAA brane is simply the canonical section of the Hitchin fibration. This statement is compatible with S-duality: the S-dual of the maximal Nahm pole boundary condition is a pure Neumann boundary condition, whose BBB image is the structure sheaf $\mathcal{O}$ of the whole Hitchin moduli space. T-duality along the fibers of the Hitchin fibration indeed maps $\mathcal{O}$ to the canonical section of the Hitchin fibration.

The S-dual of more general Nahm pole boundary conditions or pure Dirichlet boundary conditions involves non-trivial boundary degrees of freedom.

The BBB image of Nahm pole boundary conditions lies somewhat outside the regime of validity of the sigma model description, as it involves a singular boundary condition for the scalar fields in the 2d gauge multiplet. It is a variant of the skyscraper sheaf at singular loci in the Hitchin moduli space where the Higgs bundle structure group reduces to G_H.

We can promote Nahm pole boundary conditions to interfaces between G and $G' = Z_\rho(G)$ gauge theories.

On the BAA side that means promoting the Slodowy slice constraints to a Lagrangian correspondence: E_H is identified with the bundle E' of the G' Higgs bundle (E', Φ'), while Φ' is set to equal the projection of Φ onto the trivial $\mathfrak{su}(2)$ representation.

On the BBB side, that means considering a sheaf supported on the graph of the embedding of $\mathcal{M}_H(C, G')$ into $\mathcal{M}_H(C, G)$. Again, the sigma model description is not quite adequate because of the Nahm pole.

Acknowledgements

We thank Kevin Costello and Edward Witten for many illuminating conversations and explanations. The research of DG was supported by the Perimeter Institute for Theoretical Physics. Research at Perimeter Institute is supported by the Government of Canada through Industry Canada and by the Province of Ontario through the Ministry of Economic Development & Innovation.

[10]An alternative description of the slice involves a global section s of $\mathrm{Hom}(E_H, E)$ with the constraint that s, Φs, $\Phi^2 s$, etc. span flags of a specific form.

References

[1] A. Kapustin and E. Witten, *Electric-magnetic duality and the geometric Langlands program*, Commun. Number Theory Phys. **1** (2007), no. 1, 1–236. MR2306566

[2] E. Witten, *More on gauge theory and geometric langlands*, 1506.04293.

[3] M. Bershadsky, A. Johansen, V. Sadov, and C. Vafa, *Topological reduction of 4D SYM to 2D σ-models*, Nuclear Phys. B **448** (1995), no. 1-2, 166–186. MR1352404

[4] D. Gaiotto and E. Witten, *Supersymmetric boundary conditions in $\mathcal{N} = 4$ super Yang-Mills theory*, J. Stat. Phys. **135** (2009), no. 5-6, 789–855. MR2548595

[5] D. Gaiotto and E. Witten, *S-duality of boundary conditions in $\mathcal{N} = 4$ super Yang-Mills theory*, Adv. Theor. Math. Phys. **13** (2009), no. 3, 721–896. MR2610576

[6] M. Henningson, *Boundary conditions for geometric-Langlands twisted N=4 supersymmetric Yang-Mills theory*, *Phys. Rev.* **D86** (2012) 085003, [1106.3845].

[7] D. Gaiotto, *Twisted compactifications of 3d $n = 4$ theories and conformal blocks*, 1611.01528.

[8] N. J. Hitchin, *The self-duality equations on a Riemann surface*, Proc. London Math. Soc. (3) **55** (1987), no. 1, 59–126. MR887284

[9] E. Witten, *Mirror symmetry, Hitchin's equations, and Langlands duality*, The many facets of geometry, Oxford Univ. Press, Oxford, 2010, pp. 113–128. MR2681689

[10] A. Strominger, S.-T. Yau and E. Zaslow, *Mirror symmetry is T duality*, *Nucl. Phys.* **B479** (1996) 243–259, [hep-th/9606040].

[11] S. Gukov and E. Witten, *Branes and quantization*, Adv. Theor. Math. Phys. **13** (2009), no. 5, 1445–1518. MR2672467

[12] S. Gukov, *Quantization via mirror symmetry*, Jpn. J. Math. **6** (2011), no. 2, 65–119. MR2861769

[13] D. Gaiotto, G. W. Moore, and A. Neitzke, *Wall-crossing in coupled 2d-4d systems*, J. High Energy Phys. **12** (2012), 082, front matter + 166. MR3045271

[14] N. Hitchin, *Spinors, Lagrangians and rank 2 Higgs bundles*, Proc. Lond. Math. Soc. (3) **115** (2017), no. 1, 33–54. MR3669932

[15] J. Blaavand, *The dirac-higgs bundle*, *Thesis* (2015) .

[16] B. Assel and J. Gomis, *Mirror symmetry and loop operators*, J. High Energy Phys. **11** (2015), 055, front matter+90. MR3455039

[17] S. Gukov and E. Witten, *Gauge theory, ramification, and the geometric Langlands program*, Current developments in mathematics, 2006, Int. Press, Somerville, MA, 2008, pp. 35–180. MR2459305

[18] D. Gaiotto and M. Rapcak, *Boundary line defects and the geometric langlands program*, *in preparation* .

[19] L. Rozansky and E. Witten, *HyperKahler geometry and invariants of three manifolds*, *Selecta Math.* **3** (1997) 401–458, [hep-th/9612216].

[20] A. Braverman and D. Gaitsgory, *Geometric Eisenstein series*, Invent. Math. **150** (2002), no. 2, 287–384. MR1933587

[21] N. Hitchin, *Higgs bundles and characteristic classes*, Arbeitstagung Bonn 2013, Progr. Math., vol. 319, Birkhäuser/Springer, Cham, 2016, pp. 247–264. MR3618052

[22] S. Cecotti and C. Vafa, *Topological–anti-topological fusion*, Nuclear Phys. B **367** (1991), no. 2, 359–461. MR1139739

[23] S. Cecotti, D. Gaiotto, and C. Vafa, *tt^* geometry in 3 and 4 dimensions*, J. High Energy Phys. **5** (2014), 055, front matter+110. MR3223223

[24] E. Witten, *Talk at the workshop "Symplectic duality and Gauge Theory"* (April 2016) .

[25] E. Witten, *Geometric Langlands duality and the equations of Nahm and Bogomolny*, Proc. Roy. Soc. Edinburgh Sect. A **140** (2010), no. 4, 857–895. MR2672073

Perimeter Institute for Theoretical Physics, Waterloo, Ontario, N2L 2Y5, Canada
Email address: dgaiotto@gmail.com

Proceedings of Symposia in Pure Mathematics
Volume **98**, 2018
http://dx.doi.org/10.1090/pspum/098/01727

Pure SU(2) gauge theory partition function and generalized Bessel kernel

P. Gavrylenko and O. Lisovyy

Abstract. We show that the dual partition function of the pure $\mathcal{N} = 2$ SU (2) gauge theory in the self-dual Ω-background (a) is given by Fredholm determinant of a generalized Bessel kernel and (b) coincides with the tau function associated to the general solution of the Painlevé III equation of type D_8 (radial sine-Gordon equation). In particular, the principal minor expansion of the Fredholm determinant yields Nekrasov combinatorial sums over pairs of Young diagrams.

1. Introduction

The study of quantitative aspects of the isomonodromy/CFT correspondence [**SMJ, K89, M90, T11**] has been initiated in the work [**GIL12**], where the general tau function of the sixth Painlevé equation was conjectured to coincide with the Fourier tranform of the 4-point $c = 1$ Virasoro conformal block

(1.1) $$\tau_{\rm VI}\big(t \,|\, \sigma, \eta, \vec{\theta}\big) = \sum_{n\in\mathbb{Z}} e^{2\pi i n\eta}$$ 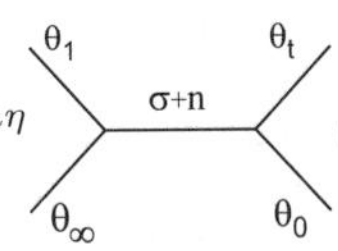 (t).

This proposal was later proved in [**ILTe, BSh1**] by CFT methods. The parameters $\vec{\theta} = (\theta_0, \theta_t, \theta_1, \theta_\infty)$ represent local monodromy exponents on the Painlevé side, and are related to external conformal dimensions of primaries in the conformal block by $\Delta_\nu = \theta_\nu^2$. The intermediate dimension is $\Delta = (\sigma + n)^2$.

As is well-known, the AGT correspondence [**AGT**] relates Virasoro 4-point conformal blocks to partition functions of the $\mathcal{N} = 2$ supersymmetric 4D gauge theories with the gauge group SU (2) and $N_f = 4$ matter multiplets, regularized by an appropriate deformation (the Ω-background) with two parameters ϵ_1, ϵ_2. The $c = 1$ case corresponds to the self-dual Ω-background ($\epsilon_1 + \epsilon_2 = 0$). Expanding conformal blocks around $t = 0$ corresponds to the weak coupling expansion in the gauge theory, explicitly computed in [**Nek**].

The Painlevé VI is the most general equation in the Painlevé family. All the others can be obtained from it by appropriate degeneration limits. In [**GIL13**], some of these limits have been computed at the level of solutions. This produces explicit formulas for Painlevé V and all three types (D_6, D_7 and D_8) of Painlevé III functions in the form of power series. From the gauge theory point of view, such

2010 *Mathematics Subject Classification.* Primary 33E17; Secondary 34E05.

degenerations correspond to decoupling of the massive fields, which means that Painlevé V and III's are related to $N_f < 4$ gauge theories, and explicit formulas for the tau functions are known in their weak coupling regions. On the CFT side, these cases are related to conformal blocks involving Whittaker vectors [**G09**, **BMT**, **GT**]. In contrast to the $N_f = 4$ case, there are interesting situations for $N_f < 4$ where explicit (asymptotic) series representations of solutions are not known: they correspond to strong coupling regions on the gauge theory side, and to conformal blocks with irregular vertex operators in the CFT framework. The present work is concerned with the most degenerate case of Painlevé III equation of type D_8 corresponding to the pure gauge theory.

It is interesting to note that an avatar of the Painlevé III (D_8) tau function was already studied by Nekrasov and Okounkov in [**NO**], although at the time the relevant object had not yet been related to isomonodromy nor to CFT. The equation of interest is of 2nd order and contains no parameters; its tau function is given by

$$\tau_{\mathrm{III}}\left(t\,|\,\sigma,\eta\right)=\sum_{n\in\mathbb{Z}}e^{4\pi i n\eta}\mathcal{Z}_{\mathrm{SU}(2)}\left(t\,|\,\sigma+n\right), \tag{1.2}$$

where (σ, η) represent the initial data. The right side of (1.2) was dubbed in [**NO**] the dual partition function of the pure gauge theory. The first reason to consider it was purely technical: it is convenient to introduce a Lagrange multiplier to control (in the non-ϵ-deformed limit) the value of σ, the vacuum expectation value of the scalar field. A second reason is the existence of a fermionic representation for the dual partition function, presented in [**NO**] in the special case $\tau_{\mathrm{III}}\left(t\,|\,\frac{1}{4},\eta\right)$. Setting in addition $\eta = 0$ or $\eta = \frac{1}{4}$, we obtain elementary solutions of PIII:

$$\tau_{\mathrm{III}}\left(t\,\middle|\,\frac{1}{4},\frac{1\pm 1}{8}\right)=t^{\frac{1}{16}}e^{\pm 4\sqrt{t}}. \tag{1.3}$$

They are related to twisted representations in the intermediate channel [**Z86**, **AZ**] generated by the realization of the Virasoro algebra in terms of one Ramond boson [**BSh2**].

In order to get a physically interesting result, namely the partition function without ϵ-deformation, one has to consider the asymptotics of the dual partition function as $\eta \to i\infty$. In this case the sum can be computed in a saddle-point approximation[1]. Different quantities scale as follows:

$$\begin{gathered}\eta=\epsilon^{-1}\tilde{\eta},\qquad \sigma=\epsilon^{-1}\tilde{\sigma},\qquad t=\epsilon^{-4}\tilde{t},\\ \mathcal{Z}_{\mathrm{SU}(2)}\left(\epsilon^{-4}\tilde{t}\,|\,\epsilon^{-1}\tilde{\sigma}\right)\sim\exp\left\{\epsilon^{-2}\mathcal{F}_0(\tilde{t}\,|\,\tilde{\sigma})+\mathcal{F}_1(\tilde{t}\,|\,\tilde{\sigma})+\ldots\right\},\end{gathered} \tag{1.4}$$

which means that the saddle point is defined by the equation $\partial_{\tilde{\sigma}}\mathcal{F}_0\left(\tilde{\sigma}\,|\,\tilde{t}\right) = -4\pi i\tilde{\eta}$. One of the main results of [**NO**] is the statement that the Seiberg-Witten prepotential [**SW**] — the function encoding the low-energy behaviour of the $\mathcal{N} = 2$ pure SU (2) gauge theory — coincides with $\mathcal{F}_0\left(\tilde{\sigma}\,|\,\tilde{t}\right)$, which confirms the Seiberg-Witten solution at the microscopic level.

A related procedure was used in [**BLMST**] to identify the Painlevé I–V tau functions also with the dual partition functions of *strongly* coupled gauge theories, including the Argyres-Douglas theories of type H_0, H_1 and H_2. Specifically, it has

[1]This is the original proposal from [**NO**, Eq. (5.5)]. The actual answer for the dual partition function also contains non-perturbative corrections (in ϵ) of crucial importance which we are going to study in a future work.

been checked that the long-distance (irregular type) tau function expansions match various magnetic and dyonic strong coupling expansions on the gauge side. A CFT counterpart of this correspondence has been suggested in [**Nag1**, **Nag2**], where some of the long-distance asymptotic series for Painlevé V and IV were conjecturally related to Fourier transforms of conformal blocks with irregular vertex operators.

In a recent paper [**GL16**], we have developed a method of representing the isomonodromic tau functions of Fuchsian systems as block Fredholm determinants. The construction is based on the Riemann-Hilbert approach. The main input is given by monodromy of a connection $\partial_z - A(z)$ with simple poles together with a pants decomposition of the appropriate punctured Riemann sphere. The relevant integral operators act on vector-valued functions defined on a collection of circles (internal boundary components of pants). Their kernels are expressed in terms of solutions of Fuchsian systems associated to different pairs of pants and having only 3 regular singular points. In rank 2, where the isomonodromy equations are equivalent to the Garnier system containing Painlevé VI as the simplest case, Fredholm determinant representations become completely explicit as the kernels have hypergeometric expressions. Furthermore, the principal minor expansion of the determinant written in the Fourier basis coincides with the combinatorial evaluation [**Nek**] of the dual partition function of the 4D $\mathcal{N} = 2$ linear quiver $\mathrm{U}(2)$ gauge theory. This yields in particular a rigorous proof of the series representation of the Painlevé VI tau function, which bypasses the use of the AGT correspondence and does not rely on CFT arguments such as crossing symmetry, null vector decoupling equations, etc.

While it is in principle clear that the approach of [**GL16**] may be extended to at least some classes of irregular isomonodromic systems, its practical implementation within the Riemann-Hilbert framework is not obvious. Our main goal in this paper is to work out the details for Painlevé III (D_8) equation which exhibits most of the subtleties of the irregular case and at the same time keeps the notational fuss to a minimum. We hope that the Fredholm determinant representation of $\tau_{\mathrm{III}}(t\,|\,\sigma,\eta)$ obtained here, besides producing a combinatorial series at weak coupling, may also turn out to be useful for the analysis of the strongly coupled regime. Let us mention that a different (?) Fredholm determinant representation for the special tau function $\tau_{\mathrm{III}}(t\,|\,\sigma,0)$ has recently appeared in the proof [**BGT1**] of a 4D version of the conjecture of [**GHM**] relating topological strings and spectral theory (see also [**BGT2**] for higher-rank generalizations). Our results could also provide some insight in this context.

A useful guideline for our work is provided by the geometric Painlevé confluence diagram proposed in [**CM**, **CMR**]. In this picture, the monodromy manifolds of different Painlevé equations are interpreted as moduli spaces of Riemann spheres with cusped boundaries. One is then tempted to replace the usual decomposition of the Painlevé VI four-holed sphere into two pairs of pants by cutting the Painlevé III (D_8) decorated cylinder into two, each of them having one regular and one 1-cusped puncture, see Fig. 1.
Furthermore, the number of cusps at a particular hole was heuristically related [**CMR**, Appendix A] to the number of Stokes rays at the corresponding irregular singular point, and to the pole order of the quadratic differential $\det A(z)\,dz^2$. As we will see, the former interpretation turns out to be the most adapted to our purposes, cf e.g. the Riemann-Hilbert contour in Fig. 3.

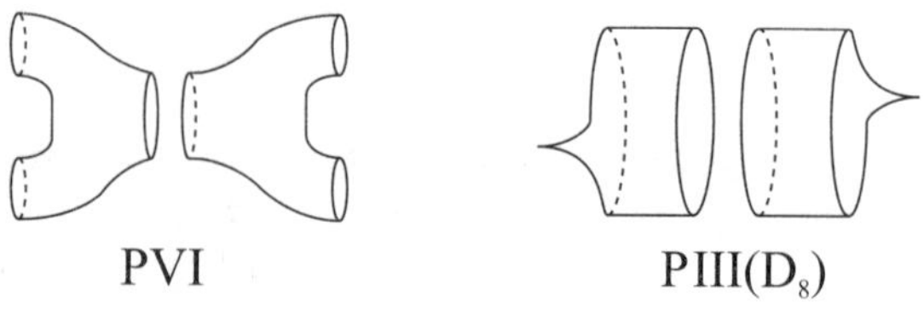

FIGURE 1. Pants decomposition for Painlevé VI and Painlevé III (D_8).

The paper is organized as follows. In Section 2, we introduce an irregular linear system leading to Painlevé III (D_8), describe its generalized monodromy, and explain the "decorated pants decomposition" of the associated Riemann-Hilbert problem. In Section 3, it is shown that the PIII (D_8) tau function admits a Fredholm determinant representation with a generalized Bessel kernel, the main result being Theorem 3.2. Section 4 is devoted to derivation of the series over pairs of Maya/Young diagrams and its identification with the dual partition function of the pure gauge theory (Theorems 4.1 and 4.4).

Acknowledgements. We would like to thank M. Bershtein, N. Iorgov, and A. Marshakov for useful discussions. The present work was supported by the CNRS/PICS project "Isomonodromic deformations and conformal field theory". P. G. was partially supported by the RSF grant No. 16-11-10160 (results of section 4). He is also a Young Russian Mathematics award winner and would like to thank its sponsors and jury. P. G. would also like to thank the KdV Institute of the University of Amsterdam, where a part of this work was done, and especially G. Helminck, for warm hospitality.

2. Isomonodromy and Riemann-Hilbert setup

2.1. Associated irregular system. Our starting point is a system of linear differential equations

$$\partial_z Y = A(z)\, Y, \tag{2.1a}$$

where $A(z)$ is a given $N \times N$ matrix with rational dependence on z. The fundamental matrix solution $Y(z)$ in general has branched singularities at the poles of the 1-form $A(z)\, dz$ on the Riemann sphere $\mathbb{P}^1$. It involves no loss of generality to assume that $\operatorname{Tr} A(z) = 0$; otherwise it suffices to transform $Y \mapsto fY$ with a suitably adjusted scalar factor.

We are going to study a special class of such linear systems in rank $N = 2$ characterized by the number of singularities and their type. Specifically, assume that there are only two irregular singular points (e.g. 0 and ∞) of Poincaré rank $\frac{1}{2}$. By this we mean that

$$A(z) = A_{-2} z^{-2} + A_{-1} z^{-1} + A_0, \qquad A_k \in \operatorname{Mat}_{2\times 2}(\mathbb{C}), \quad \operatorname{Tr} A_k = 0,$$

with non-diagonalizable A_0 and A_{-2}. Using constant gauge transformations $Y \mapsto GY$, $A(z) \mapsto GA(z)G^{-1}$ and rescaling $z \mapsto \lambda z$ if necessary, it may be further assumed that either (i) $A_0 = \sigma_+$, $A_{-2} = \sigma_-$ or (ii) $A_0 = A_{-2} = \sigma_+$, where

$$\sigma_+ = \begin{pmatrix} 0 & 1 \\ 0 & 0 \end{pmatrix}, \qquad \sigma_- = \begin{pmatrix} 0 & 0 \\ 1 & 0 \end{pmatrix}.$$

In the case (ii), the remaining freedom of conjugation by upper triangular matrices with unit diagonals leaves only two nontrivial parameters in A_{-1}. The corresponding linear system does not admit isomonodromic deformations and reduces to a special case of doubly-cofluent Heun equation. We will therefore focus on the case (i) and, after suitable rescalings, parameterize $A(z)$ as

$$A(z) = q\sigma_- z^{-2} + q^{-1}\begin{pmatrix} -p & t \\ -q & p \end{pmatrix} z^{-1} - \sigma_+. \tag{2.1b}$$

The system (2.1a) with $A(z)$ given by (2.1b) is the linear problem associated to Painlevé III (D_8) equation. Among 3 parameters p, q and t, the latter plays the role of time in the associated isomonodromic problem, and the former two are coordinates on the PIII (D_8) phase space.

The system (2.1) can be put to a more convenient form using non-constant gauge transformation. Let us define a new matrix $\tilde{Y}(\xi)$ by

$$\tilde{Y}(\xi) = G(\xi)^{-1} Y(\xi^2), \qquad G(\xi) = \frac{1}{i\sqrt{2}}\begin{pmatrix} \xi^{\frac{1}{2}} & \xi^{\frac{1}{2}} \\ \xi^{-\frac{1}{2}} & -\xi^{-\frac{1}{2}} \end{pmatrix}.$$

It solves the linear system

$$\partial_\xi \tilde{Y} = \tilde{A}(\xi)\tilde{Y}, \tag{2.2a}$$

with $\tilde{A}(\xi) = 2\xi G(\xi)^{-1} A(\xi^2) G(\xi) - G(\xi)^{-1} G'(\xi)$. Computing the latter matrix explicitly, one may see that the system (2.2a) also has irregular singularities at 0 and ∞:

(2.2b)

$$\tilde{A}(\xi) = \tilde{A}_{-2}\xi^{-2} + \tilde{A}_{-1}\xi^{-1} + \tilde{A}_0,$$

$$\tilde{A}_{-2} = \left(q + \frac{t}{q}\right)\sigma_z + \left(q - \frac{t}{q}\right) i\sigma_y, \qquad \tilde{A}_{-1} = -\left(\frac{2p}{q} + \frac{1}{2}\right)\sigma_x, \qquad \tilde{A}_0 = -2\sigma_z,$$

where $\sigma_{x,y,z}$ denote the Pauli matrices. The above is by no means a generic form of 2×2 systems with 2 irregular singular points of Poincaré rank 1; one of the properties that singles out the class described by (2.2b) is a discrete $\mathbb{Z}_2$-symmetry $\tilde{A}(-\xi) = -\sigma_x \tilde{A}(\xi)\sigma_x$.

2.2. Monodromy. The fact that the transformed coefficients $\tilde{A}_{-2}$ and $\tilde{A}_0$ are diagonalisable, in contrast to their counterparts in (2.1b), allows to write formal fundamental solutions of (2.2a) at 0 and ∞ in the standard form,

$$\tilde{Y}^{(0)}_{\text{form}}(\xi) = \left(-\frac{q}{\sqrt{t}}\right)^{-\frac{\sigma_x}{2}}\left[\mathbf{1} + \sum_{k=1}^{\infty} y_k^{(0)}\xi^k\right] e^{2\sigma_z\sqrt{t}\,\xi^{-1}}, \qquad \xi \to 0, \tag{2.3a}$$

$$\tilde{Y}^{(\infty)}_{\text{form}}(\xi) = \left[\mathbf{1} + \sum_{k=1}^{\infty} y_k^{(\infty)}\xi^{-k}\right] e^{-2\sigma_z\xi}, \qquad \xi \to \infty. \tag{2.3b}$$

The $\mathbb{Z}_2$-symmetry of $\tilde{A}(\xi)$ implies that formal solutions satisfy

$$\tilde{Y}^{(\nu)}_{\text{form}}(-\xi) = \sigma_x \tilde{Y}^{(\nu)}_{\text{form}}(\xi)\sigma_x.$$

The expansion coefficients $y_k^{(\nu)}$ can be computed in a straightforwad way to any finite order using (2.2). In what follows, the only explicit expression we need concerns the first such coefficient in (2.3a), namely,

$$y_1^{(0)}=-\frac{1}{\sqrt{t}}\left[\left(\frac{p^2}{q^2}+\frac{p}{2q}-q-\frac{t}{q}+\frac{1}{16}\right)\sigma_z+\left(\frac{p}{2q}+\frac{1}{8}\right)i\sigma_y\right]. \tag{2.4}$$

The actual solutions of the unfolded system (2.2) can only be asymptotic to $\tilde{Y}^{(\nu)}_{\text{form}}(\xi)$ inside the Stokes sectors $\tilde{\mathcal{S}}^{(\nu)}_k$ $(k=1,2,3)$ defined by

$$\tilde{\mathcal{S}}^{(0)}_k=\left\{\xi\in\mathbb{C}\,|\,\frac{\arg t-3\pi}{2}+k\pi<\arg\xi<\frac{\arg t+\pi}{2}+k\pi,\ |\xi|<R\right\},$$

$$\tilde{\mathcal{S}}^{(\infty)}_k=\left\{\xi\in\mathbb{C}\,|\,-\frac{3\pi}{2}+k\pi<\arg\xi<\frac{\pi}{2}+k\pi,\ |\xi|>R\right\}.$$

Furthermore, the requirement that $\tilde{Y}^{(\nu)}_k\simeq\tilde{Y}^{(\nu)}_{\text{form}}$ inside $\mathcal{S}^{(\nu)}_k$ as $\xi\to\nu$ fixes the solutions $\tilde{Y}^{(\nu)}_k$ uniquely. As is well-known, such canonical solutions associated to the same point are related by constant (i.e. independent of ξ) Stokes matrices

$$S^{(\nu)}_k=\tilde{Y}^{(\nu)}_k(\xi)^{-1}\tilde{Y}^{(\nu)}_{k+1}(\xi),\qquad \nu=0,\infty,\qquad k=1,2. \tag{2.5}$$

Constant connection matrix $E=\tilde{Y}^{(0)}_1(\xi)^{-1}\tilde{Y}^{(\infty)}_1(\xi)$ relates the canonical solutions at 0 and ∞. Stokes and connection matrices describe global asymptotic behavior of solutions of (2.1), (2.2) and constitute the relevant set of generalized monodromy data.

The discrete symmetry mentioned above and interlacing structure of dominant/recessive solutions imply that Stokes matrices can be written as

$$S^{(0)}_1=\sigma_x S^{(0)}_2\sigma_x=\begin{pmatrix}1&\alpha\\0&1\end{pmatrix},\qquad S^{(\infty)}_1=\sigma_x S^{(\infty)}_2\sigma_x=\begin{pmatrix}1&0\\\beta&1\end{pmatrix}. \tag{2.6a}$$

Similar constraints are also valid for the connection matrix,

$$\sigma_x E\sigma_x=S^{(0)}_1{}^{-1}E\,S^{(\infty)}_1=S^{(0)}_2E\,S^{(\infty)}_2{}^{-1}. \tag{2.6b}$$

The relations (2.6) imply that, in general, the monodromy data $\{S^{(\nu)}_k\}$, E can be parameterized by a pair of complex parameters (σ,η) in the following way:

$$\begin{aligned}&E=\frac{1}{\sin 2\pi\sigma}\begin{pmatrix}\sin 2\pi\eta & -i\sin 2\pi(\eta+\sigma)\\ i\sin 2\pi(\eta-\sigma) & \sin 2\pi\eta\end{pmatrix},\qquad \sigma\notin\mathbb{Z}/2,\\ &S^{(0)}_1=S^{(0)T}_2=S^{(\infty)T}_1=S^{(\infty)}_2\equiv S=\begin{pmatrix}1&-2i\cos 2\pi\sigma\\0&1\end{pmatrix}.\end{aligned} \tag{2.7}$$

It can be furthermore assumed that σ and η belong to the strips $-\frac12\leq\Re\sigma\leq 0$ and $-\frac12<\Re\eta\leq\frac12$. Note that the counterclockwise monodromy matrix $\tilde{M}_0$ of $\tilde{Y}^{(0)}_1(z)$ around 0 can be expressed as

$$\tilde{M}_0^{-1}=SS^T=ES^TSE^{-1}. \tag{2.8}$$

Let us finally comment on how to recover monodromy of the initial system (2.1) from the Stokes data of the unfolded equation (2.2). Introduce the solutions $Y^{(\nu)}(z)=G(\sqrt{z})\,\tilde{Y}^{(\nu)}_1(\sqrt{z})$, uniquely defined by their asymptotic behavior

$$Y^{(\nu)}(z)\simeq G(\sqrt{z})\,\tilde{Y}^{(\nu)}_{\text{form}}(\sqrt{z}),$$

as $z\to\nu$ inside the sectors $\arg t-\pi<\arg z<\arg t+3\pi$ (for $\nu=0$) and $-\pi<\arg z<3\pi$ (for $\nu=\infty$). The monodromy matrix in $Y^{(0)}(ze^{2\pi i})=Y^{(0)}(z)\,M_0$

can be computed using the Stokes matrix connecting unfolded solutions $\tilde{Y}^{(0)}_{1,2}(\xi)$ together with the symmetry properties $G\left(\xi e^{i\pi}\right) = iG(\xi)\,\sigma_x$ and $\sigma_x \tilde{Y}^{(0)}_{\text{form}}(-\xi)\,\sigma_x = \tilde{Y}^{(0)}_{\text{form}}(\xi)$. The result reads

$$M_0 = i\sigma_x S^{-1} = \begin{pmatrix} 0 & i \\ i & -2\cos 2\pi\sigma \end{pmatrix} = U^{-1} e^{2\pi i \mathfrak{S}} U, \tag{2.9a}$$

$$\mathfrak{S} = \left(\sigma + \frac{1}{2}\right)\sigma_z, \qquad U = \frac{1}{\sqrt{2\sin 2\pi\sigma}} \begin{pmatrix} e^{-i\pi\left(\sigma+\frac{1}{4}\right)} & e^{i\pi\left(\sigma+\frac{1}{4}\right)} \\ e^{i\pi\left(\sigma+\frac{1}{4}\right)} & -e^{-i\pi\left(\sigma+\frac{1}{4}\right)} \end{pmatrix}. \tag{2.9b}$$

As expected, the monodromy matrices of $Y^{(0)}$ and $\tilde{Y}^{(0)}_1$ are related by $\tilde{M}_0 = -M_0^2$. In the same way, the monodromy of $Y^{(\infty)}(z)$ around 0 is given by $E^{-1}M_0E = \sigma_x M_0 \sigma_x$.

2.3. Deformation equations and tau function. The usual construction of isomonodromic family of systems (2.2) involves varying the "time" parameter t appearing in the exponentials in (2.3), while keeping the data $\left\{S^{(\nu)}_k\right\}$, E fixed. The latter requirement implies that the matrix $\partial_t \tilde{Y} \cdot \tilde{Y}^{-1}$ is meromorphic on $\mathbb{P}^1$ with poles only possible at 0 and ∞. Analyzing the local behavior of this quantity with the help of expansions of formal solutions and recasting the result in terms of $Y(z)$, one finds that

$$\partial_t Y = B(z)\,Y, \qquad B(z) = \begin{pmatrix} 0 & -q^{-1} \\ -\frac{q}{tz} & 0 \end{pmatrix}.$$

The compatibility of this isomonodromy constraint with the system (2.1) yields the zero-curvature condition $\partial_t A - \partial_z B + [A, B] = 0$, which is equivalent to a pair of scalar equations

$$\begin{cases} tq_t = 2p + q, \\ tp_t = \dfrac{2p^2}{q} + p + q^2 - t, \end{cases} \tag{2.10}$$

or, equivalently, to a single 2nd order ODE:

$$q_{tt} = \frac{q_t^2}{q} - \frac{q_t}{t} + \frac{2q^2}{t^2} - \frac{2}{t}. \tag{2.11}$$

This is the most degenerate Painlevé III equation (of type D_8). In applications, it usually appears in the form of the radial sine-Gordon equation

$$u_{rr} + \frac{u_r}{r} + \sin u = 0, \tag{2.12}$$

which is obtained from (2.11) after the change of variables $q\left(2^{-12}r^4\right) = -2^{-6}r^2 e^{iu(r)}$. The isomonodromic provenance of these equations implies that the quantities (σ, η) introduced above to parameterize the Stokes data provide a pair of conserved quantities for (2.11) and (2.12).

Let us define the tau function $\tau_{\text{III}}(t)$ of PIII (D_8) by the logarithmic derivative

$$\zeta(t) = t\partial_t \ln \tau_{\text{III}} = \frac{(tq_t - q)^2}{4q^2} - q - \frac{t}{q}. \tag{2.13}$$

Conversely, one can express $q = -t\zeta'$. The function $\zeta(t)$ essentially coincides with the time-dependent Hamiltonian of PIII (D_8) and satisfies the equation

$$(t\zeta_{tt})^2 = 4(\zeta_t)^2(\zeta - t\zeta_t) - 4\zeta_t. \tag{2.14}$$

The tau function plays a crucial role in the rest of this note. We are going to express it in terms of monodromy, thereby providing explicit formulas for the general solution of Painlevé III (D_8).

2.4. Riemann-Hilbert problem. It is convenient to replace the linear system (2.2) by an equivalent Riemann-Hilbert problem (RHP). It will be defined by a pair (Γ, J) where Γ is an oriented contour on $\mathbb{P}^1$ and $J : \Gamma \to \mathrm{SL}(2,\mathbb{C})$ is a jump matrix. The relevant contour $\Gamma = \ell^{[0]} \cup \ell^{[\infty]} \cup C_E$ is represented by solid lines in Fig. 2 where it is assumed for simplicity that $\arg t = 0$. The segments $\ell^{[0]}$ and $\ell^{[\infty]}$ correspond to portions of anti-Stokes rays at 0 and ∞. They are close relatives of the cusps of the Chekhov-Mazzocco-Rubtsov geometric confluence diagram [**CM, CMR**] in the Riemann-Hilbert setting.

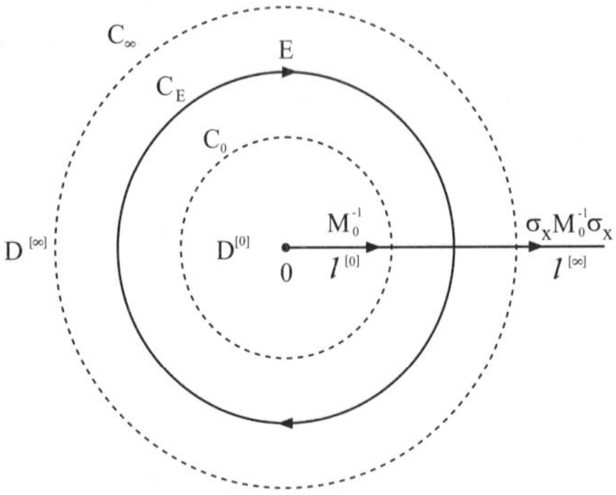

FIGURE 2. Contour Γ of the PIII (D_8) Riemann-Hilbert problem.

The Riemann sphere is decomposed by Γ into 2 connected open domains $D^{[0]}$ and $D^{[\infty]}$. The relevant RHP is to find a 2×2 matrix $\Psi(z)$ holomorphic and invertible inside each of these domains such that

(i) its boundary values on the positive and negative side of Γ satisfy $\Psi_+ = \Psi_- J$, where the piecewise constant jump matrix is given by

$$J\big|_{\ell^{[0]}} = M_0^{-1}, \qquad J\big|_{\ell^{[\infty]}} = \sigma_x M_0^{-1}\sigma_x, \qquad J\big|_{C_E} = E.$$

(ii) at 0 and ∞, the function $\Psi(z)$ behaves as

$$\Psi(z) \simeq G\left(\sqrt{z}\right)\begin{cases} Q\left[\mathbf{1} + O\left(\sqrt{z}\right)\right] e^{2\sigma_z\sqrt{t/z}}, & z \to 0, \\ \left[\mathbf{1} + O\left(\frac{1}{\sqrt{z}}\right)\right] e^{-2\sigma_z\sqrt{z}}, & z \to \infty, \end{cases}$$

where $\arg z \in]-\pi, \pi[$ and Q is a constant invertible matrix such that $[Q, \sigma_x] = 0$, cf (2.3).

The unique solution of this RHP is related to the fundamental solution of the irregular system (2.1) by

$$Y^{(0)}(z) = \begin{cases} \Psi(z), & z \in D^{[0]}, \\ \Psi(z)\, E^{-1}, & z \in D^{[\infty]}. \end{cases}$$

Let $\mathcal{A}$ be the open annulus bounded by two circles C_0 and C_∞ as shown in Fig. 2. Consider a piecewise analytic function $\hat{\Psi}(z)$ defined by

$$\hat{\Psi}(z) = \begin{cases} \Psi(z), & z \notin \bar{\mathcal{A}}, \\ Y^{(0)}(z)\, U^{-1} z^{-\mathfrak{S}}, & z \in \mathcal{A}, \end{cases} \tag{2.15}$$

where $\mathfrak{S} = \left(\sigma + \frac{1}{2}\right)\sigma_z$ and U are defined by (2.9b). This new function solves a Riemann-Hilbert problem defined by the pair $\left(\hat{\Gamma}, \hat{J}\right)$, where the contour $\hat{\Gamma}$ and the relevant jump matrices are represented in Fig. 3. The transformed RHP is of course equivalent to the initial one. The function $\hat{\Psi}(z)$ has been designed so that it has no jumps inside $\mathcal{A}$ and coincides with $\Psi(z)$ inside C_0 and outside C_∞. Cancellation of the jumps inside $\mathcal{A}$ can only be done at the expense of introducing new jumps on the circles C_0 and C_∞; as we will see in a moment, their choice above models regular singularities at ∞ and 0, respectively.

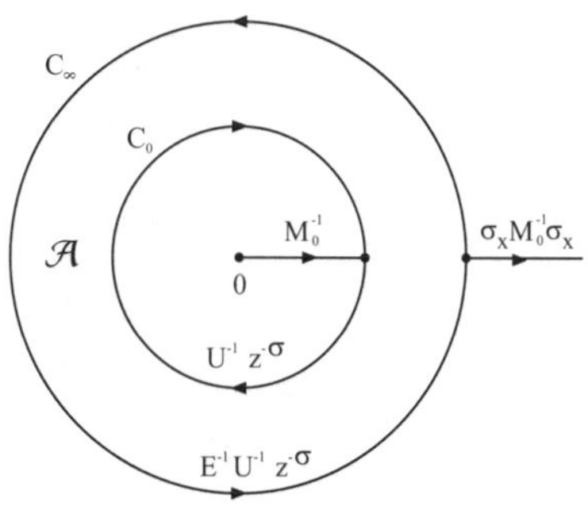

FIGURE 3. Contour $\hat{\Gamma}$ and associated jumps.

There is a natural decomposition $\hat{\Gamma} = \hat{\Gamma}^{[0]} \cup \hat{\Gamma}^{[\infty]}$, where $\hat{\Gamma}^{[0]}$ (and $\hat{\Gamma}^{[\infty]}$) consist of C_0 (resp. C_∞) and the part of the positive real axis contained inside C_0 (resp. outside C_∞). Denoting

$$\hat{J}^{[0]} = \hat{J}\Big|_{\hat{\Gamma}^{[0]}}, \qquad \hat{J}^{[\infty]} = \hat{J}\Big|_{\hat{\Gamma}^{[\infty]}},$$

we can assign to the original RHP two simpler RHPs for functions $\hat{\Psi}^{[0]}(z)$ and $\hat{\Psi}^{[\infty]}(z)$ defined by the pairs $\left(\hat{\Gamma}^{[0]}, \hat{J}^{[0]}\right)$ and $\left(\hat{\Gamma}^{[\infty]}, \hat{J}^{[\infty]}\right)$. The latter correspond to two rank 2 Fuchsian systems having one regular singular point and one irregular singular point of Poincaré rank $\frac{1}{2}$ which can be expicitly solved in terms of Bessel functions.

Let us also remark that (i) only $\hat{\Psi}^{[0]}(z)$ depends on PIII (D_8) independent variable t (via the asymptotic condition at $z = 0$); (ii) the initial RHP may also be rewritten as a RHP on a single circle inside $\mathcal{A}$ with the jump $\hat{\Psi}^{[0]}(z)\,\hat{\Psi}^{[\infty]}(z)^{-1}$. The study of an equivalent RHP on a circle is the main tool used in [**Nil**] for the asymptotic analysis of PIII (D_8).

2.5. Building block solutions. Consider a model differential system

$$\partial_z Y^{[\infty]} = A^{[\infty]}(z)\, Y^{[\infty]}, \qquad A^{[\infty]}(z) = -\sigma_+ + \begin{pmatrix} \sigma + \frac{1}{2} & 0 \\ -1 & -\sigma - \frac{1}{2} \end{pmatrix} z^{-1}. \tag{2.16}$$

Such an ansatz is inspired by the following: we would like to have an irregular singularity of Poincaré rank $\frac{1}{2}$ at $z=\infty$ and a regular singularity at $z=0$ with local monodromy exponents given by $\pm\left(\sigma+\frac{1}{2}\right)$, cf (2.9a).

Choose the fundamental matrix solution of (2.16) as

(2.17)
$$Y_{\infty}^{[\infty]}(z)=\frac{1}{i\sqrt{2\pi}}\begin{pmatrix} 2\sqrt{z}\,K_{2\sigma}\left(2\sqrt{z}\right) & 2\pi\sqrt{z}\,I_{2\sigma}\left(2\sqrt{z}\right) \\ 2K_{2\sigma+1}\left(2\sqrt{z}\right) & -2\pi I_{2\sigma+1}\left(2\sqrt{z}\right)\end{pmatrix}\begin{pmatrix} 1 & -ie^{2\pi i\sigma} \\ 0 & 1\end{pmatrix},$$

where $I_\nu(x)$, $K_\nu(x)$ denote the modified Bessel functions of the 1st and 2nd kind. This fundamental solution is defined in the domain $\arg z\in]0,2\pi[$ where it has the asymptotics

$$Y_{\infty}^{[\infty]}(z)\simeq G\left(\sqrt{z}\right)\left[\mathbf{1}+\sum_{k=1}^{\infty}y_k^{[\infty]}z^{-\frac{k}{2}}\right]e^{-2\sigma_z\sqrt{z}},\qquad z\to\infty.$$

In the vicinity of $z=0$, it becomes convenient to rewrite it as

(2.18a)
$$Y_{\infty}^{[\infty]}(z)=Y_0^{[\infty]}(z)\,e^{2\pi i\eta\sigma_z}z^{\mathfrak{S}}UE,$$
(2.18b)
$$Y_0^{[\infty]}(z)=i\sqrt{\frac{\pi}{\sin 2\pi\sigma}}\begin{pmatrix} z^{-\sigma}I_{2\sigma}\left(2\sqrt{z}\right) & -z^{\sigma+1}I_{-2\sigma}\left(2\sqrt{z}\right) \\ -z^{-\sigma-\frac{1}{2}}I_{2\sigma+1}\left(2\sqrt{z}\right) & z^{\sigma+\frac{1}{2}}I_{-2\sigma-1}\left(2\sqrt{z}\right)\end{pmatrix}e^{-i\pi\left(\sigma-\frac{1}{4}\right)\sigma_z}.$$

In fact, we used in (2.17) the same σ as in the parameterization of Stokes data precisely to achieve (2.18a). The matrix function $Y_0^{[\infty]}(z)$ is holomorphic and invertible in the entire complex plane, and normalized as to have unit determinant. Therefore, the solution $\hat{\Psi}^{[\infty]}(z)$ of the exterior auxiliary RHP may be written as

$$\hat{\Psi}^{[\infty]}(z)=\begin{cases} Y_{\infty}^{[\infty]}(z), & z \text{ outside } C_\infty, \\ Y_0^{[\infty]}(z)\,e^{2\pi i\eta\sigma_z}, & z \text{ inside } C_\infty.\end{cases}\tag{2.19}$$

Similarly, the function

$$Y_0^{[0]}(z)=\frac{1}{i\sqrt{2\pi}}\begin{pmatrix} 2\pi I_{-2\sigma-1}\left(\frac{2}{\sqrt{z}}\right) & 2K_{-2\sigma-1}\left(\frac{2}{\sqrt{z}}\right) \\ \frac{2\pi}{\sqrt{z}}I_{-2\sigma}\left(\frac{2}{\sqrt{z}}\right) & -\frac{2}{\sqrt{z}}K_{-2\sigma}\left(\frac{2}{\sqrt{z}}\right)\end{pmatrix}\begin{pmatrix} 1 & 0 \\ -ie^{2\pi i\sigma} & 1\end{pmatrix}\tag{2.20}$$

defines a fundamental matrix solution of the linear system

$$\partial_z Y^{[0]}=A^{[0]}(z)\,Y^{[0]},\qquad A^{[0]}(z)=-\sigma_- z^{-2}+\begin{pmatrix}\sigma+\frac{1}{2} & -1 \\ 0 & -\sigma-\frac{1}{2}\end{pmatrix}z^{-1}.\tag{2.21}$$

It is characterized by the asymptotic behavior

$$Y_0^{[0]}(z)\simeq G\left(\sqrt{z}\right)\left[\mathbf{1}+\sum_{k=1}^{\infty}y_k^{[0]}z^{\frac{k}{2}}\right]e^{2\sigma_z/\sqrt{z}},\tag{2.22}$$

as $z \to 0$ inside the sector $\arg z \in]0, 2\pi[$. In the neighborhood of $z = \infty$, this model solution $Y_0^{[0]}(z)$ can be suitably rewritten as

(2.23a)

$$Y_0^{[0]}(z) = Y_\infty^{[0]}(z)\, z^{\mathfrak{S}} U,$$

(2.23b)

$$Y_\infty^{[0]}(z) = i\sqrt{\frac{\pi}{\sin 2\pi\sigma}} \begin{pmatrix} z^{-\sigma-\frac{1}{2}} I_{-2\sigma-1}\left(\frac{2}{\sqrt{z}}\right) & z^{\sigma+\frac{1}{2}} I_{2\sigma+1}\left(\frac{2}{\sqrt{z}}\right) \\ z^{-\sigma-1} I_{-2\sigma}\left(\frac{2}{\sqrt{z}}\right) & z^{\sigma} I_{2\sigma}\left(\frac{2}{\sqrt{z}}\right) \end{pmatrix} e^{-i\pi\left(\sigma+\frac{1}{4}\right)\sigma_z},$$

Taking into account that the matrix ratio $G\left(\sqrt{z}\right)^{-1} G\left(\sqrt{\frac{z}{t}}\right) = t^{-\frac{\sigma_x}{4}}$ is independent of z, the solution $\hat{\Psi}^{[0]}(z)$ of the interior auxiliary RHP may now be expressed as

$$\hat{\Psi}^{[0]}(z) = \begin{cases} Y_0^{[0]}\left(\frac{z}{t}\right), & z \text{ inside } C_0, \\ Y_\infty^{[0]}\left(\frac{z}{t}\right) t^{-\mathfrak{S}}, & z \text{ outside } C_0. \end{cases} \tag{2.24}$$

The parameterization of Stokes data introduced in Subsection 2.2 now becomes more transparent. The variable σ encodes the spectrum of the single nontrivial monodromy matrix M_0 whose eigenvalues are given by $-e^{\pm 2\pi i\sigma}$, cf (2.9a). The 2nd parameter η measures a relative twist of local parametrices $Y_0^{[\infty]}(z)$, $Y_\infty^{[0]}\left(\frac{z}{t}\right)$ in the full solution $\hat{\Psi}(z)$.

3. Fredholm determinant representation

3.1. Boundary spaces. Let $V(C)$ be the space of smooth functions on a circle C which will be sometimes identified with the space of holomorphic functions in an annulus containing C. Also, define the space $H(C) = \mathbb{C}^2 \otimes V(C)$ whose elements will be represented as 2-rows of elements of $V(C)$. The subspaces of $V(C)$ and $H(C)$ that consist of functions with only positive or negative Fourier modes will be denoted by $V_\pm(C)$ and $H_\pm(C)$.

In relation with the previously discussed RHP for the function $\hat{\Psi}(\xi)$, introduce the spaces

$$\mathcal{H} = \mathcal{H}_+ \oplus \mathcal{H}_-, \qquad \mathcal{H}_\pm = H_\pm(C_0) \oplus H_\mp(C_\infty). \tag{3.1}$$

Observe that each of the subspaces $\mathcal{H}_\pm$ can be identified in a natural way with the space of vector-valued holomorphic functions on the annulus $\mathcal{A}$. We are now going to consider two operators acting on $\mathcal{H}$ *from the right* and generalizing the usual projections on positive and negative modes.

(1) The first operator, to be denoted by $\mathcal{P}_\Sigma$, is defined by

$$(f\mathcal{P}_\Sigma)(z) = \frac{1}{2\pi i}\int_{C_0\cup C_\infty} \frac{f(z')\,\hat{\Psi}_+(z')^{-1}\hat{\Psi}_+(z)\,dz'}{z-z'}, \qquad z \in C_0 \cup C_\infty.$$

The convention used to interpret the singularity at $z' = z$ is to slightly deform the integration contour so that it goes clockwise around this point.

(2) The second operator, $\mathcal{P}_\oplus$, is constructed in a similar way with the help of elementary building block solutions $\hat{\Psi}^{[0]}(z)$ and $\hat{\Psi}^{[\infty]}(z)$,

$$(f\mathcal{P}_\oplus)(z) = \frac{1}{2\pi i}\int_{C_k} \frac{f(z')\,\hat{\Psi}_+^{[k]}(z')^{-1}\,\hat{\Psi}_+^{[k]}(z)\,dz'}{z-z'}, \qquad z\in C_k, \quad k=0,\infty.$$

The absence of jumps of $\hat{\Psi}(z)\,\hat{\Psi}_+^{[0]}(z)^{-1}$ (and $\hat{\Psi}(z)\,\hat{\Psi}_+^{[\infty]}(z)^{-1}$) inside C_0 (resp. outside C_∞), and systematic application of residue theorem/collapsing the contours imply the following properties:

- $\mathcal{P}_\Sigma^2 = \mathcal{P}_\Sigma$ and $\mathcal{P}_\oplus^2 = \mathcal{P}_\oplus$, i.e. the operators $\mathcal{P}_\Sigma$, $\mathcal{P}_\oplus$ are projections.
- $\ker \mathcal{P}_\oplus = \mathcal{H}_-$, $\ker \mathcal{P}_\Sigma \supseteq \mathcal{H}_\mathcal{A}$, where $\mathcal{H}_\mathcal{A}$ is the space of boundary values of functions holomorphic on $\mathcal{A}$.
- $\mathcal{P}_\Sigma\mathcal{P}_\oplus = \mathcal{P}_\Sigma$, $\mathcal{P}_\oplus\mathcal{P}_\Sigma = \mathcal{P}_\oplus$; this means that $\mathcal{P}_\Sigma$ and $\mathcal{P}_\oplus$ have the same range, to be denoted by $\mathcal{H}_\mathcal{T}$.

Loosely speaking, the space $\mathcal{H}_\mathcal{T}$ consists of functions on $C_0\cup C_\infty$ whose continuations outside $\mathcal{A}$ share the global monodromy properties of the fundamental matrix solution of (2.2). The operators $\mathcal{P}_\Sigma$, $\mathcal{P}_\oplus$ project on $\mathcal{H}_\mathcal{T}$ along $\mathcal{H}_\mathcal{A}$ and $\mathcal{H}_-$, respectively, which may be denoted as $\mathcal{P}_\Sigma = \mathcal{H} \xrightarrow{\mathcal{H}_\mathcal{A}} \mathcal{H}_\mathcal{T}$, $\mathcal{P}_\oplus = \mathcal{H} \xrightarrow{\mathcal{H}_-} \mathcal{H}_\mathcal{T}$.

According to the decomposition (3.1), write $f\in\mathcal{H}$ as

$$f = \begin{pmatrix} f_+^{[0]} & f_-^{[\infty]} \end{pmatrix} \oplus \begin{pmatrix} f_-^{[0]} & f_+^{[\infty]} \end{pmatrix}.$$

The action of $\mathcal{P}_\oplus$ is then given by

$$f\mathcal{P}_\oplus = \begin{pmatrix} f_+^{[0]} & f_-^{[\infty]} \end{pmatrix} \oplus \begin{pmatrix} f_+^{[0]}\mathsf{d} & f_-^{[\infty]}\mathsf{a} \end{pmatrix}, \tag{3.2}$$

where the matrix integral operators $\mathsf{a}: H_-(C_\infty)\to H_+(C_\infty)$ and $\mathsf{d}: H_+(C_0)\to H_-(C_0)$ are expressed in terms of elementary solutions $\hat{\Psi}^{[0]}(z)$, $\hat{\Psi}^{[\infty]}(z)$ and have integrable form:

(3.3)

$$(f\mathsf{a})(z) = \frac{1}{2\pi i}\oint_{C_\infty} f(z')\,\mathsf{a}(z',z)\,dz', \qquad (f\mathsf{d})(z) = -\frac{1}{2\pi i}\oint_{C_0} f(z')\,\mathsf{d}(z',z)\,dz',$$

$$\mathsf{a}(z',z) = \frac{\hat{\Psi}_+^{[\infty]}(z')^{-1}\,\hat{\Psi}_+^{[\infty]}(z) - \mathbf{1}}{z-z'}, \qquad \mathsf{d}(z',z) = \frac{\mathbf{1} - \hat{\Psi}_+^{[0]}(z')^{-1}\,\hat{\Psi}_+^{[0]}(z)}{z-z'}.$$

The minus sign is introduced into the definition of $\mathsf{d}(z',z)$ to absorb the opposite orientation of the circles $C_{0,\infty}$ in some computations below. Let us note in passing that the action of a and d may be extended from the boundary circles to vector-valued functions holomorphic on $\mathcal{A}$.

The result (3.2) suggests that $f_+^{[0]}$, $f_-^{[\infty]}$ provide convenient coordinates on $\mathcal{H}_\mathcal{T}$. We are going to use this basis to describe the operator $\mathcal{P}_\Sigma$ involving the solution $\hat{\Psi}(\xi)$. Given $f\in\mathcal{H}$, write $f = g+h$ with $g\in\mathcal{H}_\mathcal{T}$ and $h\in\mathcal{H}_\mathcal{A}$. These conditions translate into

$$g = \begin{pmatrix} g_+^{[0]} & g_-^{[\infty]} \end{pmatrix} \oplus \begin{pmatrix} g_+^{[0]}\mathsf{d} & g_-^{[\infty]}\mathsf{a} \end{pmatrix}, \qquad h = \begin{pmatrix} h_+^{[0]} & h_-^{[\infty]} \end{pmatrix} \oplus \begin{pmatrix} h_-^{[\infty]} & h_+^{[0]} \end{pmatrix}$$

Expressing $h_+^{[0]}$, $h_-^{[\infty]}$ in terms of $g_+^{[0]}$, $g_-^{[\infty]}$, one obtains the equation

$$\begin{pmatrix} g_+^{[0]} & g_-^{[\infty]} \end{pmatrix}(\mathbf{1}-K) = \begin{pmatrix} f_+^{[0]} - f_+^{[\infty]} & f_-^{[\infty]} - f_-^{[0]} \end{pmatrix}, \qquad K = \begin{pmatrix} 0 & \mathsf{a} \\ \mathsf{d} & 0 \end{pmatrix}.$$

Below we assume invertibility of $\mathbf{1}-K$, which ensures the existence of a unique splitting $\mathcal{H}=\mathcal{H}_{\mathcal{T}}\oplus\mathcal{H}_{\mathcal{A}}$. Computing the action of $\mathcal{P}_\Sigma$ on $\mathcal{H}$ (essentially equivalent to solving the original RHP) thereby amounts to finding the inverse $(\mathbf{1}-K)^{-1}$.

Consider the restrictions $\mathcal{P}_{\oplus,+}=\mathcal{P}_\oplus\big|_{\mathcal{H}_+}$, $\mathcal{P}_{\Sigma,+}=\mathcal{P}_\Sigma\big|_{\mathcal{H}_+}$. We have just seen that in the previously described basis $\mathcal{P}_{\oplus,+}$ is given by the identity matrix, whereas $\mathcal{P}_{\Sigma,+}$ coincides with $(\mathbf{1}-K)^{-1}$.

DEFINITION 3.1. The tau function of the Riemann-Hilbert problem for $\hat{\Psi}(\xi)$ is defined as

$$\tau(t)=\det\left(\mathcal{H}_+\xrightarrow{\mathcal{H}_-}\mathcal{H}_{\mathcal{T}}\xrightarrow{\mathcal{H}_{\mathcal{A}}}\mathcal{H}_+\right)=\det\left(\mathcal{P}_{\oplus,+}\mathcal{P}_{\Sigma,+}{}^{-1}\right)=\det(\mathbf{1}-K). \tag{3.4}$$

The first two expressions of $\tau(t)$ are "coordinate-free" while the last Fredholm determinant corresponds to the choice of a specific basis. Our next task is to understand the relation between the last definition and the tau function of Painlevé III (D_8) equation introduced in (2.13).

3.2. Relation to $\tau_{\mathrm{III}}(t)$. Let t_0 be a constant parameter close to Painlevé III (D_8) independent variable t and consider the ratio

$$\begin{aligned}\frac{\tau(t)}{\tau(t_0)}&=\det\left(\mathcal{H}_+\xrightarrow{\mathcal{H}_-}\mathcal{H}_{\mathcal{T}}(t)\xrightarrow{\mathcal{H}_{\mathcal{A}}}\mathcal{H}_+\xrightarrow{\mathcal{H}_{\mathcal{A}}}\mathcal{H}_{\mathcal{T}}(t_0)\xrightarrow{\mathcal{H}_-}\mathcal{H}_+\right)\\&=\det\left(\mathcal{H}_{\mathcal{T}}(t_0)\xrightarrow{\mathcal{H}_-}\mathcal{H}_{\mathcal{T}}(t)\xrightarrow{\mathcal{H}_{\mathcal{A}}}\mathcal{H}_{\mathcal{T}}(t_0)\right)\\&=\det\left(\mathcal{P}_\oplus(t)\big|_{\mathcal{H}_{\mathcal{T}}(t_0)}\mathcal{P}_\Sigma(t_0)\big|_{\mathcal{H}_{\mathcal{T}}(t)}\right).\end{aligned}$$

Since for $\nu=\oplus,\Sigma$ we can express the inverses as $\left(\mathcal{P}_\nu(t)\big|_{\mathcal{H}_{\mathcal{T}}(t_0)}\right)^{-1}=\mathcal{P}_\nu(t_0)\big|_{\mathcal{H}_{\mathcal{T}}(t)}$, the logarithmic derivative of $\tau(t)$ may be written as

$$\begin{aligned}&-\partial_t\ln\tau(t)\\&=\mathrm{Tr}_{\mathcal{H}_{\mathcal{T}}(t_0)}\left\{\mathcal{P}_\oplus(t)\big|_{\mathcal{H}_{\mathcal{T}}(t_0)}\mathcal{P}_\Sigma(t_0)\big|_{\mathcal{H}_{\mathcal{T}}(t)}\partial_t\left(\mathcal{P}_\Sigma(t)\big|_{\mathcal{H}_{\mathcal{T}}(t_0)}\mathcal{P}_\oplus(t_0)\big|_{\mathcal{H}_{\mathcal{T}}(t)}\right)\right\}\\&=\mathrm{Tr}_{\mathcal{H}}\left\{\mathcal{P}_\oplus(t)\,\mathcal{P}_\Sigma(t_0)\,\partial_t\mathcal{P}_\Sigma(t)\,\mathcal{P}_\oplus(t_0)\right\}\\&=\mathrm{Tr}_{\mathcal{H}}\left\{\mathcal{P}_\oplus(t)\,\partial_t\mathcal{P}_\Sigma(t)\right\}.\end{aligned}\tag{3.5}$$

Here the middle line is obtained by using that $\operatorname{ran}\mathcal{P}_\nu(t)=\mathcal{H}_{\mathcal{T}}(t)$. The last line follows from the transversality of $\mathcal{H}_{\mathcal{T}}(t)$ and $\mathcal{H}_{\mathcal{A}}$ (as well as $\mathcal{H}_{\mathcal{T}}(t)$ and $\mathcal{H}_-$) in $\mathcal{H}$, which implies that

$$\left(\mathcal{H}\xrightarrow{\mathcal{H}_{\mathcal{T}}(t_0)}\mathcal{H}_{\mathcal{A}}\xrightarrow{\mathcal{H}_{\mathcal{A}}}\mathcal{H}_{\mathcal{T}}(t)\right)=\left(\mathcal{H}\xrightarrow{\mathcal{H}_{\mathcal{T}}(t_0)}\mathcal{H}_-\xrightarrow{\mathcal{H}_-}\mathcal{H}_{\mathcal{T}}(t)\right)=0,$$

i.e. the corresponding compositions of projections are equal to zero.

The next task is to compute the trace in the right side of (3.5). Collapsing the contours and computing residues as in Step 2 of the proof of Theorem 2.9 in [**GL16**], we arrive at

$$\begin{aligned}&\mathrm{Tr}_{\mathcal{H}}\left\{\mathcal{P}_\oplus(t)\,\partial_t\mathcal{P}_\Sigma(t)\right\}\\&=\sum_{\nu=0,\infty}\frac{1}{2\pi i}\oint_{C_\nu}\mathrm{Tr}\left\{\partial_z\left(\hat{\Psi}_+(z)\,\hat{\Psi}_+^{[\nu]}(z)^{-1}\right)\hat{\Psi}_+^{[\nu]}(z)\,\partial_t\left(\hat{\Psi}_+(z)^{-1}\right)\right\}dz.\end{aligned}$$

Recall that $\hat{\Psi}$ has the same jumps as $\hat{\Psi}^{[0]}$ inside C_0 and as $\hat{\Psi}^{[\infty]}$ outside C_∞. Therefore the "+"-indices in the above expression are redundant, the contours $C_{0,\infty}$ can be replaced by small circles around 0 and ∞, and the resulting integrals may be computed by residues. On these circles, the integrand may be represented by series involving only integer (but not half-integer) powers of z, which can be shown using once again the symmetry properties such as $G\left(\xi e^{i\pi}\right) = iG\left(\xi\right)\sigma_x$ and $\sigma_x \tilde{Y}^{(0)}_{\text{form}}\left(-\xi\right)\sigma_x = \tilde{Y}^{(0)}_{\text{form}}\left(\xi\right)$. Furthermore, the series at ∞ has the form $\sum_{k\geq 0} f_k z^{-k-2}$, hence the corresponding residue vanishes.

On the other hand, the residue at 0 reads (note the negative orientation of C_0)

$$\frac{\left(y_1^{(0)}\right)_{11} - \left(y_1^{(0)}\right)_{12}}{\sqrt{t}} - \frac{\left(y_1^{[0]}\right)_{11} - \left(y_1^{[0]}\right)_{12}}{t},$$

where $y_1^{(0)}$ is the first nontrivial coefficient of the formal solution (2.3a) and $y_1^{[0]}$ is its counterpart in the expansion (2.22) of the model solution $Y_0^{[0]}\left(z\right)$. The former quantity is explicitly given by (2.4), while the latter is readily deduced from (2.20):

$$y_1^{[0]} = \left(\sigma + \frac{1}{4}\right)\frac{i\sigma_y}{2} - \left(\sigma + \frac{1}{4}\right)^2 \sigma_z.$$

Combining the two results with (3.5) yields

$$\partial_t \ln \tau\left(t\right) = -\operatorname{Tr}_{\mathcal{H}}\left\{\mathcal{P}_{\oplus}\left(t\right)\partial_t \mathcal{P}_{\Sigma}\left(t\right)\right\} = \frac{1}{t}\left[\frac{p^2}{q^2} - q - \frac{t}{q} - \left(\sigma + \frac{1}{2}\right)^2\right].$$

The Fredholm determinant $\tau\left(t\right)$ from the Definition 3.1 may therefore be identified with the usual Painlevé III (D_8) tau function $\tau_{\text{III}}\left(t\right)$ defined by (2.13):

$$\tau_{\text{III}}\left(t\right) = \text{const} \cdot t^{\left(\sigma+\frac{1}{2}\right)^2} \tau\left(t\right). \tag{3.6}$$

In combination with explicit solutions (2.19), (2.24) of auxiliary RHPs which appear in the definition (3.3) of operators a and d, this yields the following result.

THEOREM 3.2. *Let $(\sigma, \eta) \in \mathbb{C}^2$ with $\sigma \notin \mathbb{Z}/2$ be the coordinates on the generic stratum of the space of the Stokes data of the linear system (2.1), introduced in Subsection 2.2. The corresponding Painlevé III (D_8) tau function $\tau_{\text{III}}\left(t\right) = \tau_{\text{III}}\left(t\,|\,\sigma,\eta\right)$ can be expressed as Fredholm determinant*

$$\tau_{\text{III}}\left(t\right) = \text{const} \cdot t^{\left(\sigma+\frac{1}{2}\right)^2} \det\left(\mathbf{1} - K\right), \qquad K = \begin{pmatrix} 0 & \mathsf{a} \\ \mathsf{d} & 0 \end{pmatrix}. \tag{3.7}$$

Here the operators a, d act on vector-valued functions $f \in H\left(C\right)$ on a circle C centered at the origin and oriented counterclockwise,

$$\left(f\mathsf{a}\right)\left(z\right) = \frac{1}{2\pi i}\oint_C f\left(z'\right)\mathsf{a}\left(z',z\right)dz', \qquad \left(f\mathsf{d}\right)\left(z\right) = \frac{1}{2\pi i}\oint_C f\left(z'\right)\mathsf{d}\left(z',z\right)dz', \tag{3.8}$$

and the integral kernels $\mathsf{a}\left(z',z\right)$, $\mathsf{d}\left(z',z\right)$ are explicitly given by

$$\mathsf{a}\left(z',z\right) = e^{i\pi(\sigma-2\eta)\sigma_z}\,\frac{\mathbb{J}_\sigma\left(z',z\right) - \mathbf{1}}{z - z'}\,e^{i\pi(2\eta-\sigma)\sigma_z}, \tag{3.9a}$$

$$\mathsf{d}\left(z',z\right) = t^{\mathfrak{S}} e^{i\pi\sigma\sigma_z}\sigma_y\,\frac{\mathbf{1} - \mathbb{J}_\sigma\left(\frac{t}{z'},\frac{t}{z}\right)}{z - z'}\,\sigma_y e^{-i\pi\sigma\sigma_z} t^{-\mathfrak{S}}, \tag{3.9b}$$

(3.9c) $$\frac{\sin 2\pi\sigma}{\pi}\mathbb{J}_\sigma(z',z) = \begin{pmatrix} z'j_{\sigma+\frac12}(z)j_{-\sigma}(z') - j_\sigma(z)j_{-\sigma-\frac12}(z') & iz'j_{-\sigma-\frac12}(z)j_{-\sigma}(z') - izj_{-\sigma}(z)j_{-\sigma-\frac12}(z') \\ ij_{\sigma+\frac12}(z)j_\sigma(z') - ij_\sigma(z)j_{\sigma+\frac12}(z') & zj_{-\sigma}(z)j_{\sigma+\frac12}(z') - j_{-\sigma-\frac12}(z)j_\sigma(z') \end{pmatrix},$$

with $j_\sigma(z) = z^{-\sigma}I_{2\sigma}(2\sqrt{z}) = \dfrac{{}_0F_1(2\sigma+1;z)}{\Gamma(2\sigma+1)}$ *and* $\mathfrak{S} = \left(\sigma+\frac12\right)\sigma_z$.

REMARK 3.3. The kernels $\mathsf{a}(z',z)$, $\mathsf{d}(z',z)$ are not singular at $z=z'$. That $j_\sigma(z)$ are holomorphic in the entire complex plane is a signature of the fact that $\operatorname{ran}\mathsf{a} \subseteq H_+(C) \subseteq \ker\mathsf{a}$ and $\operatorname{ran}\mathsf{d} \subseteq H_-(C) \subseteq \ker\mathsf{d}$. This implies, in particular, that $\mathsf{a}^2 = \mathsf{d}^2 = 0$ and $\operatorname{Tr}\mathsf{a}(\mathsf{da})^k = \operatorname{Tr}\mathsf{d}(\mathsf{ad})^k = 0$ for $k\in\mathbb{Z}_{\geq 0}$. It follows that $\operatorname{Tr}(\mathsf{a}+\mathsf{d})^{2k+1} = 0 = \operatorname{Tr}K^{2k+1}$ and $\operatorname{Tr}(\mathsf{a}+\mathsf{d})^{2k} = 2\operatorname{Tr}(\mathsf{ad})^k = \operatorname{Tr}K^{2k}$. The Fredholm determinant may therefore be rewritten as

$$\det(\mathbf{1}-K) = \det(\mathbf{1}+\mathsf{a}+\mathsf{d}).$$

The latter form may seem more compact while the integral kernel of $\mathsf{a}+\mathsf{d}$ is still integrable. However it turns out to be beneficial for our purposes to work with the block structure of K in (3.7).

REMARK 3.4. Let us note that the tau function (2.13) differs from [**GIL13**, Eq. (2.14)] or [**ILT14**, Eq. (2.15)] by a $\mathbb{Z}_2$-Bäcklund transformation. This discrete symmetry becomes most explicit at the level of the sine-Gordon equation (2.12) where it corresponds to the mapping $u\mapsto -u$. The relevant monodromy parameters transform as $\sigma\mapsto\frac12-\sigma$, $\eta\mapsto-\eta$ which should be taken into account before comparing (3.7)–(3.9) with (1.2). An interesting representation-theoretic interpretation of this symmetry has been recently suggested in [**BSh2**].

REMARK 3.5. The monodromy data $\{S_k^{(\nu)}\}$, E in (2.7) are invariant with respect to integer shifts $\sigma\mapsto\sigma+1$. The Painlevé III (D_8) tau function should thus be quasi-periodic in σ, namely,

$$\tau_{\rm III}(t\,|\,\sigma+1,\eta) = \mathrm{const}\cdot\tau_{\rm III}(t\,|\,\sigma,\eta),$$

where the constant expression depends on the choice of normalization of $\tau_{\rm III}(t\,|\,\sigma,\eta)$. This quasi-periodicity is not obvious at all at the level of Fredholm determinant representation (3.7)–(3.9) but will be made manifest in the next section.

Upon truncation of the Taylor expansion of the right side of (3.9b) in t, the operator d becomes finite rank so that the corresponding $t\to 0$ asymptotics of $\tau_{\rm III}(t)$ is given by a finite determinant, cf [**GL16**, Theorem 2.11]. From the point of view of this asymptotic analysis, the most efficient choice of σ is to set $-1<\Re\sigma\leq 0$. The leading asymptotic terms obtained by such procedure coincide with the known results [**Jim, IN, Nov, FIKN, Nil**]. It is an instructive exercise to check that the subleading asymptotic terms derived from the Fredholm determinant reproduce [**ILT14**, Eqs. (3.3)–(3.5)].

REMARK 3.6. The Painlevé III (D_8) isomonodromic RHP is usually formulated in the literature for the unfolded system (2.2), see e.g. [**FIKN, IP, Nil**]. While such formulation has a number of technical advantages, the correspondence with CMR confluence diagram is not manifest therein. Furthermore, the analog of the Fredholm determinant (3.4) does *not* coincide with the tau function (2.13). Instead, it gives the tau function of a special case of PIII (D_6) equation ($N_f = 2$ in the gauge theory language), related to PIII (D_8) by a quadratic transformation.

REMARK 3.7. As mentioned in the Introduction, there exists Fredholm determinant representation of PIII (D_8) solutions corresponding to special initial conditions [**MTW**, **Z94**]. The statement is as follows. Consider the spectral determinant

$$D(t|\lambda) := \det\left(\mathbf{1} + \lambda\tilde{K}\right), \tag{3.10a}$$

where $\tilde{K}$ is an integral operator acting on $L^2(\mathbb{R}_{>0})$ with the kernel

$$\tilde{K}(x,y) = \frac{\exp\left\{-\frac{r}{4}\left(x + x^{-1} + y + y^{-1}\right)\right\}}{x+y}, \qquad t = 2^{-12}r^4. \tag{3.10b}$$

This determinant can be written as

$$D(t|\pm\lambda) = \exp\left\{\pm\frac{\psi(r)}{2} + \frac{1}{4}\int_r^\infty s\left[\sinh^2\psi(s) - (\psi_s(s))^2\right]ds\right\}, \tag{3.11}$$

where $\psi(r)$ solves the radial sinh-Gordon equation

$$\psi_{rr} + \frac{\psi_r}{r} = \frac{1}{2}\sinh 2\psi,$$

subject to boundary condition $\psi(r \to +\infty) \simeq 2\lambda K_0(r)$. It becomes equivalent to Painlevé III (D_8) in the form (2.12) after identification $u(r) = -2i\psi(r) + \pi$. It is also known [**MTW**] that for $0 < \pi\lambda < 1$ the solution $\psi(r)$ has asymptotics

$$\psi\left(r \to 0^+\right) \simeq -\bar{\sigma}\ln\frac{r}{8} + \ln\frac{\Gamma\left(\frac{1+\bar{\sigma}}{2}\right)}{\Gamma\left(\frac{1-\bar{\sigma}}{2}\right)} + o(1), \tag{3.12}$$

with $\bar{\sigma} = \frac{2}{\pi}\arcsin\pi\lambda$. The representation (3.11) implies that

$$\tau_\pm(t) := t^{\frac{1}{16}}e^{-4\sqrt{t}}D(t|\pm\lambda)$$

are Painlevé III (D_8) tau functions related by Bäcklund transformation mentioned in Remark 3.4. The asymptotics (3.12) allows to identify the relevant monodromy parameters as $\sigma_\pm = -\frac{1\pm\bar{\sigma}}{4}$, $\eta_\pm = 0$.

It is natural to wonder whether it is possible to deduce the determinant (3.10) by specializing the general solution (3.7)–(3.9). We believe there is no direct way to do this. The two determinants differ by a factor depending on t (in a simple way) and also involving rather nontrivial connection constant computed in [**T91**]. They are well-adapted to the description of different asymptotic regimes ($t \to 0$ and $t \to +\infty$), which suggests that there may exist another Fredholm determinant representation of the general solution associated to different decomposition of the PIII (D_8) Riemann-Hilbert problem into "irregular pants". Further technical evidence supporting this hypothesis is that the value $\eta = 0$ has no particular meaning in the short-distance regime but corresponds to highly non-generic long-distance asymptotics of $\tau_{\text{III}}(t)$, exhibiting exponential decay instead of exponential growth. Let us finally mention that $\tilde{K}^2$ is related to the scalar integrable Macdonald kernel by a similarity transformation [**GIL13**, Subsection 5.2.1].

4. Series over Young diagrams

4.1. Cauchy matrix representations. Let us now express the operators a and d from (3.8) in the basis of Fourier modes, where they are given by semi-infinite matrices. Denoting $\mathbb{Z}'=\mathbb{Z}-\frac12$, $\mathbb{Z}'_+=\mathbb{N}-\frac12$, write

$$\mathsf{a}\left(z',z\right)=\sum_{p,q\in\mathbb{Z}'_+}\mathsf{a}_{-q}^{\;\,p}\,z'^{-\frac12+p}z^{-\frac12+q},\tag{4.1a}$$

$$\mathsf{d}\left(z',z\right)=\sum_{p,q\in\mathbb{Z}'_+}\mathsf{d}^{-p}_{\;\;q}\,z'^{-\frac12-p}z^{-\frac12-q},\tag{4.1b}$$

where $z',z\in\mathbb{C}^*$. The mode operators $\mathsf{a}_{-q}^{\;\,p},\mathsf{d}^{-p}_{\;\;q}$ are 2×2 matrices whose elements will be represented as $\mathsf{a}_{-q;s}^{\;\,p;s'}$, $\mathsf{d}^{-p;s'}_{\;\;q;s}$, with "color" indices $s',s\in\{+,-\}$. Our convention is that "+" and "−" correspond to the first and second row/column.

In order to compute these matrix elements explicitly, let us return to the original definition (3.3) of a and d. Recall that inside the annulus $\mathcal{A}$ we have $\hat{\Psi}^{[\infty]}(z)=Y_\infty^{[\infty]}(z)\,E^{-1}U^{-1}z^{-\mathfrak{S}}$ and $\hat{\Psi}^{[0]}(z)=Y_0^{[0]}\left(\frac{z}{t}\right)U^{-1}z^{-\mathfrak{S}}$, where $Y_\infty^{[\infty]}(z)$, $Y_0^{[0]}(z)$ solve the linear systems (2.16), (2.21). These relations may be used to differentiate the kernels $\mathsf{a}(z',z)$, $\mathsf{d}(z',z)$ with respect to their arguments. In particular, for $z',z\in\mathcal{A}$ one has

$$\begin{aligned}
&\left(z\partial_z+z'\partial_{z'}+1\right)\mathsf{a}\left(z',z\right)-\left[\mathfrak{S},\mathsf{a}\left(z',z\right)\right]=\\
&=\hat{\Psi}^{[\infty]}\left(z'\right)^{-1}\frac{zA^{[\infty]}(z)-z'A^{[\infty]}\left(z'\right)}{z-z'}\hat{\Psi}^{[\infty]}(z)=-\hat{\Psi}^{[\infty]}\left(z'\right)^{-1}\sigma_+\hat{\Psi}^{[\infty]}(z)=\\
&=-e^{-2\pi i\eta\sigma_z}Y_0^{[\infty]}\left(z'\right)^{-1}\begin{pmatrix}1\\0\end{pmatrix}\otimes\begin{pmatrix}0&1\end{pmatrix}Y_0^{[\infty]}(z)\,e^{2\pi i\eta\sigma_z}=\\
&=-\frac{\pi}{\sin 2\pi\sigma}\begin{pmatrix}f_{-\sigma-\frac12}\left(z'\right)\\ ie^{2\pi i(2\eta-\sigma)}f_{\sigma+\frac12}\left(z'\right)\end{pmatrix}\otimes\begin{pmatrix}f_{\sigma+\frac12}(z) & ie^{2\pi i(\sigma-2\eta)}f_{-\sigma-\frac12}(z)\end{pmatrix}.
\end{aligned}$$

Substituting into the last equation the Fourier representation (4.1a) and using the factorization of the right hand side, we obtain

$$(p+q)\,\mathsf{a}_{-q}^{\;\,p}-\left[\mathfrak{S},\mathsf{a}_{-q}^{\;\,p}\right]=e^{i\pi(\sigma-2\eta)\sigma_z}\psi^p(\nu)\otimes\bar{\psi}_q(\nu)\,e^{i\pi(2\eta-\sigma)\sigma_z},\tag{4.2}$$

with

$$\psi^{p;s}(\nu)=\sqrt{\frac{\Gamma(1+2s\nu)}{\Gamma(1-2s\nu)}}\frac{e^{-i\pi s/4}}{\left(p-\frac12\right)!\left(1-2s\nu\right)_{p-\frac12}},\tag{4.3a}$$

$$\bar{\psi}_{p;s}(\nu)=\sqrt{\frac{\Gamma(1-2s\nu)}{\Gamma(1+2s\nu)}}\;\frac{e^{i\pi s/4}}{\left(p-\frac12\right)!\left(2s\nu\right)_{p+\frac12}},\tag{4.3b}$$

where $s=\pm$, $(\alpha)_k=\alpha(\alpha+1)\ldots(\alpha+k-1)$ denotes the Pochhammer symbol, and we have introduced instead of σ a shifted monodromy parameter $\nu=\sigma+\frac12$ to make the resulting expressions more symmetric. Further introducing shifted momenta

$$x_{p;s}=p-s\nu,\qquad\qquad p\in\mathbb{Z}',\ s=\pm,$$

the solution of (4.2) can be written as

$$\mathsf{a}_{-q;s}^{\;\,p;s'}=\frac{\psi^{p;s'}(\nu)\,\bar{\psi}_{q,s}(\nu)}{x_{p;s'}-x_{-q;s}}\,e^{i\pi(2\eta-\sigma)\left(s-s'\right)},\tag{4.4a}$$

where $p, q \in \mathbb{Z}'_+$, $s', s = \pm 1$. We thus conclude that in the Fourier basis the operator a is given, up to left and right diagonal factors, by a Cauchy matrix $M_{jk} = \frac{1}{x_j - y_k}$. This allows, *inter alia*, to compute any minor of a in a factorized form.

The matrix elements of d may be computed in a similar fashion, or alternatively deduced by comparison of (3.9a) and (3.9b). The result is again a Cauchy matrix,

$$\mathsf{d}^{-q;s}_{\ \ p;s'} = \frac{\psi^{q;s}(-\nu)\,\bar{\psi}_{p,s'}(-\nu)}{x_{p;s'} - x_{-q;s}}\, e^{i\pi\sigma(s-s')} t^{(s-s')\nu+p+q}, \tag{4.4b}$$

and its nontrivial part coincides with that of (4.4b) after replacement $\nu \to -\nu$. The dependence on PIII variable t is isolated in the diagonal factors; cf Remark 3.5.

4.2. Maya and Young diagrams. Given a matrix $A \in \mathrm{Mat}_{m\times m}(\mathbb{C})$, the von Koch's formula

$$\det(\mathbf{1} + A) = \sum_{n=0}^{\infty} \sum_{i_1<\ldots<i_n} \det\left(A_{i_j i_k}\right)_{j,k=1}^{n}$$

expresses the determinant $\det(\mathbf{1}+A)$ as the sum of principal minors of A. While this series of course terminates at $n = m$, the formula has a straightforward generalization to infinite matrices. If A is indexed by elements of a discrete set $\mathfrak{X}$ instead of $\{1, \ldots, m\}$, then

$$\det(\mathbf{1} + A) = \sum_{\mathfrak{Y}\in 2^{\mathfrak{X}}} \det A_{\mathfrak{Y}}, \tag{4.5}$$

where the sum is taken over all subsets $\mathfrak{Y}$ of $\mathfrak{X}$ and $A_{\mathfrak{Y}}$ is the principal minor of A obtained by choosing the rows and columns labeled by $\mathfrak{Y}$.

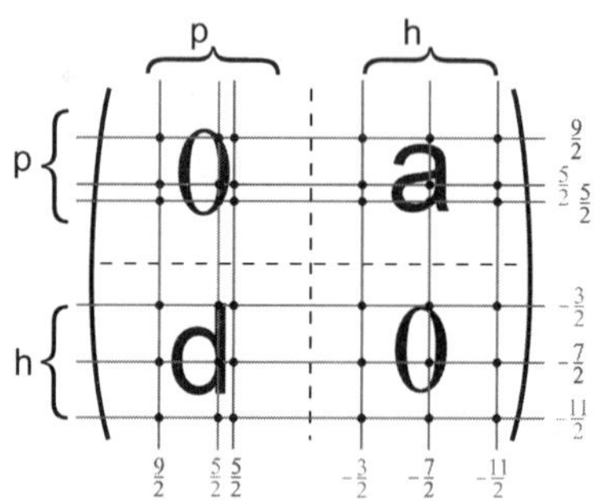

FIGURE 4. Labeling of principal minors of K by positions (p, h) of particles and holes of color $+$ (red) and $-$ (blue). Here $\mathsf{p}^+ = \left\{\frac{5}{2}\right\}$, $\mathsf{h}^+ = \left\{-\frac{3}{2}, -\frac{11}{2}\right\}$, $\mathsf{p}^- = \left\{\frac{9}{2}, \frac{5}{2}\right\}$, $\mathsf{h}^- = \left\{-\frac{7}{2}\right\}$, so that $\mathsf{m}^+ = \left\{\frac{5}{2}, -\frac{3}{2}, -\frac{11}{2}\right\}$, $\mathsf{m}^- = \left\{\frac{9}{2}, \frac{5}{2}, -\frac{7}{2}\right\}$ and $Q(\mathsf{m}^+) = -Q(\mathsf{m}^-) = -1$.

We are going apply the last formula to the Fredholm determinant (3.7) with K written in the Fourier basis. Represent appropriate subsets as $\mathfrak{Y} = (\mathsf{p}, \mathsf{h})$, where p and h correspond, respectively, to the first and second block of K, see Fig. 4. The sum in (4.5) may be restricted to (p, h) with $\sharp(\mathsf{p}) = \sharp(\mathsf{h})$, as otherwise the corresponding minors obviously vanish. It follows that

$$\det(\mathbf{1} - K) = \sum_{(\mathsf{p},\mathsf{h})\,:\,\sharp(\mathsf{p})=\sharp(\mathsf{h})} (-1)^{\sharp(\mathsf{p})} \det \mathsf{a}^{\mathsf{p}}_{\mathsf{h}} \det \mathsf{d}^{\mathsf{h}}_{\mathsf{p}}, \tag{4.6}$$

where e.g. $\mathsf{a}_{\mathsf{h}}^{\mathsf{p}}$ is a square $\sharp(\mathsf{p}) \times \sharp(\mathsf{p})$ matrix obtained by restricting a to rows p and columns h. Let us now take a closer look at the structure of subsets (p, h) labeling different contributions to (4.6):

- The set p has the form $\mathsf{p}^+ \sqcup \mathsf{p}^-$, where
$$\mathsf{p}^+ = \left\{p_1^+, \ldots, p_L^+\right\}, \quad \mathsf{p}^- = \left\{p_1^-, \ldots, p_M^-\right\}, \quad \text{and} \quad p_j^\pm \in \mathbb{Z}'_+$$
are Fourier indices of elements of p of color $\pm$. Similarly, $\mathsf{h} = \mathsf{h}^+ \sqcup \mathsf{h}^-$, where $\mathsf{h}^+ = \left\{-q_1^+, \ldots, -q_{L'}^+\right\}$, $\mathsf{h}^- = \left\{-q_1^-, \ldots, -q_{M'}^-\right\}$ with $q_j^\pm \in \mathbb{Z}'_+$ consist of Fourier indices of elements of h of colors $+$ and $-$. The elements of $\mathsf{p}^\pm$ and $\mathsf{h}^\pm$ are thus distinct positive (resp. negative) half-integers.
- Let us consider the combinations $\mathsf{m}^\pm = \mathsf{p}^\pm \cup \mathsf{h}^\pm$. Both $\mathsf{m}^\pm$ are finite subsets of $\mathbb{Z}'$ and can be represented in the usual way by Maya diagrams; $\mathsf{p}^\pm$ and $\mathsf{h}^\pm$ are positions of particles and holes of color $\pm$ in the Dirac sea, see Fig. 4 and bottom part of Fig. 5. Given a Maya diagram m, the difference $Q(\mathsf{m}) = \sharp(\text{particles}) - \sharp(\text{holes})$ is called the charge of m. The constraint $\sharp(\mathsf{p}) = \sharp(\mathsf{h})$ is then nothing but the neutrality condition $Q(\mathsf{m}^+) + Q(\mathsf{m}^-) = 0$.
- On the other hand, the set $\mathbb{M}$ of Maya diagrams can be bijectively mapped to the set $\mathbb{Y} \times \mathbb{Z}$ of charged Young diagrams/partitions. This correspondence is represented graphically in Fig. 5. The profile of the Young diagram $\mathsf{Y} \in \mathbb{Y}$ associated to a Maya diagram $\mathsf{m} \in \mathbb{M}$ is obtained by starting far away on the NW-axis and going south-east above each filled circle and north-east above each empty circle of m. The charge corresponds to relative position of the bottom boundary of Y and the NE-axis, and coincides with $Q(\mathsf{m})$.

Different contributions to (4.6) may therefore be labeled (i) by positions of particles $\mathsf{p}^\pm \in 2^{\mathbb{Z}'_+}$ and holes $\mathsf{h}^\pm \in 2^{-\mathbb{Z}'_+}$ of two colors $\pm$ satisfying the balance condition $\sharp(\mathsf{p}^+) + \sharp(\mathsf{p}^-) = \sharp(\mathsf{h}^+) + \sharp(\mathsf{h}^-)$; (ii) by pairs $(\mathsf{m}^+, \mathsf{m}^-) \in \mathbb{M}^2$ of Maya diagrams of zero total charge; and also (iii) by pairs $(\mathsf{Y}^+, \mathsf{Y}^-) \in \mathbb{Y}^2$ of Young diagrams corresponding to m^+ and m^-, and an integer $Q \equiv Q(\mathsf{m}^+)$.

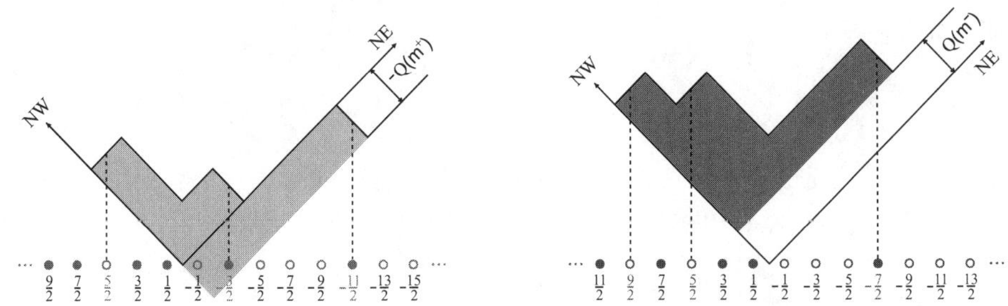

FIGURE 5. Young diagrams Y^+ and Y^- (shaded regions) associated to $\mathsf{m}^+ = \left\{\frac{5}{2}, -\frac{3}{2}, -\frac{11}{2}\right\}$ and $\mathsf{m}^- = \left\{\frac{9}{2}, \frac{5}{2}, -\frac{7}{2}\right\}$.

The individual contributions can be readily computed using the Cauchy matrix representations (4.4).

THEOREM 4.1. *The Painlevé III (D_8) tau function $\tau_{\rm III}(t) = \tau_{\rm III}(t\,|\,\sigma,\eta)$ from Theorem 3.2 admits the following series representation:*

$$\tau_{\rm III}(t) = \sum_{(\mathsf{p},\mathsf{h})\,:\,\sharp(\mathsf{p})=\sharp(\mathsf{h})} e^{-4\pi i\eta Q}\,\Xi_{\mathsf{p},\mathsf{h}}(\nu)\,\Delta^2_{\mathsf{p},\mathsf{h}}(\nu)\,t^{(\nu-Q)^2+|\mathsf{Y}^+|+|\mathsf{Y}^-|}, \tag{4.7}$$

where $|\mathsf{Y}|$ denotes the total number of boxes in $\mathsf{Y}\in\mathbb{Y}$ and

$$\Xi_{\mathsf{p},\mathsf{h}}(\nu) = (-1)^Q\frac{\Gamma^{2Q}(1+2\nu)}{\Gamma^{2Q}(1-2\nu)} \tag{4.8a}$$
$$\times\left[\prod_{(p,s')\in\mathsf{p}}\left(p-\tfrac12\right)!\,(1-2s'\nu)_{p-\frac12}\prod_{(-q,s)\in\mathsf{h}}\left(q-\tfrac12\right)!\,(2s\nu)_{q+\frac12}\right]^{-2},$$

$$\Delta_{\mathsf{p},\mathsf{h}}(\nu) = \frac{\prod\limits_{(p,s')<(\bar p,\bar s')\in\mathsf{p}}(x_{p;s'}-x_{\bar p;\bar s'})\prod\limits_{(-q,s)<(-\bar q,\bar s)\in\mathsf{h}}(x_{-\bar q;\bar s}-x_{-q;s})}{\prod\limits_{(p;s')\in\mathsf{p}}\prod\limits_{(-q;s)\in\mathsf{h}}(x_{p;s'}-x_{-q;s})}. \tag{4.8b}$$

PROOF. From (4.4) it follows that

$$\begin{aligned}(-1)^{\sharp(\mathsf{p})}\det\mathsf{a}^{\mathsf{p}}_{\mathsf{h}}\det\mathsf{d}^{\mathsf{h}}_{\mathsf{p}} &= (-1)^{\sharp(\mathsf{p})}\prod_{(p,s')\in\mathsf{p}}\psi^{p;s'}(\nu)\,\bar\psi_{p;s'}(-\nu)\prod_{(-q,s)\in\mathsf{h}}\psi^{q;s}(-\nu)\,\bar\psi_{q;s}(\nu)\\ &\times\left(t^\nu e^{2\pi i\eta}\right)^{\sum_{(-q,s)\in\mathsf{h}}s-\sum_{(p,s')\in\mathsf{p}}s'}\,t^{\sum_{(-q,s)\in\mathsf{h}}q+\sum_{(p,s')\in\mathsf{p}}p}\,\Delta^2_{\mathsf{p},\mathsf{h}}(\nu).\end{aligned} \tag{4.9}$$

The power of $t^\nu e^{2\pi i\eta}$ can be further transformed as

$$\sum_{(-q,s)\in\mathsf{h}}s-\sum_{(p,s')\in\mathsf{p}}s' = \sharp(\mathsf{h}^+)-\sharp(\mathsf{h}^-)-\sharp(\mathsf{p}^+)+\sharp(\mathsf{p}^-) = Q(\mathsf{m}^-)-Q(\mathsf{m}^+) = -2Q.$$

It may be also easily shown (see Fig. 13 in [**GL16**]) that

$$\sum_{(-q,s)\in\mathsf{h}^\pm}q+\sum_{(p,s')\in\mathsf{p}^\pm}p = \frac{Q^2}{2}+|\mathsf{Y}^\pm|,$$

so that the power of t in the second line of (4.9) becomes $Q^2+|\mathsf{Y}^+|+|\mathsf{Y}^-|$. The prefactor $\Xi_{\mathsf{p},\mathsf{h}}(\nu)$ is obtained from the diagonal products in the first line by simple algebra. □

4.3. Nekrasov functions. In this subsection we rewrite the factorized expressions $\Xi_{\mathsf{p},\mathsf{h}}(\nu)\,\Delta^2_{\mathsf{p},\mathsf{h}}(\nu)$ for the coefficients of the tau function expansion in a notation close to gauge theory. The main tool we need is a technical statement that can be found, for example, in [**GL16, GM**]. In order to formulate it, let $(\mathsf{Y}^s,Q^s)\in\mathbb{Y}\times\mathbb{Z}$ (with $s=\pm$) be two charged Young diagrams, not necessarily the same as above. Denote by $\mathsf{m}^s\in\mathbb{M}$ the associated Maya diagrams. At this point we do not need to assume that $Q^++Q^-=0$.

Introduce the following three quantities:

(1) An explicit factorized function

$$\tilde{Z}_{\mathsf{bif}}\left(\nu\,|\,\mathsf{Y}^+,Q^+;\mathsf{Y}^-,Q^-\right) \tag{4.10}$$
$$= \prod_{-q\in\mathsf{h}^+}(-\nu)_{q+\frac12}\prod_{-q\in\mathsf{h}^-}(\nu+1)_{q-\frac12}\prod_{p\in\mathsf{p}^-}(-\nu)_{p+\frac12}\prod_{p\in\mathsf{p}^+}(\nu+1)_{p-\frac12}$$
$$\times\frac{\prod_{-q\in\mathsf{h}^+}\prod_{p\in\mathsf{p}^-}(\nu-q-p)\prod_{-q\in\mathsf{h}^-}\prod_{p\in\mathsf{p}^+}(\nu+p+q)}{\prod_{-q'\in\mathsf{h}^-}\prod_{-q\in\mathsf{h}^+}(\nu-q+q')\prod_{p'\in\mathsf{p}^-}\prod_{p\in\mathsf{p}^+}(\nu+p-p')}.$$

which, as we will see in a moment, constitutes the main building block of $\Xi_{\mathsf{p},\mathsf{h}}(\nu)\,\Delta^2_{\mathsf{p},\mathsf{h}}(\nu)$.

(2) Another factorized expression, representing the Nekrasov bifundamental contribution:

(4.11)
$$Z_{\mathsf{bif}}\left(\nu\,|\,\mathsf{Y}^+,\mathsf{Y}^-\right):=\prod_{\Box\in\mathsf{Y}^+}\left(\nu+1+a_{\mathsf{Y}^+}(\Box)+l_{\mathsf{Y}^-}(\Box)\right)\prod_{\Box\in\mathsf{Y}^-}\left(\nu-1-a_{\mathsf{Y}^-}(\Box)-l_{\mathsf{Y}^+}(\Box)\right),$$

where $\mathsf{Y}^\pm\in\mathbb{Y}$ and the notation for Young diagrams follows Fig. 6. The expressions $a_{\mathsf{Y}}(\Box)$, $l_{\mathsf{Y}}(\Box)$ and $h_{\mathsf{Y}}(\Box)$ represent the arm-, leg-, and hook length of the box $\Box$ in $\mathsf{Y}\in\mathbb{Y}$. In the case where $\Box=(i,j)$ does not belong to Y, the definition of the former two quantities is extended by $a_{\mathsf{Y}}(\Box)=\mathsf{Y}_i-j$ and $l_{\mathsf{Y}}(\Box)=\mathsf{Y}'_j-i$. In particular, we have

$$Z_{\mathsf{bif}}\left(-\nu\,|\,\mathsf{Y}^-,\mathsf{Y}^+\right)=(-1)^{|\mathsf{Y}^+|+|\mathsf{Y}^-|}Z_{\mathsf{bif}}\left(\nu\,|\,\mathsf{Y}^+,\mathsf{Y}^-\right), \tag{4.12}$$
$$Z_{\mathsf{bif}}\left(0\,|\,\mathsf{Y},\mathsf{Y}\right)=(-1)^{|\mathsf{Y}|}\prod_{\Box\in\mathsf{Y}}h^2_{\mathsf{Y}}(\Box).$$

(3) For $Q\in\mathbb{Z}$, define the "structure constant" $\Upsilon(\nu\,|\,Q)$ by

$$\Upsilon(\nu\,|\,Q)=\frac{\Gamma^Q(1+\nu)\,G(1+\nu)}{G(1+\nu+Q)}. \tag{4.13}$$

Here $G(z)$ denotes the Barnes G-function satisfying the relation $G(z+1)=\Gamma(z)\,G(z)$. Note that $\Upsilon(\nu\,|\,Q)$ is actually a rational function of ν.

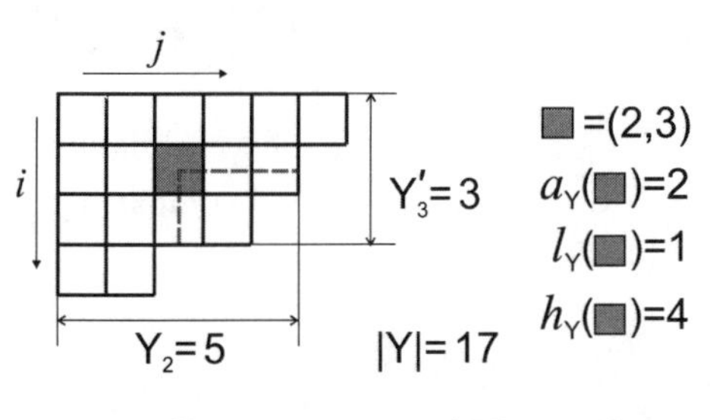

FIGURE 6. Characteristics of Young diagrams.

LEMMA 4.2. *We have*

$$\tilde{Z}_{\mathsf{bif}}\left(\nu\middle|\,\mathsf{Y}^+,Q^+;\mathsf{Y}^-,Q^-\right)=\pm\frac{Z_{\mathsf{bif}}\left(\nu+Q^+-Q^-\,\middle|\,\mathsf{Y}^+,\mathsf{Y}^-\right)}{\Upsilon\left(\nu\,\middle|\,Q^+-Q^-\right)},\tag{4.14}$$

where $\pm$ *means that the equality holds up to an overall sign.*

PROOF. See [**GL16**, Appendix A]. □

Lemma 4.2 thus relates certain products over boxes of Young diagrams to some explicit functions of particle and hole coordinates in the relevant Maya diagrams. Let us now use it to identify the corresponding Nekrasov functions in (4.1). We first prove

LEMMA 4.3. *We have*

$$\Xi_{\mathsf{p},\mathsf{h}}\left(\nu\right)\Delta^2_{\mathsf{p},\mathsf{h}}\left(\nu\right)=\frac{\Gamma^{2Q}(1+2\nu)}{\Gamma^{2Q}(1-2\nu)}\frac{\Upsilon\left(2\nu\,|\,-2Q\right)\Upsilon\left(-2\nu\,|\,2Q\right)}{\prod_{s,s'=\pm1}Z_{\mathsf{bif}}\left((Q-\nu)(s'-s)\,|\,\mathsf{Y}^{s'},\mathsf{Y}^{s}\right)}.\tag{4.15}$$

PROOF. First of all, one may further decompose $\Delta_{\mathsf{p},\mathsf{h}}$:

$$\Delta_{\mathsf{p},\mathsf{h}}(\nu)=\Delta^{++}_{\mathsf{p},\mathsf{h}}\Delta^{+-}_{\mathsf{p},\mathsf{h}}(\nu)\Delta^{--}_{\mathsf{p},\mathsf{h}},$$

where

$$\Delta^{\pm\pm}_{\mathsf{p},\mathsf{h}}=\frac{\prod_{p,\bar p\in\mathsf{p}^\pm:p<\bar p}(p-\bar p)\prod_{-q,-\bar q\in\mathsf{h}^\pm:q>\bar q}(q-\bar q)}{\prod_{p\in\mathsf{p}^\pm}\prod_{-q\in\mathsf{h}^\pm}(p+q)},$$

$$\Delta^{+-}_{\mathsf{p},\mathsf{h}}(\nu)=\frac{\prod_{p_+\in\mathsf{p}^+}\prod_{p_-\in\mathsf{p}^-}(-2\nu+p_+-p_-)\prod_{-q_+\in\mathsf{h}^+}\prod_{-q_-\in\mathsf{h}^-}(-2\nu-q_++q_-)}{\prod_{p_+\in\mathsf{p}^+}\prod_{-q_-\in\mathsf{h}^-}(-2\nu+p_++q_-)\prod_{p_-\in\mathsf{p}^-}\prod_{-q_+\in\mathsf{h}^+}(2\nu+p_-+q_+)}.$$

Comparing (4.10) with the above formulas, we can write

$$\begin{aligned}\prod_{s,s'=\pm1}\tilde{Z}_{\mathsf{bif}}&\left(\nu\,(s-s')\,|\,\mathsf{Y}^{s'},Qs';\mathsf{Y}^{s},Qs\right)=\pm\left[\Delta^{++}_{\mathsf{p},\mathsf{h}}\Delta^{+-}_{\mathsf{p},\mathsf{h}}(\nu)\Delta^{--}_{\mathsf{p},\mathsf{h}}\right]^{-2}\\
&\times\prod_{-q\in\mathsf{h}^-}(-2\nu)_{q+\frac12}\prod_{-q\in\mathsf{h}^+}(2\nu+1)_{q-\frac12}\prod_{p\in\mathsf{p}^+}(-2\nu)_{p+\frac12}\prod_{p\in\mathsf{p}^-}(2\nu+1)_{p-\frac12}\\
&\times\prod_{-q\in\mathsf{h}^+}(2\nu)_{q+\frac12}\prod_{-q\in\mathsf{h}^-}(-2\nu+1)_{q-\frac12}\prod_{p\in\mathsf{p}^+}(2\nu)_{p+\frac12}\prod_{p\in\mathsf{p}^+}(-2\nu+1)_{p-\frac12}\\
&\times\left[\prod_{-q\in\mathsf{h}^-}\left(q-\frac12\right)!\prod_{-q\in\mathsf{h}^+}\left(q-\frac12\right)!\prod_{p\in\mathsf{p}^-}\left(p-\frac12\right)!\prod_{p\in\mathsf{p}^+}\left(p-\frac12\right)!\right]^2.\end{aligned}\tag{4.16}$$

Using the identity $(z)_{q+\frac12}=z\cdot(z+1)_{q-\frac12}$ for the Pochhammer's symbol, the balance condition $\sharp\,(\mathsf{h}^+)+\sharp\,(\mathsf{h}^-)=\sharp\,(\mathsf{p}^+)+\sharp\,(\mathsf{p}^-)$, and comparing the last three lines with (4.8a), we can rewrite (4.16) as

$$\prod_{s,s'=\pm1}\tilde{Z}_{\mathsf{bif}}\left(\nu\,(s-s')\,|\,\mathsf{Y}^{s'},Qs';\mathsf{Y}^{s},Qs\right)=\pm\frac{\Gamma^{2Q}(1+2\nu)}{\Gamma^{2Q}(1-2\nu)}\,\Xi^{-1}_{\mathsf{p},\mathsf{h}}\left(\nu\right)\Delta^{-2}_{\mathsf{p},\mathsf{h}}\left(\nu\right).\tag{4.17}$$

Combining this result with (4.14), we immediately obtain (4.15) up to an overall sign. It suffices to check it for real $\nu\notin\mathbb{Z}$. For the left side this sign is obviously equal to $(-1)^Q$. From the identities (4.12) it follows that the right side of (4.15) may be rewritten as

$$\frac{G(1+2\nu)\,G(1-2\nu)}{G(1+2\nu-2Q)\,G(1-2\nu+2Q)}$$
$$\times\left[Z_{\mathsf{bif}}\left(2\nu-2Q\,|\,\mathsf{Y}^-,\mathsf{Y}^+\right)\prod_{\square\in\mathsf{Y}^+}h_{\mathsf{Y}^+}(\square)\prod_{\square\in\mathsf{Y}^-}h_{\mathsf{Y}^-}(\square)\right]^{-2}.$$

Its sign is therefore determined by the Barnes function prefactor in the last expression, and can be easily shown to be $(-1)^Q$. □

We can now formulate our final result.

THEOREM 4.4. *Let $\mathcal{Z}_{\mathrm{SU}(2)}(t\,|\,\nu)$ be the Nekrasov instanton partition function of the pure gauge theory, defined as a double sum over partitions:*

(4.18)
$$\mathcal{Z}_{\mathrm{SU}(2)}(t\,|\,\nu)=C(\nu)\sum_{\mathsf{Y}^+,\mathsf{Y}^-\in\mathbb{Y}}\frac{t^{\nu^2+|\mathsf{Y}^+|+|\mathsf{Y}^-|}}{\prod_{s,s'=\pm1}Z_{\mathsf{bif}}\left(\nu(s-s')\,|\,\mathsf{Y}^{s'},\mathsf{Y}^{s}\right)},$$

with $C(\nu)=\left[\prod_{s=\pm1}G(1+2s\nu)\right]^{-1}$ and $Z_{\mathsf{bif}}(\nu\,|\,\mathsf{Y}',\mathsf{Y})$ defined by (4.11). *The dual partition function*

(4.19)
$$\mathcal{Z}^{\mathrm{dual}}_{\mathrm{SU}(2)}(t\,|\,\nu,\eta)=\sum_{n\in\mathbb{Z}}e^{4\pi i n\eta}\mathcal{Z}_{\mathrm{SU}(2)}(t\,|\,\nu+n)$$

admits Fredholm determinant representation

(4.20)
$$\mathcal{Z}^{\mathrm{dual}}_{\mathrm{SU}(2)}(t\,|\,\nu,\eta)=C(\nu)\,t^{\nu^2}\det\left(\mathbf{1}-K\right),$$

where K is the generalized Bessel kernel from Theorem 3.2 *(with $\sigma=\nu-\frac{1}{2}$), and thereby coincides with the general tau function of the Painlevé III (D_8) equation.*

References

[AGT] L. F. Alday, D. Gaiotto, and Y. Tachikawa, *Liouville correlation functions from four-dimensional gauge theories*, Lett. Math. Phys. **91** (2010), no. 2, 167–197. MR2586871

[AZ] S. A. Apikyan and A. B. Zamolodchikov, *Conformal blocks, related to conformally invariant Ramond states of a free scalar field* (Russian), Zh. Èksper. Teoret. Fiz. **92** (1987), no. 1, 34–45; English transl., Soviet Phys. JETP **65** (1987), no. 1, 19–24. MR922674

[BSh1] M. A. Bershtein and A. I. Shchechkin, *Bilinear equations on Painlevé τ functions from CFT*, Comm. Math. Phys. **339** (2015), no. 3, 1021–1061. MR3385990

[BSh2] M. A. Bershtein and A. I. Shchechkin, *Bäcklund transformation of Painlevé* III(D_8) τ *function*, J. Phys. A **50** (2017), no. 11, 115205, 31. MR3622582

[BGT1] G. Bonelli, A. Grassi, and A. Tanzini, *Seiberg-Witten theory as a Fermi gas*, Lett. Math. Phys. **107** (2017), no. 1, 1–30. MR3598873

[BGT2] G. Bonelli, A. Grassi, and A. Tanzini, *New results in $\mathcal{N}=2$ theories from non-perturbative string*, Ann. Henri Poincaré **19** (2018), No. 3, 743–774. DOI: 10.1007/s00023-017-0643-5.

[BLMST] G. Bonelli, O. Lisovyy, K. Maruyoshi, A. Sciarappa, and A. Tanzini, *On Painlevé/gauge theory correspondence*, Lett. Math. Phys. **107** (2017), no. 12, 2359–2413. MR3719644

[BMT] G. Bonelli, K. Maruyoshi, and A. Tanzini, *Wild quiver gauge theories*, J. High Energy Phys. **2** (2012), 031, front matter+30. MR2996123

[CM] L. Chekhov and M. Mazzocco, *Colliding holes in Riemann surfaces and quantum cluster algebras*, Nonlinearity **31** (2018), No. 54. DOI:10.1088/1361-6544/aa9729.

[CMR] L. Chekhov, M. Mazzocco, and V. Rubtsov, *Painlevé monodromy manifolds, decorated character varieties and cluster algebras*, Int. Math. Res. Not. **2017** (2017), 7639–7691. DOI:10.1093/imrn/rnw219.

[FIKN] A. S. Fokas, A. R. Its, A. A. Kapaev, and V. Yu. Novokshenov, *Painlevé transcendents*, Mathematical Surveys and Monographs, vol. 128, American Mathematical Society, Providence, RI, 2006. The Riemann-Hilbert approach. MR2264522

[G09] D. Gaiotto, *Asymptotically free $\mathcal{N} = 2$ theories and irregular conformal blocks*, arXiv:0908.0307 [hep-th].

[GT] D. Gaiotto and J. Teschner, *Irregular singularities in Liouville theory and Argyres-Douglas type gauge theories*, J. High Energy Phys. **12** (2012), 050, front matter + 76. MR3045283

[GIL12] O. Gamayun, N. Iorgov, and O. Lisovyy, *Conformal field theory of Painlevé VI*, J. High Energy Phys. **10** (2012), 038, front matter + 24. MR3033880

[GIL13] O. Gamayun, N. Iorgov, and O. Lisovyy, *How instanton combinatorics solves Painlevé VI, V and IIIs*, J. Phys. A **46** (2013), no. 33, 335203, 29. MR3093009

[GL16] P. Gavrylenko, O. Lisovyy, *Fredholm determinant and Nekrasov sum representations of isomonodromic tau functions*, (2016); arXiv:1608.00958 [math-ph].

[GM] P. G. Gavrilenko and A. V. Marshakov, *Free fermions, W-algebras, and isomonodromic deformations* (Russian, with Russian summary), Teoret. Mat. Fiz. **187** (2016), no. 2, 232–262; English transl., Theoret. and Math. Phys. **187** (2016), no. 2, 649–677. MR3507535

[GHM] A. Grassi, Y. Hatsuda, and M. Mariño, *Topological strings from quantum mechanics*, Ann. Henri Poincaré **17** (2016), no. 11, 3177–3235. MR3556519

[ILTe] N. Iorgov, O. Lisovyy, and J. Teschner, *Isomonodromic tau-functions from Liouville conformal blocks*, Comm. Math. Phys. **336** (2015), no. 2, 671–694. MR3322384

[ILT14] A. Its, O. Lisovyy, and Y. Tykhyy, *Connection problem for the sine-Gordon/Painlevé III tau function and irregular conformal blocks*, Int. Math. Res. Not. IMRN **18** (2015), 8903–8924. MR3417698

[IN] A. R. Its and V. Yu. Novokshenov, *The isomonodromic deformation method in the theory of Painlevé equations*, Lecture Notes in Mathematics, vol. 1191, Springer-Verlag, Berlin, 1986. MR851569

[IP] A. Its and A. Prokhorov, *Connection problem for the tau-function of the sine-Gordon reduction of Painlevé-III equation via the Riemann-Hilbert approach*, Int. Math. Res. Not. IMRN **22** (2016), 6856–6883. MR3632069

[Jim] M. Jimbo, *Monodromy problem and the boundary condition for some Painlevé equations*, Publ. Res. Inst. Math. Sci. **18** (1982), no. 3, 1137–1161. MR688949

[JMU] M. Jimbo, T. Miwa, and K. Ueno, *Monodromy preserving deformation of linear ordinary differential equations with rational coefficients. I. General theory and τ-function*, Phys. D **2** (1981), no. 2, 306–352. MR630674

[K89] V. G. Knizhnik, *Multiloop amplitudes in the theory of quantum strings and complex geometry* (Russian), Uspekhi Fiz. Nauk **159** (1989), no. 3, 401–453; English transl., Soviet Phys. Uspekhi **32** (1989), no. 11, 945–971 (1990). MR1050755

[MTW] B. M. McCoy, C. A. Tracy, and T. T. Wu, *Painlevé functions of the third kind*, J. Mathematical Phys. **18** (1977), no. 5, 1058–1092. MR0473322

[M90] G. Moore, *Geometry of the string equations*, Comm. Math. Phys. **133** (1990), no. 2, 261–304. MR1090426

[Nag1] H. Nagoya, *Irregular conformal blocks, with an application to the fifth and fourth Painlevé equations*, J. Math. Phys. **56** (2015), no. 12, 123505, 24. MR3436751

[Nag2] H. Nagoya, *Conformal blocks and Painlevé functions*, arXiv:1611.08971 [math-ph].

[Nek] N. A. Nekrasov, *Seiberg-Witten prepotential from instanton counting*, Adv. Theor. Math. Phys. **7** (2003), no. 5, 831–864. MR2045303

[NO] N. A. Nekrasov and A. Okounkov, *Seiberg-Witten theory and random partitions*, The unity of mathematics, Progr. Math., vol. 244, Birkhäuser Boston, Boston, MA, 2006, pp. 525–596. MR2181816

[Nil] D. G. Niles, *The Riemann-Hilbert-Birkhoff inverse monodromy problem and connection formulae for the third Painleve transcendents*, ProQuest LLC, Ann Arbor, MI, 2009. Thesis (Ph.D.)–Purdue University. MR2941337

[Nov] V. Yu. Novokshënov, *The asymptotic behavior of the general real solution of the third Painlevé equation* (Russian), Dokl. Akad. Nauk SSSR **283** (1985), no. 5, 1161–1165. MR802144

[SMJ] M. Sato, T. Miwa, and M. Jimbo, *Holonomic quantum fields. I*, Publ. Res. Inst. Math. Sci. **14** (1978), no. 1, 223–267. MR499666

[SW] N. Seiberg and E. Witten, *Electric-magnetic duality, monopole condensation, and confinement in $N = 2$ supersymmetric Yang-Mills theory*, Nuclear Phys. B **426** (1994), no. 1, 19–52. MR1293681

[T11] J. Teschner, *Quantization of the Hitchin moduli spaces, Liouville theory and the geometric Langlands correspondence I*, Adv. Theor. Math. Phys. **15** (2011), no. 2, 471–564. MR2924236

[T91] C. A. Tracy, *Asymptotics of a τ-function arising in the two-dimensional Ising model*, Comm. Math. Phys. **142** (1991), no. 2, 297–311. MR1137066

[Z86] Al. B. Zamolodchikov, *Conformal scalar field on the hyperelliptic curve and critical Ashkin-Teller multipoint correlation functions*, Nuclear Phys. B **285** (1987), no. 3, 481–503. MR897030

[Z94] Al. B. Zamolodchikov, *Painlevé III and 2D polymers*, Nuclear Phys. B **432** (1994), no. 3, 427–456. MR1306209

Center for Advanced Studies, Skolkovo Institute of Science and Technology, 143026 Moscow, Russia; and National Research University Higher School of Economics, Department of Mathematics and International Laboratory of Representation Theory and Mathematical Physics, 119048 Moscow, Russia; and Bogolyubov Institute for Theoretical Physics, 03680 Kyiv, Ukraine

Email address: `pasha145@gmail.com`

Laboratoire de Mathématiques et Physique Théorique CNRS/UMR 7350, Université de Tours, Parc de Grandmont, 37200 Tours, France

Email address: `lisovyi@lmpt.univ-tours.fr`

Proceedings of Symposia in Pure Mathematics
Volume **98**, 2018
http://dx.doi.org/10.1090/pspum/098/01724

Reduction for $SL(3)$ pre-buildings

Ludmil Katzarkov, Pranav Pandit, and Carlos Simpson

ABSTRACT. Given an $SL(3)$ spectral curve over a simply connected Riemann surface, we describe in detail the reduction steps necessary to construct the core of a pre-building with versal harmonic map whose differential is given by the spectral curve.

1. Introduction

Let X be a Riemann surface with a spectral covering $\Sigma \subset T^*X$ for the group $SL(3)$. In [**15**] we proposed in general terms a reduction process that would construct a versal Σ-harmonic map to an $SL(3)$ pre-building. It was conjectured that if the Σ-spectral network [**6**–**8**] has no BPS states then the reduction process should be well-defined.

This conjecture would lead to a precise calculation of the WKB exponents for singular perturbations whose spectral curve has no BPS states, generalizing the known picture [**5**] for quadratic differentials and $SL(2)$.

The purpose of the present paper is to provide more details on the reduction process, particularly about the combinatorial structure of the singularities that occur and how they are arranged at each reduction step. We will show (although the proofs are sometimes only sketches) that the reduction steps are well-defined if there are no BPS states. It is left for later to show that the process finishes in finitely many steps.

Assume X is complete and simply connected. For the present one should think of it as being the complex plane with Σ a spectral covering similar to the one considered in our original example [**14**]. The case of the universal covering of a compact Riemann surface would pose additional problems of non-finiteness of the set of singularities, and finding the right notion of convergence.

The process and results will be summed up in §8 and the reader is referred there.

2010 *Mathematics Subject Classification.* Primary 14H30; Secondary 34E20, 51E24.

Key words and phrases. Spectral curve, spectral network, building, WKB exponent, BPS state.

The authors were partially supported by by a Simons research grant, ERC Gemis, NSF DMS-150908, DMS-1265230, DMS-1201475, FWF P24572-N25, P20778, P27784-N25, OISE-1242272, PASI, ANR 933R03/13ANR002SRAR, Simons collaborative Grant - HMS, Laboratory of Mirror Symmetry NRU HSE, RF government grant ag. No. 14.641.31.000, and ERC Advanced Grant 320593.

In the remainder of the introduction, we review the motivation for the reduction process considered here. A building B for the group $SL(3)$ is a piecewise linear cell complex that is covered by copies of the standard apartment $A = \mathbb{R}^2$, and indeed any two points of B are contained in a common apartment. One of the main properties characterizing a building is that it is negatively curved.

A *harmonic map* $h : X \to B$ is a continuous map such that any point $x \in X$, except for a discrete set of singularities, admits a neighborhood $x \in U \subset X$ such that there is an apartment $A \subset B$ with $h : U \to A \cong \mathbb{R}^2$ being a harmonic map.

The differential dh is naturally a triple of real 1-forms (η_1, η_2, η_3) with $\eta_1 + \eta_2 + \eta_3 = 0$. These are well-defined up to permutation. Now, these real harmonic forms are real parts of holomorphic 1-forms $\eta_i = \Re\phi_i$ and the collection $\{\phi_1, \phi_2, \phi_3\}$ defines the spectral curve $\Sigma \subset T^*X$. We say that h is a Σ-harmonic map.

For a given spectral curve we would like to understand the Σ-harmonic maps to buildings. We conjectured in [**14**, **15**] that, under a certain genericity hypothesis, there should be an essentially uniquely defined map

$$h_\phi : X \to B_\phi^{\text{pre}}$$

depending only on the spectral curve $\Sigma = \{\phi_1, \phi_2, \phi_3\}$ with the following properties:

(1) the pre-building B_ϕ^{pre} is a negatively curved complex built out of enclosures in A [**15**], and

(2) any Σ-harmonic map to a building $h : X \to B$ factors through an embedding $B_\phi^{\text{pre}} \to B$ isometric for the Finsler and vector distances.

The conjectured genericity hypothesis for existence of h_ϕ is that the spectral network associated to Σ should not have any BPS states [**6**–**8**].

Before getting to the proposed method for constructing B_ϕ^{pre}, let us consider the implications for exponents of WKB problems. There are several different ways of getting harmonic mappings to buildings, such as Gromov-Schoen's theory [**9**]. Parreau interpreted boundary points of the character variety as actions on buildings [**22**]. In [**14**] we extended Parreau's theory slightly for the situation of WKB problems, getting a control on the differential. Suppose ∇_t is a singular perturbation of flat connections, for t a large parameter. There are two typical ways of getting ∇_t, the Riemann-Hilbert situation

$$\nabla_t = \nabla_0 + t\varphi$$

or by solution of Hitchin's equations for the Higgs bundle $(E, t\varphi)$. In either case, there is an associated limiting Higgs bundle (E, φ) and we let Σ be its spectral curve.

For $P, Q \in X$ let $T_{PQ}(t) : E_P \to E_Q$ denote the transport for ∇_t. For an ultrafilter ω on $t \to \infty$ define the *exponent*

$$\nu_{PQ}^{\omega} := \lim_{\omega} \frac{1}{t} \log \|T_{PQ}(t)\| .$$

There is a similar *vector exponent* [**14**] that is a point in the positive Weyl chamber of A.

The groupoid version of Parreau's theory [**14**, **22**] gives a map to a building $h_\omega : X \to B_\omega$ such that the Finsler distance (resp. vector distance) between $h_\omega(P), h_\omega(Q)$ is the exponent ν_{PQ}^{ω} (resp. the vector exponent).

We showed in [**14**] for the Riemann-Hilbert situation that h_ω is a Σ-harmonic map. Mochizuki [**20**] showed this for the Hitchin WKB problem.

If there exists a Σ-harmonic map h_ϕ satisfying the properties (1), (2) above, then it follows that ν^{ω}_{PQ} is calculated as the Finsler distance between $h_\phi(P), h_\phi(Q)$. In particular, it depends only on the spectral curve Σ. Independence of the choice of ultrafilter means that the ultrafilter limit used to define the exponent is actually a limit, and we obtain a calculation of the WKB exponents for our singular perturbation.

Turn now to the reduction process for constructing $B^{\rm pre}_\phi$. A conjectural yet detailed picture of this construction was set out in [**15**] and readers are referred there for a full explanation. The first step was to make an initial construction. That is essentially what we shall be calling $Z^{\rm init}$ below, although the initial construction for the pre-building has to include additional small parallelogram-shaped regions corresponding to the folded pieces $\widetilde{Q}_i$ that we'll meet in §3 below. Our $Z^{\rm init}$ has these trimmed off.

The main problem of the initial construction is that it contains points of positive curvature, referred to as 4-fold points or $\mathbf{4}_2$ points below. These have to collapse in some way under any harmonic map to a building B since B is negatively curved. The construction of the pre-building consists of successively doing such a collapsing operation.

Unfortunately, the direction of collapsing at a 4-fold point is not well-defined, rather there are two possibilities. For that reason, the notion of *scaffolding* was introduced in [**15**]. The initial scaffolding formalizes the existence of small neighborhoods U that map, in an unfolded way, into apartments of the building. It follows that U should not be folded in the map h_ϕ to $B^{\rm pre}_\phi$. The edges gotten by gluing together sectors are, on the other hand, to be folded further, and this collection of data is complete: every edge is either marked as folded or unfolded (open) in the scaffolding.

The scaffolding tells us which direction to fold at a $\mathbf{4}_2$ point. The remaining difficulty is to propagate the information of the scaffolding into the new constructions obtained by collapsing. It was conjectured in [**15**] that this should be possible, under the hypothesis of absence of BPS states in the original spectral network.

The purpose of the present paper is to prove this conjecture on propagation of the scaffolding. We show how to make a series of reduction steps and how to propagate the scaffolding and other required information so that the series of reduction steps is well-defined.

The present work does not yet result in a full construction of $B^{\rm pre}_\phi$. Notably missing is a convergence statement saying that the process stops in a (locally) finite number of steps. Some parts of our arguments are also sketches rather than full proofs, and we don't provide here a justification for the choice of initial construction (Principle 5.2).

It was observed in [**15**] that each step of the reduction process towards the pre-building, could be accompanied by a trimming of the just glued-together parallelograms. If one does that, then the sequence of constructions is a sequence of 2-dimensional surfaces. This point of view will be useful for the program of generalizing Bridgeland-Smith's work on stability conditions, discussed briefly in §9 at the end. Also, the combinatorics of the reduction steps are all contained in this sequence of surfaces. Therefore, in the present paper we shall include the trimming operation as part of our reduction steps. In order to construct the pre-building

$B_\phi^{\rm pre}$ one should put back in the pieces that were trimmed off—the procedure for doing that was explained in [**15**].

2. Spectral networks in X

Start by considering spectral networks in X. Away from the branch points, the spectral curve Σ consists of three holomorphic 1-forms ϕ_1, ϕ_2, ϕ_3 and setting $\phi_{ij} := \phi_i - \phi_j$ these define *foliation lines* f_{ij} where $\Re\phi_{ij} = 0$. Assume the ramification of Σ consists of simple branch points. At a branch point, two indices are interchanged, picking out one of the foliations $\Re\phi_{ij} = 0$ whose singular leaves starting from the branch point are the initial edges of the spectral network.

Away from the *caustics* where the three foliation lines are tangent, the coordinates $\Re\phi_i$ define a flat structure on X by local identification with the standard apartment

$$A := \{(x_1, x_2, x_3) \in \mathbb{R}^3, \ \ x_1 + x_2 + x_3 = 0\} \cong \mathbb{R}^2$$

for buildings of the group $SL(3)$. We use this flat structure when speaking of angles.

We are going to add some extra singularities, so suppose given a finite subset $S \subset X$ and for each $p \in S$ one of the three foliations f_p. A *spectral network graph* is a map from a trivalent graph G, that can have endpoints as well as at most one infinite end, to X that sends endpoints of G to elements of S, that sends edges to foliation lines, and that sends trivalent vertices to *collisions* [**7**], points where foliation lines f_{ij}, f_{jk}, f_{ik} meet at $120°$ angles. We require that an endpoint of the graph going to $p \in S$ has adjoining edge going to a foliation line for the given foliation f_p.

A foliation line in X is an *SN-line* if there exists a spectral network graph such that the given foliation line is in the image of an edge of the graph that is either adjacent to an endpoint, or is an infinite end.

A *BPS state* [**6–8**] is a compact spectral network graph. Our main hypothesis will be that these don't exist.

We now note how to add singularities while conserving this hypothesis.

The original set S_0 of singular points is equal to the set of ramification points of the spectral curve, assumed to be simple ramifications. The foliation line f_p at a ramification point is the one determined by the two sheets of the spectral curve that come together at that point.

The following proposition will allow us successively to add points to the set of singularities.

PROPOSITION 2.1. *Suppose given a set of singularities S_{i-1} such that the resulting spectral network doesn't have BPS states. Choose a point p_i in general position along the interior of a caustic curve, let f_{p_i} be one of the foliation lines at p_i and let $S_i := S_{i-1} \cup \{p_i\}$. Then the spectral network associated to S_i also doesn't have any BPS states.*

PROOF. Suppose we are given a spectral network graph $\beta : G \to X$ that is a BPS state for S_i. Consider a nearby point $p_i(\epsilon)$ obtained by moving it a small distance ϵ in one direction along the caustic curve (the caustic is generically transverse to the foliation lines otherwise it would constitute a BPS state itself). We may assume that the BPS state follows to a nearby one $\beta(\epsilon) : G \to X$. The foliation lines are defined by differential forms $\Re\phi_{ij}$ so this gives us a way of measuring transverse distances; in these terms it makes sense to talk about the distance between one of the foliation lines and an adjacent one for the same foliation. We should choose for each edge of the graph a sign for the form in question. Now, label edges of the graph $e \in \text{edge}(G)$ by integers $k(e)$ such that the edge e moves by $k(e)\epsilon$. An edge e' adjoining an endpoint of the graph that goes to a previous singularity $q \in S_{i-1}$ has to be labeled with $k(e') = 0$, since q doesn't move with ϵ. There is a balancing condition on the labels at the collision points. The edges that end in p_i from one direction have to be labeled by $k = 1$ whereas the edges that end in p_i from the other direction have to be labeled by $k = -1$. However, the balancing condition at the trivalent vertices plus the condition that all the other endpoint labels are zero, means that the number of $k = 1$ labels and the number of $k = -1$ labels at p_i have to be equal. Therefore, we can pair up edges coming in from one direction with edges going out in the other, glueing these edges together pairwise. This results in a new graph $G' \to X$ that is a BPS state for the previous set of singularities S_{i-1}. □

Boundedness—Let us mention a boundedness hypothesis that will be useful. Suppose that there exists a compact subset $K \subset X$ whose boundary ∂K consists of foliation lines, such that the corners of ∂K are convex in the sense that each corner consists of foliation lines separated by two sectors inside K. And, we suppose that all ramification points of the spectral curve, and all caustics, are contained in the interior of K. Since the additional singularities were chosen on caustics, it follows that $S \subset K$.

With this hypothesis, a spectral network graph has no collisions outside of K, and any edge that leaves K continues as an infinite end. Indeed, all caustics are contained in the interior of K, so $X - K$ has a flat structure modeled on the standard apartment A. Suppose $x \in X - K$ were a collision point, say the first one outside of K. Then the two incoming edges must exit from K, but two distinct foliation lines that exit from K and meet, must exit from the same edge because of the convexity hypothesis on the corners. If they exit from the same edge and meet, then they must meet at a $60°$ angle, contradicting the hypothesis that they form incoming edges of a collision. Thus, such a collision cannot occur.

Referring to the reduction steps that will be discussed later, any regions that are to be folded together have to stay inside K by the same kind of considerations. Therefore, we may follow our compact subset along into the sequence of constructions Z that will occur; the compact subset will contain all modifications, and $X - K$ remains untouched as a flat space contained in each of our series of constructions.

Although we don't discuss the question of convergence in the present paper, the existence of a series of compact subsets that contain all the modifications means that one could envision an argument using decrease of the area to conclude termination of the reduction process. That would require bounding from below the size of the reduction steps.

3. The initial construction

The *initial construction* is a construction Z^{init} obtained by a first process of folding together certain regions in X and trimming away the resulting pieces. We consider a collection of regions $Q_i \subset X$ covering the caustics, such that Q_i are bounded by foliation lines meeting in two points q_i, q_i' symmetrical with respect to the caustic. In any ϕ-harmonic map to a building $h : X \to B$, q_i and q_i' have to map to the same point, and indeed $h|_{Q_i}$ has to factor through a map $\widetilde{Q}_i \to B$ where $\widetilde{Q}_i$ is Q_i folded in two along the caustic [**14, 15**].

Here is a picture of the regions Q_i in X

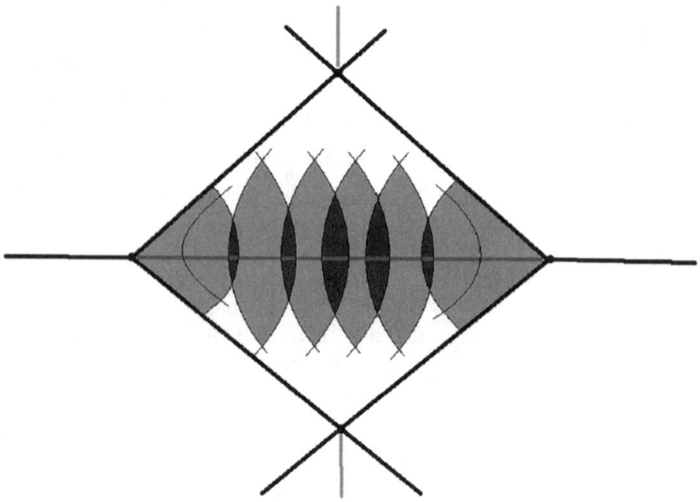

and here is what their images look like in any harmonic map to a building:

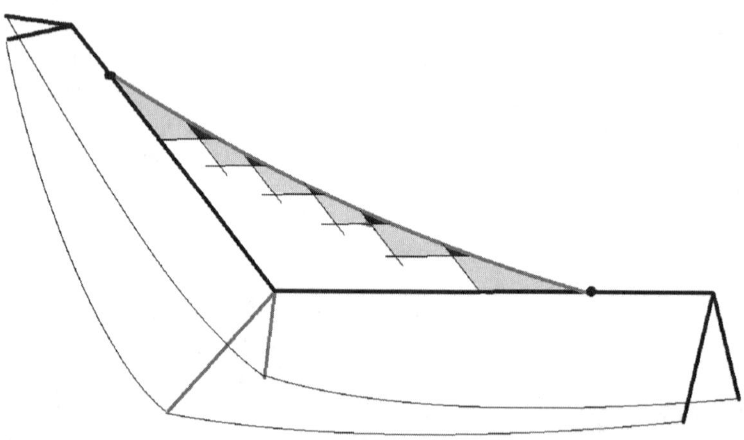

We assume that the boundaries of the regions Q_i are formed by new spectral network lines gotten after adding singularities along the caustics according to Proposition 2.1. That will be used for the refracting property in §5.

Let $\widetilde{X}$ denote the quotient of X by the equivalence relation induced by the quotients $Q_i \to \widetilde{Q}_i$. Now let $X^{\rm init}$ be the result of trimming off the folded-together pieces, in other words it is the closure in $\widetilde{X}$ of $X - \bigcup Q_i$, or equivalently the image of $X - \bigcup Q_i^o$.

The space $X^{\rm init}$ is a topological surface, and we have cut out the caustics. Thus, $X^{\rm init}$ is provided with a geometric structure locally modeled on the standard apartment A; the three foliation lines at any point correspond to the three standard directions in A. In particular $X^{\rm init}$ has a flat metric. The conformal structure for this flat metric is different from the original structure of Riemann surface on X. It has singular points of positive and negative curvature, namely 8-fold points of negative curvature whose total angle is 480° and 4-fold points of positive curvature whose total angle is 240°.

View $X^{\rm init}$ as corresponding to a construction, i.e. a presheaf on the site of enclosures as discussed in [**15**]. More precisely we have a construction $Z^{\rm init}$ whose usual set of points is $Z^{\rm init}(p) = X^{\rm init}$. It is provided with a map

$$h^{\rm init} : X - \bigcup Q_i^o \to Z^{\rm init}(p)$$

that sends foliation lines in X to segments in $Z^{\rm init}$.

This construction will be the starting point of our reduction process. It is provided with a *scaffolding* [**15**]: certain edges (the image of $\partial \bigcup Q_i^o$) are designated as "fold edges" while all the other edge germs are designated as "open" or "unfolded". If $h : X \to B$ is any ϕ-harmonic map to a building, then its restriction to $X - \bigcup Q_i^o$ factors through a map $Z^{\rm init} \to B$ that respects the scaffolding, in the sense that fold edges are folded and open edges are unfolded.

4. Structures

Let Z be a construction. We say that Z is *complete* if the associated metric space is complete. It is *normal* if the link at any point is connected. We say that Z is *ecarinate* if, at any edge there are two half-planes. Notice that $Z^{\rm init}$ satisfies these conditions, and our reduction process will conserve them, so let us consider only complete normal ecarinate and simply connected constructions Z. In particular the set of usual points $Z(p)$ is a complete 2-manifold.

If $z \in Z(p)$ is a point then the link Z_z is a connected graph such that each vertex (corresponding to a germ of edge at z) is contained in two edges (corresponding to germs of sectors at z). Therefore, the link is a polygon. By the parity property the polygon has an even number of edges. We assume that the number of edges in the link at any point is 4, 6 or 8. Most points are *flat*, meaning that their links are hexagons. The other points are called 4-fold, or 8-fold respectively. The 4-fold points are points of positive curvature, and the 8-fold points are points of negative curvature.

A *scaffolding* consists of the following structures:

(1) A marking of each edge in Z as either *open* (o) or *folded* (f), such that the fold edges are those from a discrete collection $\mathbf{F}$ of straight edges in Z.

(2) An orientation assigned to each fold edge.

(3) A marking of some subset of the fold edges said to be *refracting*.

Our initial construction $Z^{\rm init}$ contains a scaffolding in which all the fold edges are already marked as refracting, see §5 below, and our reduction process will preserve the refracting condition so we henceforth consider only fully refracting scaffoldings.

We assume that $Z^{\rm init}$ and its scaffolding have the property that there exists a harmonic mapping to a building such that the fold edges are folded and the open edges are unfolded. This constrains the local type of singularities. However, the number of possibilities is still rather large.

We describe here a list standard examples that will be sufficient for our reduction process—the statement that we remain within this standard list is indeed one of the main conclusions of our treatment in the present paper.

The notation will consist of a boldface number saying how many sectors there are, and a subscript saying how many folded lines in the scaffolding there are. Arrows in the pictures indicate the orientations of the scaffolding edges.

For example, points of type $\mathbf{6}_0$ and $\mathbf{8}_0$ are respectively 6-fold and 8-fold points with no adjoining fold edges. A point of type $\mathbf{6}_2$ is a 6-fold point with a single straight fold edge (it is nonsingular); a $\mathbf{6}_3$ point has three folded edges separated by 120°, and a $\mathbf{6}_4$ point is the same with an additional folded edge. These may be pictured as follows:

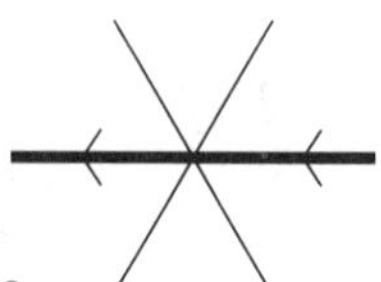
$\mathbf{6}_2$

$\mathbf{6}_3$

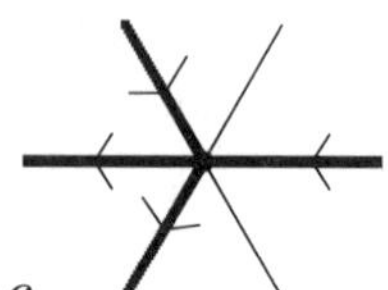
$\mathbf{6}_4$

Next we picture the 8-fold and 4-fold points. Note that the pictures cannot be conformally correct for the angles; all drawn sectors represent sectors of 60° in Z.

The 8-fold points have at most 2 fold edges; and if there are 2 of them, they are separated by either 1 or 4 sectors. The fold edge orientations go outward, so after the $\mathbf{8}_0$ picture there are three possibilities:

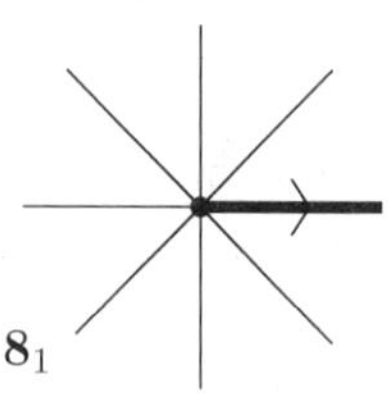
$\mathbf{8}_1$

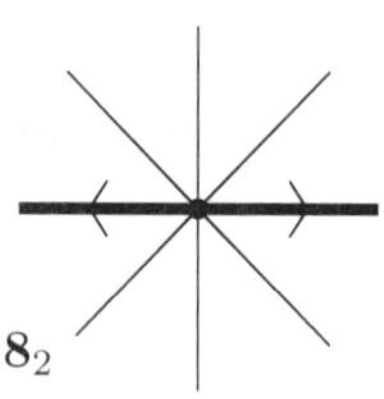
$\mathbf{8}_2$

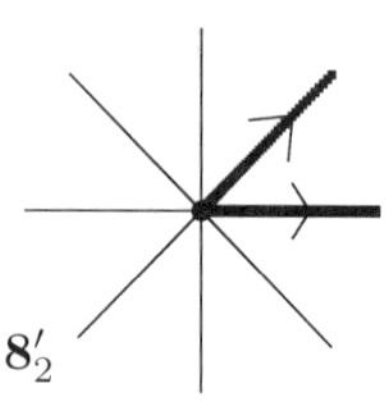
$\mathbf{8}'_2$

Recall **[15]** that at a 4-fold point, at least two of the four edges are folded, and if an edge is folded then so is the opposite one. Our standard type is when only two edges are folded, and the edge orientations are inwards towards the singularity, so it has the following picture:

$\mathbf{4}_2$

DEFINITION 4.1. We say that the singularities of the scaffolding are *initial* if at any point of Z the picture is either $\mathbf{6}_0$ (a smooth point not on the scaffolding), $\mathbf{6}_2$ (an interior point of an edge of the scaffolding), an 8-fold point of the form $\mathbf{8}_1, \mathbf{8}_2, \mathbf{8}'_2$, or a 4-fold point $\mathbf{4}_2$. We say that the singularities of the scaffolding are *standard* if at any point the local picture is one of the ones we have described, namely:

$$\mathbf{6}_0,\ \mathbf{6}_2,\ \mathbf{6}_3,\ \mathbf{6}_4,\ \mathbf{8}_0,\ \mathbf{8}_1,\ \mathbf{8}_2,\ \mathbf{8}_2',\ \text{or}\ \mathbf{4}_2$$

shown above.

Included in the above definitions are compatibility of the orientations of the fold lines with the singularity types as drawn in the pictures. Recall that edges are oriented outward at 8-fold points and inward at 4-fold points. For initial scaffoldings we obtain the following characterization:

REMARK 4.2. Given a scaffolding with initial singularities, the collection of fold lines decomposes into a disjoint union of piecewise linear curves called the *post-caustics*, whose endpoints are of type $\mathbf{8}_1$, and along which the 4-fold points alternate with 8-fold points. In particular, the number of 8-fold points on a connected post-caustic is 1 more than the number of 4-fold points.

5. The refraction property

Using Proposition 2.1 to add spectral network lines along the boundaries of the regions Q_i used to define the initial construction, will give that the scaffolding of the initial construction $Z^{\rm init}$ is a fully refracting scaffolding. The "refracting properties" with respect to the spectral network are:

(R1) At singular points of the scaffolding, all unfolded edges are initial edges of the spectral network; and

(R2) When a spectral network line crosses over a fold edge, it can continue on the other side in either one of the two available directions.

Suppose we are given a construction Z and a fully refracting scaffolding. A *spectral network graph for the refracting spectral network* is a map from a graph $G \to Z$ satisfying the following properties:

—edges of G go to segments in Z that are straight except when they cross the fold edges;

—edges do not go along fold edges of the scaffolding;

—when edges cross over the fold edges they can "refract", that is to say they go out of the fold edge on the other side in either one of the two directions;

—trivalent vertices of the graph go to collisions at $\mathbf{6}_0$ points i.e. points not on fold edges of the scaffolding; and

—endpoints of the graph go to singular points of the scaffolding or $\mathbf{8}_0$ points of Z, with the adjoining edges going outward in any non-fold directions.

Before choosing the regions $Q_i \subset X$ that will be folded and trimmed to get $X^{\rm init}$, add points to the set of singularities of our spectral network, using Proposition 2.1. Add these points along the caustic curves at the places where the boundaries of the Q_i meet the caustics, so that the boundary curves of the Q_i become spectral network curves. Notice furthermore that at the corners q_i, q_i', all three directions outward to the rest of X are spectral network lines, two from continuing the boundary curves and the third by collision of the two boundary curves.

Our above somewhat heuristic discussion leads to the following principle.

PRINCIPLE 5.1. If the original spectral network of X had no BPS states, then the refracting spectral network of $Z^{\rm init}$ has no BPS states.

We also need to know something about the arrangement of singularities of the scaffolding for Z^{init}. We state this as another principle:

PRINCIPLE 5.2. The initial construction Z^{init} may be chosen to have only initial singularities (Definition 4.1). The fold edges of the scaffolding are thus arranged into post-caustics as described in Remark 4.2.

We don't give here a formal justification for the possibility of choosing the initial construction in this way, but note that it is what happens in the pictures we have considered. For example, see [**15**] for a picture leading to an $\mathbf{8}'_2$ point.

6. Reduction

This section begins the discussion of a step in the reduction process. The first question is to show the existence of a $\mathbf{4}_2$ point about which some collapsing can be done.

PROPOSITION 6.1. *Suppose given a construction with a refracting scaffolding, such that there are no BPS states in the resulting spectral network. Suppose that the fold edges of the scaffolding are oriented and the singularities are all from our standard list. Make a directed graph using the singularities as vertices, except that we separate a* $\mathbf{6}_4$ *singularity into two vertices. The edges of the graph are the fold edges with their orientations; at a* $\mathbf{6}_4$ *point the "spine" (consisting of the two fold edges that are opposite) goes to one of the vertices and the other two edges go to the other one. This directed graph has no directed loops.*

PROOF. (Sketch)—Consider a path parallel to a directed loop, just to one side of it. This will satisfy the collision and refracting conditions, so taken together with the appropriate initial SN lines coming from singularities, it constitutes a BPS state. □

COROLLARY 6.2. *If there are no BPS states and if the set of fold lines in the scaffolding is nonempty, then there must be a* $\mathbf{4}_2$ *point.*

PROOF. In the directed graph described in Proposition 6.1, our hypothesis that there are no BPS states implies that there are no directed loops. Therefore, the orientations of edges make the graph into a poset; since it is finite it has a minimal vertex. The only type of point on the scaffolding that has all fold edges pointed inward is a $\mathbf{4}_2$ point, therefore there exists a $\mathbf{4}_2$ point. □

We now consider an extended collapsing operation at a $\mathbf{4}_2$ point a. For this, consider two pieces R_1 and R_2 that match up, and share a common vertex a and common edges f and f'. Let b (resp. b') denote the endpoint of f (resp. f') different from a. The R_i are assumed to be constructions that can be considered as isomorphic to a subset R in an abstract parallelogram P. Let us also label the corresponding vertex a and the corresponding edges f, f' in P. The embeddings $R_i \cong R \subset P$ preserve a, f, f'. We can now describe the configuration of R: the vertex a is an obtuse vertex of P, and f, f' are the full edges of P meeting a. Hence the points also labelled $b, b' \in P$ are the two acute vertices. We assume that R is a union of finitely many sub-parallelograms $R(2j-1) \subset P$ (see Figure 1 for the numbering) such that $R(2j-1)$ contains a as obtuse vertex, that is to say the edges of $R(2j-1)$ will be segments in f, f' starting at a. Let $t(2j-1)$ denote the vertices of $R(2j-1)$ opposite to a. We assume that the order $R(1), R(3), \ldots, R(2k+1)$ is

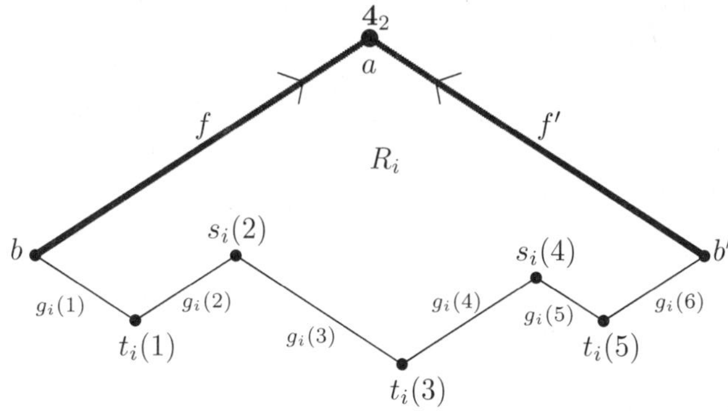

FIGURE 1. Maximal collapsing region

such that $t(1)$ is on the same edge of P as b, and they go in a consecutive sequence until $t(2k+1)$ is on the same edge of P as b'. The $t(1), t(3), \ldots, t(2k+1)$ are the other convex corners of R after a, b, b'. Let $s(2), s(4), \ldots, s(2k)$ denote the concave corners of R in between them, so that $s(2j)$ lies between $t(2j-1)$ and $t(2j+1)$. Let $g(1)$ be the edge from b to $t(1)$ and let $g(2k+2)$ be the edge from $t(2k+1)$ to b'. Let $g(2j)$ be the edge from $t(2j-1)$ to $s(2j)$ and $g(2j+1)$ be the edge from $s(2j)$ to $t(2j+1)$. Thus, $g(2j)$ and f are parallel, and $g(2j-1)$ and f' are parallel.

Now consider the same points in the regions R_i, indicated as $s_i(2j)$ and $t_i(2j-1)$, with edges $g_i(j)$. The configuration of this maximal collapsing pair of regions may be pictured as in Figure 1.

We assume that R is a maximal such region with corresponding regions $R_i \subset Z$, such that the following conditions are satisfied:

—there are no singularities in the interior of R_i;

—the only singularities on the interior of the edges f, f' are $\mathbf{6}_4$ points where f or f' is the straight fold edge (spine).

PROPOSITION 6.3. *Under the above maximality conditions, we have the following properties:*

1. The vertices b, b' are singularities, and furthermore these singularities are not $\mathbf{6}_4$ points with edge f or f' on the spine;

2. For each j there is exactly one of the $s_1(2j), s_2(2j)$ that is an 8-fold singularity, and the other is nonsingular;

3. There are no 8-fold singularities in the interiors of the edges $g_i(j)$;

4. If q_1 is a 4-fold or singular 6-fold point on an edge $g_1(j)$ of R_1 then the corresponding point q_2 on R_2 is not a singularity, and vice-versa, and this also applies for corners $t_i(2j-1)$.

PROOF. Notice in general that there can't be a fold edge going from a point on one of the edges $g_i(j)$ into R_i in the middle direction between the directions of f and f' (vertical in Figure 1); that would have to meet the 4-fold point making it of type $\mathbf{4}_4$, or meet an edge in a $\mathbf{6}_4$ point but oriented in the wrong direction.

It follows that two corresponding points $q_1 \in R_1$ and $q_2 \in R_2$ can't both be singularities, for they are joined by a common foliation line that refracts at an edge f, f'; we have seen that it can't be a fold line so it would be an initial SN-line from both singularities resulting in a BPS state. We get 4 and part of 2.

For 1, if b (resp. b') were nonsingular, or a $\mathbf{6}_4$ point with edge f (resp. f') along the spine, then one could continue the regions R_i further along the corresponding edge.

For 2, if both points are nonsingular then the regions are not maximal, so one must be singular, say on R_1. However, if it were a 6-fold point then both lines going into R_1 parallel to f and f' would be fold lines; these would have to meet f and f' in $\mathbf{6}_3$ or $\mathbf{6}_4$ points, resulting in fold lines going back in the opposite direction in R_2. These two would meet in the corresponding point of R_2 but that would have to be singular, contradicting the statement above.

For 3, notice that if some segments of one of these edges are folded, then the orientations of the fold segments are all the same (it follows from the consideration of the first paragraph). But if we had an 8-fold point on the interior of one of these edges, it would send out either a fold line or an SN-line, and that would contradict the existence of the 8-fold singularity given in part 2. □

We now consider how fold lines, SN lines and singularities can be arranged along a pair of edges $g_1(j), g_2(j)$. Each edge $g_i(j)$ has two endpoints, $q_i^{(0)}$ that is either $s_i(j'')$ or b, b', and $q_i^{(m_{ij}+1)} = t_i(j')$.

LEMMA 6.4. *One of the two edges $g_1(j)$ or $g_2(j)$ is an SN line, and has no singularities in its interior (so $m_{ij} = 0$), or at the endpoint $q_i^{(1)}$. It points towards the endpoint $q_i^{(1)} = t_i(j')$.*

PROOF. One of the two $q_i^{(0)}$ is an 8-fold point. At our admissible possibilities $\mathbf{8}_0, \mathbf{8}_1, \mathbf{8}_2, \mathbf{8}_2'$ there are never two fold edges separated by three sectors. Therefore, at this point either the outgoing edge along $g_i(j)$, or the edge that goes into R_i in the opposite direction (that is to say separated by three sectors interior to R_i), are SN-lines. If $g_i(j)$ is an SN-line then we obtain the desired conclusion. Notably, there are no singular points along this edge otherwise that would create a BPS state. In the other case, the SN line reflects at f or f' and comes back in the other region $R_{i'}$, eventually joining the edge $g_{i'}(j)$ so that edge is an SN line. Again in that case it has no singularities. The SN line points away from $q_i^{(0)}$ in both cases, so towards $q_i^{(1)}$. □

On the edge $g_i(j)$ not concerned by the above lemma, let $q_i^{(1)}, \ldots, q_i^{(m_{ij})}$ denote the singularities in order along the interior of $g_i(j)$, starting from the nearest to $q_i^{(0)}$ (recall that was $s_i(j'')$ or b, b'). The internal singularities are 4 or 6-fold points.

LEMMA 6.5. *All the segments $q_i^{(u-1)}q_i^{(u)}$ are fold lines for $1 \le u \le m_{ij}$. Furthermore, for any $1 \le u < m_{ij}$ the singularity $q_i^{(u)}$ is a $\mathbf{6}_4$ point with the connecting segments of the edge g_i being its spine. These are oriented in the direction away from $q_i^{(0)}$.*

PROOF. If any of these segments were SN lines that would create a BPS state. Note that if $q_i^{(0)}$ is not the singular one of the two points $s_1(j''), s_2(j'')$, then an SN line from $q_i^{(1)}$ to $q_i^{(0)}$ would continue to one of the edges f, f' and reflect and hit the corresponding point $s_{i'}(j')$ on the other piece, that is then a singularity by part 2 of Proposition 6.3.

If $1 \leq u < m_i$ then both segments $q_i^{(u-1)}q_i^{(u)}$ and $q_i^{(u)}q_i^{(u+1)}$ are fold edges. The only type of singularity with two fold edges separated by three sectors is a $\mathbf{6}_4$ point and these edges are the spine. For the orientation, note that all the segments from $q_i^{(0)}$ up to $q_i^{(u+1)}$ are fold edges, and $q_i^{(0)}q_i^{(1)}$ is oriented outwards from $q_i^{(0)}$. By the rule for orientations at $\mathbf{6}_4$ it follows inductively that all the segments are oriented the same way. □

When we do our glue and trim operation, the combination of any edge with a folded edge has the same marking as the first edge. So, for the problem of deciding which edges are in the scaffolding, the only indeterminacy is whether the edge $q_i^{(m_{ij})}q_i^{(m_{ij}+1)}$ is a fold edge or an SN edge. This could mean the entire edge $g_i(j)$ if $m_{ij} = 0$.

The following lemma says that the spectral network or fold lines created in the glued and trimmed construction will satisfy the refracting property.

LEMMA 6.6. *If the edge $q_i^{(m_{ij})}q_i^{(m_{ij}+1)}$ is an SN line, then it points in the direction towards the endpoint $q_i^{(m_{ij}+1)} = t_i(j')$. As a corollary, if a spectral network line crosses into R_1 from anywhere across the edge g_1, it is reflected from f or f' and creates two SN lines going out from R_2 at the corresponding point in both outward directions. Same with $1, 2$ interchanged.*

PROOF. The point $q_i^{(m_{ij})}$ is a singularity so all outgoing non-folded edges are initial for the spectral network. □

7. The new construction

Let us now proceed with the collapsing operation of glueing together R_1 and R_2, and trimming off the resulting piece. After doing this we obtain the new construction Z^{new}. It is again ecarinate, complete, normal and simply connected.

Let $g(j)$ denote the edges in Z^{new} obtained by identifying $g_1(j) \subset R_1$ with $g_2(j) \subset R_2$. Similarly let $s(j)$ (resp. $t(j)$) denote the points resulting from the identification of $s_1(j)$ and $s_2(j)$ (resp. $t_1(j)$ and $t_2(j)$). The images of b, b' are again denoted b, b'. We identify $Z - (R_1 \cup R_2)$ with $Z^{\mathrm{new}} - \bigcup g(j)$.

A point $s(j)$ is obtained by gluing together an 8-fold point and a 6-fold point, having removed 4 sectors from each one. It follows that $s(j)$ is a 6-fold point.

If say $t_1(j)$ is a $\mathbf{4}_2$ then $t_2(j)$ is a $\mathbf{6}_0$ point, indeed the edges $g_1(j)$ and $g_1(j+1)$ must be folded so by Lemma 6.4 both edges $g_2(j)$ and $g_2(j+1)$ are SN lines. It follows in this case that $t(j)$ is a $\mathbf{6}_0$ point.

If $t_i(j)$ are both 6-fold points then $t(j)$ is an 8-fold point.

Suppose that an edge say $g_1(j)$ contains some singularities $q_1^{(1)}, \ldots, q_1^{(m)}$. As we have seen, $q_1^{(1)}, \ldots, q_1^{(m-1)}$ are $\mathbf{6}_4$ points with spine along $q_1(j)$. One can see that these become $\mathbf{6}_2$ points in Z^{new}. The segments of $g_1(j)$ that are folded, become unfolded segments in Z^{new} since the opposite edge $g_2(j)$ is an SN line and a folded segment glued to an unfolded one yields an unfolded segment in the new construction.

In order to determine the scaffolding of Z^{new} it remains to see what becomes of the last segment $q_1^{(m)}q_1^{(m+1)}$, in case $g_1(j)$ contains some singularities, or of the whole of $g(j)$ in case neither side contains singularities in the interior. This depends on the singularities at $t_i(j')$ and on the last singular point $q_1^{(m)}$ if it is there, or else on $s_i(j'')$.

One must do an analysis of cases. The conclusion is that the labeling of the resulting segment of $g(j)$ as folded or unfolded, is determined by these singularities. In some cases the answer is indeterminate at $t(j')$ but determined by the other endpoint. The determinations of fold/unfold coming from the two ends of the segment must be the same, because we are supposing the existence of some harmonic map to a building compatible with the scaffolding.

If a segment becomes folded, then it is oriented outwards from the 8-fold point $t(j')$ (if it is a folded segment then we were not in the case where one $t_i(j')$ is a 4-fold point, treated above).

We have seen in Lemma 6.6 that such a folded segment will satisfy the refracting condition for reflection of SN lines. We have determined the scaffolding of Z^{new}, satisfying the refracting condition, and with orientations of edges.

Notice that the two edges meeting in $t(j)$ cannot both have singular points in the interior, as that would have meant that there was an intersection of fold lines in the interior of R_1 or R_2.

What remains to be verified is that the new points $s(j)$, $t(j)$, and the images of the $q_i^{(m)}$ if they are there, fall into the standard list of possible singularities; that the required spectral network lines emanating from these points exist; and that the orientations of fold lines are compatible with the allowable orientations at these new singularities.

Discuss first the compatibilities of orientations of fold lines. An 8-fold point $s_i(j)$ becomes a 6-fold point $s(j)$. Any fold lines not on the edges $g(j), g(j+1)$ will stay oriented in the outgoing direction. There may be one or two new fold lines on the edges $g(j), g(j+1)$, but these will now orient inwards towards $s(j)$. Indeed, this only happens if $g(j)$ (resp. $g(j+1)$) has no interior singularities and $t(j')$ is an 8-fold point. See Table 1 below for compatibility.

Consider singularities interior to the edges. Suppose say $g_1(j)$ contains a sequence of singular points with last one $q_1^{(m)}$. Let $q^{(m)}$ denote its image in Z^{new}. The segment $t(j')q^{(m)}$ might or might not be folded. If it is folded, it is oriented outward from $t(j')$ hence inward towards $q^{(m)}$. We should check that this is compatible with the singularity type of $q^{(m)}$. If $q_1^{(m)}$ is a 4-fold point then so is $q^{(m)}$ and this compatibility holds. If our new segment is folded it means that the previous segment $q_1^{(m)}t_1(j')$ had to be unfolded. Recall that $q_1^{(m-1)}q_1^{(m)}$ is folded and oriented towards $q_1^{(m)}$. The direction into R_1 that is parallel to f or f' is folded, and this fold line hits one of the edges f or f' at a $\mathbf{6}_4$ point getting reflected back into R_2 and eventually forming a fold line that will participate in the local picture at the point $q^{(m)}$.

Using these facts there are two possibilities. First, $q_1^{(m)}$ could be a $\mathbf{6}_3$ point whose fold edge along $g_1(j)$, the segment $q_1^{(m-1)}q_1^{(m)}$, is oriented inwards. The new segment $t(j')q^{(m)}$ is not folded, and the resulting point $q^{(m)}$ is a nonsingular $\mathbf{6}_2$ point with compatible orientations of fold edges (see the middle two lines of Table 2).

The other possibility is that $q_1^{(m)}$ be a $\mathbf{6}_4$ point. In this case the new segment $t(j')q^{(m)}$ must be folded, and oriented inwards towards $q^{(m)}$ since $t(j')$ is an 8-fold point, whereas $q^{(m-1)}q^{(m)}$ is unfolded. The resulting point $q^{(m)}$ is a $\mathbf{6}_4$ point. The inward oriented edge of its spine comes from the inward oriented edge of the spine of $q_1^{(m)}$ reflected into R_2. See the next-to-last line of Table 2 below for the other fold edges and their orientations.

This completes the discussion of compatibility of the orientations of new scaffolding edges.

We need to verify that the new singular types are all contained in the list that we are using. This will be done by writing down tables of the possibilities. The cases of the points b, b' are similar and are left to the reader.

Let us assume that we have a singular point on R_1. Also, for singularities inside edges, we consider edges $g_1(j)$ that are parallel to f. The other cases are the same by symmetry.

It will be convenient to establish a convention for speaking of directions in R_1, R_2 at a singularity s, t, q. Number them as follows: $\langle 0\rangle$ corresponds to the direction from our singularity, towards the interior of R_i, not parallel to f or f' ("up" in the picture). Then $\langle 1\rangle, \langle 2\rangle, \ldots$ are the directions obtained by turning clockwise $1, 2, \ldots$ sectors. On the other side, $\langle -1\rangle, \langle -2\rangle, \ldots$ are the directions obtained by turning counterclockwise that many sectors. These join up: at a 6-fold point $\langle 3\rangle = \langle -3\rangle$ while at an 8-fold point $\langle 4\rangle = \langle -4\rangle$.

The edge f is oriented $\langle 1\rangle$, and f' is oriented $\langle -1\rangle$.

In our tables below we will be listing the folded directions at singular points. In this case, a notation $\langle i\rangle$ means the edge germ emanating from the singularity in the specified direction. With this notation, in the orientation of our scaffolding, the fold edge is said to be oriented outwards. The same edge oriented inwards towards the singularity will be denoted $\overline{\langle i\rangle}$.

We now consider a singularity of the form $s_1(j)$. It is an 8-fold point, and the corresponding $s_2(j)$ is a nonsingular 6-fold point (either $\mathbf{6}_0$ or $\mathbf{6}_2$). These glue together to form a point $s(j) \in Z^{\text{new}}$. Table 1 gives the structure of the scaffolding at $s(j)$ as a function of the structure at $s_1(j)$.

In order to fill in the table, recall that four sectors are removed from the neighborhood of $s_1(j)$, as well as from the neighborhood of $s_2(j)$; then the remaining two sectors from $s_2(j)$ are glued back in to give the neighborhood of s. We make the convention that edge germs at $s(j)$ are numbered starting with the middle edge of the two sectors from $s_2(j)$ being $\langle 0\rangle$; the two indeterminate lines are $\langle 1\rangle, \langle -1\rangle$, then remaining $\langle 2\rangle, \langle 3\rangle, \langle -2\rangle$. These latter correspond to the directions $\langle 3\rangle, \langle 4\rangle, \langle -3\rangle$ respectively at $s_1(j)$.

We include a column in the table to say what is happening at the point $s_2(j)$. Note that it is a nonsingular 6-fold point, hence either $\mathbf{6}_0$ or $\mathbf{6}_2$. If it is $\mathbf{6}_2$ then the direction of the fold line comes from the direction of the fold line at $s_1(j)$ that goes in direction either $\langle 1\rangle$ or $\langle -1\rangle$. This extra column will be most useful in our third table below.

We don't include configurations that are obtained by symmetry (changing $\langle i\rangle$ to $\langle -i\rangle$) from ones that were already included, and we don't include configurations (such as $\mathbf{8}_1, \langle 0\rangle$) that are ruled out.

Along an edge $g_1(j)$ parallel to f suppose given singularities $q_1^{(1)}, \ldots, q_1^{(m)}$. We have seen that for $1 \leq u < m$ the $q_1^{(u)}$ have to be $\mathbf{6}_4$ points with spine along $g_1(j)$ resulting in a nonsingular $\mathbf{6}_2$ point in Z^{new} (as shows up in the first lines of the next table).

Let us consider now the configurations for $q_1^{(m)}$ and resulting configurations for $q^{(m)}$ in Z^{new}, shown in Table 2. Recall that in this case, the three sectors of R_1

TABLE 1. Structure at $s(j)$

$s_1(j)$	fold edges	$s_2(j)$	$s(j)$	new fold edges
$\mathbf{8}_0$	none	$\mathbf{6}_0$	$\mathbf{6}_0$	none
$\mathbf{8}_1$	$\langle 1\rangle_1$	$\mathbf{6}_2\ \overline{\langle 1\rangle}_2, \langle -2\rangle_2$	$\mathbf{6}_0$	none
$\mathbf{8}_1$	$\langle 2\rangle_1$	$\mathbf{6}_0$	$\mathbf{6}_0$	none
$\mathbf{8}_1$	$\langle 3\rangle_1$	$\mathbf{6}_0$	$\mathbf{6}_2$	$\langle 2\rangle, \overline{\langle -1\rangle}$
$\mathbf{8}_1$	$\langle 4\rangle_1$	$\mathbf{6}_0$	$\mathbf{6}_3$	$\langle 3\rangle, \overline{\langle 1\rangle}, \overline{\langle -1\rangle}$
$\mathbf{8}_2$	$\langle 1\rangle_1, \langle -3\rangle_1$	$\mathbf{6}_2\ \overline{\langle 1\rangle}_2, \langle -2\rangle_2$	$\mathbf{6}_2$	$\langle -2\rangle, \overline{\langle 1\rangle}$
$\mathbf{8}_2$	$\langle 2\rangle_1, \langle -2\rangle_1$	$\mathbf{6}_0$	$\mathbf{6}_0$	none
$\mathbf{8}'_2$	$\langle 1\rangle_1, \langle 2\rangle_1$	$\mathbf{6}_2\ \overline{\langle 1\rangle}_2, \langle -2\rangle_2$	$\mathbf{6}_0$	none
$\mathbf{8}'_2$	$\langle 2\rangle_1, \langle 3\rangle_1$	$\mathbf{6}_0$	$\mathbf{6}_2$	$\langle 2\rangle, \overline{\langle -1\rangle}$
$\mathbf{8}'_2$	$\langle 3\rangle_1, \langle 4\rangle_1$	$\mathbf{6}_0$	$\mathbf{6}_4$	$\langle 2\rangle, \langle 3\rangle, \overline{\langle 1\rangle}, \overline{\langle -1\rangle}$

and the three sectors of R_2 are cut out, and the remaining ones are glued together. We number the edges at the new point $q^{(m)}$ as follows: the edges $\langle 2\rangle, \langle 3\rangle$ exterior to R_1 keep the same numbers, whereas $\langle 2\rangle, \langle 3\rangle$ exterior to R_2 become respectively $\langle 0\rangle, \langle -1\rangle$ (in practice an edge $\langle -1\rangle$ at $q_1^{(m)}$ reflects becoming $\langle 2\rangle$ at the nonsingular point $q_2^{(m)} \in R_2$ opposite $q_1^{(m)}$ hence $\langle 0\rangle$ at $q^{(m)}$); the directions $\langle 1\rangle$ and $\langle -2\rangle$ correspond to the edge $g(j)$. Recall that the direction $\overline{\langle 1\rangle}$ is by hypothesis folded on $g_1(j)$ so the new edge $\langle 1\rangle$ is unfolded. If $\langle -2\rangle$ is folded at $q_1^{(m)}$ then it becomes unfolded at $q^{(m)}$, whereas if it is unfolded then it can become either.

TABLE 2. Structure at $q(m)$

$q_1^{(m)}$	fold edges	$q_2^{(m)}$	$q^{(m)}$	new fold edges
$\mathbf{6}_4$	$\overline{\langle 1\rangle}_1, \overline{\langle -1\rangle}_1, \langle -2\rangle_1, \langle 3\rangle_1$	$\mathbf{6}_2\ \langle -1\rangle_2, \overline{\langle 2\rangle}_2$	$\mathbf{6}_2$	$\langle 3\rangle, \overline{\langle 0\rangle}$
$\mathbf{6}_4$	$\overline{\langle 1\rangle}_1, \langle -1\rangle_1, \langle -2\rangle_1, \overline{\langle 3\rangle}_1$	$\mathbf{6}_2\ \overline{\langle -1\rangle}_2, \langle 2\rangle_2$	$\mathbf{6}_2$	$\langle 0\rangle, \overline{\langle 3\rangle}$
$\mathbf{6}_3$	$\overline{\langle 1\rangle}_1, \overline{\langle -1\rangle}_1, \langle 3\rangle_1$	$\mathbf{6}_2\ \langle -1\rangle_2, \overline{\langle 2\rangle}_2$	$\mathbf{6}_2$	$\langle 3\rangle, \overline{\langle 0\rangle}$
$\mathbf{6}_3$	$\overline{\langle 1\rangle}_1, \langle -1\rangle_1, \overline{\langle 3\rangle}_1$	$\mathbf{6}_2\ \overline{\langle -1\rangle}_2, \langle 2\rangle_2$	$\mathbf{6}_2$	$\langle 0\rangle, \overline{\langle 3\rangle}$
$\mathbf{6}_4$	$\overline{\langle 1\rangle}_1, \overline{\langle -1\rangle}_1, \langle 2\rangle_1, \langle 3\rangle_1$	$\mathbf{6}_2\ \langle -1\rangle_2, \overline{\langle 2\rangle}_2$	$\mathbf{6}_4$	$\langle 2\rangle, \langle 3\rangle, \overline{\langle 0\rangle}, \overline{\langle -2\rangle}$
$\mathbf{4}_2$	$\overline{\langle 1\rangle}_1, \overline{\langle -1\rangle}_1$	$\mathbf{6}_2\ \langle -1\rangle_2, \overline{\langle 2\rangle}_2$	$\mathbf{4}_2$	$\overline{\langle 0\rangle}, \overline{\langle -2\rangle}$

We now turn to the case of the singular points $t_1(j)$ glueing to the nonsingular $t_2(j)$ to yield $t(j)$. In this case, two sectors are removed from the neighborhood of $t_1(j)$, two sectors removed from the neighborhood of $t_2(j)$, and the remaining sectors are put back together. There are four remaining sectors from $t_2(j)$. We make the following labeling conventions, with subscripts indicating sectors coming from neighborhoods of $t_1(j)$ or $t_2(j)$:

$$\langle 2\rangle_1 \mapsto \langle 3\rangle, \quad \langle 3\rangle_1 \mapsto \langle 4\rangle, \langle -2\rangle_1 \mapsto \langle -3\rangle,$$

$$\langle 2\rangle_2 \mapsto \langle 1\rangle, \quad \langle 3\rangle_2 \mapsto \langle 0\rangle, \langle -2\rangle_2 \mapsto \langle -1\rangle,$$

and the indeterminate ones

$$\langle 1\rangle_1 \text{ or } \langle 1\rangle_2 \mapsto \langle 2\rangle, \quad \langle -1\rangle_1 \text{ or } \langle -1\rangle_2 \mapsto \langle -2\rangle.$$

In this case the structure of $t_2(j)$ is not determined by that of $t_1(j)$ since it could depend on the singularities along the adjacent edges, so it is included in the table. The possibilities are unfolded, $\mathbf{6}_2$ folded in direction $\overline{\langle 1\rangle}, \langle -2\rangle$ or $\mathbf{6}_2$ folded in direction $\overline{\langle -1\rangle}, \langle 2\rangle$. If for example there is a fold in direction $\overline{\langle 1\rangle}, \langle -2\rangle$ then it came from a fold line in direction $\langle 1\rangle$ at the singularity $s_1(j)$ that was reflected on the edge f', and in this case the edge $g_1(j)$ is unfolded, in particular the direction $\langle 1\rangle$ at $t_1(j)$ is unfolded. Similarly in the other direction. Again we omit cases that can be obtained by symmetry.

TABLE 3. Structure at $t(j)$

$t_1(j)$	fold edges	$t_2(j)$	$t(j)$	new fold edges
$\mathbf{6}_0$	none	$\mathbf{6}_0$	$\mathbf{8}_0, \mathbf{8}_1, \mathbf{8}_2$	$\langle 2\rangle?, \langle -2\rangle?$
$\mathbf{6}_0$	none	$\mathbf{6}_2$ $\overline{\langle -1\rangle}_2, \langle 2\rangle_2$	$\mathbf{8}_1, \mathbf{8}'_2$	$\langle 1\rangle, \langle 2\rangle?$
$\mathbf{6}_2$	$\overline{\langle 1\rangle}_1, \langle -2\rangle_1$	$\mathbf{6}_0$	$\mathbf{8}_1, \mathbf{8}'_2$	$\langle -3\rangle, \langle -2\rangle?$
$\mathbf{6}_2$	$\overline{\langle 1\rangle}_1, \langle -2\rangle_1$	$\mathbf{6}_2$ $\overline{\langle -1\rangle}_2, \langle 2\rangle_2$	$\mathbf{8}_2$	$\langle -3\rangle, \langle 1\rangle$
$\mathbf{6}_2$	$\overline{\langle 1\rangle}_1, \langle -2\rangle_1$	$\mathbf{6}_2$ $\overline{\langle -1\rangle}_2, \langle 2\rangle_2$	$\mathbf{8}_2$	$\langle -3\rangle, \langle 1\rangle$
$\mathbf{6}_3$	$\overline{\langle 1\rangle}_1, \overline{\langle -1\rangle}_1, \langle 3\rangle_1$	$\mathbf{6}_0$	$\mathbf{8}_1$	$\langle 4\rangle$
$\mathbf{6}_4$	$\overline{\langle 1\rangle}_1, \overline{\langle -1\rangle}_1, \langle 2\rangle_1, \langle 3\rangle_1$	$\mathbf{6}_0$	$\mathbf{8}'_2$	$\langle 3\rangle, \langle 4\rangle$
$\mathbf{4}_2$	$\overline{\langle 1\rangle}_1, \overline{\langle -1\rangle}_1$	$\mathbf{6}_0$	$\mathbf{6}_0$	none

In some rows of the table, the answer is not determined by the information local to $t_i(j)$. In those cases we have included the various possibilities. Notice that the marking of edges $\langle 2\rangle, \langle -2\rangle$ will be determined from what happens in the two previous tables at the adjacent singularities on these segments.

COROLLARY 7.1. *The scaffolding of Z^{new} is well-defined, with orientations of the fold edges. From the tables, the types of local pictures of the scaffolding for Z^{new} are in our standard list $\mathbf{4}_2, \mathbf{6}_3, \mathbf{6}_4, \mathbf{8}_0, \mathbf{8}_1, \mathbf{8}_2, \mathbf{8}'_2$. The orientations of the fold edges at these singularities are compatible with the allowable configurations.*

PROPOSITION 7.2. *Define the refracting spectral network of Z^{new} to be generated by an initial SN line going in every non-fold direction from each of the singularities, and closed under collisions as well as refraction upon crossing fold lines of the new scaffolding. Then, any SN-line of this new spectral network, outside of the edges $g(j)$, is contained in the previous spectral network of Z. SN lines along non-folded segments of the $g(j)$ are with reversed orientation with respect to those of Z. Under the assumption that there were no BPS states in the refracting spectral network of Z, then there are no BPS states in the refracting spectral network of Z^{new}.*

PROOF. (Sketch)—We verify in each of the cases contained in the tables, that there are SN lines in $Z-(R_1 \cup R_2)$ corresponding to all non-folded outward directions from singular points. In the case of non-folded edges that are segments of the $g(j)$, there were SN lines in the non-folded segments of $g_i(j)$ going in the opposite direction. Switch the directions of these, and when these new SN lines meet a $\mathbf{6}_2$ point, notice that it came from a singular point and the refracted directions are among the directions containing SN lines of Z (or we continue along the next segment of $g(j)$ to the next $\mathbf{6}_2$ point).

At an 8-fold singularity obtained from the third table, the SN lines in all directions are generated by the SN lines along the segments of $g_i(j)$ and $g_i'(j)$, sometimes by using the collision process in Z.

We should check that there are no BPS states in the new spectral network. Concerning lines not on the edges $g(j)$ this comes from the inclusion into the previous spectral network in $Z-(R_1\cup R_2)$, and the existence of SN lines going outwards from any new singularities as noted above.

We therefore need to consider new SN lines along segments of $g(j)$. For this, let us notice that in Table 1, whenever a singular $\mathbf{6}_3$ or $\mathbf{6}_4$ is created, the edges going along $g_(j)$ and $g(j+1)$, there denoted $\langle 1\rangle$ and $\langle -1\rangle$, are folded. Furthermore, whenever a $\mathbf{6}_2$ is created, one of those two edges is folded so a BPS state between the two adjacent $s(j), s(j+1)$ is not created. Similarly, in 2 when a $\mathbf{6}_4$ or point is created, the segment after it on $g(j)$, denoted there $\overline{\langle -2\rangle}$, is folded.

This has only been a sketch of proof, a more detailed discussion is needed in order to follow through all possible SN lines that might start with directions along the edges $g(j)$. □

8. Scholium

We now review what has been done (or sketched) above. From X we created an initial complete normal ecarinate construction Z^{init} and we are assuming that this is done following Principles 5.1 and 5.2.

Thus Z^{init} is provided with a fully refracting scaffolding, whose associated spectral network has no BPS states. Initially the scaffolding has only $\mathbf{8}_1, \mathbf{8}_2, \mathbf{8}_2', \mathbf{4}_2$ singularities, so the endpoints of the post-caustics are $\mathbf{8}_1$ points, and the 8-fold and 4-fold points alternate.

The reduction process will consist of a sequence of reduction steps starting with Z^{init}. Let us denote by Z the construction obtained after a certain number of steps. Our goal is to describe the next reduction step.

Our construction Z is again complete, ecarinate, normal, and it is provided with a refracting scaffolding with oriented fold edges. Our assumptions are as follows:

—that the refracting spectral network generated by the scaffolding has no BPS states;

—and that the types of points in the scaffolding are in the standard list of Definition 4.1, taking account orientations of edges.

Suppose that there are some remaining fold edges. By Corollary 6.2, there exists a $\mathbf{4}_2$ point. The reduction step will be to collapse a neighborhood of this $\mathbf{4}_2$ point in a good way.

Choose a maximal collapsing pair R_1, R_2 at this vertex. These regions are arranged with singularities and edges satisfying the properties of Proposition 6.3 and the subsequent discussion.

We then glue together R_1 and R_2, and trim away the resulting piece (except for the union of edges $g(j)$), to get a new construction Z^{new}. This is the result of a single step of our reduction process.

The main point is to verify that Z^{new} is provided with a well-defined refracting scaffolding that still satisfies the required properties. We have seen that the configurations at singularities in Z determine the fold edges at the new points in Z^{new}. Indeed, this is the case at points of the form $s(j)$ and $q^{(m)}$, and the only indeterminacy at points $t(j)$ is along segments that will be connected either to points $q^{(m)}$ or $s(j)$ so the fold edges are determined. We have also assigned orientations to the fold edges.

Specific analysis of each case allows to fill in Tables 1, 2, 3. These show that the new singularities are only of the types in our standard list. Furthermore, we see here that the types of singularities are compatible with the assigned orientations of the fold edges in the new scaffolding.

We noted along the way that the fold edges of the new scaffolding have the required refracting effect on spectral network lines. The sketch of proof of Proposition 7.2 shows how the new spectral network is a subset of the previous one, apart from the spectral network lines that might have switched directions along the unfolded segments of the edges $g(j)$, and this spectral network doesn't have any BPS states.

This completes the verification that our new construction Z^{new} has the required structures and satisfies the required properties so it can be used as the starting point in a next step of the reduction process.

9. Further questions

We have described a single step of the reduction process. The main question will now be to obtain a convergence statement saying that the process ends in finitely many steps with a construction Z^{core} whose scaffolding has an empty set of edges. Suppose it does end. The only singularities of Z^{core} are $\mathbf{8}_0$ points of negative curvature.

This reduced construction will be the *core* of the pre-building that we are conjecturing to exist in [**15**]. In order to get the pre-building, the steps of putting back in the pieces that have been trimmed off, need to be done as described in [**15**].

The construction of the core Z^{core} may be seen as a 2-dimensional analogue of the Stallings core graph [**23, 24**]. It should be interesting to compare these combinatorics to the ones of [**1, 2, 11**].

In current work with Fabian Haiden, we hope to apply this operation

$$X \mapsto Z^{\mathrm{init}} \mapsto Z^{\mathrm{core}}$$

in order to generalize the work of Bridgeland and Smith constructing stability conditions [**3**] from $SL(2)$ to $SL(3)$.

The core Z^{core} has a natural flat structure with geometry modeled on the standard apartment A for $SL(3)$. This geometric structure carries with it a natural cyclic 3-fold spectral covering with ramification at the 8-fold singular points. The 8-fold singular points are negatively curved conical points for the flat metric, with angles of 480°. The flat metric determines a conformal and hence complex structure; this is a modification of the complex structure of the original Riemann surface X, and the cyclic covering corresponds to a cubic differential. This modification looks to be a discrete or possibly "tropical" analogue of Labourie's result [**17**, Conjectures 1.6, 1.7] [**18**], replacing a general spectral curve by a cubic differential.

The first conjecture is that a minimization process provides a stability condition corresponding to the cubic differential, on the category of sections of the perverse Schober with fiber A_2^{CY2} corresponding to the cyclic spectral covering.

If we can do that, then a procedure for defining the stability condition for a general $SL(3)$-spectral curve Σ will be to define the categories $D_{\leq\theta}$ of objects of phase $\leq \theta$, for any θ by considering the core $Z^{\mathrm{core}}(e^{i\theta}\Sigma)$ and using the stability condition given by the conjectured minimization process of the previous paragraph.

These t-structures should each satisfy the required axioms for taking part in a stability condition. Therefore, taken together for all phases they should define a stability condition. We are just beginning to work on this program.

Acknowledgments

The authors wish to express their gratitude to D. Auroux, F. Haiden, M. Kapovich, M. Kontsevich and A. Noll for their help. The third author would like to thank the University of Miami for hospitality during the completion of this work.

References

[1] T. Aoki, T. Kawai, and Y. Takei, *New turning points in the exact WKB analysis for higher-order ordinary differential equations*, Analyse algébrique des perturbations singulières, I (Marseille-Luminy, 1991), Travaux en Cours, vol. 47, Hermann, Paris, 1994, pp. xiii, xv, 69–84. MR1296472

[2] T. Aoki, T. Kawai, and Y. Takei, *On the exact steepest descent method: a new method for the description of Stokes curves*, J. Math. Phys. **42** (2001), no. 8, 3691–3713. MR1845214

[3] T. Bridgeland and I. Smith, *Quadratic differentials as stability conditions*, Publ. Math. Inst. Hautes Études Sci. **121** (2015), 155–278. MR3349833

[4] B. Collier and Q. Li, *Asymptotics of Higgs bundles in the Hitchin component*, Adv. Math. **307** (2017), 488–558. MR3590524

[5] D. Dumas, *Holonomy limits of complex projective structures*, Adv. Math. **315** (2017), 427–473. MR3667590

[6] D. Gaiotto, G. W. Moore, and A. Neitzke, *Wall-crossing, Hitchin systems, and the WKB approximation*, Adv. Math. **234** (2013), 239–403. MR3003931

[7] D. Gaiotto, G. W. Moore, and A. Neitzke, *Spectral networks*, Ann. Henri Poincaré **14** (2013), no. 7, 1643–1731. MR3115984

[8] D. Gaiotto, G. W. Moore, and A. Neitzke, *Spectral networks and snakes*, Ann. Henri Poincaré **15** (2014), no. 1, 61–141. MR3147409

[9] M. Gromov and R. Schoen, *Harmonic maps into singular spaces and p-adic superrigidity for lattices in groups of rank one*, Inst. Hautes Études Sci. Publ. Math. **76** (1992), 165–246. MR1215595

[10] F. Haiden, L. Katzarkov, M. Kontsevich. *Stability in Fukaya categories of surfaces.* Preprint arXiv:1409.8611 (2014).

[11] K. Iwaki and T. Nakanishi, *Exact WKB analysis and cluster algebras*, J. Phys. A **47** (2014), no. 47, 474009, 98. MR3280000

[12] I. Kapovich and A. Myasnikov, *Stallings foldings and subgroups of free groups*, J. Algebra **248** (2002), no. 2, 608–668. MR1882114

[13] M. Kapranov, V. Schechtman. *Perverse Schobers.* Preprint arXiv: 1411.2772 (2014).

[14] L. Katzarkov, A. Noll, P. Pandit, and C. Simpson, *Harmonic maps to buildings and singular perturbation theory*, Comm. Math. Phys. **336** (2015), no. 2, 853–903. MR3322389

[15] L. Katzarkov, A. Noll, P. Pandit, C. Simpson. Constructing buildings and harmonic maps. *Algebra, Geometry, and Physics in the 21st Century (Kontsevich Festschrift)*, D. Auroux, L. Katzarkov, T. Pantev, Y. Soibelman, Y. Tschinkel, eds., PROGRESS IN MATH. **324**, Birkhäuser (2017), 203-260.

[16] M. Kontsevich and Y. Soibelman, *Wall-crossing structures in Donaldson-Thomas invariants, integrable systems and mirror symmetry*, Homological mirror symmetry and tropical geometry, Lect. Notes Unione Mat. Ital., vol. 15, Springer, Cham, 2014, pp. 197–308. MR3330788

[17] F. Labourie, *Anosov flows, surface groups and curves in projective space*, Invent. Math. **165** (2006), no. 1, 51–114. MR2221137

[18] F. Labourie, *Flat projective structures on surfaces and cubic holomorphic differentials*, Pure Appl. Math. Q. **3** (2007), no. 4, Special Issue: In honor of Grigory Margulis., 1057–1099. MR2402597

[19] R. Mazzeo, J. Swoboda, H. Weiss, and F. Witt, *Ends of the moduli space of Higgs bundles*, Duke Math. J. **165** (2016), no. 12, 2227–2271. MR3544281

[20] T. Mochizuki, *Asymptotic behaviour of certain families of harmonic bundles on Riemann surfaces*, J. Topol. **9** (2016), no. 4, 1021–1073. MR3620459

[21] J. W. Morgan and P. B. Shalen, *Valuations, trees, and degenerations of hyperbolic structures. I*, Ann. of Math. (2) **120** (1984), no. 3, 401–476. MR769158

[22] A. Parreau, *Compactification d'espaces de représentations de groupes de type fini*, Math. Z. **272** (2012), no. 1-2, 51–86. MR2968214

[23] O. Parzanchevski and D. Puder, *Stallings graphs, algebraic extensions and primitive elements in* F_2, Math. Proc. Cambridge Philos. Soc. **157** (2014), no. 1, 1–11. MR3211804

[24] J. R. Stallings, *Topology of finite graphs*, Invent. Math. **71** (1983), no. 3, 551–565. MR695906

Department of Mathematics, University of Miami, 1365 Memorial Drive, Coral Gables, Florida 33146; and National Research University Higher School of Economics, Moscow, 101000, Russia; and Fakultät für Mathematik, Universität Wien, Oskar-Morgenstern-Platz 1, 1090 Wien, Austria
Email address: lkatzarkov@gmail.com

Fakultät für Mathematik, Universität Wien, Oskar-Morgenstern-Platz 1, 1090 Wien, Austria
Email address: pranav.pandit@univie.ac.at

Université Côte d'Azur, CNRS, LJAD, 06108, Nice Cedex 2, France
Email address: carlos@unice.fr

Proceedings of Symposia in Pure Mathematics
Volume **98**, 2018
http://dx.doi.org/10.1090/pspum/098/01749

Conformal nets are factorization algebras

André Henriques

ABSTRACT. We prove that conformal nets of finite index are an instance of the notion of a factorization algebra. This result is an ingredient in our proof that, for $G = SU(n)$, the Drinfel'd center of the category of positive energy representations of the based loop group is equivalent to the category of positive energy representations of the free loop group.

1. Introduction

In this note, we prove that conformal nets of finite index (Definitions 1.1 and 3.1 in [**2**]) form an instance of the notion of a factorization algebra. Our main result, Theorem 2, is a key ingredient in the proof, announced in [**8**], that the category of solitons of a finite index conformal net is a bicommutant category.

Our main theorem is an analog, within the coordinate-free setup of [**2**], of the additivity property of conformal nets. Let $\mathcal{A}$ be a conformal net on S^1, let $I \subset S^1$ be a closed interval, and let $\{I_i \subset I\}$ be collection of closed intervals whose interiors cover that of I. *Additivity* is the statement that the von Neumann algebras $\mathcal{A}(I_i)$ then generate a dense subalgebra in $\mathcal{A}(I)$:

$$\bigcup \mathring{I}_i = \mathring{I} \quad \Rightarrow \quad \bigvee \mathcal{A}(I_i) = \mathcal{A}(I). \qquad \text{(additivity)}$$

The additivity property of chiral conformal nets was proven in [**5**]. If one takes finitely many intervals I_i whose union is I, then the corresponding property is called *strong additivity*:

$$\bigcup I_i = I \quad \Rightarrow \quad \bigvee \mathcal{A}(I_i) = \mathcal{A}(I). \qquad \text{(strong additivity)}$$

It is a result of Longo–Xu that chiral conformal nets of finite index satisfy strong additivity [**12**, §5].

Let now I be an abstract interval, and $\{I_i \subset I\}$ a finite collection of multi-intervals (a multi-interval is a finite disjoint union of intervals) satisfying $\bigcup I_i \times I_i = I \times I$. Equivalently, this is the requirement that for every pair of points $p, q \in I$ there exists an element of the cover that contains both p and q. In Theorem 2, we prove that for every coordinate free conformal net $\mathcal{A}$ of finite index, not necessarily

2010 *Mathematics Subject Classification.* Primary 81T40.

This project has received funding from the European Research Council (ERC) under the European Union's Horizon 2020 research and innovation programme (grant agreement No 674978).

chiral[1], we have:

$$\bigcup I_i \times I_i = I \times I \quad \Rightarrow \quad \operatorname{colim} \mathcal{A}(I_i) = \mathcal{A}(I). \qquad \text{(factorization algebra)}$$

The colimit which appears in the right hand side, informally denoted $\operatorname{colim} \mathcal{A}(I_i)$, is that of a diagram involving the algebras $\mathcal{A}(I_i)$ and $\mathcal{A}(I_i \cap I_j)$. (The colimit is defined by a universal property in the category of von Neumann algebras and normal $*$-homomorphisms.) That diagram is written out in the left hand side of equation (1), below. Here, the equation "$\operatorname{colim} \mathcal{A}(I_i) = \mathcal{A}(I)$" is the statement that the natural inclusions $\mathcal{A}(I_i) \to \mathcal{A}(I)$ extend to an isomorphism $\operatorname{colim} \mathcal{A}(I_i) \to \mathcal{A}(I)$.

REMARK. The category of von Neumann algebras and normal $*$-homomorphisms is cocomplete [**11**, Prop. 5.7] (see also [**7**, §7]).

2. Factorization algebras

Let Man^n be the category whose objects are n-dimensional manifolds and whose morphisms are embeddings. We equip it with the symmetric monoidal structure given by disjoint union. A collection of open subsets $\{U_i \subset M\}$ of a manifold M is a *Weiss cover* if for every finite subset $S \subset M$, there exists an index i such that $S \subset U_i$ [**4**, Chapt. 6]. Equivalently, being a Weiss cover means that for every $n \in \mathbb{N}$, the condition $\bigcup U_i^n = M^n$ is satisfied. Let C be a symmetric monoidal category.

DEFINITION ([**4**, Chapt. 6]). *An n-dimensional C-valued factorization algebra is a symmetric monoidal functor $\mathcal{A} : \mathsf{Man}^n \to \mathsf{C}$ which is a co-sheaf with respect to Weiss covers.*

Here, being a co-sheaf with respect to Weiss covers means that, for every Weiss cover $\{U_i \subset M\}$, the natural map

$$\operatorname{colim} \left(\begin{array}{ccc} \mathcal{A}(U_1 \cap U_2) & & \\ \mathcal{A}(U_1 \cap U_3) & & \mathcal{A}(U_1) \\ \mathcal{A}(U_2 \cap U_3) & \rightrightarrows & \mathcal{A}(U_2) \\ \mathcal{A}(U_1 \cap U_4) & & \mathcal{A}(U_3) \\ \mathcal{A}(U_2 \cap U_4) & & \mathcal{A}(U_4) \\ \vdots & & \vdots \end{array} \right) \longrightarrow \mathcal{A}(M) \tag{1}$$

is an isomorphism. For later notational convenience, we abbreviate the left hand side of (1) as $\operatorname{colim}(\{\mathcal{A}(U_i \cap U_j)\} \rightrightarrows \{\mathcal{A}(U_i)\})$.

In this paper, we are interested in 1-dimensional factorization algebras (or rather, a small variant of the notion of 1-dimensional factorization algebra) with values in the category of von Neumann algebras and normal $*$-homomorphisms.

An *interval* is an oriented 1-manifold diffeomorphic to $[0, 1]$. A *multi-interval* is a finite disjoint union of intervals. Let INT^* be the category whose objects are multi-intervals and whose morphisms are orientation preserving embeddings, and let $\mathsf{INT} \subset \mathsf{INT}^*$ be its full subcategory of intervals. Let VN denote the category of von Neumann algebras and normal $*$-homomorphisms, equipped with the

[1] A coordinate-free conformal net is called *chiral* if the action of the rotation group on its vacuum sector has positive energy, and if the latter admits a unique, cyclic, $PSL(2, \mathbb{R})$-invariant vacuum vector.

symmetric monoidal structure given by spatial tensor product. By the split property [**2**, Def. 1.1], a conformal net can be viewed as a symmetric monoidal functor $\mathcal{A} : \mathsf{INT}^* \to \mathsf{VN}$.

We introduce a variant of the notion of Weiss cover that accounts for the fact that morphisms in INT^* are not open but rather closed inclusions. Given a topological space X, a *Weiss c-cover* is a family of closed subsets $\{V_i \subset X\}$ that satisfies $\bigcup \mathring{V}_i^n = X^n$ for every $n \in \mathbb{N}$. Here, $\mathring{V}_i$ denotes the *relative* interior of V_i inside X, i.e., the largest subset of V_i which is open in X. (For example, the relative interior of $[0,1]$ inside $[0,2]$ is the half-open interval $[0,1[$.)

Throughout this paper, all conformal nets are assumed irreducible, i.e., all the algebras $\mathcal{A}(I)$ are assumed to be factors (we work with conformal nets in the sense of [**2**, Def. 1.1]). The following statement expresses the idea that conformal nets are factorization algebras:

THEOREM 1. *Let $\mathcal{A} : \mathsf{INT}^* \to \mathsf{VN}$ be a conformal net of finite index. Then $\mathcal{A}$ is a co-sheaf with respect to Weiss c-covers. Namely, for every multi-interval I, and every Weiss c-cover of I by multi-intervals $I_i \subset I$, the natural map*

$$q : \operatorname{colim}\big(\{\mathcal{A}(I_i \cap I_j)\} \rightrightarrows \{\mathcal{A}(I_i)\}\big) \;\to\; \mathcal{A}(I)$$

is an isomorphism.

REMARK. Here, it is crucial to use covers by *closed* multi-intervals. For a chiral conformal net $\mathcal{A}$ on S^1, the functor that sends an open multi-interval U to the algebra $\mathcal{A}(\bar{U})$ is never a factorisation algebra, unless $\mathcal{A}$ is trivial ($\bar{U}$ denotes the closure of U in S^1), because the map

$$A := \operatorname*{colim}_{\varepsilon \to 0} \mathcal{A}([\varepsilon, 1-\varepsilon]) \to \mathcal{A}([0,1])$$

is not an isomorphism. Let $I_\varepsilon \subset S^1$ be the image of $[\varepsilon, 1-\varepsilon]$ under the exponential map $t \mapsto e^{2\pi i t} : \mathbb{R} \to S^1$. The obvious isomorphisms $\mathcal{A}([\varepsilon, 1-\varepsilon]) \to \mathcal{A}(I_\varepsilon)$, followed by the standard actions of $\mathcal{A}(I_\varepsilon)$ on the vacuum Hilbert space of $\mathcal{A}$ equip the latter with the structure of an irreducible A-module. However, $\mathcal{A}([0,1])$ is a III_1 factor, which admits no irreducible representations [**6**, Thm. 2.13].

We can sharpen the above result a little bit. Given a compact topological space X, a collection $\{V_i \subset X\}_{i \in \mathcal{I}}$ of closed subsets is called a *2-cover* if there exists a finite subset $\mathcal{I}' \subset \mathcal{I}$ such that $\bigcup_{i \in \mathcal{I}'} V_i^2 = X^2$. Any Weiss c-cover is a 2-cover, and Theorem 1 is a formal consequence of the following stronger result:

THEOREM 1′. *Let $\mathcal{A}$ be a conformal net of finite index, let I be a multi-interval, and let $\{I_i \subset I\}$ be a 2-cover by multi-intervals. Assume furthermore that there exists an element of the 2-cover that contains ∂I. Then the natural map*

$$q : \operatorname{colim}\big(\{\mathcal{A}(I_i \cap I_j)\} \rightrightarrows \{\mathcal{A}(I_i)\}\big) \;\to\; \mathcal{A}(I)$$

is an isomorphism.

When I is connected, the statement of Theorem 1′ simplifies:

THEOREM 2. *Let $\mathcal{A}$ be a conformal net of finite index, let I be an interval, and let $\{I_i \subset I\}$ be a 2-cover by multi-intervals. Then the natural map*

$$q : \operatorname{colim}\big(\{\mathcal{A}(I_i \cap I_j)\} \rightrightarrows \{\mathcal{A}(I_i)\}\big) \;\to\; \mathcal{A}(I)$$

is an isomorphism.

3. Proofs

In this section, we present the proofs of the above theorems. We first prove Theorem 2. We then prove Theorem 1′ by a slight variation of the argument. Theorem 1 is then a formal consequence of Theorem 1′. We begin with some lemmas. We first note that, when working with multi-intervals, a 2-cover induces a cover in the usual sense:

LEMMA 3. *Let I be a multi-interval and let $\{I_i \subset I\}_{i\in\mathcal{I}}$ be a 2-cover by multi-intervals. Then $\bigcup \mathring{I}_i = \mathring{I}$.*

PROOF. By definition, $\bigcup_{i\in\mathcal{I}'} I_i^2 = I^2$ for some finite subset $\mathcal{I}' \subset \mathcal{I}$. Given a point $p \in \mathring{I}$, pick sequences (x_n) and (y_n) in I converging to p and satisfying $x_n < p < y_n$. For every n, there exists an index $i \in \mathcal{I}'$ such that x_n and y_n are both in I_i. The set $\mathcal{I}'$ being finite, there exists an I_i that contains infinitely many x_n's and y_n's. Since I_i is a multi-interval, it contains p in its interior. □

The next lemma is technical in nature. It is a generalisation of [**2**, Lem. 1.9]. Let $\mathcal{A}$ be a conformal net (not necessarily of finite index) and let I be a multi-interval:

LEMMA 4. *Let $\mathcal{I} = \{I_i \subset I\}$ be a collection of multi-intervals satisfying $\bigcup \mathring{I}_i = \mathring{I}$. Let $\varphi \in \mathrm{Diff}(I)$ be a diffeomorphism in the connected component of the identity, and let $\hat{I} := \varphi(I_0)$ for some $I_0 \in \mathcal{I}$. Let H be a Hilbert space equipped with actions $\rho_i : \mathcal{A}(I_i) \to B(H)$ satisfying*

(1) $\rho_i|_{\mathcal{A}(I_i\cap I_j)} = \rho_j|_{\mathcal{A}(I_i\cap I_j)} : \mathcal{A}(I_i \cap I_j) \to B(H)$.
(2) *For every $I_j, I_k \in \mathcal{I}$ and every intervals $J \subset I_j$, $K \subset I_k$ with disjoint interiors, the algebras $\rho_j(\mathcal{A}(J))$ and $\rho_k(\mathcal{A}(K))$ commute.*

Then the actions $\rho_i|_{\mathcal{A}(\hat{I}\cap I_i)}$ of $\mathcal{A}(\hat{I} \cap I_i)$ on H extend (uniquely) to an action $\hat{\rho} : \mathcal{A}(\hat{I}) \to B(H)$.

PROOF. We write ρ_0 for the action of $\mathcal{A}(I_0)$ on H. We may assume without loss of generality that φ fixes a neighbourhood of ∂I. Provided that is the case, we can write φ as a product of diffeomorphisms $\varphi = \varphi_1 \circ \ldots \circ \varphi_n$ with $\mathrm{supp}(\varphi_s) \subset \mathring{I}_{i_s}$ for some $I_{i_s} \in \mathcal{I}$. Let $u_s \in \mathcal{A}(I_{i_s})$ be unitaries s.t. $\mathrm{Ad}(u_s) = \mathcal{A}(\varphi_s)$ [**2**, Def. 1.1 (iv)]. Identifying the elements u_s with their images in $B(H)$, we set

$$\hat{\rho}(a) := u_1 \ldots u_n \rho_0\big(\mathcal{A}(\varphi^{-1})(a)\big) u_n^* \ldots u_1^*.$$

For every $I_\ell \in \mathcal{I}$ and every sufficiently small interval $K \subset \hat{I} \cap I_\ell$, we will show that

$$\hat{\rho}|_{\mathcal{A}(K)} = \rho_\ell|_{\mathcal{A}(K)}. \tag{2}$$

Here, 'sufficiently small' means that the intervals $K_s := \varphi_s^{-1}(\ldots(\varphi_1^{-1}(K)))$ are contained in I_{k_s} for some $I_{k_s} \in \mathcal{I}$, and that for every $s' \le n$ either $K_s \subset I_{i_{s'}}$ or $K_s \cap \mathrm{supp}(\varphi_{s'}) = \emptyset$.

For every $s \le n$, we claim that

$$u_1 \ldots u_s \rho_{k_s}\big(\mathcal{A}(\varphi_s^{-1} \circ \ldots \circ \varphi_1^{-1})(a)\big) u_s^* \ldots u_1^* = \rho_\ell(a) \qquad \forall a \in \mathcal{A}(K). \tag{3}$$

Equation (2) is the special case $s = n$. We prove (3) by induction on s. The base case ($s = 0$, $k_0 = \ell$) is trivial. The induction step reduces to the equation

$$\rho_{k_s}\big(\mathcal{A}(\varphi_s^{-1})(b)\big) = u_s^* \rho_{k_{s-1}}(b) u_s,$$

with $b = \mathcal{A}(\varphi_{s-1}^{-1} \circ \ldots \circ \varphi_1^{-1})(a)$. Recall that $b \in \mathcal{A}(K_{s-1})$, $u_s \in \mathcal{A}(I_{i_s})$ and that, by assumption, either $K_{s-1} \subset I_{i_s}$ or $K_{s-1} \cap \operatorname{supp}(\varphi_s) = \emptyset$. In the first case, we have

$$u_s^* \rho_{k_{s-1}}(b) u_s = u_s^* \rho_{i_s}(b) u_s = \rho_{i_s}(u_s^* b u_s) = \rho_{i_s}\big(\mathcal{A}(\varphi_s^{-1})(b)\big) = \rho_{k_s}\big(\mathcal{A}(\varphi_s^{-1})(b)\big).$$

In the second case, the elements $\rho_{k_{s-1}}(b)$ and u_s commute:

$$u_s^* \rho_{k_{s-1}}(b) u_s = \rho_{k_{s-1}}(b) = \rho_{k_s}(b) = \rho_{k_s}\big(\mathcal{A}(\varphi_s^{-1})(b)\big),$$

where the last equality holds since $b \in A(K_{s-1})$ and φ_s acts trivially on K_{s-1}. This finishes the proof of (3) and hence of (2). Finally, by strong additivity (which is one of the axioms in [**2**, Def. 1.1]), it follows from (2) that $\hat{\rho}(a) = \rho_\ell(a)$ for every $a \in \mathcal{A}(\hat{I} \cap I_\ell)$. □

We now establish some assumptions under which the hypotheses of Lemma 4 are satisfied:

LEMMA 5. *Let I be a multi-interval, and let $\mathcal{I} = \{I_i \subset I\}$ be a 2-cover. Let $\rho_i : \mathcal{A}(I_i) \to B(H)$ be actions satisfying $\rho_i|_{\mathcal{A}(I_i \cap I_j)} = \rho_j|_{\mathcal{A}(I_i \cap I_j)}$. Then, for every $I_j, I_k \in \mathcal{I}$ and every intervals $J \subset I_j$, $K \subset I_k$ with disjoint interiors, we have*

$$[\rho_j(\mathcal{A}(J)), \rho_k(\mathcal{A}(K))] = 0.$$

PROOF. We assume without loss of generality that the 2-cover is finite. The finite set $S := \bigcup_{i \in \mathcal{I}} \partial I_i$ decomposes J and K into a finitely many intervals: $J = J_1 \cup \ldots \cup J_n$ and $K = K_1 \cup \ldots \cup K_m$. For each pair J_r, K_s of above intervals, we will argue that

$$[\rho_j(\mathcal{A}(J_r)), \rho_k(\mathcal{A}(K_s))] = 0. \tag{4}$$

Pick interior points $x \in \mathring{J}_r$ and $y \in \mathring{K}_s$. Since $\mathcal{I}$ is a 2-cover, there exists an $i \in \mathcal{I}$ such that $\{x, y\} \subset I_i$. It follows that $J_r \cup K_s \subset I_i$. The actions of $\mathcal{A}(J_r)$ and $\mathcal{A}(K_s)$ on H factor through that of $\mathcal{A}(I_i)$, so equation (4) follows.

Equation (4) being true for every pair J_r, K_s as above, by strong additivity, it follows that the algebras $\rho_j(\mathcal{A}(J)) = \bigvee_r \rho_j(\mathcal{A}(J_r))$ and $\rho_k(\mathcal{A}(K)) = \bigvee_s \rho_k(\mathcal{A}(K_s))$ commute with each other. □

The following lemma contains the main argument of the proof of Theorem 2. Let $\mathcal{A}$ be a conformal net of finite index:

LEMMA 6. *Let I be an interval and let $\mathcal{I} = \{I_i \subset I\}$ be a 2-cover. Let H be a Hilbert space equipped with actions $\rho_i : \mathcal{A}(I_i) \to B(H)$ satisfying $\rho_i|_{\mathcal{A}(I_i \cap I_j)} = \rho_j|_{\mathcal{A}(I_i \cap I_j)}$. Then those maps extend to an action of $\mathcal{A}(I)$.*

PROOF. We may assume, without loss of generality, that the 2-cover is closed under taking subsets: ($I_i \in \mathcal{I}$ and $J \subset I_i$, J a multi-interval) $\Rightarrow$ ($J \in \mathcal{I}$).

By Lemmas 3 and 5, we are in a situation to apply Lemma 4. The latter implies that for every interval $J \subsetneq I$, the actions of $\mathcal{A}(I_i \cap J)$ extend (uniquely) to an action of $\mathcal{A}(J)$. We may therefore assume without loss of generality that $I = [0, 5]$, and that the 2-cover contains the multi-intervals $[0, 2] \cup [3, 5]$ and $[1, 4]$ as elements.

Recall that $L^2(-)$ is the unit for the operation $\boxtimes$ of Connes fusion. We have $H \cong L^2\mathcal{A}([1, 4]) \boxtimes_{\mathcal{A}([1,4])} H$, both as $\mathcal{A}([1, 4])$-modules and as $\mathcal{A}([0, 1] \cup [4, 5])$-modules. By [**3**, Cor. 2.9], the vacuum sector $L^2\mathcal{A}([1, 4])$ is isomorphic to

$$L^2\mathcal{A}([2, 3]) \boxtimes_{\mathcal{A}([2,3] \cup_{\{2,3\}} \overline{[2,3]})} \big(L^2\mathcal{A}([1, 4]) \boxtimes_{\mathcal{A}([1,2] \cup [3,4])} L^2\mathcal{A}([1, 4])\big)$$

as an $\mathcal{A}([1,4])$-$\mathcal{A}([1,4])$-bimodule (this is where we use the assumption that $\mathcal{A}$ has finite index). Here, $\overline{[2,3]}$ denotes the interval $[2,3]$ equipped with the opposite orientation, and the algebra $\mathcal{A}([2,3]\cup_{\{2,3\}}\overline{[2,3]})$ associated to the circle $[2,3]\cup_{\{2,3\}}\overline{[2,3]}$ is described in [**3**, Prop. 1.25]. The action of $\mathcal{A}([2,3]\cup_{\{2,3\}}\overline{[2,3]})$ on $L^2\mathcal{A}([1,4])\boxtimes_{\mathcal{A}([1,2]\cup[3,4])}L^2\mathcal{A}([1,4])$ comes from the left action of $\mathcal{A}([2,3])\subset\mathcal{A}([1,4])$ on the second copy of $L^2\mathcal{A}([1,4])$ and the right action of $\mathcal{A}([2,3])\subset\mathcal{A}([1,4])$ on the first copy of $L^2\mathcal{A}([1,4])$.

Let us abbreviate $\mathcal{A}([a,b])$ by $\mathcal{A}_{ab}$, $\mathcal{A}([a,b]\cup[c,d])$ by $\mathcal{A}_{ab\cup cd}$, and $\mathcal{A}([a,b]\cup_{\{a,b\}}\overline{[a,b]})$ by $\mathcal{A}_{\widehat{ab}}$. We have:

$$H\cong L^2\mathcal{A}_{14}\boxtimes_{\mathcal{A}_{14}}H \qquad\text{and}\qquad L^2\mathcal{A}_{14}\cong L^2\mathcal{A}_{23}\boxtimes_{\mathcal{A}_{\widehat{23}}}\big(L^2\mathcal{A}_{14}\boxtimes_{\mathcal{A}_{12\cup34}}L^2\mathcal{A}_{14}\big).$$

Combining those two facts, one gets

$$\begin{aligned}
H &\cong L^2\mathcal{A}_{14}\boxtimes_{\mathcal{A}_{14}}H\\
&\cong \big(L^2\mathcal{A}_{23}\boxtimes_{\mathcal{A}_{\widehat{23}}}(L^2\mathcal{A}_{14}\boxtimes_{\mathcal{A}_{12\cup34}}L^2\mathcal{A}_{14})\big)\boxtimes_{\mathcal{A}_{14}}H\\
&\cong L^2\mathcal{A}_{23}\boxtimes_{\mathcal{A}_{\widehat{23}}}\big(L^2\mathcal{A}_{14}\boxtimes_{\mathcal{A}_{12\cup34}}L^2\mathcal{A}_{14}\boxtimes_{\mathcal{A}_{14}}H\big)\\
&\cong L^2\mathcal{A}_{23}\boxtimes_{\mathcal{A}_{\widehat{23}}}\big(L^2\mathcal{A}_{14}\boxtimes_{\mathcal{A}_{12\cup34}}H\big).
\end{aligned}\tag{5}$$

Using that $H\cong L^2\mathcal{A}_{02\cup35}\boxtimes_{\mathcal{A}_{02\cup35}}H$ and the existence of a (non-canonical) isomorphism

$$L^2\mathcal{A}_{14}\boxtimes_{\mathcal{A}_{12\cup34}}L^2\mathcal{A}_{02\cup35}\cong L^2\mathcal{A}_{02}\boxtimes_{\mathcal{A}_{12}^{op}}L^2\mathcal{A}_{14}\boxtimes_{\mathcal{A}_{34}}L^2\mathcal{A}_{35}\cong L^2\mathcal{A}_{05}$$

which is compatible with the left actions of $\mathcal{A}_{14}$ and $\mathcal{A}_{01\cup45}$ and the right actions of $\mathcal{A}_{02\cup35}$ and $\mathcal{A}_{23}$ ([**2**, Cor. 1.33] and [**3**, Lem. A.4]), we get the following sequence of isomorphisms of $\mathcal{A}_{14}$- and $\mathcal{A}_{01\cup45}$-modules:

$$\begin{aligned}
H &\cong L^2\mathcal{A}_{23}\boxtimes_{\mathcal{A}_{\widehat{23}}}\big(L^2\mathcal{A}_{14}\boxtimes_{\mathcal{A}_{12\cup34}}H\big)\\
&\cong L^2\mathcal{A}_{23}\boxtimes_{\mathcal{A}_{\widehat{23}}}\big(L^2\mathcal{A}_{14}\boxtimes_{\mathcal{A}_{12\cup34}}L^2\mathcal{A}_{02\cup35}\boxtimes_{\mathcal{A}_{02\cup35}}H\big)\\
&\cong L^2\mathcal{A}_{23}\boxtimes_{\mathcal{A}_{\widehat{23}}}\big(L^2\mathcal{A}_{05}\boxtimes_{\mathcal{A}_{02\cup35}}H\big).
\end{aligned}\tag{6}$$

The actions of $\mathcal{A}_{14}$ and of $\mathcal{A}_{01\cup45}$ on

$$L^2\mathcal{A}_{23}\boxtimes_{\mathcal{A}_{\widehat{23}}}\big(L^2\mathcal{A}_{05}\boxtimes_{\mathcal{A}_{02\cup35}}H\big)$$

extend to an action of $\mathcal{A}_{05}$ because they both act on $L^2\mathcal{A}_{05}$. The actions of $\mathcal{A}_{14}$ and of $\mathcal{A}_{01\cup45}$ on H therefore also extend to an action of $\mathcal{A}_{05}$. □

To help the reader digest the argument in the above proof, we include a graphical rendering of the isomorphisms which appear in (5) and (6):

With Lemma 6 in place, the proof of Theorem 2 is now easy:

PROOF OF THEOREM 2. We first note that, by the strong additivity property of conformal nets [**2**, Def. 1.1], the map q has dense image. It is therefore surjective, as any morphism of von Neumann algebras whose image is dense is automatically surjective [**13**, Chapt. III, Prop. 3.12].

To show that q is injective, pick a faithful representation

$$\pi:\operatorname{colim}\big(\{\mathcal{A}(I_i\cap I_j)\}\rightrightarrows\{\mathcal{A}(I_i)\}\big)\to B(H)$$

and let $\rho_i := \pi|_{\mathcal{A}(I_i)}$. By Lemma 6, this extends to an action $\rho : \mathcal{A}(I) \to B(H)$. As π is injective and $\pi = \rho \circ q$, the map q is also injective. $\square$

The proof of Theorem 1′ follows closely that of Theorem 2.

PROOF OF THEOREM 1′. Let I be a multi-interval, and let $\mathcal{I} = \{I_i \subset I\}$ be a 2-cover one of whose elements contains ∂I. As in the proof of Theorem 2, it is enough to argue that if H is a Hilbert space equipped with actions $\rho_i : \mathcal{A}(I_i) \to B(H)$ satisfying $\rho_i|_{\mathcal{A}(I_i \cap I_j)} = \rho_j|_{\mathcal{A}(I_i \cap I_j)}$, then those extend to an action of $\mathcal{A}(I)$.

Without loss of generality, we may assume that the 2-cover is closed under taking subsets: $(I_i \in \mathcal{I}$ and $J \subset I_i) \Rightarrow (J \in \mathcal{I})$. In particular, we may assume that the 2-cover admits an element which doesn't intersect its boundary, and which has exactly one connected component in each connected component on I.

By Lemmas 3 and 5, we are in a situation to apply Lemma 4. We can therefore assume, without loss of generality, that $I = \bigsqcup_{k=1}^n [0,5]$, and that $\mathcal{I}$ contains the multi-intervals $\bigsqcup_{k=1}^n ([0,2] \cup [3,5])$ and $\bigsqcup_{k=1}^n [1,4]$ as elements.

Following the structure of the proof of Lemma 6, we have isomorphisms:

$$\begin{aligned}
H &\cong L^2(\mathcal{A}_{14})^{\otimes n} \boxtimes_{\mathcal{A}_{14}^{\otimes n}} H \\
&\cong \big(L^2(\mathcal{A}_{23})^{\otimes n} \boxtimes_{\mathcal{A}_{\widehat{23}}^{\otimes n}} (L^2(\mathcal{A}_{14})^{\otimes n} \boxtimes_{\mathcal{A}_{12\cup 34}^{\otimes n}} L^2(\mathcal{A}_{14})^{\otimes n})\big) \boxtimes_{\mathcal{A}_{14}^{\otimes n}} H \\
&\cong L^2(\mathcal{A}_{23})^{\otimes n} \boxtimes_{\mathcal{A}_{\widehat{23}}^{\otimes n}} \big(L^2(\mathcal{A}_{14})^{\otimes n} \boxtimes_{\mathcal{A}_{12\cup 34}^{\otimes n}} L^2(\mathcal{A}_{14})^{\otimes n} \boxtimes_{\mathcal{A}_{14}^{\otimes n}} H\big) \\
&\cong L^2(\mathcal{A}_{23})^{\otimes n} \boxtimes_{\mathcal{A}_{\widehat{23}}^{\otimes n}} \big(L^2(\mathcal{A}_{14})^{\otimes n} \boxtimes_{\mathcal{A}_{12\cup 34}^{\otimes n}} H\big). \\
&\cong L^2(\mathcal{A}_{23})^{\otimes n} \boxtimes_{\mathcal{A}_{\widehat{23}}^{\otimes n}} \big(L^2(\mathcal{A}_{14})^{\otimes n} \boxtimes_{\mathcal{A}_{12\cup 34}^{\otimes n}} L^2(\mathcal{A}_{02\cup 35})^{\otimes n} \boxtimes_{\mathcal{A}_{02\cup 35}^{\otimes n}} H\big) \\
&\cong L^2(\mathcal{A}_{23})^{\otimes n} \boxtimes_{\mathcal{A}_{\widehat{23}}^{\otimes n}} \big(L^2(\mathcal{A}_{05})^{\otimes n} \boxtimes_{\mathcal{A}_{02\cup 35}^{\otimes n}} H\big)
\end{aligned}$$

of $\mathcal{A}(\bigsqcup^n([0,2] \cup [3,5]))$- and $\mathcal{A}(\bigsqcup^n[1,4])$-modules.

The actions of $\mathcal{A}(\bigsqcup^n([0,2] \cup [3,5]))$ and of $\mathcal{A}(\bigsqcup^n[1,4])$ on the Hilbert space $L^2(\bigotimes^n \mathcal{A}_{23}) \boxtimes_{\mathcal{A}_{\widehat{23}}^{\otimes n}} \big(L^2(\bigotimes^n \mathcal{A}_{05}) \boxtimes_{\bigotimes^n \mathcal{A}_{02\cup 35}} H\big)$ visibly extend to an action of the von Neumann algebra $\mathcal{A}(\bigsqcup^n[0,5]) = \mathcal{A}(I)$. They therefore also extend to an action of $\mathcal{A}(I)$ on H. $\square$

4. An application

In our recent preprint [**8**], we introduced higher categorical analogs of von Neumann algebras called *bicommutant categories*. A bicommutant category is a tensor category which is equivalent to its bicommutant inside $\mathrm{Bim}(R)$. (The latter is the category of all bimodules over a hyperfinite factor; it plays the role of the algebra of bounded operators on a Hilbert space.) A bicommutant category is also equipped with a higher categorical analog of a $*$-structure, called a bi-involutive structure [**9**, Def. 2.3].

In [**8**], we made the following announcement: for G the group $SU(n)$ and for k a positive integer, the category of positive energy representations of the based loop group of G at level k is a bicommutant category. Moreover, its Drinfel'd center is the category of positive energy representations of the free loop group of G:

$$Z\big(\mathrm{Rep}^{k}_{\substack{pos.\\en.}}(\Omega G)\big) = \mathrm{Rep}^{k}_{\substack{pos.\\en.}}(LG). \tag{7}$$

We then argued that this result provides good evidence for our claim that the tensor category of positive energy representation of the based loop group is the value of Chern-Simons theory on a point.

REMARK 7. The tensor category of positive energy representations of LG, as defined using conformal nets (see, e.g., [**6, 15**]), has, to our knowledge, not been compared to the corresponding tensor category defined using affine Lie algebras (or vertex algebras, or quantum groups). The right hand side of (7) refers to the tensor category defined in [**15**].

REMARK 8. We expect the relation (7) to hold true for every compact connected Lie group G and every level $k \in H^4_+(BG, \mathbb{Z})$. It is conjectured by many people that all chiral WZW conformal nets have finite index (see [**6**][**14**][**15**][**2**, §4.C][**10**, §8] for the definition of these conformal nets in various degrees of generality). The finite index property is known when $G = SU(n)$ [**15, 16**], and in a few other cases. Our proof of (7) relies crucially on the fact that the chiral WZW conformal nets associated to G have finite index. However, our dependence on this result is the only place where we use that G is the group $SU(n)$.

We can generalize (7) to arbitrary conformal nets of finite index. The analog of the tensor category of positive energy representations of ΩG is the tensor category $\mathcal{T}_\mathcal{A}$ of *solitons* of the conformal net $\mathcal{A}$ [**12**, §3.0.1]. The claim is then that the Drinfel'd center of the tensor category of solitons of $\mathcal{A}$ is the braided tensor category of representations of $\mathcal{A}$.

By definition, a soliton is a Hilbert space equipped with compatible actions of the algebras $\mathcal{A}(I)$, where I ranges over all subintervals of the standard circle whose interior does not contain the base point $1 \in S^1$. Equivalently, it ranges over all subintervals $I \subsetneq S^1_{\text{cut}}$, where S^1_{cut} is the manifold obtained from the standard circle by removing its base point and replacing it by two points:

S^1 : 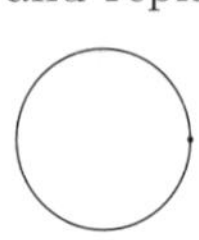 S^1_{cut} : 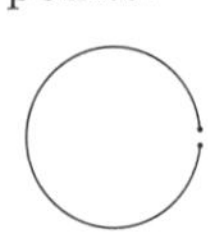

The monoidal structure on $\mathcal{T}_\mathcal{A}$ is defined as follows. Let H and K be two solitons. Let I_+ be the upper half of S^1, and let I_- be its lower half. Precomposing the left action of $\mathcal{A}(I_-)$ on H by the map

$$\mathcal{A}(z \mapsto \bar{z} : I_+ \to I_-) \ : \ \mathcal{A}(I_+)^{op} \to \mathcal{A}(I_-)$$

yields a right action of $\mathcal{A}(I_+)$ on H. We let

$$H \boxtimes K := H \boxtimes_{\mathcal{A}(I_+)} K.$$

Here, $\mathcal{A}(I_+)$ acts on K in the usual way, and acts on H on the right via the action described above. The left actions of $\mathcal{A}(I_+)$ on H and of $\mathcal{A}(I_-)$ on K induce corresponding actions on $H \boxtimes K$. Given an interval $J \subset S^1$, $1 \notin J$, $-1 \in \mathring{J}$, then, by the same argument as in [**2**, Def. 1.31], the actions of $\mathcal{A}(J \cap I_+)$ and $\mathcal{A}(J \cap I_-)$ on $H \boxtimes K$ extend to an action of $\mathcal{A}(J)$. By Lemma 4, it follows that for *every* interval $J \subsetneq S^1_{\text{cut}}$, the actions of $\mathcal{A}(J \cap I_+)$ and $\mathcal{A}(J \cap I_-)$ on $H \boxtimes K$ extend to an action

$$\rho_J : \mathcal{A}(J) \to B(H \boxtimes K).$$

All together, these actions equip $H \boxtimes K$ with the structure of a soliton.

Given a soliton H, with actions $\rho_I : \mathcal{A}(I) \to B(H)$ for $I \subsetneq S^1_{\text{cut}}$, its *conjugate* $\overline{H}$ is the complex conjugate Hilbert space equipped with the actions

$$\mathcal{A}(I) \xrightarrow{\mathcal{A}(z\mapsto\bar{z})} \mathcal{A}(\bar{I})^{op} \xrightarrow{\;*\;} \overline{\mathcal{A}(\bar{I})} \xrightarrow{\overline{\rho_{\bar{I}}}} \overline{B(H)} = B(\overline{H}).$$

Here, $\bar{I}$ denotes the image of $I \subset S^1$ under the complex conjugation map $S^1 \to S^1$. The conjugation on $\mathcal{T}_\mathcal{A}$ squares to the identity and satisfies $\overline{H \boxtimes K} \cong \overline{K} \boxtimes \overline{H}$.

DEFINITION ([**9**, Def. 5.3]). *An object Ω of a tensor category $(\mathcal{T}, \otimes)$ is called absorbing if it is non-zero and satisfies*

$$(X \neq 0) \Rightarrow (X \otimes \Omega \cong \Omega \cong \Omega \otimes X) \qquad \forall X \in \mathcal{T}.$$

REMARK. If $\mathcal{T}$ admits a conjugation, it is a little bit easier to check that an object is absorbing. $\Omega \in \mathcal{T}$ is absorbing if it is non-zero and if $X \otimes \Omega \cong \Omega$ for every $X \neq 0$; see the comments after [**9**, Def. 5.3] for a proof.

The next result, about the existence of absorbing objects, is a key ingredient in the proof, announced in [**8**], that the category of solitons of a conformal net with finite index is a bicommutant category, and that its Drinfel'd center is the category of representations of $\mathcal{A}$. The proof relies on Lemma 6 (which is essentially equivalent to Theorem 2).

Given a non-trivial conformal net $\mathcal{A}$, let

$$\Omega_\mathcal{A} := L^2(\mathcal{A}(S^1_{\text{cut}})) \in \mathcal{T}_\mathcal{A},$$

with actions of $\mathcal{A}(I)$ for $I \subsetneq S^1_{\text{cut}}$ provided by the obvious inclusion $\mathcal{A}(I) \to \mathcal{A}(S^1_{\text{cut}})$ followed by the left action of $\mathcal{A}(S^1_{\text{cut}})$ on its L^2-space.

Alternatively, the soliton $\Omega_\mathcal{A}$ can be described as follows. We let its underlying Hilbert space be the vacuum Hilbert space H_0 of the conformal net $\mathcal{A}$. Given an interval $I = [e^{ia}, e^{ib}] \subset S^1$ with $0 \leq a < b \leq 2\pi$, let $\sqrt{I} := [e^{ia/2}, e^{ib/2}]$. The square root function induces an isomorphism $\mathcal{A}(I) \to \mathcal{A}(\sqrt{I})$. For $I \subset S^1$, $1 \notin \mathring{I}$, the action of $\mathcal{A}(I)$ on H_0 is the composite $\mathcal{A}(I) \to \mathcal{A}(\sqrt{I}) \to B(H_0)$ of the above isomorphism with the standard action of $\mathcal{A}(\sqrt{I})$ on H_0.

The equivalence between the above two descriptions of $\Omega_\mathcal{A}$ is provided by the linear map $L^2(\mathcal{A}(\sqrt{\cdot}\,)) : L^2(\mathcal{A}(S^1_{\text{cut}})) \to L^2(\mathcal{A}([e^{i0}, e^{i\pi}])) = H_0$.

THEOREM 9. *Let $\mathcal{A}$ be a non-trivial conformal net. Then $\Omega_\mathcal{A} \in \mathcal{T}_\mathcal{A}$ is characterized up to isomorphism by the following three properties:*

(*a*) *it is non-zero,*

(*b*) $\Omega_\mathcal{A} \oplus \Omega_\mathcal{A} \cong \Omega_\mathcal{A}$, *and*

(*c*) *the actions of $\mathcal{A}(I)$ for $I \subset S^1$, $1 \notin \mathring{I}$, factor through an action of $\mathcal{A}(S^1_{\text{cut}})$.*

Moreover, if $\mathcal{A}$ has finite index, then $\Omega_\mathcal{A}$ is an absorbing object.

PROOF. We first check that $\Omega_\mathcal{A}$ satisfies the above three properties. The first one is obvious. The third one holds by construction. The second property is a consequence of Lemma 10 below: since $\mathcal{A}(S^1_{\text{cut}})$ is an infinite factor, the Hilbert spaces $L^2(\mathcal{A}(S^1_{\text{cut}})) \oplus L^2(\mathcal{A}(S^1_{\text{cut}}))$ and $L^2(\mathcal{A}(S^1_{\text{cut}}))$ are isomorphic as left $\mathcal{A}(S^1_{\text{cut}})$-modules. It follows that $\Omega_\mathcal{A} \oplus \Omega_\mathcal{A} \cong \Omega_\mathcal{A}$.

Let Ω' be another soliton that satisfies the same three properties. Then Ω' is naturally an $\mathcal{A}(S^1_{\text{cut}})$-module. By the classification of modules over factors, there is a unique non-zero $\mathcal{A}(S^1_{\text{cut}})$-module (up to isomorphism) that satisfies $\Omega' \oplus \Omega' \cong \Omega'$. It follows that $\Omega' \cong \Omega_\mathcal{A}$.

Let us now assume that $\mathcal{A}$ has finite index. Given a non-zero soliton X, we need to show that $X \boxtimes \Omega_\mathcal{A} \cong \Omega_\mathcal{A}$. Equivalently, we need to show that $X \boxtimes \Omega_\mathcal{A}$ satisfies the three properties listed above. The first property, $X \boxtimes \Omega_\mathcal{A} \neq 0$, holds because fusing over a factor sends non-zero Hilbert spaces to non-zero Hilbert spaces (see, e.g., [**1**, Prop. 5.2]). The second property, $X \boxtimes \Omega_\mathcal{A} \oplus X \boxtimes \Omega_\mathcal{A} \cong X \boxtimes \Omega_\mathcal{A}$, is an immediate consequence of the corresponding property of $\Omega_\mathcal{A}$. The third property is more tricky and its verification will occupy the rest of this proof.

Let $\mathcal{I}_0$ be the collection of all subintervals of S^1 whose interior does not contain the base point $1 \in S^1$. Equivalently, $\mathcal{I}_0$ is the collection of all subintervals $I \subsetneq S^1_{\mathrm{cut}}$. By definition, a soliton is a Hilbert space equipped with compatible actions of all the algebras $\mathcal{A}(I)$ for $I \in \mathcal{I}_0$. Let $I_1 := [e^{i0}, e^{i\pi/2}]$ and $I_4 := [e^{i3\pi/2}, e^{i2\pi}]$ be the first and fourth quadrants of the standard circle, and let $I_{14} := I_1 \sqcup I_4 \subset S^1_{\mathrm{cut}}$ be their disjoint union (whereas I_1 and I_4 are not disjoint in S^1, these intervals *are* disjoint when viewed as subsets of S^1_{cut}). The collection $\mathcal{I}_0$ is not a 2-cover of S^1_{cut}, because no element of $\mathcal{I}_0$ contains $\partial S^1_{\mathrm{cut}}$. But

$$\mathcal{I} := \mathcal{I}_0 \cup \{I_{14}\}$$

is a 2-cover.

Recall that, by definition, $X \boxtimes \Omega_\mathcal{A} = X \boxtimes_{\mathcal{A}(I_+)} \Omega_\mathcal{A}$. By the split property, the actions of $\mathcal{A}(I_+)$ and of $\mathcal{A}(I_4)$ on $\Omega_\mathcal{A}$ extend to an action of their spatial tensor product. Equivalently, there exists an intermediate type I factor:

$$\mathcal{A}(I_4) \subset N \subset \mathcal{A}(I_+)' \qquad \text{(commutant inside } B(\Omega_\mathcal{A})).$$

The action of N on $\Omega_\mathcal{A}$ commutes with that of $\mathcal{A}(I_+)$ and thus induces an action on $X \boxtimes_{\mathcal{A}(I_+)} \Omega_\mathcal{A}$. The latter commutes with the action of $\mathcal{A}(I_1)$ coming from X, so we get an intermediate type I factor:

$$\mathcal{A}(I_4) \subset N \subset \mathcal{A}(I_1)' \qquad \text{(commutant inside } B(X \boxtimes_{\mathcal{A}(I_+)} \Omega_\mathcal{A})).$$

Equivalently, the actions of $\mathcal{A}(I_1)$ and $\mathcal{A}(I_4)$ on $X \boxtimes_{\mathcal{A}(I_+)} \Omega_\mathcal{A}$ extend to their spatial tensor product $\mathcal{A}(I_{14})$.

The above action of $\mathcal{A}(I_{14})$ on $X \boxtimes \Omega_\mathcal{A}$, together with the actions of $\mathcal{A}(I)$ for $I \in \mathcal{I}_0$ coming from the fact that $X \boxtimes \Omega_\mathcal{A}$ is a soliton, assemble to a compatible family of actions

$$\rho_I : \mathcal{A}(I) \to B(X \boxtimes \Omega_\mathcal{A}) \qquad \forall I \in \mathcal{I}.$$

Finally, by Lemma 6, since $\mathcal{I}$ is a 2-cover, these extend to an action of $\mathcal{A}(S^1_{\mathrm{cut}})$. This finishes the proof of condition (c). □

5. Appendix

For classical conformal nets, it is well known that, unless $\mathcal{A}(I) = \mathbb{C}$ for all I, the algebras $\mathcal{A}(I)$ are hyperfinite III_1 factors [**6**, Thm. 2.13]. Hyperfiniteness is a formal consequence of the split property, and holds equally well in the coordinate free setup (i.e., for conformal net as in [**2**, Def. 1.1]). Indeed, given an interval I, write $\mathring{I} = \bigcup I_n$ with $I_n \subset \mathring{I}_{n+1}$. By the split property, there exist intermediate type I subfactors $\mathcal{A}(I_n) \subset N_n \subset \mathcal{A}(I_{n+1})$, and so $\mathcal{A}(I) = \bigvee N_n$ is hyperfinite.

We do not know how to prove the type III_1 property in the coordinate free setup. The following lemma is the best we can offer:

LEMMA 10. *Let $\mathcal{A}$ be a non-trivial conformal net. Then the algebras $\mathcal{A}(I)$ are infinite factors (they are infinite dimensional, and they are not of type II_1).*

PROOF. The algebra $\mathcal{A}(I)$ is infinite dimensional as it contains infinitely many non-trivial commuting subalgebras.

Let I_0 be the upper half of the standard circle, so that the vacuum sector H_0 is $L^2(\mathcal{A}(I_0))$. Assume by contradiction that the algebra $\mathcal{A}(I_0)$ is of type II_1. Then the von Neumann dimension of H_0 as an $\mathcal{A}(I_0)$-module is equal to 1. By diffeomorphism covariance, for every interval $I \subset S^1$, the dimension of H_0 as an $\mathcal{A}(I)$-module is also 1. Given two intervals $I \subsetneq J \subset S^1$, we have

$$\dim_{\mathcal{A}(I)}(H_0) = [\mathcal{A}(J) : \mathcal{A}(I)] \cdot \dim_{\mathcal{A}(J)}(H_0).$$

It follows that $[\mathcal{A}(J) : \mathcal{A}(I)] = 1$. The inclusion $\mathcal{A}(I) \to \mathcal{A}(J)$ is therefore an isomorphism, a contradiction. □

Acknowledgement

We thank Claudia Scheimbauer for useful comments on this manuscript.

References

[1] Arthur Bartels, Christopher L. Douglas, and André Henriques, *Dualizability and index of subfactors*, Quantum Topol. **5** (2014), no. 3, 289–345. MR3342166
[2] Arthur Bartels, Christopher L. Douglas, and André Henriques, *Conformal nets I: Coordinate-free nets*, Int. Math. Res. Not. IMRN **13** (2015), 4975–5052. MR3439097
[3] Arthur Bartels, Christopher L. Douglas, and André Henriques, *Conformal nets I: Coordinate-free nets*, Int. Math. Res. Not. IMRN **13** (2015), 4975–5052. MR3439097
[4] Kevin Costello and Owen Gwilliam, *Factorization algebras in quantum field theory. Vol. 1*, New Mathematical Monographs, vol. 31, Cambridge University Press, Cambridge, 2017. MR3586504
[5] Klaus Fredenhagen and Martin Jörß, *Conformal Haag-Kastler nets, pointlike localized fields and the existence of operator product expansions*, Comm. Math. Phys. **176** (1996), no. 3, 541–554. MR1376431
[6] Fabrizio Gabbiani and Jürg Fröhlich, *Operator algebras and conformal field theory*, Comm. Math. Phys. **155** (1993), no. 3, 569–640. MR1231644
[7] Alain Guichardet, *Sur la catégorie des algèbres de von Neumann* (French), Bull. Sci. Math. (2) **90** (1966), 41–64. MR0201989
[8] A. Henriques. What Chern-Simons theory assigns to a point. *arXiv:1503.06254* (2015).
[9] André Henriques and David Penneys, *Bicommutant categories from fusion categories*, Selecta Math. (N.S.) **23** (2017), no. 3, 1669–1708. MR3663592
[10] André Henriques, *The classification of chiral WZW models by $H^4_+(BG, \mathbb{Z})$*, Lie Algebras, Vertex Operator Algebras, and Related Topics, Contemp. Math., vol. 695, Amer. Math. Soc., Providence, RI, 2017, pp. 99–121. MR3709708
[11] A. Kornell. Quantum collections. *arXiv:1202.2994* (2012).
[12] Roberto Longo and Feng Xu, *Topological sectors and a dichotomy in conformal field theory*, Comm. Math. Phys. **251** (2004), no. 2, 321–364. MR2100058
[13] M. Takesaki, *Theory of operator algebras. I*, Encyclopaedia of Mathematical Sciences, vol. 124, Springer-Verlag, Berlin, 2002. Reprint of the first (1979) edition; Operator Algebras and Non-commutative Geometry, 5. MR1873025
[14] V. Toledano Laredo. Fusion of positive energy representations of $LSpin_{2n}$. Ph.D. thesis, St. John's College, Cambridge, 1997.
[15] Antony Wassermann, *Operator algebras and conformal field theory. III. Fusion of positive energy representations of* LSU(N) *using bounded operators*, Invent. Math. **133** (1998), no. 3, 467–538. MR1645078
[16] Feng Xu, *Jones-Wassermann subfactors for disconnected intervals*, Commun. Contemp. Math. **2** (2000), no. 3, 307–347. MR1776984

MATHEMATICAL INSTITUTE UNIVERSITY OF OXFORD, ANDREW WILES BUILDING, RADCLIFFE OBSERVATORY QUARTER, WOODSTOCK ROAD, OXFORD OX2 6GG, UNITED KINGDOM
Email address: andre.henriques@maths.ox.ac.uk

Proceedings of Symposia in Pure Mathematics
Volume **98**, 2018
http://dx.doi.org/10.1090/pspum/098/01725

Contracting the Weierstrass locus to a point

Alexander Polishchuk

ABSTRACT. We construct an open substack $\mathcal{U} \subset \mathcal{M}_{g,1}$ with the complement of codimension ≥ 2 and a morphism from $\mathcal{U}$ to a weighted projective stack, which sends the Weierstrass locus $\mathcal{W} \cap \mathcal{U}$ to a point, and maps $\mathcal{M}_{g,1} \setminus \mathcal{W}$ isomorphically to its image. The construction uses alternative birational models of $\mathcal{M}_{g,1}$ and $\mathcal{M}_{g,2}$ studied earlier by the author.

Introduction

Let $\mathcal{W} \subset \mathcal{M}_{g,1}$ denote the locus in the moduli stack of smooth one-pointed curves of genus g, consisting of (C,p) such that p is a Weierstrass point on C, i.e., $h^1(gp) \neq 0$. It is well known that $\mathcal{W}$ is an irreducible divisor. In this paper we construct a rational map from $\mathcal{M}_{g,1}$ to a proper DM-stack with projective coarse moduli space, which contracts $\mathcal{W}$ to a single point and maps $\mathcal{M}_{g,1} \setminus \mathcal{W}$ isomorphically to its image (see Theorem A below). This is partly motivated by the question whether the class of the closure of W in $\overline{M}_{g,1}$ generates an extremal ray (we do not solve this; however, see Prop. 2.4.6, Rem. 2.4.7 and the discussion below). Note that for small g some pointed Brill-Noether divisors were shown to generate extremal rays in the effective cone of $\overline{M}_{g,1}$ in [**R01**], [**J13**] and [**J10**].

The construction involves certain moduli stacks studied in [**P17**]. Namely, in [**P17**] we introduced and studied the moduli stack of curves with marked points $(C, p_1, \ldots, p_n)$, where C is a reduced projective curve of arithmetic genus g, such that $h^1(a_1p_1 + \ldots + a_np_n) = 0$ for fixed integer weights $a_i \geq 0$ such that $a_1 + \ldots + a_n = g$ (we assume that the marked points are smooth and distinct). We denote this stack by $\mathcal{U}^{ns}_{g,n}(a_1, \ldots, a_n)$. We showed that $\mathcal{U}^{ns}_{g,n}(a_1, \ldots, a_n)$ can be realized as a quotient of an affine scheme by a torus action and studied the related GIT picture which leads to interesting projective birational models of $M_{g,n}$. In particular, for $n = 1$ and $a_1 = g$ there is a unique nonempty GIT quotient stack $\overline{\mathcal{U}}^{ns}_{g,1}(g)$, obtained from $\mathcal{U}^{ns}_{g,1}(g)$ by deleting one point corresponding to the most singular cuspidal curve. Furthermore, $\overline{\mathcal{U}}^{ns}_{g,1}(g)$ is a closed substack in a weighted projective stack (see Sec. 1.1 for details).

We start by considering the natural rational map

$$\text{for}_2 : \mathcal{U}^{ns}_{g,2}(g-1, 1) \dashrightarrow \overline{\mathcal{U}}^{ns}_{g,1}(g) \tag{0.0.1}$$

2010 *Mathematics Subject Classification.* Primary 14H10; Secondary 14H20.

Supported in part by the NSF grant DMS-1400390 and by the Russian Academic Excellence Project '5-100'.

given by forgetting the second marked point (more precisely, the map for_2 is regular on a certain open substack which is dense in the component corresponding to smoothable curves). Our main technical result is that (0.0.1) is regular on the open substack of (C, p_1, p_2) such that $h^1((g+1)p_1) = 0$, and that the divisor, defined by the condition $h^1(gp_1) \neq 0$, gets contracted to a point (see Prop. 1.2.2). Furthermore, we show that this point has trivial group of automorphisms. We derive from this the following result.

Theorem A. *Assume that $g \geq 2$. The natural open embedding of stacks*

$$\mathcal{M}_{g,1} \setminus \mathcal{W} \hookrightarrow \overline{\mathcal{U}}^{ns}_{g,1}(g)$$

extends to a regular morphism

$$\phi = \phi_g : \mathcal{U} \to \overline{\mathcal{U}}^{ns}_{g,1}(g),$$

for some open substack $\mathcal{U} \subset \mathcal{M}_{g,1}$ containing $\mathcal{M}_{g,1} \setminus \mathcal{W}$ and such that $\mathcal{M}_{g,1} \setminus \mathcal{U}$ has codimension ≥ 2 in $\mathcal{M}_{g,1}$. Furthermore, ϕ contracts $\mathcal{U} \cap \mathcal{W}$ to a single point, which has no nontrivial automorphisms.

More precisely, the open substack $\mathcal{U}$ in the above Theorem consists of (C, p) such that $h^1((g+1)p) = 0$ and $h^0((g-1)p) = 1$.

We study the case $g = 2$ in more detail. In this case we get a more precise result involving a certain modular compactification of $\mathcal{M}_{2,1}$.

Recall that Smyth introduced in [**Sm13**] the notion of an extremal assignment, which is a rule associating to each stable curve of given arithmetic genus some of its irreducible components (this rule should be stable under degenerations). For each extremal assignment $\mathcal{Z}$, Smyth considered the moduli stack $\overline{\mathcal{M}}_{g,n}(\mathcal{Z})$ of $\mathcal{Z}$-stable curves, i.e., pointed curves C for which there exists a stable curve C' and a map of pointed curves $C' \to C$, contracting precisely the components of C', assigned by $\mathcal{Z}$, in a certain controlled way. In this paper we consider only one extremal assignment which associates to every stable curve all of its unmarked components (see [**Sm13**, Ex. 1.12]), so when we say $\mathcal{Z}$-stable we always mean this particular extremal assignment.

We prove that the map ϕ_2 extends to a regular morphism of stacks

$$\phi_2 : \overline{\mathcal{M}}_{2,1}(\mathcal{Z}) \to \overline{\mathcal{U}}^{ns}_{2,1}(2)$$

contracting the closure of $\mathcal{W}$ to one point (see Theorem 2.4.5). Furthermore, we identify the point $\phi_2(\mathcal{W})$ explicitly as a certain cuspidal curve C_0 (see Definition 2.2.2), and show that ϕ_2 induces an isomorphism of the complement of $\mathcal{W}$ to the complement of $\phi_2(\mathcal{W})$.

We also prove that the natural rational map of the coarse moduli spaces $\overline{M}_{2,1} \dashrightarrow \overline{U}^{ns}_{2,1}(2)$ is a birational contraction with the exceptional divisors $\overline{W}$ and Δ_1 (see Proposition 2.4.6). One can expect that the rational map $\overline{M}_{g,1} \dashrightarrow \overline{U}^{ns}_{g,1}(g)$ is still a birational contraction for $g > 2$ (see Remark 2.4.7 for further discussion).

In addition, in Sec. 2.1 we obtain an isomorphism

$$\overline{\mathcal{U}}^{ns}_{2,1}(2) \simeq \mathbb{P}(2,3,4,5,6),$$

where the right-hand side is the weighted projective stack.

Conventions. In Sec. 2.1 we work over $\mathbb{Z}[1/6]$. Everywhere else we work over $\mathbb{C}$. By a curve we mean a connected reduced projective curve. By the genus of a curve we always mean arithmetic genus. For DM-stacks whose notation involves calligraphic letters $\mathcal{M}, \mathcal{U}$ and $\mathcal{W}$, we denote their coarse moduli spaces by replacing these letters by M, U and W.

1. Rational maps for_2 and ϕ

1.1. Moduli spaces of curves with non-special divisors. We start by recalling some results from [**P17**] about the stacks $\mathcal{U}^{ns}_{g,n}(\mathbf{a})$, where $\mathbf{a} = (a_1, \dots, a_n)$ and a_i are non-negative integers with $a_1 + \dots + a_n = g$. We denote by $\widetilde{\mathcal{U}}^{ns}_{g,n}(\mathbf{a})$ the $\mathbb{G}_m^n$-torsor over $\mathcal{U}^{ns}_{g,n}(\mathbf{a})$, corresponding to choices of nonzero tangent vectors at the marked points. It is proved in [**P17**] that $\widetilde{\mathcal{U}}^{ns}_{g,n}(\mathbf{a})$ is an affine scheme of finite type. In this paper we only need the case when all a_i are positive, so we assume this is the case.

The key result we will use is that for each $i = 1, \dots, n$, and each $(C, p_\bullet, v_\bullet)$ in $\widetilde{\mathcal{U}}^{ns}_{g,n}(\mathbf{a})$ (where v_i is a nonzero tangent vector at p_i), there is a canonical formal parameter t_i on C at p_i, such that $\langle v_i, dt_i \rangle = 1$, which is defined as follows. Given a formal parameter t_i, for each $m > a_i$ there is unique, up to adding a constant, rational function $f_i[-m] \in H^0(C, \mathcal{O}(mp_i + \sum_{j \neq i} a_j p_j))$ with the Laurent expansion in t_i of the form

$$f_i[-m] = t_i^{-m} + \sum_{q \geq -a_i} \alpha_i[-m, q] t_i^q. \tag{1.1.1}$$

The canonical parameter is uniquely characterized by the conditions that $\langle v_i, dt_i \rangle = 1$ and $\alpha_i[-m, -a_i] = 0$ for every $m > a_i$. Using these formal parameters we can consider for every pair (i, j) and $m > a_i$ the expansion of $f_i[-m]$ at p_j:

$$f_i[-m] = \delta_{ij} t_j^{-m} + \sum_{q \geq -a_j} \alpha_{ij}[-m, q] t_j^q$$

(note that $\alpha_i[-m, q] = \alpha_{ii}[-m, q]$). Now we can view the coefficients $\alpha_{ij}[-m, q]$ as functions on $\widetilde{\mathcal{U}}^{ns}_{g,n}(\mathbf{a})$, where we fix the ambiguity in adding a constant to $f_i[-m]$ by requiring that $\alpha_i[-m, 0] = 0$. It follows from the results of [**P17**] that these functions are all expressed in terms of a finite number of them, which gives a closed embedding of $\widetilde{\mathcal{U}}^{ns}_{g,n}(\mathbf{a})$ into an affine space.

The rescaling of the tangent vectors (v_i) defines an action of $\mathbb{G}_m^n$ on $\widetilde{\mathcal{U}}^{ns}_{g,n}(\mathbf{a})$, so that the weight of the function $\alpha_{ij}[-m, q]$ is $m\mathbf{e}_i + q\mathbf{e}_j$, where $(\mathbf{e}_i)$ is the standard basis in the character lattice of $\mathbb{G}_m^n$.

There is a special point in $\widetilde{\mathcal{U}}^{ns}_{g,n}(\mathbf{a})$ which is a unique point invariant under the action of $\mathbb{G}_m^n$: it is the point where all the functions $\alpha_{ij}[-m, q]$ vanish, i.e., it corresponds to the origin in the ambient affine space. The underlying curve is the union of n rational cuspidal curves $C^{\mathrm{cusp}}(a_i)$, glued transversally at the cusp. Here $C^{\mathrm{cusp}}(a)$ is the projective curve with the affine part given by $\mathrm{Spec}(k \cdot 1 + x^{a+1} k[x])$, and with one smooth point at infinity (see [**P17**, Sec. 2.1]). In [**P17**] we also studied the GIT picture for the $\mathbb{G}_m^n$-action on $\widetilde{\mathcal{U}}^{ns}_{g,n}(\mathbf{a})$. In general we have stability conditions depending on a character χ of $\mathbb{G}_m^n$. In the case $n = 1$, i.e., for $\widetilde{\mathcal{U}}^{ns}_{g,1}(g)$ there is a unique nonempty stability condition, so that the unique unstable point in $\widetilde{\mathcal{U}}^{ns}_{g,1}(g)$ is the origin, i.e., the point corresponding to the curve $C^{\mathrm{cusp}}(g)$. We denote this point by $[C^{\mathrm{cusp}}(g)]$. Then the functions $\alpha_{ij}[-m,q]$ identify the corresponding GIT quotient stack,

$$\overline{\mathcal{U}}^{ns}_{g,1}(g) := (\widetilde{\mathcal{U}}^{ns}_{g,1}(g) \setminus [C^{\mathrm{cusp}}(g)])/\mathbb{G}_m,$$

with a closed substack in the weighted projective stack.

For two collections of weights as above, $\mathbf{a}$ and $\mathbf{a}'$, we denote by $\widetilde{\mathcal{U}}^{ns}_{g,n}(\mathbf{a},\mathbf{a}')$ the intersection of the stacks $\widetilde{\mathcal{U}}^{ns}_{g,n}(\mathbf{a})$ and $\widetilde{\mathcal{U}}^{ns}_{g,n}(\mathbf{a}')$. In other words, we impose both conditions, $h^1(\sum a_ip_i)=0$ and $h^1(\sum a_i'p_i)=0$, on the marked points.

1.2. The forgetful map. The rational map (0.0.1) corresponds to a regular morphism

$$\text{(1.2.1)} \qquad \mathrm{for}_2 : \widetilde{\mathcal{U}}^{ns}_{g,2}((g-1,1),(g,0)) \to \widetilde{\mathcal{U}}^{ns}_{g,1}(g),$$

which is given as the composition of the open embedding $\widetilde{\mathcal{U}}^{ns}_{g,2}((g-1,1),(g,0)) \hookrightarrow \widetilde{\mathcal{U}}^{ns}_{g,2}(g,0)$ followed by the forgetful map

$$\mathrm{for}_2 : \widetilde{\mathcal{U}}^{ns}_{g,2}(g,0) \to \widetilde{\mathcal{U}}^{ns}_{g,1}(g)$$

defined in [**P17**, Thm. A]. The latter map sends (C,p_1,p_2,v_1,v_2), with C irreducible, to (C,p_1,v_1) (if C is reducible then it gets replaced by a certain curve $\overline{C}$, such that $C \to \overline{C}$ is a contraction of the component containing p_2).

Let $Z \subset \widetilde{\mathcal{U}}^{ns}_{g,2}((g-1,1),(g,0))$ be the closed subscheme given as the preimage of the origin under (1.2.1). Then there is a regular morphism

$$\text{(1.2.2)} \qquad \widetilde{\mathcal{U}}^{ns}_{g,2}((g-1,1),(g,0)) \setminus Z \to \overline{\mathcal{U}}^{ns}_{g,1}(g)$$

induced by (1.2.1). Note that (C,p_1,p_2,v_1,v_2) with irreducible C belongs to Z if and only if (C,p_1) is the cuspidal curve $C^{\mathrm{cusp}}(g)$ (with the marked point at infinity).

Let us denote by

$$\widetilde{\mathcal{U}}^{ns}_{g,2}((g-1,1),(g+1,0)) \subset \widetilde{\mathcal{U}}^{ns}_{g,2}(g-1,1)$$

the open subset given by the condition $h^1((g+1)p_1)=0$. Let also

$$\widetilde{\mathcal{W}} \subset \widetilde{\mathcal{U}}^{ns}_{g,2}((g-1,1),(g+1,0))$$

denote the closed locus given by the condition $h^1(gp_1)\neq 0$, so that

$$\widetilde{\mathcal{U}}^{ns}_{g,2}((g-1,1),(g,0)) = \widetilde{\mathcal{U}}^{ns}_{g,2}((g-1,1),(g+1,0)) \setminus \widetilde{\mathcal{W}}.$$

Recall that we have sections $f_1[-m] \in H^0(C,\mathcal{O}(mp_1+p_2))$, where C is the universal curve over $\widetilde{\mathcal{U}}^{ns}_{g,2}(g-1,1)$, for $m\geq g$, with expansions at p_1 of the form (1.1.1) (with $i=1$) with $\alpha_1[-m,-g+1]=\alpha_1[-m,0]=0$.

LEMMA 1.2.1. *Let us set $\alpha=\alpha_{12}[-g,-1]$, $\beta=\alpha_{12}[-g-1,-1]$. Then the open subset*

$$\widetilde{\mathcal{U}}^{ns}_{g,2}((g-1,1),(g,0)) \subset \widetilde{\mathcal{U}}^{ns}_{g,2}(g-1,1)$$

is given by the condition $\alpha\neq 0$. Similarly, the open subset

$$\widetilde{\mathcal{U}}^{ns}_{g,2}((g-1,1),(g+1,0)) \subset \widetilde{\mathcal{U}}^{ns}_{g,2}(g-1,1)$$

is the locus where either $\alpha\neq 0$ or $\beta\neq 0$.

Proof. Recall that the open subset $\widetilde{\mathcal{U}}^{ns}_{g,2}((g-1,1),(g,0))$ is characterized by the condition $h^1(gp_1)=0$. Since $h^1(gp_1+p_2)=0$, the long exact sequence of cohomology associated with the exact sequence of sheaves

$$0\to\mathcal{O}(gp_1)\to\mathcal{O}(gp_1+p_2)\to\mathcal{O}(p_2)/\mathcal{O}\to 0$$

shows that $h^1(gp_1)\neq 0$ precisely for those curves for which $f_1[-g]$ is regular at p_2. But this is equivalent to the vanishing of α, since α is the coefficient of t_2^{-1} in the expansion of $f_1[-g]$ at p_2.

The case of $\widetilde{\mathcal{U}}^{ns}_{g,2}((g-1,1),(g+1,0))$ is similar: now we consider the exact sequence

$$0\to\mathcal{O}((g+1)p_1)\to\mathcal{O}((g+1)p_1+p_2)\to\mathcal{O}(p_2)/\mathcal{O}\to 0$$

which shows that $h^1((g+1)p_1)\neq 0$ when both $f_1[-g]$ and $f_1[-g-1]$ are regular at p_2, i.e., both α and β vanish. □

The following Proposition is a crucial step in proving Theorem A.

PROPOSITION 1.2.2. *The subset* $Z\subset\widetilde{\mathcal{U}}^{ns}_{g,2}((g-1,1),(g+1,0))\setminus\widetilde{\mathcal{W}}$ *is closed in* $\widetilde{\mathcal{U}}^{ns}_{g,2}((g-1,1),(g+1,0))$. *There exists a regular morphism*

$$\overline{\mathrm{for}}_2:\widetilde{\mathcal{U}}^{ns}_{g,2}((g-1,1),(g+1,0))\setminus Z\to\overline{\mathcal{U}}^{ns}_{g,1}(g), \tag{1.2.3}$$

extending the morphism (1.2.2) *and sending* $\widetilde{\mathcal{W}}$ *to a point. Furthermore, this point has no nontrivial automorphisms.*

Proof. Let C' denote the universal curve over the open subset $\widetilde{\mathcal{U}}^{ns}_{g,2}((g-1,1),(g,0))$. To calculate explicitly the map (1.2.1), we need to find the sections

$$f[-m]\in H^0(C',\mathcal{O}(mp_1)),$$

for $m\geq g+1$, and a modified formal parameter u at p_1, such that $f[-m]$ would have expansions of the form

$$f[-m]=u^{-m}+\alpha[-m,-g+1]u^{-g+1}+\alpha[-m,-g+2]u^{-g+2}+\dots, \tag{1.2.4}$$

where $\alpha[-m,q]$ are some rational expressions of the coordinates on $\widetilde{\mathcal{U}}^{ns}_{g,2}(g-1,1)$ with only powers of α in the denominator.

As the first approximation let us set for $m\geq g+1$,

$$\widetilde{f}[-m]=f_1[-m]-\frac{\alpha_{12}[-m,-1]}{\alpha}f_1[-g].$$

The constant is chosen in such a way that the poles at p_2 cancel out, so we have $\widetilde{f}[-m]\in H^0(\mathcal{O}(mp_1))$, while the expansion of $\widetilde{f}[-m]$ at p_1 has form

$$\widetilde{f}[-m]=t_1^{-m}-\frac{\alpha_{12}[-m,-1]}{\alpha}\cdot t_1^{-g}+\dots,$$

where t_1 is the canonical parameter at p_1 on C'.

Now we need to change the canonical parameter to $u=t_1+c_1t_1^2+\dots$, and to add to each $\widetilde{f}[-m]$ a linear combination of $\widetilde{f}[-m']$ with $m'<m$, to get the expansions of the required form (1.2.4). We want to know only the highest order polar parts of the functions $\alpha[-m,q]$, i.e., those with the highest power of α (prescribed below) in the denominator, so we introduce the following filtration F_n on the space of formal Laurent series in t_1 with coefficients in $R=\mathcal{O}(\widetilde{\mathcal{U}}^{ns}_{g,2}((g-1,1),(g,0)))$. By definition,

a Laurent series belongs to F_n if it can be written in the form $\sum_i a_i\alpha^{-i-n}t_1^i$, where each a_i extends to a regular function on $\widetilde{\mathcal{U}}^{ns}_{g,2}(g-1,1)$.

It will be enough for us to keep track only of $f[-m] \bmod F_{m-1}$. It is easy to see that the change of variables $t_1 \mapsto t_1 + c_1t_1^2 + c_2t_1^3 + \ldots$, where for each i, $\alpha^i c_i$ extends to a regular function on $\widetilde{\mathcal{U}}^{ns}_{g,2}(g-1,1)$, preserves the filtration (F_n). Since to go from t_1 to u we will only use the changes of variables of this form, it suffices for us to know that

$$\widetilde{f}[-g-1] \equiv t_1^{-g-1} - \lambda t_1^{-g} \bmod F_g, \tag{1.2.5}$$

where $\lambda = \frac{\alpha_{12}[-g-1,-1]}{\alpha} = \frac{\beta}{\alpha}$, while

$$\widetilde{f}[-m] \equiv t_1^{-m} \bmod F_{m-1} \quad \text{for } m > g+1. \tag{1.2.6}$$

We claim that there exist rational constants $(r_{m,j})$, $1 \le j < m-g$, and (r_i), $i \ge 1$, such that

$$f[-m] \equiv \widetilde{f}[-m] + \sum_{1\le j<m-g} r_{m,j}\lambda^j \widetilde{f}[-m+j] \bmod F_{m-1}, \tag{1.2.7}$$

for each $m \ge g+1$, and

$$t_1 \equiv u + r_1\lambda u^2 + r_2\lambda^2u^3 + \ldots \bmod F_{-2}. \tag{1.2.8}$$

Namely, we prove by induction on $n \ge 1$ that (1.2.7) holds for all m with $m \le g+n$, and that the required relation between t_1 and u holds modulo $t_1^n R[[t_1]] + F_{-2}$.

Let us recall the recursive construction of $(f[-g-n])$ and of formal parameters u_n such that $u_n \equiv u \bmod t_1^{n+1}R[[t_1]]$, where u is the canonical parameter (cf. [**FP14**, Lem. 4.1.3]). For $n=1$ we have $f[-g-1] = \widetilde{f}[-g-1]$ and $u_1 = t_1$. Assume $f[-g-n']$ are already defined for $n' < n$ and $u_{n-1} \equiv u \bmod t_1^n R[[t_1]]$ is known, so that

$$f[-g-n'] \equiv u_{n-1}^{-g-n'} \bmod t_1^{-g+1}R[[t_1]] \text{ for } n' < n-1, \text{ while}$$

$$f[-g-n+1] \equiv u_{n-1}^{-g-n+1} + c\cdot u_{n-1}^{-g} \bmod t_1^{-g+1}R[[t_1]]. \tag{1.2.9}$$

Then we set $u_n = u_{n-1} + \frac{c}{g+n-1}u_{n-1}^n$, the expansion of $f[-g-n+1]$ in u_n will take the form

$$f[-g-n+1] \equiv u_n^{-g-n+1} \bmod t_1^{-g+1}R[[t_1]],$$

and the expansions of $f[-g-n']$ for all $n' < n-1$ in u_n will still have the correct form. Now, if the expansion of $\widetilde{f}[-g-n]$ in u_n has the form

$$\widetilde{f}[-g-n] = u_n^{-g-n} + p_1u_n^{-g-n+1} + \ldots + p_{n-1}u_n^{-g-1} + \ldots, \tag{1.2.10}$$

then we set

$$f[-g-n] = \widetilde{f}[-g-n] - p_1 f[-g-n+1] - \ldots - p_{n-1}f[-g-1]. \tag{1.2.11}$$

The induction assumption implies that the function c in (1.2.9) has the leading polar term $r\lambda^{n-1}$ for some $r \in \mathbb{Q}$, so the change of variables from u_{n-1} to u_n is of the right form, as discussed above. It follows that

$$t_1 \equiv u_n + s_1\lambda u_n^2 + \ldots + s_{n-1}\lambda^{n-1}u_n^n \bmod t_1^{n+1}R[[t_1]] + F_{-2}$$

for some $s_i \in \mathbb{Q}$. Now from (1.2.6) we get that

$$\widetilde{f}[-g-n] = (u_n + s_1\lambda u_n^2 + \ldots + s_{n-1}\lambda^{n-1}u_n^n)^{-g-n} \bmod t_1^{-g}R[[t_1]] + F_{g+n-1}.$$

This implies that for $i = 1, \dots, n-1$, the leading polar term of the coefficient p_i in the expansion (1.2.10) is of the form $a_i\lambda^i$, for $a_i \in \mathbb{Q}$. Now (1.2.11) shows that (1.2.7) holds for $m = g + n$. This finishes the proof of our claim.

Now combining (1.2.5)–(1.2.8), we get that for each $m \geq g+1$ the expansion of $f[-m]$ in the canonical parameter u has form

$$f[-m] \equiv u^{-m} + \sum_{j\geq 1} s_{m,j}\lambda^{m-g+j}u^{-g+j} \operatorname{mod} F_{m-1},$$

for some rational constants $(s_{m,j})$. In other words, the functions $\alpha[-m, -g+j] \in R$, defining the map (1.2.1), have form

$$\alpha[-m, -g+j] = s_{m,j}\lambda^{m-g+j} + \dots$$

where the omitted terms have smaller powers of α in the denominator.

Finally, we need to know that not all $(s_{m,j})$ are zero, so let us compute $s_{-g-1,-g+1}$ and $s_{g-1,-g+2}$ following the above procedure (we will need to look at two coordinates to prove that the point, which is the image of $\mathcal{W}$, has no nontrivial automorphisms). Due to (1.2.5), the first change of variables is

$$t_1 = u_2 - \frac{\lambda}{g+1}u_2^2 \operatorname{mod} F_{-2}.$$

Then we get expansions

$$f[-g-1] = \widetilde{f}[-g-1] \equiv u_2^{-g-1} + \frac{2-g}{2(g+1)}\lambda^2 u_2^{-g+1} + \frac{-g^2+g+3}{3(g+1)^2}\lambda^3 u_2^{-g+2} \operatorname{mod} u_2^{-g+3}R[[u_2]] + F_g,$$

$$\widetilde{f}[-g-2] \equiv u_2^{-g-2} + \frac{g+2}{g+1}\lambda u_2^{-g-1} + \frac{(g+2)(g+3)}{2(g+1)^2}\lambda^2 u_2^{-g} \operatorname{mod} u_2^{-g+1}R[[u_2]] + F_{g+1},$$

$$\widetilde{f}[-g-3] \equiv u_2^{-g-3} + \frac{g+3}{g+1}\lambda u_2^{-g-2} + \frac{(g+3)(g+4)}{2(g+1)^2}\lambda^2 u_2^{-g-1} + \frac{(g+3)(g+4)(g+5)}{6(g+1)^3}\lambda^3 u_2^{-g} \operatorname{mod} u_2^{-g+1}R[[u_2]] + F_{g+2}.$$

Hence, the coefficient of u_2^{-g} in $f[-g-2] \operatorname{mod} F_{g+1}$ (which is the same as in $\widetilde{f}[-g-2] \operatorname{mod} F_{g+1}$) is $\frac{(g+2)(g+3)}{2(g+1)^2}\lambda^2$. Thus, the second change of variables (defined so that the coefficient of u_3^{-g} in $f[-g-2]$ is zero) is

$$u_2 = u_3 + \frac{(g+2)(g+3)}{2(g+2)(g+1)^2}\lambda^2 u_3^3 \operatorname{mod} F_{-2},$$

and we get the expansion

$$f[-g-1] = u_3^{-g-1} - \frac{2g+1}{2(g+1)}\lambda^2 u_3^{-g+1} + \frac{-g^2+g+3}{3(g+1)^2}\lambda^3 u_3^{-g+2} \operatorname{mod} u_3^{-g+3}R[[u_3]] + F_g,$$

which shows that

$$s_{g+1,1} = -\frac{2g+1}{2(g+1)}.$$

Also, we see that the coefficient of u_3^{-g} in the expansion of $\widetilde{f}[-g-3] \bmod F_{g+2}$ is equal to $-\frac{(g+3)(g^2+3g-1)}{3(g+1)^3}\lambda^3$. This dictates that the next change of variables is

$$u_3 = u_4 - \frac{(g^2+3g-1)}{3(g+1)^3}\lambda^3 u_4^4 \bmod F_{-2}.$$

Finally, we get that the coefficient of u_4^{-g+2} in the expansion of $f[-g-1] \bmod F_g$ is equal to

$$\frac{-g^2+g+3}{3(g+1)^2}\lambda^3 + \frac{(g^2+3g-1)}{3(g+1)^2}\lambda^3 = \frac{4g+2}{3(g+1)^2}\lambda^3,$$

and hence,

$$s_{g+1,2} = \frac{4g+2}{3(g+1)^2}.$$

Now let us consider the modified map

$$\alpha \cdot \mathrm{for}_2 : \widetilde{\mathcal{U}}^{ns}_{g,2}((g-1,1),(g,0)) \to \widetilde{\mathcal{U}}^{ns}_{g,1}(g) : x \mapsto \alpha(x) \cdot \mathrm{for}_2(x)$$

Since the weight of $\alpha[-m,-g+j]$ is $m-g+j$, the modifed map sends x to the point in $\widetilde{\mathcal{U}}^{ns}_{g,1}(g)$ with coordinates

$$\alpha(x)^{m-g+j}\alpha[-m,-g+j](x) = s_{m,j}\beta(x)^{m-g+j} + \alpha(x)\cdot f_{m,j}(x), \tag{1.2.12}$$

where $f_{m,j}$ are regular functions on $\widetilde{\mathcal{U}}^{ns}_{g,2}(g-1,1)$. In particular, $\alpha \cdot \mathrm{for}_2$ can be viewed as a regular map from $\widetilde{\mathcal{U}}^{ns}_{g,2}(g-1,1)$.

Recall that by Lemma 1.2.1, the open subset $\widetilde{\mathcal{U}}^{ns}_{g,2}((g-1,1),(g+1,0))$ is the locus where either $\alpha \neq 0$ or $\beta \neq 0$, and the locus $\widetilde{\mathcal{W}}$ is given by $\alpha = 0$. Thus, (1.2.12) gives for $x \in \widetilde{\mathcal{W}}$:

$$\alpha \cdot \mathrm{for}_2(x) = (s_{m,j}\beta(x)^{m-g+j}) = \beta(x)\cdot(s_{m,j}).$$

Furthermore, as we have seen above, the constants $s_{-g-1,1}$ and $s_{-g-1,2}$ are nonzero, so the corresponding coordinates in the above expression are also nonzero. Note also that the corresponding point of $\overline{\mathcal{U}}^{ns}_{g,1}(g)$ is equal to $(s_{m,j})$, so it does not depend on x.

Denoting by $U_{\beta\neq 0} \subset \widetilde{\mathcal{U}}^{ns}_{g,2}((g-1,1),(g+1,0))$ the open subset where $\beta \neq 0$, we get

$$(\alpha \cdot \mathrm{for}_2)^{-1}(0) \cap U_{\beta\neq 0} = Z \cap U_{\beta\neq 0},$$

and so $Z \cap U_{\beta\neq 0}$ is closed in $U_{\beta\neq 0}$. Since, Z is closed in the open subset $\alpha \neq 0$, we derive that Z is closed in $\widetilde{\mathcal{U}}^{ns}_{g,2}((g-1,1),(g+1,0))$.

We have a covering of $\widetilde{\mathcal{U}}^{ns}_{g,2}((g-1,1),(g+1,0)) \setminus Z$ by two open subsets: $\widetilde{\mathcal{U}}^{ns}_{g,2}((g-1,1),(g,0)) \setminus Z$ and $U_{\beta\neq 0}$. The required regular morphism (1.2.3) to $\overline{\mathcal{U}}^{ns}_{g,1}(g)$ is induced by for_2 on $\widetilde{\mathcal{U}}^{ns}_{g,2}((g-1,1),(g,0)) \setminus Z$ and by $\alpha \cdot \mathrm{for}_2$ on $U_{\beta\neq 0}$. As we have seen above, this morphism sends $\widetilde{\mathcal{W}} \subset U_{\beta\neq 0}$ to the point $(s_{m,j})$ of the weighted projective stack with two nonzero homogeneous coordinates, of weights 2 and 3. Hence, this point does not have nontrivial automorphisms. □

1.3. Proof of Theorem A. It is well known that $\mathcal{W} \subset \mathcal{M}_{g,1}$ is an irreducible divisor (see [**A74, A78**]). Now let $\mathcal{U} \subset \mathcal{M}_{g,1}$ be the open substack of (C,p) satisfying

$$h^1((g+1)p) = 0, \quad h^0((g-1)p) = 1.$$

Note also that we have an inclusion

$$\mathcal{M}_{g,1} \setminus \mathcal{W} \subset \mathcal{U}$$

since the condition $h^0(gp) = 1$ implies that $h^1((g+1)p) = 0$ and $h^0((g-1)p) = 1$. Furthermore, the complement to $\mathcal{U}$ is a proper closed substack in $\mathcal{W}$, so it has codimension ≥ 2 in $\mathcal{M}_{g,1}$. In particular, $\mathcal{U} \cap \mathcal{W}$ is dense in $\mathcal{W}$.

Note that we have a natural open inclusion

$$\mathcal{M}_{g,1} \setminus \mathcal{W} \hookrightarrow \overline{\mathcal{U}}^{ns}_{g,1}(g). \tag{1.3.1}$$

Indeed, the only unstable point in $\mathcal{U}^{ns}_{g,1}(g)$ corresponds to the singular curve $C^{\mathrm{cusp}}(g)$. We are going to show that the above morphism extends to a regular morphism

$$\mathcal{U} \to \overline{\mathcal{U}}^{ns}_{g,1}(g),$$

such that $\mathcal{U} \cap \mathcal{W}$ is mapped to a point.

Recall that by Proposition 1.2.2, we have a regular morphism

$$\overline{\mathrm{for}}_2 : \widetilde{\mathcal{U}}^{ns}_{g,2}((g-1,1),(g+1,0)) \setminus Z \to \overline{\mathcal{U}}^{ns}_{g,1}(g),$$

sending $\widetilde{\mathcal{W}}$ to a point. Let $V \subset \widetilde{\mathcal{U}}^{ns}_{g,2}((g-1,1),(g+1,0))$ be the open subset corresponding to smooth curves. Then $V \cap Z = \emptyset$ because for points of Z the underlying curve is singular. Thus, the above morphism induces a regular morphism

$$\widetilde{\phi} : V \to \overline{\mathcal{U}}^{ns}_{g,1}(g), \tag{1.3.2}$$

mapping $\widetilde{\mathcal{W}} \cap V$ to a point.

Now we claim that the natural projection $V \to \mathcal{M}_{g,1}$ induces a smooth surjective morphism $V \to \mathcal{U}$. Indeed, if $h^0((g-1)p_1 + p_2) = 1$ then $h^0((g-1)p_1) = 1$, so this projection factors through $\mathcal{U}$. Conversely, if for $(C,p_1) \in \mathcal{M}_{g,1}$ one has $h^0((g-1)p_1) = 1$ then for generic p_2 we will have $h^0((g-1)p_1 + p_2) = 1$, hence the map $V \to \mathcal{U}$ is surjective. It is smooth since V is a $\mathbb{G}_m^2$-torsor over an open substack of a universal curve over $\mathcal{U}$.

It remains to prove that the morphism (1.3.2) factors through a morphism $\phi : \mathcal{U} \to \overline{\mathcal{U}}^{ns}_{g,1}(g)$ (it will then map $\mathcal{W} \cap \mathcal{U}$ to a point, since (1.3.2) sends $\widetilde{\mathcal{W}} \cap V$ to a point). Indeed, this is true if we restrict to the open subset $\mathcal{M}_{g,1} \setminus \mathcal{W}$, by the construction. Now let us set $T := V \times_{\mathcal{U}} V$ and consider two morphisms

$$f_1 = \widetilde{\phi} \circ \pi_1, f_2 = \widetilde{\phi} \circ \pi_2 : T \to \overline{\mathcal{U}}^{ns}_{g,1}(g),$$

where π_1 and π_2 are two projections to V. We know that these two maps agree on the open subset $\pi^{-1}(\mathcal{M}_{g,1} \setminus \mathcal{W})$, where π is the projection $T = V \times_{\mathcal{U}} V \to \mathcal{U}$.

Note that the scheme T parametrizes data $(C, p_1, p_2, p_2', v_1, v_2, v_2')$ such that $h^0((g-1)p_1 + p_2) = h^0((g-1)p_1 + p_2') = 1$ and $h^1((g+1)p_1) = 0$ (and C smooth, $p_1 \neq p_2$, $p_1 \neq p_2'$). Thus, it is an open subset in a $\mathbb{G}_m^3$-torsor over the universal curve over $\mathcal{M}_{g,2}$ (via the projection to (C, p_1, p_2, p_2')), in particular, T is smooth and irreducible.

Let us consider the cartesian diagram

$$\begin{array}{ccc} T' & \longrightarrow & \overline{\mathcal{U}}^{ns}_{g,1}(g) \\ \Big\downarrow \rho & & \Big\downarrow \Delta \\ T & \xrightarrow{(f_1, f_2)} & \overline{\mathcal{U}}^{ns}_{g,1}(g) \times \overline{\mathcal{U}}^{ns}_{g,1}(g) \end{array}$$

Since the stack $\overline{\mathcal{U}}^{ns}_{g,1}(g)$ is separated, the vertical arrows are finite morphisms. Finally, we observe that a generic pointed curve (C,p) in $\mathcal{M}_{g,1}$ does not have nontrivial automorphisms (note that in the case $g=2$ this is true since we can take p not to be a Weierstrass point). Hence, the preimages of points with trivial automorphisms in $\overline{\mathcal{U}}^{ns}_{g,1}(g)$ under f_1 and f_2 are nonempty open subsets in T. Since f_1 and f_2 agree on a nonempty open subset, we deduce that there exists a nonempty open subset $X \subset T$ such that $\rho^{-1}(X) \to X$ is an isomorphism. Let $T'' \subset T'$ be an irreducible component of T', containing $\rho^{-1}(X)$, with reduced scheme structure. Then $\rho|_{T''} : T'' \to T$ is a finite birational morphism. Since T is smooth, we deduce that $\rho|_{T''}$ is an isomorphism. Hence, ρ admits a section, and so we have $f_1 = f_2$, which means that the map (1.3.2) descends to a morphism from $\mathcal{U}$. □

2. Curves of genus 2

2.1. Explicit identification of $\widetilde{\mathcal{U}}^{ns}_{2,1}(2)$.

PROPOSITION 2.1.1. *Let us work over* $\mathbb{Z}[1/6]$. *One has an isomorphism of the moduli scheme* $\widetilde{\mathcal{U}}^{ns}_{2,1}(2)$ *with the affine space* $\mathbb{A}^5$ *with coordinates* $q_1, q_{2,0}, q_{2,1}, q_{3,0}, q_{3,1}$, *so that the affine universal curve* $C \setminus \{p\}$ *is given by the following equations in the independent variables* f, h, k:

$$\begin{aligned} h^2 &= fk + q_1 h + 2q_1^2 + f(q_{2,0} + q_{2,1} f), \\ hk &= f(q_{3,0} + q_{3,1} f + f^2) - q_1 k + (q_{2,0} + q_{2,1} f)h + q_1(q_{2,0} + q_{2,1} f), \\ k^2 &= (q_{3,0} + q_{3,1} f + f^2)h + (q_{2,0} + q_{2,1} f)^2 - 2q_1(q_{3,0} + q_{3,1} f + f^2). \end{aligned} \tag{2.1.1}$$

The weights of the $\mathbb{G}_m$*-action are:*

$$\deg(q_{2,1}) = 2, \ \deg(q_{3,1}) = 3, \ \deg(q_1) = 4, \ \deg(q_{2,0}) = 5, \ \deg(q_{3,0}) = 6.$$

Hence, we get the identification of $\overline{\mathcal{U}}^{ns}_{2,1}(2)$ *with the weighted projective stack* $\mathbb{P}(2,3,4,5,6)$.

Proof. This is proved using the same method as in [**P13**, Thm. A] and [**P17**, Thm. A]. Let (C,p,v) be a point in $\widetilde{\mathcal{U}}^{ns}_{2,1}(2)$. Since $h^1(2p) = 0$, we have $h^0(np) = n-1$ for $n \geq 2$. Let t be a formal parameter at p compatible with the given tangent vector. We can find the elements $f \in H^0(C, \mathcal{O}(3p))$, $h \in H^0(C, \mathcal{O}(4p))$ and $k \in H^0(C, \mathcal{O}(5p))$ with the Laurent expansions

$$f = \frac{1}{t^3} + \dots, \quad h = \frac{1}{t^4} + \dots, \quad k = \frac{1}{t^5} + \dots,$$

where the omitted terms have poles of smaller order. Then the elements

$$f^n, f^n h, f^n k, \quad \text{for } n \geq 0, \tag{2.1.2}$$

form a linear basis on $H^0(C\setminus\{p\},\mathcal{O})$, so we can express h^2, hk and k^2 as their linear combinations. Taking into account the above Laurent expansion, we get relations of the form

$$\begin{aligned} h^2 &= p_1(f)k + q_1(f)h + c_1(f), \\ hk &= p_2(f)k + q_2(f)h + c_2(f), \\ k^2 &= p_3(f)k + q_3(f)h + c_3(f), \end{aligned} \tag{2.1.3}$$

where p_i, q_i, c_i are polynomials in f with the following restrictions:

$$\begin{aligned} &\deg p_1 = 1, \deg p_2 \le 1, \deg p_3 \le 1, \deg q_1 \le 1, \deg q_2 \le 1, \deg q_3 = 2, \\ &\deg c_1 \le 2, \deg c_2 = 3, \deg c_3 \le 3, \end{aligned}$$

and the polynomials p_1, q_3 and c_2 are monic. Note that f is defined up to adding a constant, while h and k are defined up to the transformation

$$(h,k) \mapsto (\widetilde{h} = h + A(f), \widetilde{k} = k + Bh + C(f)),$$

where A and C are linear polynomials in f and B is a constant. It is easy to check that we can fix the ambiguity in the choice of h and k by requiring that $p_3 = 0$ and $p_2 = -q_1$ is a constant, i.e., does not have a linear term in f. More precisely, we should set

$$A = -\frac{q_1 + p_2}{3}, \quad B = \frac{1}{3}(q_1' - 2p_2'), \quad C = -\frac{p_3}{2} - \frac{B^2 p_1}{2} - Bp_2 \tag{2.1.4}$$

(here q_1' and p_2' are derivatives of the linear polynomials q_1 and p_2). Note that here we use our assumption that 6 is invertible. Finally, we can fix the ambiguity in the choice of f by requiring that $p_1(f) = f$.

Now the fact that the elements (2.1.2) form a basis of $H^0(C\setminus\{p\},\mathcal{O})$ is equivalent to the condition that the relations (2.1.3) form a Gröbner basis in the ideal they generate (with respect to the degree reverse lexicographical order such that $f < h < k$, $\deg(f) = 3$, $\deg(h) = 4$, $\deg(k) = 5$). Applying the Buchberger's Criterion (see [**Eis95**, Thm. 15.8]) we compute that this condition is equivalent to the following expressions of c_1, c_2, c_3 in terms of the other variables (where in the second expression in each line we take into account the normalization $p_3 = 0$, $p_2 = -q_1$):

$$\begin{aligned} c_1 &= p_2^2 + p_1 q_2 - q_1 p_2 = 2q_1^2 + p_1 q_2, \\ c_2 &= p_1 q_3 - p_2 q_2 = p_1 q_3 + q_1 q_2, \\ c_3 &= q_2^2 + p_2 q_3 - q_1 q_3 = q_2^2 - 2q_1 q_3. \end{aligned}$$

Thus, if we set

$$q_2 = q_{2,0} + q_{2,1} f, \quad q_3 = q_{3,0} + q_{3,1} f + f^2,$$

then we see that the constants $(q_1, q_{2,0}, q_{2,1}, q_{3,0}, q_{3,1})$ determine the curve (C,p). The above process can be run in families and can be reversed (see the proofs of [**P13**, Thm. A] and [**P17**, Thm. A]), so this gives the required identification of our moduli space with $\mathbb{A}^5$. □

2.2. Special cuspidal curve C_0. Let C_0 denote the curve obtained from $\mathbb{P}^1$ by pinching the point 0 into a genus 2 cuspidal singular point, so that a regular function f near 0 descends to C_0 if and only if the expansion of f in the standard parameter t has form

$$f \equiv c_0 + c_2 \cdot t^2 \bmod(t^4). \tag{2.2.1}$$

Note that this condition depends on coordinates, i.e., the point $\infty \in C_0$ plays a special role. For example, the standard $\mathbb{G}_m$-action on $\mathbb{P}^1$, preserving 0 and ∞, descends to a $\mathbb{G}_m$-action on C_0. Also, note that $C_0 \setminus \{\infty\} = \mathrm{Spec}(\mathbb{C}[t^2, t^5])$.

The next Lemma shows that if we equip C_0 with a smooth marked point $p \neq \infty$ then we get a point of $\overline{\mathcal{U}}^{ns}_{2,1}(2)$.

LEMMA 2.2.1. *Let $p \in C_0 \setminus \{0, \infty\}$. Then $h^0(C_0, \mathcal{O}(2p)) = 1$. On the other hand, for $p = \infty$ we have $h^0(C_0, \mathcal{O}(2p)) = 2$.*

Proof. In the case $p \neq 0, \infty$ we can assume that $t(p) = 1$. Then $\mathcal{O}_{\mathbb{P}^1}(2p)$ is spanned by 1, $\frac{1}{1-t}$ and $\frac{1}{(1-t)^2}$. Looking at the expansions at $t = 0$ we see that the only sections of $\mathcal{O}_{\mathbb{P}^1}(2p)$ satisfying (2.2.1) are constants.

In the case $p = \infty$ the functions $(1, t^2)$ give a basis of $H^0(C_0, \mathcal{O}(2p))$. □

DEFINITION 2.2.2. We denote by $[C_0]$ the point of $\overline{\mathcal{U}}^{ns}_{2,1}(2)$ corresponding to (C_0, p), where $p \neq 0, \infty$.

2.3. Classification of singular irreducible curves of genus 2. The results of this section are well known to the experts (see [**Eb65**], [**St96**]).

Let C be an irreducible curve of genus 2, and let $\rho : \widetilde{C} \to C$ be the normalization. If C is singular then the genus of $\widetilde{C}$ is either 1 or 0.

If the genus of $\widetilde{C}$ is 1 then $\mathrm{coker}(\mathcal{O}_C \to \rho_* \mathcal{O}_{\widetilde{C}})$ has length 1, so it is supported at one singular point $q \in C$. If $\rho^{-1}(q)$ contains two distinct points $q_1, q_2 \in C$ then ρ factors through a morphism $C' \to C$, where C' is the nodal curve obtained by gluing q_1 and q_2 on $\widetilde{C}$. Since C' has genus 2 we should have $C \simeq C'$. If $\rho^{-1}(q)$ is one point on C then it is easy to see that C has a simple cusp at q.

In the remaining case when $\widetilde{C} = \mathbb{P}^1$ we have more possibilities. The length of the sheaf $\mathcal{F} := \mathrm{coker}(\mathcal{O}_C \to \rho_* \mathcal{O}_{\widetilde{C}})$ is now 2, so the support of $\mathcal{F}$ can consist of ≤ 2 points.

Case I: support of $\mathcal{F}$ consists of two distinct points q_1, q_2. We have the following subcases.

Case Ia: $|\rho^{-1}(q_1)| > 1$ **and** $|\rho^{-1}(q_2)| > 1$. In this case the map ρ factors through the nodal curve C' obtained by gluing two pairs of distinct points in $\mathbb{P}^1$. Since the genus of C' is 2, we should have $C \simeq C'$.

Case Ib: $|\rho^{-1}(q_1)| = 1$ **and** $|\rho^{-1}(q_2)| > 1$. In this case ρ factors through the curve C' obtained by gluing a pair of distinct point in $\mathbb{P}^1$ and pinching one extra point to a simple cusp. Again, we have that the genus of C' is 2, so $C \simeq C'$.

Case Ic: $|\rho^{-1}(q_1)| = |\rho^{-1}(q_2)| = 1$. In this case C is obtained by pinching two points of $\mathbb{P}^1$ into simple cusps.

Case II: $\mathcal{F}$ is supported at one point q.

Case IIa: $|\rho^{-1}(q)| > 2$. In this case ρ factors through the curve C' obtained by gluing transversally 3 points on $\mathbb{P}^1$ into a single point (with the coordinate cross singularity). Since the genus of C' is 2, we get $C \simeq C'$.

Case IIb: $|\rho^{-1}(q)| = 2$. Let $\rho^{-1}(q) = \{q_1, q_2\}$. Let t be a generator of the maximal ideal $\mathfrak{m}_q \subset \mathcal{O}_{C,q}$. Assume first that $t \in \mathfrak{m}_{q_1}^2$. Then ρ factors through the curve C' obtained from $\mathbb{P}^1$ by first pinching q_1 into a simple cusp and then gluing it transversally with the point q_2. Since C' has genus 2, we have $C \simeq C'$. On the other hand, if t maps to a generator of $\mathfrak{m}_{q_i}$ for $i = 1, 2$, then ρ factors through the curve C' obtained from $\mathbb{P}^1$ by gluing q_1 and q_2 into a tacnode singularity. Since such C' has genus 2, we have $C \simeq C'$.

Case IIc: $|\rho^{-1}(q)| = 1$. In this case we can identify C with $\mathbb{P}^1$ as a topological space, so that $\mathcal{O}_C$ is a subsheaf of $\mathcal{O}_{\mathbb{P}^1}$, which differs from it only at one point q, so that $\mathfrak{m}_{C,q} \subset \mathfrak{m}_{\mathbb{P}^1,q}^2$ is an embedding of codimension 1. We claim that there are two curves of this type, up to an isomorphism. If $\mathfrak{m}_{C,q} \subset \mathfrak{m}_{\mathbb{P}^1,q}^3$ then $\mathfrak{m}_{C,q} = \mathfrak{m}_{\mathbb{P}^1,q}^3$ and $C = C^{\mathrm{cusp}}(2)$ (see Sec. 1.1). Now assume that $\mathfrak{m}_{C,q} \not\subset \mathfrak{m}_{\mathbb{P}^1,q}^3$. Let t be a formal parameter near q on $\mathbb{P}^1$. Then $\hat{\mathfrak{m}}_{C,q}$ is a (non-unital) subalgebra in $t^2\mathbb{C}[[t]]$ of codimension 1, and there exists an element $f \in \hat{\mathfrak{m}}_{C,q}$ such that $f \equiv t^2 \bmod t^3\mathbb{C}[[t]]$. Changing the formal parameter we can assume that $f = t^2$. There could not be an element $h \in \hat{\mathfrak{m}}_{C,q}$ such that $h \equiv t^3 \bmod t^4\mathbb{C}[[t]]$, since then we would have $\hat{\mathfrak{m}}_{C,q} = t^2\mathbb{C}[[t]]$. Therefore,

$$\hat{\mathfrak{m}}_{C,q} = \mathbb{C} \cdot t^2 + t^4\mathbb{C}[[t]].$$

Note that the subspace in the right-hand side depends only on $t \bmod t^3\mathbb{C}[[t]]$. Now we observe that any formal parameter at q, modulo $\mathfrak{m}_{\mathbb{P}^1,q}^3$, can be obtained from a unique regular function on $\mathbb{P}^1 \setminus \{p\}$, for some $p \neq q$. Using automorphisms of $\mathbb{P}^1$ we can make $q = 0$, $p = \infty$, so that C is the curve C_0 defined before.

2.4. Comparison of stabilities for irreducible curves of genus 2.

PROPOSITION 2.4.1. *Let C be an irreducible curve of genus 2, and let p be a smooth point. Then (C, p) is $\mathcal{Z}$-stable if and only if C is not of type IIc.*

Proof. It is easy to see that a curve C of type IIc is not $\mathcal{Z}$-stable. Indeed, if there is a contracting map $C' \to C$ then C' would have a rational component with only two distinguished points, so it could not be stable. Assume now that C is not of type IIc. If (C, p) is nodal then it is stable (since C is irreducible), hence it is $\mathcal{Z}$-stable.

Next, if C is obtained by pinching a point on an irreducible nodal curve E of genus 1 into a cusp, then there is a contraction $f : E \cup E' \to C$, where $E \cup E'$ is the stable curve with E and E' glued nodally at one point. Here the marked point is placed on E and $f(E')$ is the cusp on C. This shows that (C, p) is $\mathcal{Z}$-stable. Similarly, if C is a rational curve with two cusps then there is a contraction to C from $\mathbb{P}^1$ with two elliptic tails (that get contracted into cusps).

There remains two cases for C: IIa and IIb. In the case IIa we have a contraction to C from the union of two $\mathbb{P}^1$'s, joined nodally at 2 points. In the case IIb there is a contraction to C from the curve with an elliptic bridge. In other words, we consider the union $\mathbb{P}^1 \cup E$, where E is an elliptic curve, $\mathbb{P}^1$ and E are joined nodally at 2 points, so that there are no marked points on E. It is known (see [**Sm13**, Ex. 2.5]) that there exists a contraction $\mathbb{P}^1 \cup E \to C$, mapping E to the singular point, for both types of curves occurring in the case IIb. □

COROLLARY 2.4.2. *The stack $\overline{\mathcal{M}}_{2,1}(\mathcal{Z})$ is smooth and irreducible.*

Proof. The possible singular points that can appear in $\mathcal{Z}$-stable curves of genus 2, other that nodes, are: a simple cusp, a tacnode, and a coordinate cross in 3-space. All of these have smooth versal deformation spaces and are smoothable, hence the assertion (see [**Sm11**, Lem. 2.1]). □

Using the classification from Sec. 2.3 we easily get the following codimension estimate.

LEMMA 2.4.3. *Away from a closed subset of codimension ≥ 2, for every point (C,p) in $\overline{\mathcal{M}}_{2,1}(\mathcal{Z})$ (resp., $\overline{\mathcal{U}}^{ns}_{2,1}(2)$), C is either smooth, or a nodal curve with the normalization of genus 1.*

Proof. Both $\overline{\mathcal{M}}_{2,1}(\mathcal{Z})$ and $\overline{\mathcal{U}}^{ns}_{2,1}(2)$ are irreducible of dimension 4. Now we just go through the strata described in Sec. 2.3 and see that they all have dimension ≤ 2, except when C is either smooth or nodal with the normalization of genus 1. □

We need one more simple observation.

LEMMA 2.4.4. *Let C be an irreducible curve of genus 2, and let $p \in C$ be a smooth point. Then $h^0(p) = 1$ and $h^1(3p) = 0$.*

Proof. First, if $h^0(p) = 2$ then we would get a degree 1 regular map $C \to \mathbb{P}^1$. Composing it with the normalization map $\widetilde{C} \to C$, we get that the normalization map is the inverse map $\mathbb{P}^1 \to C$, which is impossible. Hence, $h^0(p) = 1$.

If $h^1(2p) = 0$ then we also have $h^1(3p) = 0$, so it is enough to consider the case $h^1(2p) \neq 0$, i.e., $h^0(2p) = 2$. Suppose that $h^0(3p) = 3$. Then we can choose $f \in H^0(C, \mathcal{O}(2p))$ and $h \in H^0(C, \mathcal{O}(3p))$ with the Laurent expansions $f = \frac{1}{t^2} + \dots$, $h = \frac{1}{t^3} + \dots$ at p (for some formal parameter t at p). Furthermore, there is a canonical choice of f and h, such that the relation

$$h^2 = f^3 + af + b$$

holds for some constants a and b. Then the algebra $\mathcal{O}(C \setminus \{p\})$ has the linear basis (f^n), (hf^n), and is isomorphic to the algebra $A = \mathbb{C}[h, f]/(h^2 - f^3 - af - b)$. Since C is irreducible, it is isomorphic to Proj of the Rees algebra of A, which is a plane cubic, so we get that the arithmetic genus of C is equal to 1, which is a contradiction. This shows that $h^0(3p) = 2$, i.e., $h^1(3p) = 0$. □

THEOREM 2.4.5. *Let $\overline{\mathcal{W}} \subset \overline{\mathcal{M}}_{2,1}(\mathcal{Z})$ be the closure of the Weierstrass locus $\mathcal{W} \subset \mathcal{M}_{2,1}$. Then $\overline{\mathcal{W}}$ coincides with the locus where $h^1(2p) \neq 0$. There is a regular morphism*

$$\phi_2 : \overline{\mathcal{M}}_{2,1}(\mathcal{Z}) \to \overline{\mathcal{U}}^{ns}_{2,1}(2),$$

such that $\phi_2(\overline{\mathcal{W}}) = [C_0]$ and ϕ_2 induces an isomorphism

$$\overline{\mathcal{M}}_{2,1}(\mathcal{Z}) \setminus \overline{\mathcal{W}} \xrightarrow{\sim} \overline{\mathcal{U}}^{ns}_{2,1}(2) \setminus [C_0].$$

Proof. First, we observe that every irreducible component of the locus $h^1(2p) \neq 0$ has codimension 1 in $\overline{\mathcal{M}}_{2,1}(\mathcal{Z})$ (recall that the latter stack is smooth and irreducible by Corollary 2.4.2). By Lemma 2.4.3, to see that this locus coincides with $\overline{\mathcal{W}}$, it is enough to see that the locus of (C,p), such that C is nodal with normalization E of genus 1 and $h^1(2p) \neq 0$ has dimension 2 (and hence has codimension 2 in

$\overline{\mathcal{M}}_{2,1}(\mathcal{Z})$). But if C is obtained from E by identifying points $q_1 \neq q_2$ then the condition that $h^0(2p) = 2$ implies the existence of a rational function on E with pole of order 2 at p and vanishing at both q_1 and q_2. In other words, we should have a linear equivalence $2p \sim q_1 + q_2$. Thus, we have a finite number of choices for each (E, q_1, q_2), so the dimension is 2.

Next, let us denote by

$$V^{\mathcal{Z}} \subset \widetilde{\mathcal{U}}^{ns}_{2,2}(1,1)$$

the open substack consisting of (C, p_1, p_2) such that (C, p_1) is $\mathcal{Z}$-stable (in particular, C is irreducible). Lemma 2.4.4 shows that every (C, p) in $\overline{\mathcal{M}}_{2,1}(\mathcal{Z})$ satisfies $h^0(p) = 1$ and $h^1(3p) = 0$. This implies that

$$V^{\mathcal{Z}} \subset \widetilde{\mathcal{U}}^{ns}_{2,2}((1,1),(3,0))$$

and the projection $V^{\mathcal{Z}} \to \overline{\mathcal{M}}_{2,1}(\mathcal{Z})$ is surjective. Furthermore, since the curve $[C^{\text{cusp}}(2)]$ is not $\mathcal{Z}$-stable, we have the inclusion

$$V^{\mathcal{Z}} \subset \widetilde{\mathcal{U}}^{ns}_{2,2}((1,1),(3,0)) \setminus Z.$$

Thus, the restriction of the map (1.2.3) gives us a regular morphism

$$V^{\mathcal{Z}} \to \overline{\mathcal{U}}^{ns}_{2,1}(2), \tag{2.4.1}$$

contracting $\widetilde{\mathcal{W}}$ to a point.

Now, similarly to the proof of Theorem A we check that the morphism (2.4.1) factors through $\overline{\mathcal{M}}_{2,1}(\mathcal{Z})$. Note that to apply the same argument as in Theorem A we use the following facts: (i) $\overline{\mathcal{M}}_{2,1}(\mathcal{Z})$ is smooth (see Corollary 2.4.2); (ii) the projection $V^{\mathcal{Z}} \to \overline{\mathcal{M}}_{2,1}(\mathcal{Z})$ is smooth (since p_2 varies in a smooth part of a curve); and (iii) $V^{\mathcal{Z}} \times_{\overline{\mathcal{M}}_{2,1}(\mathcal{Z})} V^{\mathcal{Z}}$ is irreducible, as a $\mathbb{G}_m^3$-torsor over the moduli stack of (C, p_1, p_2, p_2') with C smoothable.

This gives us the required morphism ϕ_2 contracting to $\overline{\mathcal{W}}$ to some point in $\overline{\mathcal{U}}^{ns}_{2,1}(2)$. On the other hand, by Proposition 2.4.1, the only point in $\overline{\mathcal{U}}^{ns}_{2,1}(2)$, which is not $\mathcal{Z}$-stable is $[C_0]$ (recall that by this we mean the pointed curve (C_0, p), where $p \neq 0, \infty$, see Lemma 2.2.1). Thus, the rational map ϕ_2^{-1} is regular on $\overline{\mathcal{U}}^{ns}_{2,1}(2) \setminus [C_0]$ (and sends (C, p) to (C, p)). Also, the restriction of ϕ_2 to $\overline{\mathcal{M}}_{2,1}(\mathcal{Z}) \setminus \overline{\mathcal{W}}$, i.e., to the locus where $h^0(2p) = 1$, is an open embedding sending (C, p) to (C, p). This implies that $\phi_2(\overline{\mathcal{W}}) = [C_0]$, and ϕ_2 induces an isomorphism of $\overline{\mathcal{M}}_{2,1}(\mathcal{Z}) \setminus \overline{\mathcal{W}}$ with $\overline{\mathcal{U}}^{ns}_{2,1}(2) \setminus [C_0]$. □

Let us consider the natural birational maps of the coarse moduli spaces

$$\overline{M}_{2,1} \dashrightarrow \overline{M}_{2,1}(\mathcal{Z}) \dashrightarrow \overline{U}^{ns}_{2,1}(2).$$

Note that all these spaces are normal (for the last two this follows from Proposition 2.1.1 and Corollary 2.4.2). Note also that we only know that $\overline{M}_{2,1}(\mathcal{Z})$ is a proper algebraic space.

Let $\overline{W} \subset \overline{M}_{2,1}$ denote the closure of W, and let $\Delta_1 \subset \overline{M}_{2,1}$ be the boundary divisor, whose generic point corresponds to the union of two elliptic curves.

Proposition 2.4.6. *The natural birational morphism $f : \overline{M}_{2,1} \dashrightarrow \overline{M}_{2,1}(\mathcal{Z})$ (resp., $g : \overline{M}_{2,1} \dashrightarrow \overline{U}^{ns}_{2,1}(2)$) is a birational contraction with the exceptional divisor Δ_1 (resp., exceptional divisors Δ_1 and $\overline{W}$).*

Proof. Recall that to check that f (resp., g) is a birational contraction we need to check that the exceptional locus $\mathrm{Exc}(f^{-1})$ (resp., $\mathrm{Exc}(g^{-1})$) has codimension ≥ 2. But this immediately follows from Lemma 2.4.3. Next, the restriction of f to the complement of Δ_1 induces an isomorphism with the open subset in $\overline{M}_{2,1}(\mathcal{Z})$ consisting of (C,p) with C smooth or nodal, so we have an inclusion $\mathrm{Exc}(f) \subset \Delta_1$. On the other hand, the generic point of Δ_1 corresponds to the union of elliptic curves $E_1 \cup E_2$, with the marked point on E_1. Under the map f this curve gets replaced by the cuspidal curve $\overline{E}_1$, so that we have a contraction $E_1 \cup E_2 \to \overline{E}_1$ sending the elliptic tail E_2 to the cusp. Since this map forgets the j-invariant of E_2, this means that Δ_1 gets contracted by f. Now the fact that $\mathrm{Exc}(g) = \Delta_1 \cup \overline{W}$ follows from Theorem 2.4.5. □

REMARK 2.4.7. Let $\overline{\mathcal{V}}^{ns}_{g,1}(g) \subset \overline{\mathcal{U}}^{ns}_{g,1}(g)$ be the irreducible component consisting of smoothable curves. Theorem A implies that the natural birational map

$$\overline{\mathcal{M}}_{g,1} \dashrightarrow \overline{\mathcal{V}}^{ns}_{g,1}(g)$$

contracts $\overline{\mathcal{W}}$ to a point. Passing to the normalizations of the coarse moduli spaces we get the birational map $\phi : \overline{M}_{g,1} \dashrightarrow X$, where X is a normal projective variety, contracting $\overline{W}$ to a point. It seems plausible that ϕ is a birational contraction (which would imply that $\overline{W}$ is an extremal divisor). To check this we would need to prove that $\mathrm{Exc}(\phi^{-1})$ has codimension ≥ 2. In other words, we would need to check that the locus in $\overline{\mathcal{V}}^{ns}_{g,1}(g)$, consisting of unstable (i.e., non-nodal) curves, has codimension ≥ 2. In the case $g = 2$ we have shown this in Lemma 2.4.3. Note that the fact that the class of $\overline{W}$ generates an extremal ray in $\overline{NE}^1(\overline{M}_{g,1})$ is known for $g \leq 3$ and $g = 5$, by the works [**R01**], [**J13**] and [**J10**].

References

[A74] E. Arbarello, *Weierstrass points and moduli of curves*, Compositio Math. **29** (1974), 325–342. MR0360601

[A78] E. Arbarello, *On subvarieties of the moduli space of curves of genus g defined in terms of Weierstrass points* (English, with Italian summary), Atti Accad. Naz. Lincei Mem. Cl. Sci. Fis. Mat. Natur. Sez. Ia (8) **15** (1978), no. 1, 3–20. MR531917

[Eb65] S. Ebey, *The classification of singular points of algebraic curves*, Trans. Amer. Math. Soc. **118** (1965), 454–471. MR0176983

[Eis95] D. Eisenbud, *Commutative algebra*, Graduate Texts in Mathematics, vol. 150, Springer-Verlag, New York, 1995. With a view toward algebraic geometry. MR1322960

[FP14] R. Fisette and A. Polishchuk, *A_∞-algebras associated with curves and rational functions on $\mathcal{M}_{g,g}$. I*, Compos. Math. **150** (2014), no. 4, 621–667. MR3200671

[J13] D. Jensen, *Birational contractions of $\overline{M}_{3,1}$ and $\overline{M}_{4,1}$*, Trans. Amer. Math. Soc. **365** (2013), no. 6, 2863–2879. MR3034451

[J10] D. Jensen, *Rational fibrations of $\overline{M}_{5,1}$ and $\overline{M}_{6,1}$*, J. Pure Appl. Algebra **216** (2012), no. 3, 633–642. MR2864765

[P13] A. Polishchuk, *Moduli of curves as moduli of A_∞-structures*, Duke Math. J. **166** (2017), no. 15, 2871–2924. MR3712167

[P17] A. Polishchuk, *Moduli of curves, Gröbner bases, and the Krichever map*, Adv. Math. **305** (2017), 682–756. MR3570146

[R01] W. F. Rulla, *The birational geometry of moduli space M(3) and moduli space M(2,1)*, ProQuest LLC, Ann Arbor, MI, 2001. Thesis (Ph.D.)–The University of Texas at Austin. MR2701950

[Sm11] D. I. Smyth, *Modular compactifications of the space of pointed elliptic curves II*, Compos. Math. **147** (2011), no. 6, 1843–1884. MR2862065

[Sm13] D. I. Smyth, *Towards a classification of modular compactifications of* $\mathcal{M}_{g,n}$, Invent. Math. **192** (2013), no. 2, 459–503. MR3044128

[St96] J. Stevens, *On the classification of reducible curve singularities*, Algebraic geometry and singularities (La Rábida, 1991), Progr. Math., vol. 134, Birkhäuser, Basel, 1996, pp. 383–407. MR1395193

Department of Mathematics, University of Oregon, Eugene, Oregon 97403; and National Research University Higher School of Economics, Moscow, Russia

Email address: apolish@uoregon.edu

Proceedings of Symposia in Pure Mathematics
Volume **98**, 2018
http://dx.doi.org/10.1090/pspum/098/01722

Spectral theory and mirror symmetry

Marcos Mariño

ABSTRACT. Recent developments in string theory have revealed a surprising connection between spectral theory and local mirror symmetry: it has been found that the quantization of mirror curves to toric Calabi–Yau threefolds leads to trace class operators, whose spectral properties are conjecturally encoded in the enumerative geometry of the Calabi–Yau. This leads to a new, infinite family of solvable spectral problems: the Fredholm determinants of these operators can be found explicitly in terms of Gromov–Witten invariants and their refinements; their spectrum is encoded in exact quantization conditions, and turns out to be determined by the vanishing of a quantum theta function. Conversely, the spectral theory of these operators provides a non-perturbative definition of topological string theory on toric Calabi–Yau threefolds. In particular, their integral kernels lead to matrix integral representations of the topological string partition function, which explain some number-theoretic properties of the periods. In this paper we give a pedagogical overview of these developments with a focus on their mathematical implications.

1. Introduction

Mirror symmetry has played a fundamental rôle in the interface of theoretical physics and mathematics, and has led to beautiful developments in string theory and enumerative geometry. Although mirror symmetry was originally formulated for compact Calabi–Yau (CY) manifolds, it can be extended to the so-called local case [**KKV**, **CKYZ**], which involves toric, hence non-compact CY manifolds. Local mirror symmetry is in a sense even richer than its compact counterpart, since it is related to many other fields of mathematical physics, like Chern–Simons theory, integrable systems, supersymmetric gauge theory, and random matrix theory. Very often, these connections have been obtained by studying topological string theory, the physical theory underlying mirror symmetry. In some circumstances, this theory can be described by other, "dual" theories, which look in principle very different, and this physical equivalence leads to a non-trivial mathematical equivalence. For example, the large N duality obtained by Gopakumar and Vafa in [**GV1**] relates topological string theory on the resolved conifold to Chern–Simons theory on the three-sphere, and this equivalence eventually led to the theory of the topological

2010 *Mathematics Subject Classification.* Primary 14J33; Secondary 34K08.

The author is supported in part by the Swiss National Science Foundation, subsidies 200020-149226, 200021-156995, and by the NCCR 51NF40-141869 "The Mathematics of Physics" (SwissMAP).

vertex [**AKMV**], which gives a complete algorithm to obtain the Gromov–Witten invariants of toric CY manifolds in terms of Chern–Simons knot invariants.

The spectral theory of self-adjoint operators plays a more established rôle in physics, due to its connection to Quantum Mechanics. In recent years, there have been many interesting developments in the arena of exactly solvable spectral problems. For example, in some one-dimensional models, exact quantization conditions incorporating non-perturbative effects at all orders have been found (see for example [**ZJJ, Vo3, Al**]). Some of these solvable models are closely related to integrable systems [**DDT**] and two-dimensional CFT [**BLZ**].

The purpose of this review paper is to give a summary of some recent results in string theory which suggest a deep connection between these two separate subjects. There is now growing evidence that, given a toric CY manifold, one can associate to it trace class operators whose spectral properties are encoded in the enumerative invariants of the underlying manifold. This correspondence is quite precise and leads to conjectural, explicit formulae for the Fredholm determinants of these operators, from which one can derive exact quantization conditions for their spectrum. From the point of view of spectral theory, this connection provides a new family of solvable operators with beautiful properties. From the point of view of enumerative geometry, it gives a new meaning to the generating functionals encoding the enumerative invariants of local CY geometries: they appear as asymptotic expansions of quantities defined by the spectral theory of the corresponding quantum operators. For example, one can obtain the genus expansion of the topological string free energy as an asymptotic expansion of the "fermionic" spectral traces of these operators.

The correspondence between spectral theory and topological strings which we will describe can be regarded in many ways as a large N string/gauge duality, like those appearing in the AdS/CFT correspondence. It relates a simple quantum-mechanical problem on the real line to a string propagating on a toric CY manifold. The weak coupling regime of the quantum problem, when the Planck constant is small, corresponds to the regime in which the string coupling constant is strong. The genus expansion of the topological string emerges in a 't Hooft-like limit of the quantum mechanical problem, which can be regarded as a rigorous non-perturbative definition of topological string theory on these backgrounds.

This review paper is organized as follows. In section 2, we present a simple example of the general type of spectral problems we will deal with, and we present both the approximate solution obtained with elementary methods and the exact solution conjectured in the recent literature. In section 3 we present the general conjecture formulated in [**GHM**] connecting spectral theory and topological string theory on toric CY manifolds. The emphasis is on how string theory solves the spectral problem. In section 4 we explain how, reciprocally, the spectral problem leads to new insights on topological string theory, providing in particular a non-perturbative definition of the genus expansion. Finally, in section 5, we conclude with some interesting problems for the future.

2. A problem in spectral theory

2.1. A quantum curve. Let us start by formulating a sharp question concerning the spectrum of an interesting operator on the real line. This will hopefully provide a motivation for the type of problems that we would like to address in this overview paper.

Let x, y be Heisenberg operators on the real line, satisfying the commutation relation

(1) $$[\mathsf{x}, \mathsf{y}] = \mathrm{i}\hbar.$$

Let us consider the following operator,

(2) $$\mathsf{O}_{\mathbb{P}^2} = \mathrm{e}^{\mathsf{x}} + \mathrm{e}^{\mathsf{y}} + \mathrm{e}^{-\mathsf{x}-\mathsf{y}}.$$

Why look at this beast? The reason is that it can be obtained by "quantizing" the mirror curve to the CY threefold known as "local $\mathbb{P}^2$", which is the total space of the canonical bundle of $\mathbb{P}^2$:

(3) $$X = \mathcal{O}(-3) \to \mathbb{P}^2.$$

This is probably the most studied example among all non-compact CY manifolds. As we will briefly review later on, the mirror of this manifold is encoded in an elliptic curve in $\mathbb{C}^\star \times \mathbb{C}^\star$, which can be written in the form

(4) $$\mathrm{e}^x + \mathrm{e}^y + \mathrm{e}^{-x-y} + \kappa = 0.$$

Here, κ parametrizes the moduli space of complex structures of the curve. We can now perform the standard Weyl quantization of the first three terms of the curve (4), i.e. we promote x and y to Heisenberg operators and we use the Weyl ordering prescription. This gives the operator (2). It involves the exponentiated "coordinate" and "momentum" operators

(5) $$\mathsf{X} = \mathrm{e}^{\mathsf{x}}, \qquad \mathsf{Y} = \mathrm{e}^{\mathsf{y}},$$

which are self-adjoint and they satisfy the Weyl algebra

(6) $$\mathsf{XY} = q\mathsf{YX},$$

where

(7) $$q = \mathrm{e}^{\mathrm{i}\hbar}.$$

The domain of the operator X, $D(\mathsf{X})$, consists of functions $\psi(x) \in L^2(\mathbb{R})$ such that

(8) $$\mathrm{e}^x \psi(x) \in L^2(\mathbb{R}).$$

Similarly, the domain of Y, $D(\mathsf{Y})$, consists of functions $\psi(x) \in L^2(\mathbb{R})$ such that

(9) $$\mathrm{e}^y \widehat{\psi}(y) \in L^2(\mathbb{R}),$$

where

(10) $$\widehat{\psi}(y) = \int \frac{\mathrm{d}x}{\sqrt{2\pi\hbar}} \mathrm{e}^{-\mathrm{i}xy/\hbar} \psi(x)$$

is the wavefunction in the y representation, which is essentially given by a Fourier transform. The condition (9) can be translated into a condition on $\psi(x)$, requiring analyticity on strips. The domain of the operator $\mathsf{O}_{\mathbb{P}^2}$ should be chosen appropriately, by considering functions such that $\mathsf{O}_{\mathbb{P}^2}\psi \in L^2(\mathbb{R})$.

As we will review in the next section, it was shown in [**KM**] that the operator $\mathsf{O}_{\mathbb{P}^2}$ has an inverse which is positive-definite and of trace class in $L^2(\mathbb{R})$, so its spectrum (which depends on $\hbar$) consists of an infinite set of positive eigenvalues e^{E_n}, $n = 0, 1, \cdots$. This spectrum can be determined numerically with standard methods, as first pointed out in [**HW**]. First, one should choose an appropriate orthonormal basis of functions $\{\phi_i\}_{i=0,1,\cdots}$ of $L^2(\mathbb{R})$, in the domain of $\mathsf{O}_{\mathbb{P}^2}$ (for

TABLE 1. Numerical spectrum of the operator (2) for $n = 0, 1, \cdots, 4$, and $\hbar = 2\pi$.

n	E_n
0	2.56264206862381937
1	3.91821318829983977
2	4.91178982376733606
3	5.73573703542155946
4	6.45535922844299896

example, the eigenfunctions of the harmonic oscillator will do). Then, the e^{E_n} are the eigenvalues of the infinite-dimensional matrix

$$M_{ij} = (\phi_i, \mathsf{O}_{\mathbb{P}^2}\phi_j), \qquad i, j = 0, 1, \cdots. \tag{11}$$

They can be obtained numerically by truncating the matrix to very large sizes. Boundary effects can be partially eliminated with standard extrapolation methods, and one can find very accurate values for the E_n. For $\hbar = 2\pi$, the results for the very first eigenvalues are listed in Table 1. We can now ask the following question: can these eigenvalues be determined *analytically*? Usually, this is hopeless, since there are very few operators whose spectral properties can be determined exactly. However, a conjecture put forward in [**GHM**] affirms that in this case it can be done: there is an exact, conjectural expression for the Fredholm determinant of the inverse operator

$$\rho_{\mathbb{P}_2} = \mathsf{O}_{\mathbb{P}^2}^{-1}, \tag{12}$$

from which one can deduce an *exact quantization condition* for the spectrum[1]. Before presenting this conjecture, let us see what can be said about the spectrum of the above operator with elementary methods.

2.2. A first approach to the spectrum. Although the above spectral problem is not of the standard form found in Quantum Mechanics, it is possible to use conventional approximation techniques to gain some insight. In particular, one can try to use the WKB approximation. Let us briefly review this method. Let us assume that we have a classical Hamiltonian of the form,

$$H(x, y) = \frac{y^2}{2} + V(x), \tag{13}$$

where y is interpreted as the momentum, and $V(x)$ is a potential supporting bound states. We will assume for simplicity that $V(x)$ is a confining potential, i.e. that $V(x) \to +\infty$ for $|x| \to \infty$. After quantization, one finds a quantum Hamiltonian H with a discrete spectrum of energies E_n, $n = 0, 1, \cdots$. The Bohr–Sommerfeld quantization condition gives an *approximate* quantization condition which determines the spectrum in the limit of small $\hbar$ or large quantum numbers $n \gg 1$. It can be formulated geometrically as follows. One considers the plane curve

$$H(x, y) = E, \tag{14}$$

[1]The exact quantization condition for this particular operator and $\hbar = 2\pi$ was first written down in [**HW**] by generalizing a result of [**KaMa**] for a different CY geometry.

where E is interpreted as the energy of the system, and a differential on this curve given by

$$\lambda = y(x)\,\mathrm{d}\,x. \tag{15}$$

In simple situations, there is a B-cycle in the curve (14) which corresponds to periodic motion in the potential, between two turning points determined by the equation of the curve. The integral of λ over this cycle gives the volume of the region $\mathcal{R}(E)$ in phase space with energy less or equal than E,

$$\mathcal{R}(E) = \{(x,y) \in \mathbb{R}^2 : H(x,y) \leq E\}. \tag{16}$$

We will denote this volume by $\mathrm{vol}_0(E)$. The Bohr–Sommerfeld quantization condition reads

$$\mathrm{vol}_0(E) = \oint_B \lambda = 2\pi\hbar\left(n + \frac{1}{2}\right), \qquad n = 0, 1, \cdots. \tag{17}$$

It can be also interpreted as saying that each cell of volume $2\pi\hbar$ in $\mathcal{R}(E)$ corresponds roughly to a quantum state.

It turns out that, although our quantum operator (2) is not of the standard form (13), we can still use the Bohr–Sommerfeld condition. The classical counterpart of the operator $\mathsf{O}_{\mathbb{P}^2}$ is the function on the phase space $\mathbb{R}^2$ given by the first three terms in (4), with the symplectic form $\omega = \mathrm{d}\,x \wedge \mathrm{d}\,y$,

$$\mathcal{O}_{\mathbb{P}^2}(x,y) = \mathrm{e}^x + \mathrm{e}^y + \mathrm{e}^{-x-y}. \tag{18}$$

The analogue of the hyperelliptic curve (14) is now,

$$\mathcal{O}_{\mathbb{P}^2}(x,y) = \mathrm{e}^E, \tag{19}$$

and the function (18) defines a compact region in phase space

$$\mathcal{R}(E) = \{(x,y) \in \mathbb{R}^2 : \mathcal{O}_{\mathbb{P}^2}(x,y) \leq \mathrm{e}^E\}, \tag{20}$$

which is the analogue of (16). Note that we are implicitly interpreting $\mathcal{O}_{\mathbb{P}^2}$ as the exponential of a classical Hamiltonian. The Bohr–Sommerfeld quantization condition for this problem is again given by (17). The period of λ can be computed explicitly, but since the resulting quantization condition is only valid at large energies, it is worth to simplify it, as follows. At large E, the region $\mathcal{R}(E)$ has a natural mathematical interpretation: it is the region enclosed by the tropical limit [**BIMS**] of the curve

$$\mathrm{e}^x + \mathrm{e}^y + \mathrm{e}^{-x-y} - \mathrm{e}^E = 0, \tag{21}$$

and limited by the lines

$$x = E, \qquad y = E, \qquad x + y + E = 0, \tag{22}$$

see Fig. 1 for an illustration when $E = 15$. In this polygonal limit, we find

$$\mathrm{vol}_0(E) \approx \frac{9E^2}{2}, \tag{23}$$

and we obtain from (17) the following approximate behavior for the eigenvalues, at large quantum numbers,

$$E_n \approx \frac{2}{3}\sqrt{\pi\hbar}\, n^{1/2}, \qquad n \gg 1. \tag{24}$$

It can be seen, by direct examination of the numerical spectrum, that this rough WKB estimate gives a good approximation to the eigenvalues of the operator (2) when n is large. The estimate (24) has been rigorously proved in [**LST**].

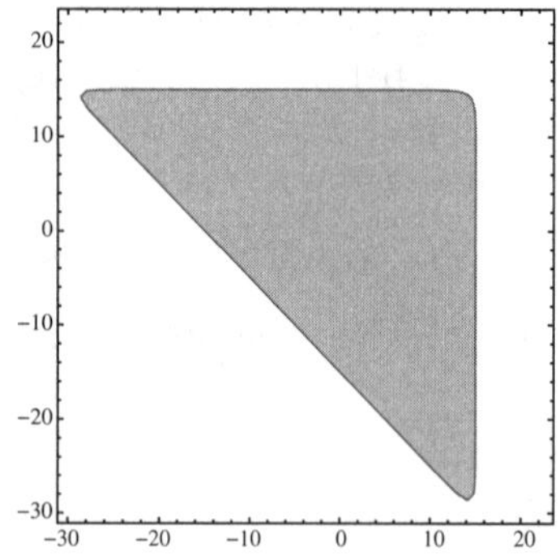

FIGURE 1. The region $\mathcal{R}(E)$ for $E = 15$.

2.3. An exact quantization condition. An important question is whether the Bohr–Sommerfeld quantization condition obtained above can be improved. One possible strategy is to consider the WKB quantization scheme to *all orders* in $\hbar$. The all-orders WKB quantization condition was derived for standard Schrödinger operators by Dunham in [**D**]. It can be adapted to operators of the form (12) by using a general formalism due to Voros [**Vo1**], or by using the extension of the WKB method to difference equations in [**DM**]. For the operator (2), this all-orders WKB method was studied in [**ACDKV**]. As we will explain in detail in the next section, the main (conjectural) result of [**ACDKV**] is that the resulting quantization condition can be expressed in terms of enumerative invariants of the CY (the so-called NS limit of the refined topological string.) However, as pointed out in [**KaMa**], this can not be the full answer to the spectral problem, since it leads to unacceptable poles for values of $\hbar$ of the form $\hbar = q\pi$, where $q \in \mathbb{Q}$. Physically, one has to take into account the contribution from quantum-mechanical instantons, which are known to correct the all-orders WKB quantization condition [**BPV**].

It turns out that in the case $\hbar = 2\pi$, the exact quantization condition is remarkably simple. Let us consider the Picard–Fuchs equation,

$$\left(\theta^3 - 3z(3\theta + 2)(3\theta + 1)\theta\right) \Pi = 0, \tag{25}$$

where

$$\theta = z\frac{\mathrm{d}}{\mathrm{d}\, z}. \tag{26}$$

This equation has three independent solutions at $z = 0$: a trivial, constant solution, a logarithmic solution $\varpi_1(z)$ and a double logarithmic solution $\varpi_2(z)$. If we introduce the power series,

$$\begin{aligned} \widetilde{\varpi}_1(z) &= \sum_{j\geq 1} 3\frac{(3j-1)!}{(j!)^3}(-z)^j, \\ \widetilde{\varpi}_2(z) &= \sum_{j\geq 1} \frac{18}{j!}\frac{\Gamma(3j)}{\Gamma(1+j)^2}\left\{\psi(3j) - \psi(j+1)\right\}(-z)^j, \end{aligned} \tag{27}$$

where $\psi(z)$ is the digamma function, we have that

$$\begin{aligned} \varpi_1(z) &= \log(z) + \widetilde{\varpi}_1(z), \\ \varpi_2(z) &= \log^2(z) + 2\widetilde{\varpi}_1(z)\log(z) + \widetilde{\varpi}_2(z). \end{aligned} \tag{28}$$

It is easy to see that the series in (27) have a radius of convergence $|z| = 1/27$. The solutions $\varpi_{1,2}(z)$ can be written in terms of hypergeometric and Meijer functions. Let us now introduce the following function,

$$\xi(E) = \frac{1}{8\pi^2}\frac{\varpi_1(z)\varpi_2'(z) - \varpi_2(z)\varpi_1'(z)}{\varpi_1'(z)}, \tag{29}$$

where

$$z = \mathrm{e}^{-3E}. \tag{30}$$

Then, the conjecture of [**GHM**] implies that the spectrum of (2), for $\hbar = 2\pi$, is determined by the condition

$$\xi(E) - \frac{1}{4} = n + \frac{1}{2}, \qquad n = 0, 1, \cdots. \tag{31}$$

It can be easily seen that the values of E_n determined in this way agree with the ones in Table 1. A graphical representation of the function $\xi(E) - 1/4$, showing the different energy levels, can be found in Fig. 2.

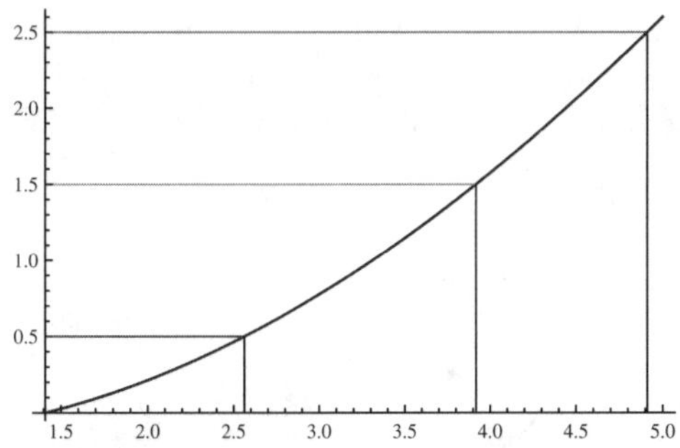

FIGURE 2. The function $\xi(E) - 1/4$, as a function of E, determining the energy spectrum of the operator (2) for $\hbar = 2\pi$.

What is the content of the quantization condition (31)? The function $\xi(E)$ has a convergent power series expansion of the form,

$$\xi(E) = \frac{9E^2}{8\pi^2} + E\sum_{n=1}^{\infty} a_n\,\mathrm{e}^{-3nE} + \sum_{n=1}^{\infty} b_n\,\mathrm{e}^{-3nE}. \tag{32}$$

When E is large, we can approximate

$$\xi(E) - \frac{1}{4} \approx \frac{9E^2}{8\pi^2}, \tag{33}$$

and we recover the Bohr–Sommerfeld estimate (24) of the previous section. The constant $1/4$ is a correction which can be easily obtained with the next-to-leading WKB method [**KaMa, HW, GHM**], and then we have an infinite series of exponentially small corrections in the energy. As we will explain later on, they have a mixed physical origin: they combine higher order, perturbative WKB corrections with non-perturbative, instanton-type corrections.

The reader familiar with local mirror symmetry has surely recognized the equation (25): it is the Picard–Fuchs equation governing the periods (28) of the mirror of local $\mathbb{P}^2$, which fully determine the genus zero Gromov–Witten invariants of this CY. The variable z is a modulus for the mirror curve, and it is related to κ in (4) by

$$z = \kappa^{-3}. \tag{34}$$

Therefore, the corrections to the Bohr–Sommerfeld quantization condition are governed by the topological string on local $\mathbb{P}^2$, whose mirror is precisely the algebraic curve (4)! As we will see in the next section, this is not a coincidence, but a general story which seems to hold for *all* toric CY threefolds: the spectral theory of the operators obtained by quantizing their mirror curves is solved by their enumerative invariants.

3. From topological strings to spectral theory

3.1. Mirror symmetry and topological strings. In this section we will make a quick summary of the relevant background of mirror symmetry and topological string theory. Useful reviews of these subjects might be found in [**CK**, **H+**].

Mirror symmetry relates pairs of CY threefolds, X, $\widehat{X}$, in such a way that the enumerative geometry of X is reformulated in terms of the deformation of complex structures of $\widehat{X}$. There are various types of enumerative invariants of X. Let us consider holomorphic maps from a Riemann surface of genus g to the CY X,

$$f : \Sigma_g \to X. \tag{35}$$

Let $[S_i] \in H_2(X, \mathbb{Z})$, $i = 1, \cdots, s$, be a basis for the two-homology of X, with $s = b_2(X)$. The maps (35) are classified topologically by the homology class

$$f_*[(\Sigma_g)] = \sum_{i=1}^{s} d_i [S_i] \in H_2(X, \mathbb{Z}), \tag{36}$$

where d_i are integers called the *degrees* of the map. We will put them together in a degree vector $\mathbf{d} = (d_1, \cdots, d_s)$. The *Gromov–Witten invariant* at genus g and degree $\mathbf{d}$, which we will denote by $N_g^{\mathbf{d}}$, "counts" (in an appropriate way) the number of holomorphic maps of degree $\mathbf{d}$ from a Riemann surface of genus g to the CY X. Due to the nature of the moduli space of maps, these invariants are in general rational, rather than integer, numbers; see for example [**CK**] for rigorous definitions and examples.

The Gromov–Witten invariants at fixed genus g but at all degrees can be put together in generating functionals $F_g(\mathbf{t})$, usually called genus g free energies. These are formal power series in e^{-t_i}, $i = 1, \cdots, s$, where t_i are the *Kähler parameters* of X. More precisely, the t_i are flat coordinates on the moduli space of Kähler structures of X. It is convenient to add to these generating functionals polynomial terms which appear naturally in the study of mirror symmetry and topological strings. In this way, we have, at genus zero,

$$F_0(\mathbf{t}) = \frac{1}{6} \sum_{i,j,k=1}^{s} a_{ijk} t_i t_j t_k + \sum_{\mathbf{d}} N_0^{\mathbf{d}} \, \mathrm{e}^{-\mathbf{d}\cdot\mathbf{t}}. \tag{37}$$

In the case of a compact CY threefold, the numbers a_{ijk} are interpreted as triple intersection numbers of two-classes in X. At genus one, one has

$$F_1(\mathbf{t}) = \sum_{i=1}^{s} b_i t_i + \sum_{\mathbf{d}} N_1^{\mathbf{d}} \, \mathrm{e}^{-\mathbf{d}\cdot\mathbf{t}} . \tag{38}$$

In the compact case, the coefficients b_i are related to the second Chern class of the CY manifold [**BCOV1**]. At higher genus one finds

$$F_g(\mathbf{t}) = C_g + \sum_{\mathbf{d}} N_g^{\mathbf{d}} \, \mathrm{e}^{-\mathbf{d}\cdot\mathbf{t}}, \qquad g \geq 2, \tag{39}$$

where C_g is a constant, called the constant map contribution to the free energy [**BCOV2**]. It turns out that these functionals have a physical interpretation as the free energies at genus g of the so-called type A topological string on X. Roughly speaking, this free energy can be computed by considering the path integral of a string theory on Riemann surfaces or "worldsheets" of genus g. This makes it possible to use a large amount of methods and ideas of physics in order to shed light on these quantities.

Although the above generating functionals are in principle formal generating functions, they have a common region of convergence near the large radius point $t_i \to \infty$. The total free energy of the topological string is formally defined as the sum,

$$F^{\mathrm{WS}}(\mathbf{t}, g_s) = \sum_{g \geq 0} g_s^{2g-2} F_g(\mathbf{t}) = F^{(\mathrm{p})}(\mathbf{t}, g_s) + \sum_{g \geq 0} \sum_{\mathbf{d}} N_g^{\mathbf{d}} \, \mathrm{e}^{-\mathbf{d}\cdot\mathbf{t}} \, g_s^{2g-2}, \tag{40}$$

where

$$F^{(\mathrm{p})}(\mathbf{t}, g_s) = \frac{1}{6 g_s^2} \sum_{i,j,k=1}^{s} a_{ijk} t_i t_j t_k + \sum_{i=1}^{s} b_i t_i + \sum_{g \geq 2} C_g g_s^{2g-2}. \tag{41}$$

The superscript "WS" refers to worldsheet instantons, which are counted by this generating functional. The variable g_s, called the *topological string coupling constant*, is in principle a formal variable keeping track of the genus. However, in string theory this constant has a physical meaning, and measures the strength of the string interaction. When g_s is very small, only Riemann surfaces of low genus contribute to a given quantum observable. On the contrary, if g_s is large, the contribution of higher genus Riemann surfaces becomes very important.

The convergence properties of the formal series (40) are less understood, but there is strong evidence that, for fixed t_i in the common region of convergence, the numerical series $F_g(\mathbf{t})$ diverges factorially, as $(2g)!$ (see [**M2**] and references therein.) Therefore, the total free energy (40) does *not* define in principle a function of g_s and t_i. In a remarkable paper [**GV2**], Gopakumar and Vafa pointed out that the series (40) can be however resummed order by order in $\exp(-t_i)$, at all orders in g_s. This resummation involves a new set of enumerative invariants, the so-called *Gopakumar–Vafa invariants* $n_g^{\mathbf{d}}$. Out of these invariants, one constructs the generating series

$$F^{\mathrm{GV}}(\mathbf{t}, g_s) = \sum_{g \geq 0} \sum_{\mathbf{d}} \sum_{w=1}^{\infty} \frac{1}{w} n_g^{\mathbf{d}} \left(2 \sin \frac{w g_s}{2} \right)^{2g-2} \mathrm{e}^{-w\mathbf{d}\cdot\mathbf{t}}, \tag{42}$$

and one has, as an equality of formal series,

$$F^{\mathrm{WS}}(\mathbf{t}, g_s) = F^{(\mathrm{p})}(\mathbf{t}, g_s) + F^{\mathrm{GV}}(\mathbf{t}, g_s)\,. \tag{43}$$

The Gopakumar–Vafa invariants turn out to be integers, in contrast to the original Gromov–Witten invariants. One can obtain one set of invariants from the other by comparing (40) to (43), but there exist direct mathematical constructions of the Gopakumar–Vafa invariants as well, see [**PT**].

The mirror manifold $\widehat{X}$ to X has s complex deformation parameters z_i, $i = 1, \cdots, s$, which are related to the Kähler parameters of X by the so-called *mirror map* $t_i(\boldsymbol{z})$. At genus zero, the theory of deformation of complex structures of $\widehat{X}$ can be formulated as a theory of periods for the holomorphic $(3,0)$ form of the CY $\widehat{X}$, which vary with the complex structure moduli z_i. As first noted in [**CGPO**], this theory is remarkably simple. In particular, one can determine a function $F_0(\boldsymbol{z})$ depending on the complex moduli, called in this context the *prepotential*, which is the generating functional of genus zero Gromov–Witten invariants (37), once the mirror map is used. The theory of deformation of complex structures also has a physical realization in terms of the so-called type B topological string. However, for general CY manifolds, a precise definition of this theory at higher genus is still lacking. In practice, one often uses the description of the B-model provided in [**BCOV2**]. This provides a set of constraints for a series of functions $F_g(\boldsymbol{z})$, $g \geq 1$, known as holomorphic anomaly equations. After using the mirror map, these functions become the generating functionals (39).

One of the consequences of mirror symmetry and the B-model is the existence of many possible descriptions of the topological string, related by symplectic transformations. The existence of these descriptions can be regarded as a generalization of electric-magnetic duality. In this way, one finds different "frames" in which the topological string amplitudes can be expressed. Although these frames are in principle equivalent, some of them might be more convenient, depending on the region of moduli space we are looking at. The topological string amplitudes $F_g(\mathbf{t})$ written down above, in terms of Gromov–Witten invariants, correspond to the so-called large radius frame in the B-model, and they are appropriate for the so-called large radius limit $\mathrm{Re}(t_i) \gg 1$. The topological string free energies in different frames are related by a formal Fourier–Laplace transform, as explained in [**ABK**].

Although mirror symmetry was formulated originally for compact CY threefolds, one can extend it to the so-called "local" case [**KKV**, **CKYZ**]. In local mirror symmetry, the CY X is taken to be a toric CY manifold, which is necessarily non-compact. The theory of deformation of complex structures of the mirror $\widehat{X}$ is encoded in an algebraic curve of the form

$$W(\mathrm{e}^x, \mathrm{e}^y) = 0, \tag{44}$$

where the variables appear naturally exponentiated. Local mirror symmetry is considerably simpler than full-fledged mirror symmetry. On the A-model side, the enumerative invariants can be computed algorithmically in various ways, either by localization [**Ko**, **GP**, **KZ**], or by using the so-called topological vertex [**AKMV**]. At the same time, the theory of deformation of complex structures of $\widehat{X}$ can be simplified very much. At genus zero, one should consider the periods of the differential

$$\lambda = y(x)\,\mathrm{d}x \tag{45}$$

on the curve (44) [**KKV, CKYZ**]. The mirror map and the genus zero free energy $F_0(\mathbf{t})$ in the large radius frame are determined by making an appropriate choice of cycles on the curve, α_i, β_i, $i = 1, \cdots, s$, and one finds

$$t_i = \oint_{\alpha_i} \lambda, \qquad \frac{\partial F_0}{\partial t_i} = \oint_{\beta_i} \lambda, \qquad i = 1, \cdots, s. \tag{46}$$

In general, $s \geq g_\Sigma$, where g_Σ is the genus of the mirror curve. Also, in the local case, the type B topological string can be formulated in a more precise way, by using the topological recursion of [**EO1**], in terms of periods and residues of meromorphic forms on the curve (44) [**M1, BKMP**]. In particular, mirror symmetry can be proved to all genera [**EO2, FLZ**] by comparing the definition of the B model in [**M1, BKMP**] with localization computations in the A model.

EXAMPLE 3.1. *Local* $\mathbb{P}^2$. In the case of the local $\mathbb{P}^2$ CY given in (3), the genus zero free energy can be obtained by mirror symmetry as follows. The complex deformation parameter in the mirror curve is the parameter κ appearing in (4), or equivalently the parameter z defined in (34). The mirror map is given by

$$t = -\varpi_1(z), \tag{47}$$

where $\varpi_1(z)$ is the power series in (28). Then, $F_0(t)$ is defined, up to a constant, by

$$\frac{\partial F_0}{\partial t} = \frac{\varpi_2(z)}{6}, \tag{48}$$

where $\varpi_2(z)$ is the other period in (28). One then finds, after fixing the integration constant appropriately,

$$F_0(t) = \frac{t^3}{18} + 3\,\mathrm{e}^{-t} - \frac{45}{8}\,\mathrm{e}^{-2t} + \frac{244}{9}\,\mathrm{e}^{-3t} + \cdots. \tag{49}$$

From here one can read the very first Gromov–Witten invariants, like for example $N_0^{d=1} = 3$, see [**CKYZ**] for more details. □

Another interesting feature of the local case is that it is possible to define a more general set of enumerative invariants, and consequently a more general topological string theory, known sometimes as the *refined* topological string. This refinement has its roots in the instanton partition functions of Nekrasov for supersymmetric gauge theories [**N**]. Different aspects of the refinement have been worked out in for example [**IKV, KW, HK**]. In particular, one can generalize the Gopakumar–Vafa invariants to the so-called *refined BPS invariants*. Precise mathematical definitions can be found in [**CKK, NO**]. These invariants, which are also integers, depend on the degrees $\mathbf{d}$ and on two non-negative half-integers, j_L, j_R, or "spins". We will denote them by $N^{\mathbf{d}}_{j_L, j_R}$. The Gopakumar–Vafa invariants are particular combinations of these refined BPS invariants, and one has the following relationship,

$$\sum_{j_L, j_R} \chi_{j_L}(q)(2j_R + 1) N^{\mathbf{d}}_{j_L, j_R} = \sum_{g \geq 0} n^{\mathbf{d}}_g \left(q^{1/2} - q^{-1/2}\right)^{2g}, \tag{50}$$

where q is a formal variable and

$$\chi_j(q) = \frac{q^{2j+1} - q^{-2j-1}}{q - q^{-1}} \tag{51}$$

is the $SU(2)$ character for the spin j. We note that the sums in (50) are well-defined, since for given degrees $\mathbf{d}$ only a finite number of j_L, j_R, g give a non-zero contribution.

Out of these refined BPS invariants, one can define the Nekrasov–Shatashvili (NS) free energy,

$$
(52)\qquad \begin{aligned} F^{\rm NS}(\mathbf{t},\hbar) &= \frac{1}{6\hbar}\sum_{i,j,k=1}^{s} a_{ijk}t_it_jt_k + \sum_{i=1}^{s} b_i^{\rm NS}t_i\hbar \\ &+ \sum_{j_L,j_R}\sum_{w,\mathbf{d}} N^{\mathbf{d}}_{j_L,j_R}\frac{\sin\frac{\hbar w}{2}(2j_L+1)\sin\frac{\hbar w}{2}(2j_R+1)}{2w^2\sin^3\frac{\hbar w}{2}}\,{\rm e}^{-w\mathbf{d}\cdot\mathbf{t}}. \end{aligned}
$$

In this equation, the coefficients a_{ijk} are the same ones that appear in (37), while $b_i^{\rm NS}$ can be obtained by using mirror symmetry as in [**KW**, **HK**]. The free energy (52) is not the most general generating functional for the refined BPS invariants, but involves a particular combination thereof, which defines the so-called *NS limit* of the refined topological string. This limit was first discussed in the context of gauge theory in [**NS**]. By expanding (52) in powers of $\hbar$, we find the NS free energies at order n, $F_n^{\rm NS}(\mathbf{t})$, as

$$
(53)\qquad F^{\rm NS}(\mathbf{t},\hbar)=\sum_{n=0}^{\infty}F_n^{\rm NS}(\mathbf{t})\hbar^{2n-1}.
$$

The expression (52) can be regarded as a Gopakumar–Vafa-like resummation of the series in (53). An important observation is that the first term in this series, $F_0^{\rm NS}(\mathbf{t})$, is equal to $F_0(\mathbf{t})$, the standard genus zero free energy. Note that the term involving the coefficients $b_i^{\rm NS}$ contributes to $F_1^{\rm NS}(\mathbf{t})$.

3.2. Quantizing curves. The semiclassical limit of many interesting problems in quantum physics turns out to be described by a plane curve, together with a choice of a meromorphic differential on it. Perhaps the simplest example is the Bohr–Sommerfeld quantization condition (17) for the one-dimensional Schrödinger equation. A more complicated example is the planar limit of random matrix models, which is typically encoded in a spectral curve which captures the information on the limiting distribution of eigenvalues (here, the semiclassical limit is the large N limit). Another example which has received much attention recently is the semi-classical limit of $\mathrm{SL}(2,\mathbb{C})$ Chern–Simons theory on the complement of a knot. In this case, the relevant algebraic curve is the A-polynomial of a knot [**G**].

The genus zero free energy of topological string theory can be formally regarded as a semiclassical limit, in which the rôle of $\hbar$ is played by the string coupling constant g_s: in the limit $g_s\to 0$, the dominant term in the genus expansion of (40) is indeed the genus zero free energy. As we explained in (46), this semiclassical limit can be obtained by computing periods of the meromorphic differential (45) on the mirror curve.

It is tempting to think that, in a situation where an algebraic curve encodes the classical limit of a quantum problem, the quantum corrections are obtained by considering a "quantum" version of the curve. This has led to a rich literature on "quantum curves," where the original curve is typically promoted to a differential operator (see for example [**No**, **GS**] for recent reviews). In the case of topological string theory, an approach based on quantum curves was first proposed in [**ADKMV**], where it was suggested that the higher genus free energies of topological string theory (which can be regarded as "quantum corrections") might be obtained by quantizing the mirror curve in an appropriate way. Building on work on supersymmetric gauge theories and quantum integrable systems [**NS**], it was later shown in [**ACDKV**] that mirror curves can be formally quantized by using the WKB approximation (this was first pointed out in [**MM**], in the context of Seiberg–Witten curves). However, the quantum corrections obtained in this way do not give the conventional higher genus Gromov–Witten invariants, but rather the NS free energy (53).

It is well-known that the quantization of a classical system is in general not unique: there can be different quantizations of the same function, leading to the same classical limit but differing when $\hbar \neq 0$. The quantization of an algebraic curve is also plagued with ambiguities, for at least three different reasons. First of all, the quantization of a curve inherits the ordering ambiguities of any quantization procedure. Second, algebraic curves are naturally defined in the complex domain, while their quantum versions typically involve a choice of reality conditions which can be made in different ways. Finally, when an algebraic curve is promoted to an operator, one has to specify the Hilbert space of wavefunctions where this operator acts, and in particular the boundary conditions satisfied by such wavefunctions. In most of the recent literature on "quantum curves," these issues are not addressed explicitly, and quantum curves are only studied at the level of formal WKB expansions, as in [**ACDKV**]. This perturbative approach has led to many interesting results, but does not give a non-perturbative formulation of the quantum problem.

In [**KaMa**] it was pointed out that the mirror curves of toric CY threefolds can be quantized in such a way that the resulting operators have a discrete spectrum (we have already seen an example in section 2, in the case of local $\mathbb{P}^2$.) It was then conjectured in [**GHM**] that *the quantization of mirror curves leads to positive definite, trace class operators on* $L^2(\mathbb{R})$. This was proved in [**KM**] in many examples. In the scheme proposed in [**GHM, KM**], the ambiguities arising in the quantization of algebraic curves are solved in a natural way: ordering ambiguities are fixed by using Weyl quantization, which is particularly well suited for exponentiated position and momentum operators. Reality conditions are chosen in such a way that the classical regions (20) in phase space are compact. Finally, the Hilbert space where the operators act is simply $L^2(\mathbb{R})$.

In the rest of this paper, we will focus for simplicity on toric (almost) del Pezzo CY threefolds. These CYs are defined as the total space of the canonical bundle on a toric (almost) del Pezzo surface S,

$$X = \mathcal{O}(K_S) \to S. \tag{54}$$

They are sometimes called "local S." For example, if $S = \mathbb{P}^2$, the total space of its canonical bundle will be called local $\mathbb{P}^2$. Examples of toric del Pezzos include, besides $\mathbb{P}^2$, the Hirzebruch surfaces $\mathbb{F}_n$, $n = 0, 1, 2$, and the blowups of $\mathbb{P}^2$ at n points, denoted by $\mathcal{B}_n$, for $n = 1, 2, 3$ (note that $\mathbb{F}_1 = \mathcal{B}_1$, and that $\mathbb{F}_0 = \mathbb{P}^1 \times \mathbb{P}^1$). The main simplifying characteristic of these manifolds is that their mirror curve has genus one. This makes their analysis much simpler. We hasten to add that there is a very interesting generalization of all the considerations in this review paper to mirror curves of higher genus [**CGM2**]. In this case, a mirror curve of genus g_Σ leads in general to g_Σ trace class operators.

By standard results in toric geometry (see for example [**HKP**, **CR**]), toric, almost del Pezzo surfaces can be classified by reflexive polyhedra in two dimensions. The polyhedron Δ_S associated to a surface S is the convex hull of a set of two-dimensional vectors

$$\nu^{(i)} = \left(\nu_1^{(i)}, \nu_2^{(i)}\right), \qquad i = 1, \cdots, s+2, \tag{55}$$

together with the origin, see Fig. 3 for the example of local $\mathbb{P}^2$. In order to construct the total space of the canonical bundle over S, we have to consider the extended vectors

$$\begin{aligned} \overline{\nu}^{(0)} &= (1,0,0), \\ \overline{\nu}^{(i)} &= \left(1, \nu_1^{(i)}, \nu_2^{(i)}\right), \qquad i = 1, \cdots, s+2. \end{aligned} \tag{56}$$

They satisfy the relations

$$\sum_{i=0}^{s+2} Q_i^{\alpha} \overline{\nu}^{(i)} = 0, \qquad \alpha = 1, \cdots, s, \tag{57}$$

where Q_i^{α} is a matrix of integers (called the charge matrix) which characterizes the geometry.

The construction of the mirror geometry to (54) goes back to Batyrev, and it has been recently reviewed in [**CR**], to which we refer for further details. In order to write down the equation for the mirror curve of (54), we note that the s complex parameters of the mirror can be divided in two types: one "true" modulus κ (for a curve of genus one) and a set of "mass" parameters ξ_i, $i = 1, \cdots, s-1$ [**HKP**, **HKRS**]. In terms of these variables, the mirror curve for a local del Pezzo CY threefold can be written as,

$$W(\mathrm{e}^x, \mathrm{e}^y) = \mathcal{O}_S(x,y) + \kappa = 0, \tag{58}$$

where

$$\mathcal{O}_S(x,y) = \sum_{i=1}^{s+2} \exp\left(\nu_1^{(i)} x + \nu_2^{(i)} y + f_i(\boldsymbol{\xi})\right), \tag{59}$$

and $f_i(\boldsymbol{\xi})$ are suitable functions of the parameters ξ_j.

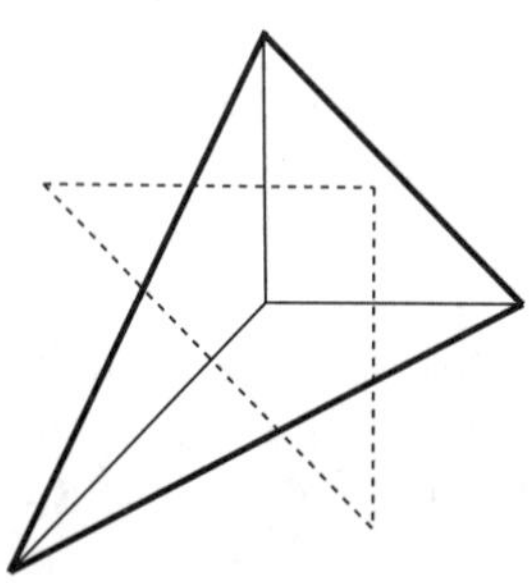

FIGURE 3. The vectors (60) defining the local $\mathbb{P}^2$ geometry, together with the polyhedron $\Delta_{\mathbb{P}^2}$ (in thick lines) and the dual polyhedron (in dashed lines).

TABLE 2. The functions $\mathcal{O}_S(x,y)$ associated to some local del Pezzo CYs.

S	$\mathcal{O}_S(x,y)$
$\mathbb{P}^2$	$e^x + e^y + e^{-x-y}$
$\mathbb{F}_0$	$e^x + \xi\, e^{-x} + e^y + e^{-y}$
$\mathbb{F}_1$	$e^x + e^y + e^{-x-y} + \xi\, e^{-x}$
$\mathbb{F}_2$	$e^x + e^y + e^{-2x-y} + \xi\, e^{-x}$
$\mathcal{B}_2$	$e^x + e^y + e^{-x-y} + \xi_1\, e^{-y} + \xi_2\, e^{-x}$
$\mathcal{B}_3$	$e^x + e^y + e^{-x-y} + \xi_1\, e^{-x} + \xi_2\, e^{-y} + \xi_3\, e^{x+y}$

EXAMPLE 3.2. The simplest case of a local del Pezzo CY is local $\mathbb{P}^2$. In this case, we have $s = 1$. The vectors (55) are given by

$$\nu^{(1)} = (1,0), \qquad \nu^{(2)} = (0,1), \qquad \nu^{(3)} = (-1,-1). \tag{60}$$

We show these vectors, together with the origin and their convex hull $\Delta_{\mathbb{P}^2}$, in Fig. 3. In this geometry there is one complex deformation parameter κ, and the function $\mathcal{O}_{\mathbb{P}^2}(x,y)$ is given by

$$\mathcal{O}_{\mathbb{P}^2}(x,y) = e^x + e^y + e^{-x-y}. \tag{61}$$

This is the curve (4). Note that, as shown in Fig. 3, the dual polyhedron to $\Delta_{\mathbb{P}^2}$ is precisely the region in Fig. 1. □

EXAMPLE 3.3. The previous example can be generalized by considering the canonical bundle over the weighted projective space $\mathbb{P}(1,m,n)$, where $m,n \in \mathbb{Z}_{>0}$. This is not a smooth manifold, but it can be analyzed by using extensions of Gromov–Witten theory, see for example [**BC**] for a study of the case $n = 1$. The vectors are in this case

$$\nu^{(1)} = (1,0), \qquad \nu^{(2)} = (0,1), \qquad \nu^{(3)} = (-m,-n), \tag{62}$$

and the function appearing in the mirror curve (58) is given by

$$\mathcal{O}_{m,n}(x,y) = e^x + e^y + e^{-mx-ny}. \tag{63}$$

Some of these geometries can arise as degeneration limits of toric del Pezzos. For example, the mirror curve to local $\mathbb{F}_2$ is characterized by the function

$$\mathcal{O}_{\mathbb{F}_2}(x,y) = e^x + e^y + e^{-2x-y} + \xi\, e^{-x}, \tag{64}$$

and when $\xi = 0$ we recover the geometry (63) with $m = 2$ and $n = 1$. □

Some further examples of functions obtained by quantization of local del Pezzos can be found in Table 2. Details on the corresponding geometries can be found in for example [**HKP**].

The "quantization" of the mirror curve (58), in the case of local del Pezzo CYs, is based on the promotion of the function $\mathcal{O}_S(x,y)$ to an operator, which will be denoted by O_S. First, we promote x, y to self-adjoint Heisenberg operators x, y on the real line, satisfying the commutation relation (1). This involves a choice of reality conditions for the complex variables x, y. Then, we apply Weyl's quantization to $\mathcal{O}_S(x,y)$. In particular, this fixes possible ordering ambiguities. We recall that, in Weyl's quantization, we have

$$\mathrm{e}^{rx+sy} \to \mathrm{e}^{r\mathsf{x}+s\mathsf{y}}. \tag{65}$$

As noted in [**GHM**], instead of focusing on the operator O_S (which is not of trace class), one should rather consider its inverse

$$\rho_S = \mathsf{O}_S^{-1} \tag{66}$$

acting on $L^2(\mathbb{R})$. It was conjectured in [**GHM**] that the operators ρ_S are of trace class and positive definite, provided appropriate positivity conditions are imposed on the mass parameters appearing in the mirror curves.

In order to motivate this assertion, note that, if e^{E_n} are the eigenvalues of O_S, the eigenvalues of ρ_S are e^{-E_n}. It is easy to extend the Bohr–Sommerfeld estimate (24) to all the operators O_S (see [**GHM**]): we can obviously define an available region in phase space, $\mathcal{R}(E)$, similar to what we did in (20), but involving this time the classical curve $\mathcal{O}_S$. The volume of this region behaves, at large E, as

$$\mathrm{vol}_0(E) \approx CE^2, \qquad E \gg 1. \tag{67}$$

The coefficient C can be easily computed by considering the tropical limit of the classical curve, where the region $\mathcal{R}(E)$ becomes a polygon (in fact, it is the dual polyhedron to Δ_S). We then find the estimate

$$E_n \approx \sqrt{\frac{2\pi\hbar}{C}} n^{1/2}, \qquad n \gg 1. \tag{68}$$

This heuristic argument indicates that the *spectral traces* of the operators ρ_S,

$$\mathrm{Tr}\, \rho_S^\ell = \sum_{n\geq 0} \mathrm{e}^{-\ell E_n}, \qquad \ell = 1, 2, \cdots \tag{69}$$

are finite. We then expect ρ_S to be positive-definite (at least for some range of the parameters) and of trace class. This was proved in [**KM**] for all the operators appearing in Table 2. Moreover, in some cases it is possible to calculate the exact integral kernel of the corresponding trace class operator [**KM, KMZ, CGuM**]. An important example in this respect is the family of operators associated to the function (63):

$$\mathsf{O}_{m,n} = \mathrm{e}^{\mathsf{x}} + \mathrm{e}^{\mathsf{y}} + \mathrm{e}^{-m\mathsf{x}-n\mathsf{y}}, \quad m, n \in \mathbb{R}_{>0}. \tag{70}$$

These operators were called three-term operators in [**KM**]. In this discussion, m, n are positive, real numbers, although in applications to the quantization of mirror curves they are often positive integers (like in the quantization of the mirror curves to local $\mathbb{P}(1, n, m)$). Let us first introduce some notation. As in [**KM**], we will denote by $\Phi_{\mathsf{b}}(x)$ Faddeev's quantum dilogarithm [**F, FK**]. We define as well

$$\Psi_{a,c}(x) = \frac{\mathrm{e}^{2\pi a x}}{\Phi_{\mathsf{b}}(x - \mathrm{i}(a+c))}. \tag{71}$$

We then have the following proposition, proved in [**KM**].

PROPOSITION 3.4. The operator on $L^2(\mathbb{R})$

$$\rho_{m,n} = \mathsf{O}_{m,n}^{-1} \tag{72}$$

is positive-definite and of trace class. Let us define normalized Heisenberg operators q, p, satisfying the normalized commutation relation

$$[\mathsf{p}, \mathsf{q}] = (2\pi \mathsf{i})^{-1}. \tag{73}$$

They are related to x, y by the linear canonical transformation,

$$\mathsf{x} \equiv 2\pi \mathsf{b} \frac{(n+1)\mathsf{p} + n\mathsf{q}}{m+n+1}, \quad \mathsf{y} \equiv -2\pi \mathsf{b} \frac{m\mathsf{p} + (m+1)\mathsf{q}}{m+n+1}, \tag{74}$$

so that $\hbar$ is related to b by

$$\hbar = \frac{2\pi \mathsf{b}^2}{m+n+1}. \tag{75}$$

In the momentum representation associated to p, the operator $\rho_{m,n}$ has the integral kernel,

$$\rho_{m,n}(p, p') = \frac{\overline{\Psi_{a,c}(p)}\, \Psi_{a,c}(p')}{2\mathsf{b} \cosh\left(\pi \frac{p-p'+\mathsf{i}(a+c-nc)}{\mathsf{b}}\right)}. \tag{76}$$

In this equation, a, c are given by

$$a = \frac{m\mathsf{b}}{2(m+n+1)}, \qquad c = \frac{\mathsf{b}}{2(m+n+1)}. \tag{77}$$

Once the trace class property has been established for the operators $\rho_{m,n}$, it can be easily established for operators ρ_S whose inverse O_S is a perturbation of a three-term operator $\mathsf{O}_{m,n}$ by a positive self-adjoint operator [**KM**]. In addition, it can be shown that the operator ρ_S for $S = \mathbb{F}_0$ is also of trace class. This proves the trace class property for a large number of operators arising from the quantization of mirror curves, including all the operators appearing in Table 2. The trace class property of $\rho_{m,n}$ and of $\rho_{\mathbb{F}_0}$ follows as well from the estimate (68), which was established rigorously in these cases in [**LST**].

3.3. Fredholm determinants from topological strings. In the previous sections, we have shown that the operators $\rho_S = \mathsf{O}_S^{-1}$ arising from mirror curves, for many toric del Pezzo threefolds, are of trace class. This means in particular that their *spectral* or *Fredholm determinant*,

$$\Xi_S(\kappa, \hbar) = \det\left(1 + \kappa \rho_S\right) \tag{78}$$

exists and is an entire function of κ [**Si**]. The Fredholm determinant has as a power series expansion around $\kappa = 0$, of the form

$$\Xi_S(\kappa, \hbar) = 1 + \sum_{N=1}^{\infty} Z_S(N, \hbar) \kappa^N, \tag{79}$$

where $Z_S(N, \hbar)$ is the *fermionic spectral trace*, given by

$$Z_S(N, \hbar) = \mathrm{Tr}\left(\Lambda^N(\rho_S)\right), \qquad N = 1, 2, \cdots \tag{80}$$

In this expression, the operator $\Lambda^N(\rho_S)$ is defined by $\rho_S^{\otimes N}$ acting on $\Lambda^N\left(L^2(\mathbb{R})\right)$. A theorem of Fredholm asserts that, if $\rho_S(p_i, p_j)$ is the kernel of ρ_S, the fermionic spectral trace can be computed as a multi-dimensional integral,

$$Z_S(N, \hbar) = \frac{1}{N!} \int \det\left(\rho_S(p_i, p_j)\right) \mathrm{d}^N p. \tag{81}$$

In turn, the logarithm of the Fredholm determinant can be regarded as a generating functional of the spectral traces of ρ_S, since

$$\mathcal{J}_S(\kappa,\hbar) = \log \Xi_S(\kappa,\hbar) = -\sum_{\ell=1}^{\infty} \frac{(-\kappa)^\ell}{\ell} \operatorname{Tr} \rho_S^\ell. \tag{82}$$

The Fredholm determinant (78) has the infinite product representation [**Si**]

$$\Xi_S(\kappa,\hbar) = \prod_{n=0}^{\infty} \left(1+\kappa\, \mathrm{e}^{-E_n}\right), \tag{83}$$

where e^{-E_n} are the eigenvalues of ρ_S. Therefore, one way of obtaining the spectrum of ρ_S is to look for the zeroes of Ξ_S, which occur at

$$\kappa = -\,\mathrm{e}^{E_n}, \qquad n \geq 0. \tag{84}$$

There are very few operators on $L^2(\mathbb{R})$ for which the Fredholm determinant can be written down explicitly. One family of examples which has been studied in some detail are Schrödinger operators with homogeneous potentials $V(x) = |x|^s$, $s = 1, 2, \cdots$ [**Vo3, DDT**]. In this case, the Fredholm determinant is defined as a regularized version of

$$D(E) = \prod_{n\geq 0} \left(1+\frac{E}{E_n}\right), \tag{85}$$

see for example [**Vo2**] for a detailed treatment. For these potentials, the function $D(E)$ is known to satisfy certain functional equations and it is captured by integral equations of the TBA type [**DDT**]. There is however no closed formula for it. One of the surprising (conjectural) results of [**GHM**] is that the Fredholm determinant of the operators ρ_S can be computed *exactly and explicitly* in terms of the enumerative invariants of X encoded in the topological string amplitudes. Moreover, the zeroes of the Fredholm determinant are determined by *exact quantization conditions.* The result stated in (31) for the spectrum of $\rho_{\mathbb{P}^2}$ at $\hbar = 2\pi$ is a particular example of the general conjecture in [**GHM**]. The resulting quantization conditions are akin to those found in conventional Quantum Mechanics (like in, for example, [**ZJJ, Vo3**]), but with an important difference: they involve *convergent* series, and in particular one does not need the apparatus of Borel–Écalle resummation in order to determine the spectrum.

The conjectural expression of [**GHM**] for the Fredholm determinant of the operator ρ_S requires the two generating functionals of enumerative invariants considered before, (42) and (52). In order to state the result, we identify the parameter κ appearing in the Fredholm determinant, with the geometric modulus of X appearing in (58). We will also write κ in terms of the "chemical potential" μ

$$\kappa = \mathrm{e}^{\mu}\,. \tag{86}$$

In addition to the modulus κ, we have the "mass parameters" ξ_j. The flat coordinates for the Kähler moduli space, t_i, are related to κ and ξ_j by the mirror map, and at leading order, in the large radius limit, we have

$$t_i \approx c_i \mu - \sum_{j=1}^{s-1} \alpha_{ij} \log \xi_j, \quad i = 1, \cdots, s, \tag{87}$$

where c_i, α_{ij} are constants. As shown in [**ACDKV**], the WKB approach makes it possible to define a *quantum mirror map* $t_i(\hbar)$, which in the limit $\hbar \to 0$ agrees with the conventional mirror map. It is a function of μ, ξ_j and $\hbar$ and it can be computed as an A-period of a quantum corrected version of the differential (45). This quantum correction is simply obtained by using the all-orders, perturbative WKB method of Dunham.

We now introduce two different functions of μ. The first one is the WKB grand potential,

$$\begin{aligned} \mathsf{J}_S^{\mathrm{WKB}}(\mu, \boldsymbol{\xi}, \hbar) = & \sum_{i=1}^{s} \frac{t_i(\hbar)}{2\pi} \frac{\partial F^{\mathrm{NS}}(\mathbf{t}(\hbar), \hbar)}{\partial t_i} + \frac{\hbar^2}{2\pi} \frac{\partial}{\partial \hbar} \left(\frac{F^{\mathrm{NS}}(\mathbf{t}(\hbar), \hbar)}{\hbar} \right) \\ & + \sum_{i=1}^{s} \frac{2\pi}{\hbar} b_i t_i(\hbar) + A(\boldsymbol{\xi}, \hbar). \end{aligned} \tag{88}$$

In this equation, $F^{\mathrm{NS}}(\mathbf{t}, \hbar)$ is given by the expression (52). In the second term of (88), the derivative w.r.t. $\hbar$ does not act on the implicit dependence of t_i. The coefficients b_i appearing in the last line are the same ones appearing in (38). The function $A(\boldsymbol{\xi}, \hbar)$ is not known in closed form for arbitrary geometries, although detailed conjectures for its form exist in various examples. It is closely related to a all-genus resummed version of the constant map contribution C_g appearing in (39). The second function is the "worldsheet" grand potential, which can be obtained from the generating functional (42),

$$\mathsf{J}_S^{\mathrm{WS}}(\mu, \boldsymbol{\xi}, \hbar) = F^{\mathrm{GV}} \left(\frac{2\pi}{\hbar} \mathbf{t}(\hbar) + \pi \mathrm{i} \mathbf{B}, \frac{4\pi^2}{\hbar} \right). \tag{89}$$

In this formula, $\mathbf{B}$ is a constant vector ("B-field") which depends on the geometry under consideration. This vector should satisfy the following requirement: for all $\mathbf{d}$, j_L and j_R such that the refined BPS invariant $N^{\mathbf{d}}_{j_L, j_R}$ is non-vanishing, we must have

$$(-1)^{2j_L + 2j_R + 1} = (-1)^{\mathbf{B} \cdot \mathbf{d}}. \tag{90}$$

For local del Pezzo CY threefolds, the existence of such a vector was established in [**HMMO**]. Note that the effect of this constant vector is to introduce a sign

$$(-1)^{w \mathbf{d} \cdot \mathbf{B}} \tag{91}$$

in the generating functional (42). An important remark is that, in (89), the topological string coupling constant g_s appearing in (40) and (42) is related to the Planck constant appearing in the spectral problem by,

$$g_s = \frac{4\pi^2}{\hbar}. \tag{92}$$

Therefore, the regime of weak coupling for the topological string coupling constant, $g_s \ll 1$, corresponds to the strong coupling regime of the spectral problem, $\hbar \gg 1$, and conversely, the semiclassical limit of the spectral problem corresponds to the strongly coupled topological string. We therefore have a *strong-weak coupling duality* between the spectral problem and the conventional topological string.

The *total grand potential* is the sum of these two functions,

$$\mathsf{J}_S(\mu, \boldsymbol{\xi}, \hbar) = \mathsf{J}_S^{\mathrm{WKB}}(\mu, \boldsymbol{\xi}, \hbar) + \mathsf{J}_S^{\mathrm{WS}}(\mu, \boldsymbol{\xi}, \hbar), \tag{93}$$

and it was first considered in [**HMMO**]. It has the structure
(94)

$$\mathsf{J}_S(\mu,\boldsymbol{\xi},\hbar) = \frac{1}{12\pi\hbar}\sum_{i,j,k=1}^{s} a_{ijk}t_it_jt_k + \sum_{i=1}^{s}\left(\frac{2\pi b_i}{\hbar} + \frac{\hbar b_i^{\rm NS}}{2\pi}\right)t_i + \mathcal{O}\left(\mathrm{e}^{-t_i}, \mathrm{e}^{-2\pi t_i/\hbar}\right),$$

where the last term stands for a formal power series in e^{-t_i}, $\mathrm{e}^{-2\pi t_i/\hbar}$, whose coefficients depend explicitly on $\hbar$. Note that the trigonometric functions appearing in (52) and (42) have double poles when $\hbar$ is a rational multiple of π. However, as shown in [**HMMO**], the poles *cancel* in the sum (93). This HMO cancellation mechanism was first discovered in [**HMO**], in a slightly different context, and it was first advocated in [**KaMa**] in the study of quantum curves.

A natural question is whether the formal power series in (93) converges, at least for some values of its arguments. Although we do not have rigorous results on this problem, the available evidence suggests that, for real $\hbar$, $\mathsf{J}_S(\mu,\boldsymbol{\xi},\hbar)$ *converges* in a neighbourhood of the large radius point $t_i \to \infty$. However, the series seems to be divergent when $\hbar$ is complex. This divergence is inherited from the generating functionals (42) and (52). Explicit calculations indicate, for example, that the Gopakumar–Vafa generating functional (42) is divergent for complex g_s, as first noted in [**HMMO**]. In the one-modulus case, if we write this series as

$$F^{\rm GV}(t,g_s) = \sum_{\ell\geq 1} a_\ell(g_s)\,\mathrm{e}^{-\ell t}, \tag{95}$$

one finds that, if $g_s \in \mathbb{C}\backslash\mathbb{R}$,

$$\log|a_\ell(g_s)| \sim \ell^2, \qquad \ell \gg 1. \tag{96}$$

Therefore, the generating functional (42) is in general ill-defined and, as it stands, it can not be used as a non-perturbative definition of the topological string free energy[2]. Finally, note that, after using the explicit expression for the t_i in terms of μ and the mass parameters, one easily finds that $\mathsf{J}_S(\mu,\boldsymbol{\xi},\hbar)$ is a cubic polynomial in μ, plus an infinite series of exponentially small corrections in $\mathrm{e}^{-\mu}$, $\mathrm{e}^{-2\pi\mu/\hbar}$. At large μ, the leading behaviour is given by,

$$\mathsf{J}_S(\mu,\boldsymbol{\xi},\hbar) \approx \frac{C}{6\pi\hbar}\mu^3, \qquad \mu \gg 1, \tag{97}$$

where C is the constant appearing in (67).

We can now state the main conjecture of [**GHM**].

CONJECTURE 3.5. The Fredholm determinant of the operator ρ_S is given by

$$\Xi_S(\kappa,\boldsymbol{\xi},\hbar) = \sum_{n\in\mathbb{Z}} \exp\left(\mathsf{J}_S(\mu+2\pi\mathrm{i}n,\boldsymbol{\xi},\hbar)\right). \tag{98}$$

The sum over n defines a *quantum theta function* $\Theta_S(\mu,\boldsymbol{\xi},\hbar)$,

$$\Xi_S(\kappa,\boldsymbol{\xi},\hbar) = \mathrm{e}^{\mathsf{J}_S(\mu,\boldsymbol{\xi},\hbar)}\,\Theta_S(\mu,\boldsymbol{\xi},\hbar). \tag{99}$$

The reason for this name is that, when $\hbar = 2\pi$, the quantum theta function becomes a classical, conventional theta function, as we will see in a moment. The vanishing locus of the Fredholm determinant, as we have seen in (84), gives the spectrum of

---

[2] In some geometries, (42) can be resummed by using instanton calculus, and one obtains a convergent function if $g_s \in \mathbb{C}\backslash\mathbb{R}$ [**BS**]. The resulting function still has a dense set of poles on the real line.

the trace class operator. Given the form of (99), this is the vanishing locus of the quantum theta function.

It would seem that the above conjecture is very difficult to test, since it gives the Fredholm determinant as a formal, infinite sum. However, it is easy to see that the r.h.s. of (98) has a series expansion at large μ in powers of $\mathrm{e}^{-\mu}$, $\mathrm{e}^{-2\pi\mu/\hbar}$. In addition, it leads to an integral representation for the fermionic spectral trace which is very useful in practice: if we write $Z_S(N,\hbar)$ as a contour integral around $\kappa=0$, simple manipulations give the expression [**HMO**, **GHM**]

$$Z_S(N,\hbar)=\frac{1}{2\pi\mathrm{i}}\int_{\mathcal{C}}\mathrm{e}^{\mathsf{J}_S(\mu,\boldsymbol{\xi},\hbar)-N\mu}\,\mathrm{d}\,\mu, \tag{100}$$

where $\mathcal{C}$ is a contour going from $\mathrm{e}^{-\mathrm{i}\pi/3}\infty$ to $\mathrm{e}^{\mathrm{i}\pi/3}\infty$. Note that this is the standard contour for the integral representation of the Airy function and it should lead to a convergent integral, since $\mathsf{J}_S(\mu,\boldsymbol{\xi},\hbar)$ is given by a cubic polynomial in μ, plus exponentially small corrections. Finally, as we will see in a moment, in some cases the r.h.s. of (98) can be written in terms of well-defined functions.

What is the interpretation of the total grand potential that we introduced in (93)? The WKB part takes into account the perturbative corrections (in $\hbar$) to the spectral problem defined by the operator ρ_S. In fact, it can be calculated order by order in the $\hbar$ expansion, by using standard techniques in Quantum and Statistical Mechanics (see for example [**MP**, **H**]). The expression (88) provides a resummation of this expansion at large μ. However, the WKB piece is insufficient to solve the spectral problem, since as we mentioned above, (88) is divergent for a dense set of values of $\hbar$ on the real line. Physically, the additional generating functional (89) contains the contribution of *complex* instantons, which are non-perturbative in $\hbar$ [**MP**, **KaMa**] and cancel the poles in the all-orders WKB contribution. Surprisingly, the full instanton contribution in this spectral problem is simply encoded in the standard topological string partition function.

The formulae (88), (89), (98) are relatively complicated, in the sense that they involve the full generating functionals (52) and (42). There is however an important case in which they simplify considerably, namely, when $\hbar=2\pi$. This was called in [**GHM**] the "maximally supersymmetric case," since in the closely related spectral problem of ABJM theory, it occurs when there is enhanced $\mathcal{N}=8$ supersymmetry [**CGM**]. In this case, as it can be easily seen from the explicit expressions for the generating functionals, many contributions vanish. For example, in the Gopakumar–Vafa generating functional, all terms involving $g\geq 2$ are zero. A simple calculation shows that the function (93) is given by

$$\mathsf{J}_S(\mu,\boldsymbol{\xi},2\pi)=\frac{1}{8\pi^2}\sum_{i,j=1}^{s}t_it_j\frac{\partial^2\widehat{F}_0}{\partial t_i\partial t_j}-\frac{1}{4\pi^2}\sum_{i=1}^{s}t_i\frac{\partial\widehat{F}_0}{\partial t_i}+\frac{1}{4\pi^2}\widehat{F}_0(\mathbf{t})+\widehat{F}_1(\mathbf{t})+\widehat{F}_1^{\mathrm{NS}}(\mathbf{t}), \tag{101}$$

where $\widehat{F}_0(\mathbf{t})$, $\widehat{F}_1(\mathbf{t})$ and $\widehat{F}_1^{\mathrm{NS}}(\mathbf{t})$ are the generating functions appearing in (37), (38) and (53), with the only difference that one has to include as well the sign (91) in the expansion in e^{-t_i}. It can be also seen that, for this value of $\hbar$, the quantum mirror map becomes the classical mirror map, up to a change of sign $\kappa\to-\kappa$. One can now use (101) to compute the quantum theta function appearing in (99). It is easy to see that, provided some integrality properties hold for the constant C in (67), the quantum theta function becomes a classical theta function. We will now illustrate this simplification in the case of local $\mathbb{P}^2$.

EXAMPLE 3.6. Local $\mathbb{P}^2$ has one single Kähler parameter, and no mass parameters. The function $A(\hbar)$ appearing in (88) has been conjectured in closed form, and it is given by

$$A(\hbar) = \frac{3A_c(\hbar/\pi) - A_c(3\hbar/\pi)}{4}, \tag{102}$$

where

$$A_c(k) = \frac{2\zeta(3)}{\pi^2 k}\left(1 - \frac{k^3}{16}\right) + \frac{k^2}{\pi^2}\int_0^\infty \frac{x}{e^{kx}-1}\log(1-e^{-2x})\, dx. \tag{103}$$

This function was first introduced in [**MP**], and determined in integral form in [**Ha+**, **HO**]. It can be obtained by an appropriate all-genus resummation of the constants C_g appearing in (39). In the "maximally supersymmetric case" $\hbar = 2\pi$, one can write the Fredholm determinant in closed form. The standard topological string genus zero free energy is given in (48), (49). The genus one free energies are given by [**KZ**, **HK**]:

$$\begin{aligned} F_1(t) &= \frac{1}{2}\log\left(-\frac{dz}{dt}\right) - \frac{1}{12}\log\left(z^7(1+27z)\right), \\ F_1^{\rm NS}(t) &= -\frac{1}{24}\log\left(\frac{1+27z}{z}\right). \end{aligned} \tag{104}$$

We identify

$$z = e^{-3\mu} = \frac{1}{\kappa^3}. \tag{105}$$

Then, it follows from the conjecture above that

(106)
$$\begin{aligned} \Xi_{\mathbb{P}^2}(-\kappa, 2\pi) &= \det(1 - \kappa\rho_{\mathbb{P}^2}) \\ &= \exp\left\{A(2\pi) + \frac{1}{4\pi^2}\left(F_0(t) - t\partial_t F_0(t) + \frac{t^2}{2}\partial_t^2 F_0(t)\right) + F_1(t) + F_1^{\rm NS}(t)\right\} \\ &\quad \times e^{\pi i/8}\,\vartheta_2\left(\xi - \frac{1}{4}, \tau\right), \end{aligned}$$

where

$$\xi = \frac{3}{4\pi^2}\left(t\partial_t^2 F_0(t) - \partial_t F_0(t)\right), \qquad \tau = \frac{2i}{\pi}\partial_t^2 F_0(t), \tag{107}$$

and $\vartheta_2(z,\tau)$ is the Jacobi theta function. The τ appearing here is the standard modulus of the genus one mirror curve of local $\mathbb{P}^2$. In particular, one has that $\mathrm{Im}(\tau) > 0$. It can be easily checked that the function ξ agrees with (29), with the identification $\mu = E$. The formulae that we have written down here are slightly different from the ones listed earlier in this section, since we are changing the sign of κ in the Fredholm determinant, but they can be easily derived from them. We can now perform a power series expansion around $\kappa = 0$, by using analytic continuation of the topological string free energies to the so-called *orbifold point*, which corresponds to the limit $z \to \infty$. This is a standard exercise in topological string theory (see for example [**AGM**, **ABK**]), and one finds [**GHM**]

$$\Xi_{\mathbb{P}^2}(\kappa, 2\pi) = 1 + \frac{\kappa}{9} + \left(\frac{1}{12\sqrt{3}\pi} - \frac{1}{81}\right)\kappa^2 + \mathcal{O}(\kappa^3), \tag{108}$$

provided some non-trivial identities are used for theta functions. This predicts the values of the very first fermionic spectral traces $Z_{\mathbb{P}^2}(N, 2\pi)$, for $N = 1, 2$. Interestingly, these traces can be calculated directly in spectral theory, since $\rho_{\mathbb{P}^2}$ is the operator $\rho_{1,1}$ defined in (72). The integral kernel of this operator is given in (76) (for $m = n = 1$), and the values appearing in the expansion (108) have been verified in this way in [**KM, OZ**].

Finally, it can be easily checked that the function (106) vanishes precisely at the values of μ given by the quantization condition (31) (since we are changing the sign of κ in this formula, the zeros occur at $\kappa = e^{E_n}$, $n = 0, 1, \cdots$). Note that this corresponds to the vanishing locus of the Jacobi theta function appearing in (106). In Fig. 4 we show a plot of the Fredholm determinant $\Xi_{\mathbb{P}^2}(\kappa, 2\pi)$ as a function of κ on the real axis. For large, positive values of κ, its behavior is dictated by the large μ behavior of $\mathsf{J}_{\mathbb{P}^2}(\mu, 2\pi)$, i.e.

$$\Xi_{\mathbb{P}^2}(\kappa, 2\pi) \approx \exp\left\{\frac{3}{8\pi^2}(\log(\kappa))^3\right\}, \qquad \kappa \gg 1. \tag{109}$$

On the other hand, for large, negative κ, one has the oscillatory behavior,

$$\Xi_{\mathbb{P}^2}(-\kappa, 2\pi) \approx \exp\left\{\frac{3}{8\pi^2}(\log(\kappa))^3\right\}\cos\left(\pi\left(\xi - \frac{1}{4}\right)\right), \qquad \kappa \gg 1, \tag{110}$$

where ξ is again given by (107). □

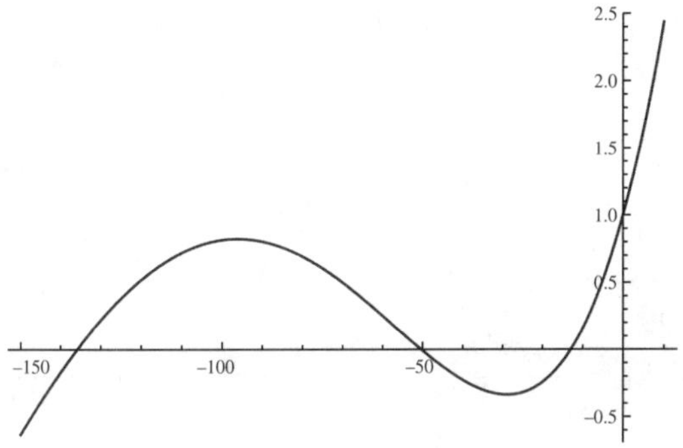

FIGURE 4. The Fredholm determinant $\Xi_{\mathbb{P}^2}(\kappa, 2\pi)$ as a function of κ, showing the first three zeroes on the negative real axis, corresponding to the first three energy levels $-e^{E_n}$, $n = 0, 1, 2$.

What happens in the general case, when $\hbar \neq 2\pi$? One can of course still compute the Fredholm determinant from our conjecture (98), but it now contains contributions from all the higher genus Gopakumar–Vafa invariants, and from the all-orders NS free energy. In order to study its properties, one possibility is to expand it in power series around the "semiclassical" case $\hbar = 2\pi$. This was done in [**Gr**], and the resulting expansion puts on a rigorous footing the $1/N$ expansion of the "non-perturbative partition function" studied in [**EM**]. One can also use (98) to determine the spectrum by looking at the vanishing locus of the Fredholm determinant for general $\hbar$. By using the large radius expansion of the grand potential, one can write an exact quantization condition defined by an explicit power series. As shown in [**GHM**] in many examples, the predictions for the spectrum obtained from this exact quantization condition are in perfect agreement with direct numerical calculations (as done for example in [**HW**]). It was found in [**WZH**] that, when the mirror curve has genus one, the quantization condition of [**GHM**] can be written down in closed form in terms solely of the NS free energy. The results of [**WZH, GHM**], when put together, imply that the generating functional (52), capturing the NS limit of the topological string, is closely related to the generating functional (42), capturing the *standard* topological string. This connection has been further explored in [**SWH**] and finally established in [**GG**] in many cases.

For a generic value of $\hbar$, it is difficult to extract analytic results for the spectral traces from (98), since the analytic continuation of the generating functionals (42), (52) to the orbifold point for arbitrary $\hbar$ is not known. However, one can still obtain very precise *numerical* results. The basis for this is the formula (100). By expanding $\mathsf{J}_S(\mu, \boldsymbol{\xi}, \hbar)$ at large μ, the fermionic spectral trace $Z_S(N, \hbar)$ can be evaluated as an infinite sum of Airy functions and their derivatives. Let us spell this out in some detail for local $\mathbb{P}^2$. One can write,

$$\mathrm{e}^{\mathsf{J}_{\mathbb{P}^2}(\mu,\hbar)} = \mathrm{e}^{\mathsf{J}^{(\mathrm{p})}_{\mathbb{P}^2}(\mu,\hbar)} \sum_{l,n} a_{l,n} \mu^n \, \mathrm{e}^{-l\mu}, \tag{111}$$

where

$$\mathsf{J}^{(\mathrm{p})}_{\mathbb{P}^2}(\mu, \hbar) = \frac{C(\hbar)}{3}\mu^3 + B(\hbar)\mu + A(\hbar). \tag{112}$$

The values of $C(\hbar)$, $B(\hbar)$ can be read off from the general expression (93) and are given by [**GHM**]

$$C(\hbar) = \frac{9}{4\pi\hbar}, \qquad B(\hbar) = \frac{\pi}{2\hbar} - \frac{\hbar}{16\pi}. \tag{113}$$

Then, assuming good convergence properties for the expansion of $\mathsf{J}_{\mathbb{P}^2}(\mu, \hbar)$, we can exchange it with integration in (100). We obtain in this way,

$$Z_{\mathbb{P}^2}(N, \hbar) = \frac{\mathrm{e}^{A(\hbar)}}{(C(\hbar))^{1/3}} \sum_{l,n} a_{l,n} \left(-\frac{\partial}{\partial N}\right)^n \mathrm{Ai}\left(\frac{N + l - B(\hbar)}{(C(\hbar))^{1/3}}\right), \tag{114}$$

where $\mathrm{Ai}(z)$ is the Airy function. Note that n takes non-negative integer values, but l is of the form $3p + 6\pi q/\hbar$, where p, q are non-negative integers. The leading behavior of (114) in the limit $N \to \infty$, $\hbar$ fixed, is given by the first term in the r.h.s., which is an Airy function

$$Z_{\mathbb{P}^2}(N, \hbar) \approx \mathrm{Ai}\left(\frac{N - B(\hbar)}{(C(\hbar))^{1/3}}\right), \qquad N \gg 1. \tag{115}$$

The additional terms in the r.h.s. of (114) give an infinite series of exponentially suppressed corrections to (115). Remarkably, this series seems to be convergent, and it should be regarded as the analogue in this theory of the Rademacher expansion in number theory. It also produces highly accurate numerical answers for the spectral traces $Z_{\mathbb{P}^2}(N, \hbar)$. Using this procedure, one obtains for example,

$$Z_{\mathbb{P}^2}\left(1, \hbar = \frac{2\pi}{3}\right) = 0.460452148172832597790488985616874708763212420 7\ldots \tag{116}$$

This can be compared to the analytic result obtained by integrating (76),

$$\text{(117)} \qquad \operatorname{Tr} \rho_{1,1}\left(\hbar = \frac{2\pi}{3}\right) = \frac{1}{3}\exp\left(\frac{V}{2\pi}\right), \qquad V = 2\,\mathrm{Im}\left(\mathrm{Li}_2\left(\mathrm{e}^{\pi\,\mathrm{i}/3}\right)\right).$$

The agreement is quite remarkable, and the precision increases with the number of terms retained in the expansion of the grand potential. Many tests of the main conjecture (98), for various local del Pezzo CYs, can be found in [**GKMR**]. In all cases, one finds that (98) captures in exquisite detail the spectrum and the spectral traces of the corresponding operator.

It follows from (115) that, at large N and $\hbar$ fixed, the spectral traces display the $N^{3/2}$ scaling

$$\text{(118)} \qquad \log Z_{\mathbb{P}^2}(N,\hbar) \approx -\frac{4\sqrt{\pi}}{9} N^{3/2}\hbar^{1/2}, \qquad N \gg 1.$$

This can be also derived as a consequence of the leading WKB scaling (24), see for example [**MP**] for a detailed explanation of this type of relation. The above behaviors, (115) and (118), are not restricted to local $\mathbb{P}^2$, but hold for all local del Pezzo CYs: the leading term in the large N expansion of the fermionic spectral trace is always an Airy function of the form (115), which leads to the $N^{3/2}$ scaling. These behaviors are typical of a theory of N M2 branes [**KT**], and they are very similar to what happens in ABJM theory [**DMP, FHM, MP**]. A universal Airy behavior for topological strings has been recently discovered in [**AYZ, CS**], and it would be interesting to understand its precise relation to what is found in this formalism.

4. From spectral theory to topological strings

4.1. Random matrices from operators and topological strings. An important consequence of the conjecture of [**GHM**] is the following. As we have seen, the Fredholm determinant involves both the WKB grand potential, which is captured by the NS free energy, and the worldsheet instanton grand potential, which is essentially given by the topological string free energy. It turns out that, due to their functional form, there are appropriate scaling limits in which only one of these ingredients contributes. We will focus on the 't Hooft-like limit

$$\text{(119)} \qquad \hbar \to \infty, \qquad \mu \to \infty, \qquad \frac{\mu}{\hbar} = \zeta \text{ fixed.}$$

For simplicity, we will also assume that the mass parameters ξ_j scale in such a way that

$$\text{(120)} \qquad m_j = \xi_j^{2\pi/\hbar}, \qquad j = 1, \cdots, s-1,$$

are fixed (although other limits are possible, see [**MZ**] for a detailed discussion). It is easy to see that, in this limit, the quantum mirror map becomes trivial, and the approximation (87) is exact. The grand potential has then the asymptotic expansion,

$$\text{(121)} \qquad \mathsf{J}_S^{\text{'t Hooft}}(\zeta, \boldsymbol{m}, \hbar) = \sum_{g=0}^{\infty} \mathsf{J}_g^S(\zeta, \boldsymbol{m})\,\hbar^{2-2g},$$

where

$$
\begin{aligned}
\mathsf{J}_0^S(\zeta,\boldsymbol{m}) &= \frac{1}{16\pi^4}\left(\widehat{F}_0(\boldsymbol{t}) + 4\pi^2 \sum_{i=1}^{s} b_i^{\rm NS} t_i + 14\pi^4 A_0(\boldsymbol{m})\right),\\
\mathsf{J}_1^S(\zeta,\boldsymbol{m}) &= A_1(\boldsymbol{m}) + \widehat{F}_1(\boldsymbol{t}),\\
\mathsf{J}_g^S(\zeta,\boldsymbol{m}) &= A_g(\boldsymbol{m}) + (4\pi^2)^{2g-2}\left(\widehat{F}_g(\boldsymbol{t}) - C_g\right), \qquad g \geq 2.
\end{aligned} \tag{122}
$$

In these equations,

$$
t_i = 2\pi c_i \zeta - \sum_{j=1}^{s-1} \alpha_{ij} \log m_j, \tag{123}
$$

and $\widehat{F}_g(\boldsymbol{t})$ are the standard topological string free energies as a function of the standard Kähler parameters $\boldsymbol{t}$, after turning on the B-field as in (91). We have also assumed that the function $A(\boldsymbol{\xi},\hbar)$ has the expansion

$$
A(\boldsymbol{\xi},\hbar) = \sum_{g=0}^{\infty} A_g(\boldsymbol{m})\hbar^{2-2g}, \tag{124}
$$

and this assumption can be tested in examples.

Therefore, in the 't Hooft limit (119), the grand potential has an asymptotic expansion which is essentially equivalent to the genus expansion of the standard topological string. According to our conjecture, this expansion gives a concrete *prediction* for the 't Hooft expansion of the fermionic spectral traces, thanks to the equality (100). Indeed, let us consider the corresponding 't Hooft limit of the fermionic spectral traces,

$$
\hbar \to \infty, \qquad N \to \infty, \qquad \frac{N}{\hbar} = \lambda \text{ fixed}. \tag{125}
$$

Then, it follows from (100) that one has the following asymptotic expansion,

$$
\log Z_S(N,\hbar) = \sum_{g\geq 0} \mathcal{F}_g(\lambda)\hbar^{2-2g}, \tag{126}
$$

where $\mathcal{F}_g(\lambda)$ can be obtained by evaluating the integral in the right hand side of (100) in the saddle-point approximation and using just the 't Hooft expansion of $\mathsf{J}_S(\mu,\boldsymbol{\xi},\hbar)$. At leading order in $\hbar$, this is just a Legendre transform, and one finds

$$
\mathcal{F}_0(\lambda) = \mathsf{J}_0^S(\zeta,\boldsymbol{m}) - \lambda\zeta, \tag{127}
$$

evaluated at the saddle-point given by

$$
\lambda = \frac{\partial \mathsf{J}_0^S}{\partial \zeta}. \tag{128}
$$

The higher order corrections can be computed systematically. In fact, the formalism of [**ABK**] gives a nice geometric description of the integral transform (100) in the saddle-point approximation: the functions $\mathcal{F}_g(\lambda)$ are simply the free energies of the topological string on X, but on a different frame. This frame corresponds to the so-called *conifold* point of the geometry. In particular, λ turns out to be a vanishing period at the conifold point.

Therefore, the conjecture (98), in its form (100), provides a precise prediction for the 't Hooft limit of the fermionic spectral traces: they are encoded in the standard topological string free energy, but evaluated at the conifold frame. It turns

out that this prediction can be tested in detail. The reason is that, at least in some cases, the fermionic spectral traces can be expressed as random matrix integrals, and their 't Hooft limit can be studied by using various techniques developed for matrix models at large N. In the case of the three-term operators (70), the explicit expression for the kernel appearing in (76) can be combined with Fredholm's theorem (81) to obtain the following expression for their fermionic spectral traces [**MZ**],
(129)

$$Z_{m,n}(N,\hbar) = \frac{1}{N!}\int_{\mathbb{R}^N} \frac{\mathrm{d}^N u}{(2\pi)^N} \prod_{i=1}^{N} \left|\Psi_{a,c}\left(\frac{\mathsf{b}u_i}{2\pi}\right)\right|^2 \frac{\prod_{i<j} 4\sinh\left(\frac{u_i-u_j}{2}\right)^2}{\prod_{i,j} 2\cosh\left(\frac{u_i-u_j}{2} + \mathrm{i}\,\pi C_{m,n}\right)}.$$

This can be regarded as an $O(2)$ matrix model [**K, EK**], with a potential depending on $\hbar$ and given by

$$V(u,\hbar) = -\hbar \log\left|\Psi_{a,c}\left(\frac{\mathsf{b}u}{2\pi}\right)\right|^2. \tag{130}$$

In [**MZ**] it is shown in detail that this potential admits an asymptotic expansion at large $\hbar$, which can be used to obtain the 't Hooft expansion of $Z_{m,n}(N,\hbar)$. In particular, when $m=n=1$, one finds a description of the all-genus topological string free energy of local $\mathbb{P}^2$ as the asymptotic 't Hooft expansion of this matrix integral. Detailed calculations show that the expansion obtained in this way agrees exactly with the predictions of (100). In fact, the conjecture "explains" many aspects of topological string theory at the conifold point, for many toric CY geometries. For example, it has been known for some time that the leading terms in the expansion of the conifold free energies are universal [**GhV**], and that they agree with the all-genus free energy of the Gaussian matrix model. Since matrix models like (129) are deformations of the Gaussian one, this observation is a consequence of our conjecture (or, equivalently, this observation must hold in order for the conjecture to be true).

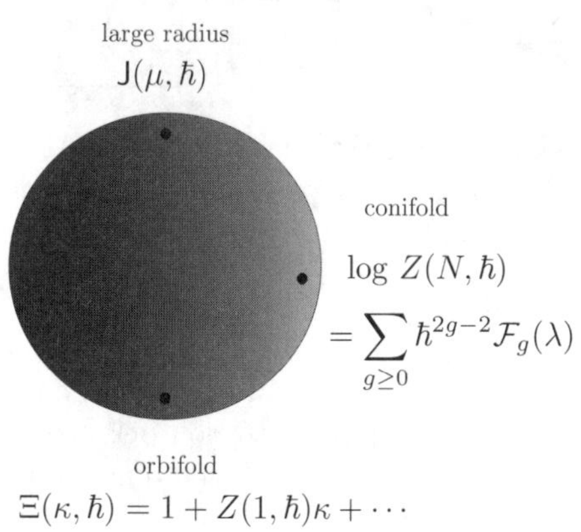

FIGURE 5. Different points in the moduli space lead to different expansions for the spectral quantities.

Note that, as shown in Fig. 5, different regions of the moduli space of a toric CY are associated to different expansions of the spectral quantities: the large radius region defines the grand potential at large μ. The orbifold point is the appropriate one to calculate the Taylor expansion of the Fredholm determinant at the origin, as in (108), while the conifold point encodes the 't Hooft limit of the fermionic spectral traces.

4.2. Non-perturbative topological strings. As we mentioned above, the total free energy of the topological string (40) is hopelessly divergent. This is also the case for the total free energy at the conifold frame. Let us consider the formal power series in $\hbar$ appearing in the r.h.s. of (126). As it happens at large radius, the functions $\mathcal{F}_g(\lambda)$ have a common, finite radius of convergence. For each λ inside this radius of convergence, the resulting series in $\hbar$ diverges factorially, like

$$\mathcal{F}_g(\lambda) \sim (2g)!(A(\lambda))^{-2g}, \qquad g \gg 1. \tag{131}$$

Therefore, the expansion (126) is asymptotic and does not define a function of λ, $\hbar$. However, we have now found a quantity, the fermionic spectral trace, with the following properties:

(1) It is rigorously well-defined. This is a consequence of the trace class property of ρ_S.
(2) Its asymptotic expansion, in the 't Hooft limit, reproduces exactly the genus expansion of the topological string in the conifold frame.

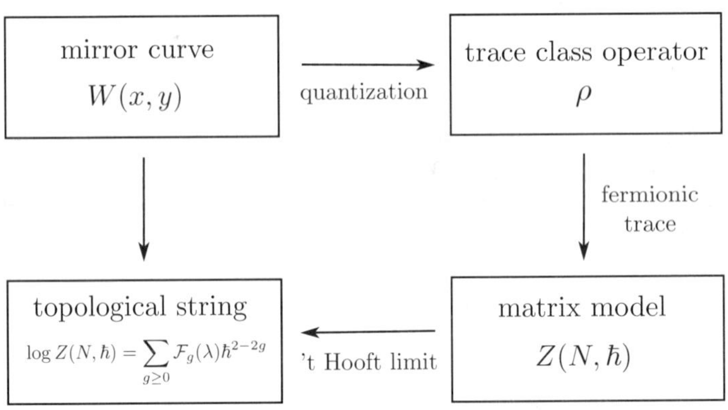

FIGURE 6. Given a toric CY threefold, the quantization of its mirror curve leads to a trace class operator ρ. The standard topological string free energy is obtained in the 't Hooft limit of its fermionic traces $Z(N,\hbar)$.

It follows that *the fermionic spectral traces provide a non-perturbative completion of the topological string*, in the conifold frame. The results of [**GHM**, **MZ**, **KMZ**] lead then to an elegant approach to non-perturbative topological strings: given a toric CY manifold X, one considers the quantization of its mirror curve, which leads to a trace class operator. The standard, all-genus topological string free energy emerges as an asymptotic expansion of the fermionic spectral traces of the operator, in a 't Hooft-like limit. This procedure is sketched in Fig. 6. Note that, from the point of view of the spectral problem, the conventional topological string emerges at *strong* coupling $\hbar \gg 1$.

Non-perturbative completions are not unique. One way to obtain well-defined answers from asymptotic series is to use Borel resummation. In favorable cases, when the series is Borel summable, this procedure is unambiguous and leads to a function whose asymptotic expansion reproduces the original asymptotic series (see for example [**M2**] for a review). In the case of topological string theory in the conifold frame, it turns out that, in many examples (like local $\mathbb{P}^2$), the genus expansion is Borel summable for real $\hbar$ and real λ. The result of the Borel resummation however turns out to be *different* from the non-perturbative completion defined by the fermionic spectral traces. For example, in the case of local $\mathbb{P}^2$, one finds that [**CMS**]

$$\log Z_{\mathbb{P}^2}^{\text{Borel}}(1, 2\pi) = -2.197217... \tag{132}$$

to be compared with the value obtained from spectral theory (see (108)),

$$\log Z_{\mathbb{P}^2}(1, 2\pi) = -\log(9) = -2.197224... \tag{133}$$

The difference between the Borel resummation and the answer from spectral theory is due to exponentially small, non-perturbative effects associated to complex instantons. In the context of the theory of resurgence, this indicates that the perturbative genus expansion has to be generalized to a *trans-series* including the non-perturbative effects in the form of additional asymptotic series. Such trans-series can be produced by using the powerful techniques of [**CESV**], which are based on a generalization of the holomorphic anomaly equation of [**BCOV2**] to the non-perturbative realm. One can now wonder whether the non-perturbative information contained in the fermionic spectral traces can be recovered from the Borel–Écalle resummation of the full trans-series. In [**CMS**], strong numerical evidence has been given that this is the case [3], relating in this way the "exact" non-perturbative approach of [**GHM**] and the resurgent analysis of non-perturbative effects in [**CESV**].

The fermionic spectral trace $Z(N, \hbar)$ provides a non-perturbative completion of the topological string in the conifold frame, and it actually promotes it to an *entire* function of N, as first noted in [**CGM**]. In the large radius frame, we can also promote the topological string partition function to a entire function on moduli space, which is nothing but the spectral determinant itself. More precisely, in the 't Hooft regime (119), we have the asymptotic behavior

$$\log \Xi_S(\kappa, \boldsymbol{\xi}, \hbar) \sim \mathsf{J}_S^{\text{'t Hooft}}(\mathbf{t}). \tag{134}$$

In this equation, the t_i are related to ζ, $\boldsymbol{m}$ as spelled out in (123). The r.h.s. is given in (121) and (122), and it involves the genus expansion of the topological string in the large radius regime. Therefore, we can promote the total topological string partition function in the large radius frame, to an entire function of moduli space. It was pointed out by Witten in [**W**] that this partition function should be thought as a wavefunction on the moduli space of the CY. In this view, the genus expansion should be regarded as a sort of WKB approximation to this wavefunction. However, the WKB solution has branch cuts and singularities which are an artifact of the approximation, and are not present in the exact answer, which is sometimes an entire function. In [**MMSS**], such a phenomenon has been argued to occur as well in string theory, where perturbative wavefunctions should be "smoothed out" by non-perturbative effects. It is tempting to interpret (134) in this light, and to

[3]Non-perturbative effects are explicitly included in the NS contribution to the grand potential, but this contribution does not have the standard form of a trans-series in the string coupling constant.

regard the spectral determinant as the *exact*, *entire* wavefunction on moduli space postulated in [**W**]. As in the WKB approximation, the branch cuts and conifold singularities appearing in the perturbative free energies of the topological strings should be regarded as artifacts of the genus expansion, and they are smoothed out in the final answer.

4.3. Number-theoretic aspects. It has been observed in [**RV**, **DK**] (see also [**MOY**]) that the values of the Kähler parameters of many toric CYs at the conifold point are given by special values of the dilogarithm function. It turns out that these number-theoretic properties of the CY periods at the conifold point are consequences of the main conjecture (98). According to this conjecture (98), the topological string amplitudes are encoded in the spectral properties of the operator obtained by quantizing the curve. On the other hand, as shown in [**KM**], this operator is closely related to the quantum dilogarithm. When these two statements are put together, the results of [**RV**, **DK**] follow.

To see this in more detail, note that (127) implies that

$$\frac{\partial \mathcal{F}_0}{\partial \lambda} = -\zeta. \tag{135}$$

It follows from (126) that the l.h.s. of this equation can be calculated by studying the large N limit of the fermionic spectral traces. For the operators $\mathsf{O}_{m,n}$, it can be computed from the matrix integral (129), and the result is [**MZ**]

$$\left(\frac{\partial \mathcal{F}_0}{\partial \lambda}\right)_{\lambda=0} = \frac{m+n+1}{2\pi^2} D(-q^{m+1}\chi_m). \tag{136}$$

In this equation,

$$q = \exp\left(\frac{\mathrm{i}\,\pi}{m+n+1}\right), \qquad \chi_k = \frac{q^k - q^{-k}}{q - q^{-1}}, \tag{137}$$

and $D(z)$ is the Bloch–Wigner function defined by,

$$D(z) = \operatorname{Im} \operatorname{Li}_2(z) + \arg(1-z)\log|z|, \tag{138}$$

where arg denotes the branch of the argument between $-\pi$ and π. Now, the point $\lambda = 0$ is the conifold point, and since ζ is related to the Kähler parameters t_i by (123), one can evaluate these parameters at the conifold point in terms of (136).

The simplest example of this situation is again local $\mathbb{P}^2$. In this case there is a single Kähler parameter t, related to ζ by

$$t = 6\pi\zeta. \tag{139}$$

The conifold point occurs at $z = -1/27$, which is at the boundary of convergence of the series in (28). The value of t at this point, which we will denote by t_c, can be in principle obtained by evaluating the mirror map (47),

$$t_c = -\log(1/27) - \widetilde{\varpi}_1(-1/27)\,. \tag{140}$$

Here, we are dropping an imaginary piece $\pm\pi\,\mathrm{i}$ coming from the log. According to (136), (140) can be evaluated in terms of the Bloch–Wigner function as

$$t_c = \frac{9}{\pi} D\left(\mathrm{e}^{\pi\,\mathrm{i}/3}\right). \tag{141}$$

This is clearly a non-trivial number-theoretic property of the period. In the framework of [**GHM**, **MZ**], it follows ultimately from the fact that the operator $\rho_{\mathbb{P}^2}$ involves the quantum dilogarithm [**KM**]. The result (141) turns out to be true, and it has been proved in [**RV**, **DK**]. Similar identities can be obtained for other CYs, and some of them seem to be new (see [**CGM2**, **CGuM**] for examples involving mirror curves of genus two.)

The fact that the spectral theory realization of the topological string "explains" some number-theoretic properties of the periods seems to indicate that this non-perturbative completion is capturing essential aspects of the theory.

5. Outlook

In this paper we have reviewed the correspondence between spectral theory and topological strings presented in [**GHM**]. This correspondence leads to a new family of exactly solvable problems in operator theory, providing explicit expressions for the corresponding Fredholm determinants. At the same time, it leads to new insights on topological strings on toric CY threefolds, and to a non-perturbative definition of these string theories in the spirit of large N string/gauge dualities.

In this review, we have focused on mirror curves of genus one. The case of genus zero is certainly special, since the corresponding operators do not have discrete spectra. However, the ideas of [**HMMO**, **GHM**] can be extended to this case in a natural way, as it has been noted recently in [**H**, **HO2**, **KrMk**]. More interesting, and more difficult, is the extension to mirror curves of higher genus. This has been done in [**CGM2**] and further explored in [**CGuM**]. When the mirror curve has genus g_Σ, its quantization leads to g_Σ different trace class operators. It is possible to define a generalized Fredholm determinant depending on g_Σ moduli. There are also natural generalizations of the fermionic spectral traces, which now depend on g_Σ different non-negative integers N_i, $i = 1, \cdots, g_\Sigma$, and provide a non-perturbative definition of the topological string in a 't Hooft-like limit. However, in contrast to the situation for quantum integrable systems, there is a single quantization condition, which determines a codimension one locus in moduli space through the vanishing of the Fredholm determinant

The results presented here have important consequences for the study of a class of integrable systems, called *cluster integrable systems*, which can be associated to local CY manifolds. These systems were constructed by Goncharov and Kenyon in [**GoKe**] and include in particular the relativistic Toda lattice [**EFS**, **Mar**, **FM**]. The number of Hamiltonians in these systems is equal to the genus of the mirror curve. In the case of curves of genus one, the operator O_S is nothing but the (single) quantum Hamiltonian of the Goncharov–Kenyon system. Therefore, the results of [**GHM**] provide a conjectural solution to the quantum integrable system in the genus one case. By using the formulation of the quantization condition of [**GHM**] presented in [**WZH**], an exact quantization condition for the spectrum of the general Goncharov–Kenyon system has been conjectured in [**HM**, **FrHaMa**].

There are clearly many avenues to explore in the future. Although we have presented some detailed results for the spectrum of the operators, one should address the construction of the eigenfunctions, which is closely related to the open string sector. A first step in this direction has been made in [**MZ2**], where a general proposal has been made in the case of genus one curves, and an exact solution has been written down explicitly in the maximally supersymmetric case of local $\mathbb{P}^1 \times \mathbb{P}^1$. However, much more work is needed in order to have a general solution for the eigenfunctions. Eventually, one should provide a rigorous proof (or at least a derivation at the physics level of rigor) of the quantization condition. This looks difficult, since, for this type of operators, we do not even have a semi-rigorous method to calculate instanton corrections. Developing these tools is an important task for the future.

On the topological string side, one limitation of the framework we have presented is that it does not accommodate the general refined theory, but only the two one-parameter specializations corresponding to the standard and the NS limits. However, one lesson we have learned from the developments reviewed in this paper is that these two one-parameter cases are not independent: in order to solve the spectral problem, we need both generating functions, (42) and (52), and each one of them can be regarded as a non-perturbative correction to the other. Maybe we should think about the refined topological string with general (ϵ_1, ϵ_2) parameters as a non-perturbative object, similar to the (p, q) string, instead of looking for a perturbative worldsheet description.

Acknowledgments

I would like to thank, first of all, my co-workers on this subject, for the enjoyable collaboration: Santiago Codesido, Ricardo Couso, Alba Grassi, Jie Gu, Yasxuyuki Hatsuda, Johan Kallen, Rinat Kashaev, Albrecht Klemm, Sanefumi Moriyama, Kazumi Okuyama, Jonas Reuter, Ricardo Schiappa, and Szabolcs Zakany. In particular, Santiago Codesido, Ricardo Couso, Alba Grassi, Yasuyuki Hatsuda, Rinat Kashaev, Ricardo Schiappa and Szabolcs Zakany read a preliminary version of this paper and made many useful comments. Finally, I would like to thank Boris Pioline and the rest of the organizers of *String-Math 2016* for the invitation to speak and for the opportunity to contribute to the proceedings.

References

[ABK] M. Aganagic, V. Bouchard, and A. Klemm, *Topological strings and (almost) modular forms*, Comm. Math. Phys. **277** (2008), no. 3, 771–819. MR2365453

[ACDKV] M. Aganagic, M. C. N. Cheng, R. Dijkgraaf, D. Krefl, and C. Vafa, *Quantum geometry of refined topological strings*, J. High Energy Phys. **11** (2012), 019, front matter + 52. MR3036500

[ADKMV] M. Aganagic, R. Dijkgraaf, A. Klemm, M. Mariño, and C. Vafa, *Topological strings and integrable hierarchies*, Comm. Math. Phys. **261** (2006), no. 2, 451–516. MR2191887

[AKMV] M. Aganagic, A. Klemm, M. Mariño, and C. Vafa, *The topological vertex*, Comm. Math. Phys. **254** (2005), no. 2, 425–478. MR2117633

[AYZ] M. Alim, S.-T. Yau, and J. Zhou, *Airy equation for the topological string partition function in a scaling limit*, Lett. Math. Phys. **106** (2016), no. 6, 719–729. MR3500418

[Al] G. Alvarez, *Langer-Cherry derivation of the multi-instanton expansion for the symmetric double well*, J. Math. Phys. **45** (2004), no. 8, 3095–3108. MR2077502

[AGM] P. S. Aspinwall, B. R. Greene, and D. R. Morrison, *Measuring small distances in $N = 2$ sigma models*, Nuclear Phys. B **420** (1994), no. 1-2, 184–242. MR1282646

[BPV] R. Balian, G. Parisi, and A. Voros, *Quartic oscillator*, Feynman path integrals (Proc. Internat. Colloq., Marseille, 1978), Lecture Notes in Phys., vol. 106, Springer, Berlin-New York, 1979, pp. 337–360. MR553093

[BLZ] V. V. Bazhanov, S. L. Lukyanov, and A. B. Zamolodchikov, *Spectral determinants for Schrödinger equation and **Q**-operators of conformal field theory*, Proceedings of the Baxter Revolution in Mathematical Physics (Canberra, 2000), J. Statist. Phys. **102** (2001), no. 3-4, 567–576. MR1832065

[BCOV1] M. Bershadsky, S. Cecotti, H. Ooguri, and C. Vafa, *Holomorphic anomalies in topological field theories*, Nuclear Phys. B **405** (1993), no. 2-3, 279–304. MR1240687

[BCOV2] M. Bershadsky, S. Cecotti, H. Ooguri, and C. Vafa, *Kodaira-Spencer theory of gravity and exact results for quantum string amplitudes*, Comm. Math. Phys. **165** (1994), no. 2, 311–427. MR1301851

[BS] M. A. Bershtein and A. I. Shchechkin, *q-deformed Painlevé τ function and q-deformed conformal blocks*, J. Phys. A **50** (2017), no. 8, 085202, 22. MR3609062

[BKMP] V. Bouchard, A. Klemm, M. Mariño, and S. Pasquetti, *Remodeling the B-model*, Comm. Math. Phys. **287** (2009), no. 1, 117–178. MR2480744

[BC] A. Brini and R. Cavalieri, *Crepant resolutions and open strings II*, arXiv:1407.2571.

[BIMS] E. Brugallé, I. Itenberg, G. Mikhalkin, and K. Shaw, *Brief introduction to tropical geometry*, Proceedings of the Gökova Geometry-Topology Conference 2014, Gökova Geometry/Topology Conference (GGT), Gökova, 2015, pp. 1–75. MR3381439

[CGPO] P. Candelas, X. C. de la Ossa, P. S. Green, and L. Parkes, *A pair of Calabi-Yau manifolds as an exactly soluble superconformal theory*, Nuclear Phys. B **359** (1991), no. 1, 21–74. MR1115626

[CKYZ] T.-M. Chiang, A. Klemm, S.-T. Yau, and E. Zaslow, *Local mirror symmetry: calculations and interpretations*, Adv. Theor. Math. Phys. **3** (1999), no. 3, 495–565. MR1797015

[CKK] J. Choi, S. Katz, and A. Klemm, *The refined BPS index from stable pair invariants*, Comm. Math. Phys. **328** (2014), no. 3, 903–954. MR3201216

[CR] B. Fang, *Mirror symmetry, constructible sheaves and toric varieties*, ProQuest LLC, Ann Arbor, MI, 2010. Thesis (Ph.D.)–Northwestern University. MR2736788

[CGM] S. Codesido, A. Grassi, and M. Mariño, *Exact results in $\mathcal{N} = 8$ Chern-Simons-matter theories and quantum geometry*, J. High Energy Phys. **7** (2015), 011, front matter+38. MR3384204

[CGM2] S. Codesido, A. Grassi, and M. Mariño, *Spectral theory and mirror curves of higher genus*, Ann. Henri Poincaré **18** (2017), no. 2, 559–622. MR3596771

[CGuM] S. Codesido, J. Gu, and M. Mariño, *Operators and higher genus mirror curves*, J. High Energy Phys. **2** (2017), 092, front matter+52. MR3637522

[CS] R. Couso-Santamaría, *Universality of the topological string at large radius and NS-brane resurgence*, Lett. Math. Phys. **107** (2017), no. 2, 343–366. MR3600919

[CESV] R. Couso-Santamaría, J. D. Edelstein, R. Schiappa, and M. Vonk, *Resurgent transseries and the holomorphic anomaly: nonperturbative closed strings in local $\mathbb{CP}^2$*, Comm. Math. Phys. **338** (2015), no. 1, 285–346. MR3345378

[CMS] R. Couso-Santamaría, M. Mariño, and R. Schiappa, *Resurgence matches quantization*, J. Phys. A **50** (2017), no. 14, 145402, 34. MR3622998

[CK] D. A. Cox and S. Katz, *Mirror symmetry and algebraic geometry*, Mathematical Surveys and Monographs, vol. 68, American Mathematical Society, Providence, RI, 1999. MR1677117

[DM] R. B. Dingle and G. J. Morgan, WKB *methods for difference equations. I, II*, Appl. Sci. Res. **18** (1967/1968), 221–237; 238–245. MR0225511

[DK] C. F. Doran and M. Kerr, *Algebraic K-theory of toric hypersurfaces*, Commun. Number Theory Phys. **5** (2011), no. 2, 397–600. MR2851155

[DDT] P. Dorey, C. Dunning, and R. Tateo, *The ODE/IM correspondence*, J. Phys. A **40** (2007), no. 32, R205–R283. MR2370538

[DMP] N. Drukker, M. Mariño, and P. Putrov, *From weak to strong coupling in ABJM theory*, Comm. Math. Phys. **306** (2011), no. 2, 511–563. MR2824480

[D] J. L. Dunham, *The Wentzel–Brillouin–Kramers method of solving the wave equation*, Phys. Rev. **41**, 713 (1932).

[EFS] R. Eager, S. Franco, and K. Schaeffer, *Dimer models and integrable systems*, J. High Energy Phys. **6** (2012), 106, front matter+24. MR3006855

[EK] B. Eynard and C. Kristjansen, *More on the exact solution of the* O(n) *model on a random lattice and an investigation of the case $|n| > 2$*, Nuclear Phys. B **466** (1996), no. 3, 463–487. MR1391455

[EM] B. Eynard and M. Mariño, *A holomorphic and background independent partition function for matrix models and topological strings*, J. Geom. Phys. **61** (2011), no. 7, 1181–1202. MR2788322

[EO1] B. Eynard and N. Orantin, *Invariants of algebraic curves and topological expansion*, Commun. Number Theory Phys. **1** (2007), no. 2, 347–452. MR2346575

[EO2] B. Eynard and N. Orantin, *Computation of open Gromov-Witten invariants for toric Calabi-Yau 3-folds by topological recursion, a proof of the BKMP conjecture*, Comm. Math. Phys. **337** (2015), no. 2, 483–567. MR3339157

[F] L. D. Faddeev, *Discrete Heisenberg-Weyl group and modular group*, Lett. Math. Phys. **34** (1995), no. 3, 249–254. MR1345554

[FK] L. D. Faddeev and R. M. Kashaev, *Quantum dilogarithm*, Modern Phys. Lett. A **9** (1994), no. 5, 427–434. MR1264393

[FLZ] B. Fang, M. C.-C. Liu, and Z. Zong, *On the remodelling conjecture for toric Calabi-Yau 3-orbifolds*, arXiv:1604.07123.

[FM] A. Marshakov, *Lie groups, cluster variables and integrable systems*, J. Geom. Phys. **67** (2013), 16–36. MR3027553

[FrHaMa] S. Franco, Y. Hatsuda, and M. Mariño, *Exact quantization conditions for cluster integrable systems*, J. Stat. Mech. Theory Exp. **6** (2016), 063107, 30. MR3522771

[FHM] H. Fuji, S. Hirano, and S. Moriyama, *Summing up all genus free energy of ABJM matrix model*, J. High Energy Phys. **8** (2011), 001, 17. MR2876085

[GK] S. Garoufalidis and R. Kashaev, *Evaluation of state integrals at rational points*, Commun. Number Theory Phys. **9** (2015), no. 3, 549–582. MR3399926

[GhV] D. Ghoshal and C. Vafa, *$c = 1$ string as the topological theory of the conifold*, Nuclear Phys. B **453** (1995), no. 1-2, 121–128. MR1358307

[GoKe] A. B. Goncharov and R. Kenyon, *Dimers and cluster integrable systems* (English, with English and French summaries), Ann. Sci. Éc. Norm. Supér. (4) **46** (2013), no. 5, 747–813. MR3185352

[GV1] R. Gopakumar and C. Vafa, *On the gauge theory/geometry correspondence*, Adv. Theor. Math. Phys. **3** (1999), no. 5, 1415–1443. MR1796682

[GV2] R. Gopakumar and C. Vafa, *M theory and topological strings. 2*, hep-th/9812127.

[GP] T. Graber and R. Pandharipande, *Localization of virtual classes*, Invent. Math. **135** (1999), no. 2, 487–518. MR1666787

[Gr] A. Grassi, *Spectral determinants and quantum theta functions*, J. Phys. A **49** (2016), no. 50, 505401, 33. MR3584406

[GG] A. Grassi and J. Gu, *BPS relations from spectral problems and blowup equations*, arXiv:1609.05914 [hep-th].

[GHM] A. Grassi, Y. Hatsuda, and M. Mariño, *Topological strings from quantum mechanics*, Ann. Henri Poincaré **17** (2016), no. 11, 3177–3235. MR3556519

[GMZ] A. Grassi, M. Mariño, and S. Zakany, *Resumming the string perturbation series*, J. High Energy Phys. **5** (2015), 038, front matter+34. MR3359359

[GKMR] J. Gu, A. Klemm, M. Mariño, and J. Reuter, *Exact solutions to quantum spectral curves by topological string theory*, J. High Energy Phys. **10** (2015), 025, front matter+68. MR3435614

[G] S. Gukov, *Three-dimensional quantum gravity, Chern-Simons theory, and the A-polynomial*, Comm. Math. Phys. **255** (2005), no. 3, 577–627. MR2134725

[GS] S. Gukov and I. Saberi, *Lectures on knot homology and quantum curves*, Physics and mathematics of link homology, Contemp. Math., vol. 680, Amer. Math. Soc., Providence, RI, 2016, pp. 59–97. MR3591643

[Ha+] M. Hanada, M. Honda, Y. Honma, J. Nishimura, S. Shiba and Y. Yoshida, *Numerical studies of the ABJM theory for arbitrary N at arbitrary coupling constant*, JHEP **1205**, 121 (2012) [arXiv:1202.5300 [hep-th]].

[H] Y. Hatsuda, *Spectral zeta function and non-perturbative effects in ABJM Fermi-gas*, J. High Energy Phys. **11** (2015), 086, front matter+32. MR3455009

[HM] Y. Hatsuda and M. Mariño, *Exact quantization conditions for the relativistic Toda lattice*, J. High Energy Phys. **5** (2016), 133, front matter+34. MR3521852

[HMMO] Y. Hatsuda, M. Mariño, S. Moriyama, and K. Okuyama, *Non-perturbative effects and the refined topological string*, J. High Energy Phys. **9** (2014), 168, front matter+41. MR3267911

[HMO] Y. Hatsuda, S. Moriyama, and K. Okuyama, *Instanton effects in ABJM theory from Fermi gas approach*, J. High Energy Phys. **1** (2013), 158, front matter + 39. MR3045482

[HO] Y. Hatsuda and K. Okuyama, *Probing non-perturbative effects in M-theory*, J. High Energy Phys. **10** (2014), 158, front matter+34. MR3325158

[HO2] Y. Hatsuda and K. Okuyama, *Resummations and non-perturbative corrections*, J. High Energy Phys. **9** (2015), 051, front matter+28. MR3429436

[H+] K. Hori, S. Katz, A. Klemm, R. Pandharipande, R. Thomas, C. Vafa, R. Vakil, and E. Zaslow, *Mirror symmetry*, Clay Mathematics Monographs, vol. 1, American Mathematical Society, Providence, RI; Clay Mathematics Institute, Cambridge, MA, 2003. With a preface by Vafa. MR2003030

[HK] M.-x. Huang and A. Klemm, *Direct integration for general Ω backgrounds*, Adv. Theor. Math. Phys. **16** (2012), no. 3, 805–849. MR3024275

[HKP] M. x. Huang, A. Klemm and M. Poretschkin, *Refined stable pair invariants for E-, M- and $[p,q]$-strings*, JHEP **1311**, 112 (2013) [arXiv:1308.0619 [hep-th]].

[HKRS] M.-x. Huang, A. Klemm, J. Reuter, and M. Schiereck, *Quantum geometry of del Pezzo surfaces in the Nekrasov-Shatashvili limit*, J. High Energy Phys. **2** (2015), 031, front matter+51. MR3321432

[HW] M.-x. Huang and X.-f. Wang, *Topological strings and quantum spectral problems*, J. High Energy Phys. **9** (2014), 150, front matter+46. MR3267929

[IKV] A. Iqbal, C. Kozçaz, and C. Vafa, *The refined topological vertex*, J. High Energy Phys. **10** (2009), 069, 58. MR2607441

[KaMa] J. Källén and M. Mariño, *Instanton effects and quantum spectral curves*, Ann. Henri Poincaré **17** (2016), no. 5, 1037–1074. MR3486428

[KM] R. Kashaev and M. Mariño, *Operators from mirror curves and the quantum dilogarithm*, Comm. Math. Phys. **346** (2016), no. 3, 967–994. MR3537342

[KMZ] R. Kashaev, M. Mariño, and S. Zakany, *Matrix models from operators and topological strings, 2*, Ann. Henri Poincaré **17** (2016), no. 10, 2741–2781. MR3546986

[KKV] S. Katz, A. Klemm, and C. Vafa, *Geometric engineering of quantum field theories*, Nuclear Phys. B **497** (1997), no. 1-2, 173–195. MR1467889

[KT] I. R. Klebanov and A. A. Tseytlin, *Entropy of near-extremal black p-branes*, Nuclear Phys. B **475** (1996), no. 1-2, 164–178. MR1412100

[KZ] A. Klemm and E. Zaslow, *Local mirror symmetry at higher genus*, hep-th/9906046.

[Ko] M. Kontsevich, *Enumeration of rational curves via torus actions*, hep-th/9405035.

[K] I. K. Kostov, *$O(n)$ vector model on a planar random lattice: spectrum of anomalous dimensions*, Modern Phys. Lett. A **4** (1989), no. 3, 217–226. MR1016518

[KW] D. Krefl and J. Walcher, *Extended holomorphic anomaly in gauge theory*, Lett. Math. Phys. **95** (2011), no. 1, 67–88. MR2764334

[KrMk] D. Krefl and R. L. Mkrtchyan, *Exact Chern-Simons/topological string duality*, J. High Energy Phys. **10** (2015), 045, front matter+26. MR3435594

[LST] A. Laptev, L. Schimmer, and L. A. Takhtajan, *Weyl type asymptotics and bounds for the eigenvalues of functional-difference operators for mirror curves*, Geom. Funct. Anal. **26** (2016), no. 1, 288–305. MR3494491

[MMSS] J. Maldacena, N. Seiberg, G. Moore, and D. Shih, *Exact vs. semiclassical target space of the minimal string*, J. High Energy Phys. **10** (2004), 020, 40. MR2116030

[M1] M. Mariño, *Open string amplitudes and large order behavior in topological string theory*, J. High Energy Phys. **3** (2008), 060, 34. MR2391060

[M2] M. Mariño, *Lectures on non-perturbative effects in large N gauge theories, matrix models and strings*, Fortschr. Phys. **62** (2014), no. 5-6, 455–540. MR3214736

[MP] M. Mariño and P. Putrov, *ABJM theory as a Fermi gas*, J. Stat. Mech. Theory Exp. **3** (2012), P03001, 53. MR2935521

[MZ] M. Mariño and S. Zakany, *Matrix models from operators and topological strings*, Ann. Henri Poincaré **17** (2016), no. 5, 1075–1108. MR3486429

[MZ2] M. Mariño and S. Zakany, *Exact eigenfunctions and the open topological string*, arXiv:1606.05297 [hep-th].

[Mar] A. Marshakov, *Lie groups, cluster variables and integrable systems*, J. Geom. Phys. **67** (2013), 16–36. MR3027553

[MM] A. Mironov and A. Morosov, *Nekrasov functions and exact Bohr-Sommerfeld integrals*, J. High Energy Phys. **4** (2010), 040, 15. MR2673092

[MOY] K. Mohri, Y. Onjo, and S.-K. Yang, *Closed sub-monodromy problems, local mirror symmetry and branes on orbifolds*, Rev. Math. Phys. **13** (2001), no. 6, 675–715. MR1841743

[N] N. A. Nekrasov, *Seiberg-Witten prepotential from instanton counting*, Adv. Theor. Math. Phys. **7** (2003), no. 5, 831–864. MR2045303

[NO] N. Nekrasov and A. Okounkov, *Membranes and sheaves*, Algebr. Geom. **3** (2016), no. 3, 320–369. MR3504535

[NS] N. A. Nekrasov and S. L. Shatashvili, *Quantization of integrable systems and four dimensional gauge theories*, XVIth International Congress on Mathematical Physics, World Sci. Publ., Hackensack, NJ, 2010, pp. 265–289. MR2730782

[No] P. Norbury, *Quantum curves and topological recursion*, String-Math 2014, Proc. Sympos. Pure Math., vol. 93, Amer. Math. Soc., Providence, RI, 2016, pp. 41–65. MR3524233

[OZ] K. Okuyama and S. Zakany, *TBA-like integral equations from quantized mirror curves*, JHEP **1603** (2016) 101 [arXiv:1512.06904 [hep-th]].

[PT] R. Pandharipande and R. P. Thomas, *Curve counting via stable pairs in the derived category*, Invent. Math. **178** (2009), no. 2, 407–447. MR2545686

[RV] F. R. Villegas, *Modular Mahler measures. I*, Topics in number theory (University Park, PA, 1997), Math. Appl., vol. 467, Kluwer Acad. Publ., Dordrecht, 1999, pp. 17–48. MR1691309

[Si] B. Simon, *Trace ideals and their applications*, 2nd ed., Mathematical Surveys and Monographs, vol. 120, American Mathematical Society, Providence, RI, 2005. MR2154153

[SWH] K. Sun, X. Wang, and M.-x. Huang, *Exact quantization conditions, toric Calabi-Yau and non-perturbative topological string*, J. High Energy Phys. **1** (2017), 061, front matter+101. MR3628591

[Vo1] A. Voros, *Asymptotic ħ-expansions of stationary quantum states* (English, with French summary), Ann. Inst. H. Poincaré Sect. A (N.S.) **26** (1977), no. 4, 343–403. MR0456138

[Vo2] A. Voros, *Spectral functions, special functions and the Selberg zeta function*, Comm. Math. Phys. **110** (1987), no. 3, 439–465. MR891947

[Vo3] A. Voros, *Corrigendum: "Exact resolution method for general 1D polynomial Schrödinger equation" [J. Phys. A* **32** *(1999), no. 32, 5993–6007; MR1721984 (2000k:81071)]*, J. Phys. A **33** (2000), no. 32, 5783–5784. MR1784724

[WZH] K. Sun, X. Wang, and M.-x. Huang, *Exact quantization conditions, toric Calabi-Yau and non-perturbative topological string*, J. High Energy Phys. **1** (2017), 061, front matter+101. MR3628591

[W] E. Witten, *Quantum background independence in string theory*, hep-th/9306122.

[ZJJ] J. Zinn-Justin and U. D. Jentschura, *Multi-instantons and exact results. I. Conjectures, WKB expansions, and instanton interactions*, Ann. Physics **313** (2004), no. 1, 197–267. MR2081095

Section de Mathématiques et Département de Physique Théorique, Université de Genève, 1211 Genève 4, Switzerland

Email address: `marcos.marino@unige.ch`

ISBN 978-1-4704-3515-8

9 781470 435158

PSPUM/98